THE CHEMISTRY OF FREE RADICAL POLYMERIZATION

Related Pergamon Titles of Interest

CARRUTHERS
Cycloaddition Reactions in Organic Synthesis

PERLMUTTER
Conjugate Addition Reactions in Organic Synthesis

SIMPKINS
Sulphones in Organic Synthesis

HASSNER & STUMER
Named Reactions and Un-named Reactions in Organic Synthesis

LEVY & TANG
The Chemistry of *C*-Glycosides

ALLEN *et al.*
Comprehensive Polymer Science & Supplements

TROST & FLEMING
Comprehensive Organic Synthesis

Related Journals
Free sample copy gladly sent on request

Polymer
European Polymer Journal
Progress in Polymer Science
Tetrahedron
Tetrahedron Letters

THE CHEMISTRY OF FREE RADICAL POLYMERIZATION

GRAEME MOAD
*CSIRO Division of Chemicals and Polymers, Bayview Avenue,
Vic Postal Address, Private Bag 10, Clayton,
Victoria 3168, Australia*

and

DAVID H. SOLOMON
*School of Chemistry, The University of Melbourne, Parkville,
Victoria 3052, Australia*

PERGAMON

U.K.   Elsevier Science Ltd, The Boulevard, Langford Lane,
            Kidlington, Oxford, OX5 1GB, U.K.

U.S.A.   Elsevier Science Inc., 660 White Plains Road, Tarrytown,
            New York 10591-5153, U.S.A.

JAPAN   Elsevier Science Japan, Tsunashima Building Annex,
            3-20-12 Yushima, Bunkyo-ku, Tokyo 113, Japan

Copyright © 1995 Elsevier Science Ltd

All Rights Reserved. No part of this publication may be reproduced, stored in a retrieval system or transmitted in any form or by any means: electronic, electrostatic, magnetic tape, mechanical, photocopying, recording or otherwise, without permission in writing from the publishers.

First Edition 1995

Library of Congress Cataloging in Publication Data
A catalog record for this book is available from the
Library of Congress

British Library Cataloguing in Publication Data
A catalogue record for this book is available from the
British Library

ISBN 0 08 042078 8 (H)
ISBN 0 08 042079 6 (F)

In order to make this volume available as economically and as rapidly as possible the authors' typescripts have been reproduced in their original forms. This method unfortunately has its typographical limitations but it is hoped that they in no way distract the reader.

Printed and bound in Great Britain by Bookcraft, Bath

Contents

CONTENTS v
INDEX TO TABLES xii
PREFACE xiv
ACKNOWLEDGMENTS xvi

1 INTRODUCTION 1

2 FREE RADICAL REACTIONS 7
2.1 INTRODUCTION 7
2.2 ADDITION TO CARBON-CARBON DOUBLE BONDS 8
    2.2.1 Steric Factors 12
    2.2.2 Polar Factors 13
    2.2.3 Bond Strengths 15
    2.2.4 Stereoelectronic Factors 15
    2.2.5 Reaction Conditions 16
        2.2.5.1 Temperature 16
        2.2.5.2 Solvent 17
    2.2.6 Theoretical Treatments 18
    2.2.7 Summary 20
2.3 HYDROGEN ATOM TRANSFER 21
    2.3.1 Bond Dissociation Energies 21
    2.3.2 Steric Factors 21
    2.3.3 Polar Factors 22
    2.3.4 Stereoelectronic Factors 23
    2.3.5 Reaction Conditions 25
    2.3.6 Abstraction *vs.* Addition 26
    2.3.7 Summary 27
2.4 RADICAL-RADICAL REACTIONS 28
    2.4.1 Pathways for Combination 28
    2.4.2 Pathways for Disproportionation 30
    2.4.3 Combination *vs.* Disproportionation 31
        2.4.3.1 Statistical factors 31
        2.4.3.2 Steric factors 32
        2.4.3.3 Polar factors 33
        2.4.3.4 Stereoelectronic and other factors 33
        2.4.3.5 Reaction conditions 34
    2.4.4 Summary 35

3 INITIATION 43
3.1 INTRODUCTION 43
3.2 THE INITIATION PROCESS 44
- 3.2.1 Reaction with Monomer 45
- 3.2.2 Fragmentation 47
- 3.2.3 Reaction with Solvents, Additives, or Impurities 48
- 3.2.4 Effects of the Reaction Medium on Radical Reactivity 49
- 3.2.5 Reaction with Oxygen 49
- 3.2.6 Initiator Efficiency 50
- 3.2.7 Cage Reaction and Initiator-Derived By-products 51
- 3.2.8 Primary Radical Termination 52
- 3.2.9 Transfer to Initiator 52
- 3.2.10 Initiation in Heterogeneous Polymerization 53

3.3 THE INITIATORS 53
- 3.3.1 Azo-Compounds 54
  - 3.3.1.1 Dialkyldiazenes 57
    - 3.3.1.1.1 Thermal decomposition 58
    - 3.3.1.1.2 Photochemical decomposition 61
    - 3.3.1.1.3 Initiator efficiency 61
    - 3.3.1.1.4 Transfer to initiator 64
  - 3.3.1.2 Hyponitrites 65
- 3.3.2 Peroxides 66
  - 3.3.2.1 Diacyl peroxides 67
    - 3.3.2.1.1 Thermal decomposition 67
    - 3.3.2.1.2 Photochemical decomposition 70
    - 3.3.2.1.3 Initiator efficiency 70
    - 3.3.2.1.4 Transfer to initiator and induced decomposition 71
    - 3.3.2.1.5 Redox reactions 72
  - 3.3.2.2 Dialkyl peroxydicarbonates 73
  - 3.3.2.3 Peresters 74
    - 3.3.2.3.1 Thermal decomposition 75
    - 3.3.2.3.2 Photochemical decomposition 76
  - 3.3.2.4 Dialkyl peroxides 77
  - 3.3.2.5 Alkyl hydroperoxides 78
  - 3.3.2.6 Inorganic peroxides 79
    - 3.3.2.6.1 Persulfate 80
    - 3.3.2.6.2 Hydrogen peroxide 81
- 3.3.3 Multifunctional Initiators 82
  - 3.3.3.1 Concerted decomposition 82
  - 3.3.3.2 Non-concerted decomposition 83
- 3.3.4 Photochemical Initiators 84
  - 3.3.4.1 Aromatic carbonyl compounds 84
    - 3.3.4.1.1 Benzoin derivatives 85

   3.3.4.1.2 Carbonyl compound-tertiary amine systems 88
  3.3.4.2 Sulfur compounds 89
 3.3.5 Redox Initiators 89
  3.3.5.1 Metal complex-organic halide systems 90
  3.3.5.2 Ceric ion systems 90
 3.3.6 Thermal Initiation 92
  3.3.6.1 Styrene homopolymerization 92
  3.3.6.2 Acrylate homopolymerization 94
  3.3.6.3 Copolymerization 95
3.4 THE RADICALS 96
 3.4.1 Carbon-Centered Radicals 96
  3.4.1.1 Alkyl radicals 97
  3.4.1.2 α-Cyanoalkyl radicals 99
  3.4.1.3 Aryl radicals 100
 3.4.2 Oxygen-Centered Radicals 101
  3.4.2.1 Alkoxy radicals 105
   3.4.2.1.1 t-Butoxy radicals 105
   3.4.2.1.2 Other t-alkoxy radicals 107
   3.4.2.1.3 Primary and secondary alkoxy radicals 108
  3.4.2.2 Acyloxy and alkoxycarbonyloxy radicals 108
   3.4.2.2.1 Benzoyloxy radicals 109
   3.4.2.2.2 Alkoxycarbonyloxy radicals 110
  3.4.2.3 Hydroxy radicals 111
  3.4.2.4 Sulfate radical anion 111
  3.4.2.5 Alkylperoxy radicals 113
 3.4.3 Other Heteroatom-Centered Radicals 114
  3.4.3.1 Sulfur- and selenium-centered radicals 114
  3.4.3.2 Phosphorous-centered radicals 115
3.5 TECHNIQUES 115
 3.5.1 Radical Trapping 116
  3.5.1.1 Spin traps 116
  3.5.1.2 Transition metal salts 118
  3.5.1.3 Metal hydrides 119
  3.5.1.4 Nitroxides 120
 3.5.2 Direct Detection of End Groups 122
  3.5.2.1 Infra-red and UV-visible spectroscopy 123
  3.5.2.2 Nuclear magnetic resonance spectroscopy 123
  3.5.2.3 Electron paramagnetic resonance spectroscopy 124
  3.5.2.4 Mass spectrometry 124
  3.5.2.5 Chemical methods 124
 3.5.3 Labeling Techniques 125
  3.5.3.1 Radiolabeling 125
  3.5.3.2 Stable isotopes and nuclear magnetic resonance 126

4 PROPAGATION 145
4.1 INTRODUCTION 145
4.2 STEREOSEQUENCE ISOMERISM - TACTICITY 146
4.2.1 Terminology and Mechanisms...146
4.2.2 Experimental Methods 150
4.2.3 Tacticities of Common Polymers 151
4.3 REGIOSEQUENCE ISOMERISM - HEAD VS. TAIL ADDITION 152
4.3.1 Monoene Polymers 152
4.3.1.1 Poly(vinyl acetate) 154
4.3.1.2 Poly(vinyl chloride) 156
4.3.1.3 Fluoro-olefin polymers 157
4.3.1.4 Allyl polymers 158
4.3.1.5 Acrylic polymers..158
4.3.2 Conjugated Diene Polymers..159
4.3.2.1 Polybutadiene..161
4.3.2.2 Polychloroprene, polyisoprene 161
4.4 STRUCTURAL ISOMERISM - REARRANGEMENT 162
4.4.1 Cyclopolymerization 162
4.4.1.1 1,6-Dienes...163
4.4.1.2 Triene monomers 168
4.4.1.3 1,4- and 1,5-dienes 169
4.4.1.4 1,7- and higher 1,n-dienes..169
4.4.1.5 Cyclo-copolymerization..170
4.4.2 Ring-Opening Polymerization 171
4.4.2.1 Vinyl substituted cyclic compounds..172
4.4.2.2 Methylene substituted cyclic compounds.................................175
4.4.2.3 Double ring-opening polymerization 182
4.4.3 Intramolecular Atom Transfer 184
4.4.3.1 Polyethylene...184
4.4.3.2 Vinyl polymers...185
4.4.3.3 "Addition-abstraction" polymerization 186
4.5 PROPAGATION KINETICS AND THERMODYNAMICS 187
4.5.1 Polymerization Thermodynamics 187
4.5.2 Measurement of Propagation Rate Constants 190
4.5.3 Chain Length Dependence of Propagation Rate Constants................192

5 TERMINATION 207
5.1 INTRODUCTION 207
5.2 RADICAL-RADICAL TERMINATION 208
5.2.1 Termination Kinetics 209
5.2.1.1 Homopolymerization..210
5.2.1.2 Copolymerization 214
5.2.1.2.1 Chemical control model 215
5.2.1.2.2 Classical diffusion control model 217

5.2.2 Disproportionation *vs.* Combination - General Considerations..........217
5.2.3 Disproportionation *vs.* Combination - Model Studies 218
    5.2.3.1 Polystyrene and derivatives 219
    5.2.3.2 Poly(alkyl methacrylates)..222
    5.2.3.3 Poly(methacrylonitrile) 224
    5.2.3.4 Polyethylene...226
    5.2.3.5 Poly(styrene-*co*-methyl methacrylate) 226
    5.2.3.6 Poly(styrene-*co*-methacrylonitrile) 227
    5.2.3.7 Other copolymers...227
5.2.4 Disproportionation *vs.* Combination - Polymerization 228
    5.2.4.1 Polystyrene..229
    5.2.4.2 Poly(alkyl methacrylates)..230
    5.2.4.3 Poly(methacrylonitrile) 231
    5.2.4.4 Poly(alkyl acrylates)...231
    5.2.4.5 Poly(acrylonitrile) 231
    5.2.4.6 Poly(vinyl acetate)..231
    5.2.4.7 Poly(vinyl chloride) 232
    5.2.4.8 Poly(styrene-*co*-methyl methacrylate) 232
5.2.5 Disproportionation *vs.* Combination - Summary 233
5.3 CHAIN TRANSFER 234
  5.3.1 Mechanisms..237
    5.3.1.1 Atom or group transfer 237
    5.3.1.2 Addition-elimination 238
    5.3.1.3 Measurement of Transfer Constants 238
  5.3.2 Transfer Agents..240
    5.3.2.1 Thiols...240
    5.3.2.2 Disulfides..241
    5.3.2.3 Monosulfides 243
    5.3.2.4 Halocarbons 243
    5.3.2.5 Solvents and other reagents...244
    5.3.2.6 Unsaturated compounds..246
    5.3.2.7 Cobalt complexes 249
  5.3.3 Transfer to Monomer 251
    5.3.3.1 Styrene 251
    5.3.3.2 Vinyl acetate..252
    5.3.3.3 Vinyl chloride 253
    5.3.3.4 Allyl monomers 254
  5.3.4 Transfer to Polymer...254
    5.3.4.1 Polyethylene..256
    5.3.4.2 Poly(alkyl methacrylates)..256
    5.3.4.3 Poly(alkyl acrylates)...257
    5.3.4.4 Poly(vinyl acetate)..257
    5.3.4.5 Poly(vinyl chloride) 259

          5.3.4.6 Poly(vinyl fluoride) 259
5.4 INHIBITION AND RETARDATION 260
    5.4.1 'Stable' Radicals 260
    5.4.2 Oxygen 262
    5.4.3 Captodative olefins 262
    5.4.4 Phenols 263
    5.4.5 Quinones 264
    5.4.6 Nitrones, nitro- and nitroso-compounds 265
    5.4.7 Transition Metal Salts 266

6 COPOLYMERIZATION 277
6.1 INTRODUCTION 277
6.2 STATISTICAL COPOLYMERIZATION 278
    6.2.1 Copolymerization Mechanisms 279
        6.2.1.1 Terminal model 280
        6.2.1.2 Penultimate model 284
        6.2.1.3 Models involving monomer complexes 288
        6.2.1.4 Copolymerization with depropagation 290
        6.2.1.5 Chain statistics 291
    6.2.2 Estimation of Reactivity Ratios 293
        6.2.2.1 Composition data 293
        6.2.2.2 Monomer sequence distribution 295
    6.2.3 Prediction of Reactivity Ratios 295
        6.2.3.1 Q-e scheme 295
        6.2.3.2 Patterns of reactivity scheme 297
6.3 BLOCK & GRAFT COPOLYMERIZATION 297
    6.3.1 End-Functional Polymers 298
        6.3.1.1 Functional initiators 298
        6.3.1.2 Functional transfer agents 300
        6.3.1.3 Functional monomers 301
        6.3.1.4 Functional inhibitors 302
    6.3.2 Block and Graft Copolymer Synthesis 302
        6.3.2.1 Polymeric and multifunctional initiators 302
        6.3.2.2 Transformation reactions 303
        6.3.2.3 Macromonomers 306

7 CONTROLLING POLYMERIZATION 315
7.1 INTRODUCTION 315
7.2 CONTROLLING STRUCTURAL IRREGULARITIES 316
    7.2.1 "Defect Structures" in Polystyrene 316
    7.2.2 "Defect Structures" in Poly(methyl methacrylate) 318
7.3 CONTROLLING PROPAGATION 321
    7.3.1 Solvent 323
        7.3.1.1 Homopolymerization 324

    7.3.1.2 Copolymerization..325
  7.3.2 Lewis Acids and Inorganics 327
    7.3.2.1 Homopolymerization 327
    7.3.2.2 Copolymerization..328
  7.3.3 Template Polymers..329
    7.3.3.1 Non-covalently bonded templates 329
    7.3.3.2 Covalently bonded templates..330
7.4 COMPOSITIONAL HETEROGENEITY IN COPOLYMERS 332
7.5 AGENTS FOR CONTROLLING TERMINATION...335
  7.5.1 Organosulfur iniferters 336
  7.5.2 Hexasubstituted ethanes and azo compounds 339
  7.5.3 Alkoxyamines and related species..341
  7.5.4 Organocobalt complexes 345

ABBREVIATIONS 353

AUTHOR INDEX..355

SUBJECT INDEX 381

Index to Tables

Table 2.1	Relative Rate Constants and Regiospecificities for Addition of Radicals to Halo-Olefins	10
Table 2.2	Relative Rate Constants for Reactions of Radicals with Alkyl-Substituted Acrylate Esters	10
Table 2.3	Hammett ρ and ρ+ Parameters for Reactions of Radicals	14
Table 2.4	Specificity of Intramolecular Hydrogen abstraction	24
Table 2.5	Bond Dissociation Energies	26
Table 2.6	Values of k_{td}/k_{tc} for the Cross-Reaction between Fluoromethyl and Ethyl Radicals (25 °C)	33
Table 2.7	Values of k_{td}/k_{tc} for t-Butyl Radicals (25 °C)	35
Table 3.1	Guide-lines to Properties of Polymerization Initiators	55
Table 3.2	Selected Kinetic Data for Decomposition of Azo-Compounds	59
Table 3.3	Solvent Dependence of Rate Constants for AIBMe Decomposition	60
Table 3.4	Zero-Conversion Initiator Efficiency for AIBMe under Various Reaction Conditions	63
Table 3.5	Selected Kinetic Data for Decomposition of Peroxides	68
Table 3.6	Rate Data for Reactions of Carbon-Centered Radicals	98
Table 3.7	Selected Rate Data for Reactions of Oxygen-Centered Radicals	102
Table 3.8	Specificity Observed in the Reactions of Oxygen-Centered Radicals with Various Monomers at 60°C	103
Table 3.9	Selected Rate Data for Reactions of Heteroatom-Centered Radicals	114
Table 3.10	Radical Trapping Agents for Studying Initiation	116
Table 4.1	Tacticities of Some Common Homopolymers	152
Table 4.2	Temperature Dependence of Head *vs.* Tail Addition for Fluoro-olefin Monomers	157
Table 4.3	Microstructure of Poly(chloroprene) *vs.* Temperature	161
Table 4.4	Symmetrical 1,6-Diene Monomers for Cyclopolymerization	166
Table 4.5	Extents of Ring-opening During Polymerizations of 2-Methylene-1,3-dioxolane and Related Species	176
Table 4.6	Extent of Ring-Opening During Polymerizations of 4-Methylene-1,3-dioxolane and 2-Methylene-1,4-dioxane Derivatives	180
Table 4.7	Extents of Ring-Opening During Polymerizations of 2-Methylenetetrahydrofuran and Related Compounds	181
Table 4.8	Extents of Double Ring-Opening During Polymerization of Polycyclic Monomers	183
Table 4.9	Heats of Polymerization for Selected Monomers ($CH_2=CRX$)	188

Table 4.10	Kinetic Parameters for Propagation in Selected Free Radical Polymerizations	191
Table 5.1	Values of k_{td}/k_{tc} for Styryl Radical Model Systems	221
Table 5.2	Values of k_{td}/k_{tc} for Methacryate Ester Model Systems	223
Table 5.3	Values of k_{td}/k_{tc} for Reactions of Cyanoisopropyl Radicals	225
Table 5.4	Determinations of k_{td}/k_{tc} for MMA Polymerization	230
Table 5.5	Identity of Chain End Units Involved in Radical-Radical Termination in MMA-S Copolymerization	232
Table 5.6	Chain Length Dependence of Transfer Constants (C_S)	236
Table 5.7	Transfer Constants (60°C, bulk) for Thiols (RSH) With Various Monomers	240
Table 5.8	Transfer Constants for Disulfides (R-S-S-R) With Various Monomers	242
Table 5.9	Transfer Constants (80°C, bulk) for Halocarbons With Various Monomers	243
Table 5.10	Transfer Constants (60°C, bulk) for Selected Solvents and Additives With Various Monomers	245
Table 5.11	Transfer Constants for Some Benzyloxy Ether and Allylic Sulfide Transfer Agents	248
Table 5.12	Transfer Constants to Monomer	251
Table 5.13	Transfer Constants to Polymer	255
Table 5.14	Absolute Rate Constants for the Reaction of Carbon-Centered Radicals with Some Common Inhibitors	260
Table 6.1	Reactivity Ratios for Some Common Monomer Pairs	282
Table 6.2	Relative Rates for Addition of Substituted Propyl Radicals to AN and S	285
Table 6.3	Q-e values for Some Common Monomers	296
Table 7.1	Solvent Effect on Propagation Rate Constants at 30°C	325
Table 7.2	Solvent Dependence of Reactivity Ratios for MMA-MAA Copolymerization at 70°C	325
Table 7.3	Solvent Dependence of Penultimate Model Reactivity Ratios for S-AN Copolymerization at 60°C	326

Preface

In recent years, the study of radical polymerization has gone through something of a renaissance. This has seen significant changes in our understanding of the area and has led to major advances in our ability to control and predict the outcome of polymerization processes. Two major factors may be judged responsible for bringing this about and for spurring an intensified interest in all aspects of radical chemistry:

Firstly, the classical theories on radical reactivity and polymerization mechanism do not adequately explain the rate and specificity of simple radical reactions. As a consequence, they can not be used to predict the manner in which polymerization rate parameters and details of polymer microstructure depend on reaction conditions, conversion and molecular weight distribution.

Secondly, new techniques have been developed which allow a more detailed characterization of both polymer microstructures and the kinetics and mechanism of polymerizations. This has allowed mechanism-structure-property relationships to be more rigorously established.

The new knowledge and understanding of radical processes has resulted in new polymer structures and in new routes to established materials; many with commercial significance. For example, radical polymerization is now used in the production block copolymers, narrow polydispersity homopolymers, and other materials of controlled architecture that were previously available only by more demanding routes. These commercial developments have added to the resurgence of studies on radical polymerization.

We believe it is now timely to review the recent developments in radical polymerization placing particular emphasis on the organic and physical-organic chemistry of the polymerization process. In this book we critically evaluate the findings of the last few years, where necessary reinterpreting earlier work in the light of these ideas, and point to the areas where current and future research is being directed. The overall aim is to provide a framework for further extending our understanding of free radical polymerization and create a definable link between synthesis conditions and polymer structure and properties. The end result should be polymers with predictable and reproducible properties.

The book commences with a general introduction outlining the basic concepts. This is followed by a chapter on radical reactions that is intended to lay the theoretical ground-work for the succeeding chapters on initiation, propagation, and termination. Because of its importance, radical copolymerization is treated in a separate chapter. We then consider some of the implications of these chapters by discussing the prospects for controlling the polymerization process and structure-property relationships. In each chapter we describe some of the techniques that have been employed to characterize polymers and polymerizations and which have led to breakthroughs in our understanding of radical polymerization. Emphasis is placed on recent developments.

This book will be of major interest to researchers in industry and in academic institutions as a reference source on the factors which control radical polymerization and as an aid in designing polymer syntheses. It is also intended to serve as a text for graduate students in the broad area of polymer chemistry. The book places an emphasis on reaction mechanisms and the organic chemistry of polymerization. It also ties in developments in polymerization kinetics and physical chemistry of the systems to provide a complete picture of this most important subject.

Graeme Moad
David H Solomon

Dr Graeme Moad
CSIRO, Division of Chemicals and Polymers
Private Bag 10, Rose bank MDC, Clayton, Victoria 3169
AUSTRALIA

Email: g.moad@chem.csiro.au

Professor David H. Solomon
Melbourne University, School of Chemistry
Parkville, Victoria 3052
AUSTRALIA

Email: david solomon@muwayf.unimelb.edu.au

Acknowledgments

We gratefully acknowledge the contribution of the following for their assistance in the preparation of the manuscript.

Simmi Abrol
Debra Bednarek
Anne Bramfit
Mary-Kathleen Hodgkin
Margaret James
Fiona Kerr
Julia Krstina
Marlene McKnight
Catherine Moad
Danny Naug
Asfia Qureshi
Laile Strong
Dominica Tannock
San Thang
Ellana Wakeman
Sheela Veluayitham

1
Introduction

From an industrial stand-point, a major virtue of radical polymerizations is that they can typically be carried out under relatively undemanding conditions. In marked contrast to ionic or coordination polymerizations, they exhibit a tolerance of trace impurities. A consequence of this is that high molecular weight polymers can often be produced without removal of the stabilizers present in commercial monomers, in the presence of trace amounts of oxygen, or in solvents that have not been rigorously dried. Indeed, radical polymerizations are remarkable amongst chain polymerizations in that they can be conveniently conducted in aqueous media.

It is the apparent simplicity of radical polymerization that has led to this technique being widely adopted for both industrial and laboratory scale polymer syntheses. Today, the bulk of commercial polymer production involves radical chemistry during some stage of the synthesis or the subsequent processing. These factors have, in turn, provided the driving force for extensive research efforts directed towards more precisely defining the mechanisms of radical polymerizations. The aim of these studies has been to define the parameters necessary for predictable and reproducible polymer syntheses and give better understanding of the properties of the polymeric materials produced.

Staudinger[1] was the first to propose the concept of a chain polymerization and define the basic structure of the polymer molecules produced by such mechanisms. He proposed that the monomer residues (1) were connected, head to tail, by covalent linkages in large cyclic structures (2).

The basic mechanism of free-radical polymerization as we know it today[2] (Scheme 1), was laid out in the 1940s and 50s.[3,4] The essential features of this mechanism are initiation and propagation steps, which involve radicals adding to the less substituted end of the double bond ("tail addition"), and a termination

The Chemistry of Free Radical Polymerization

step, which involves disproportionation or combination between two growing chains.

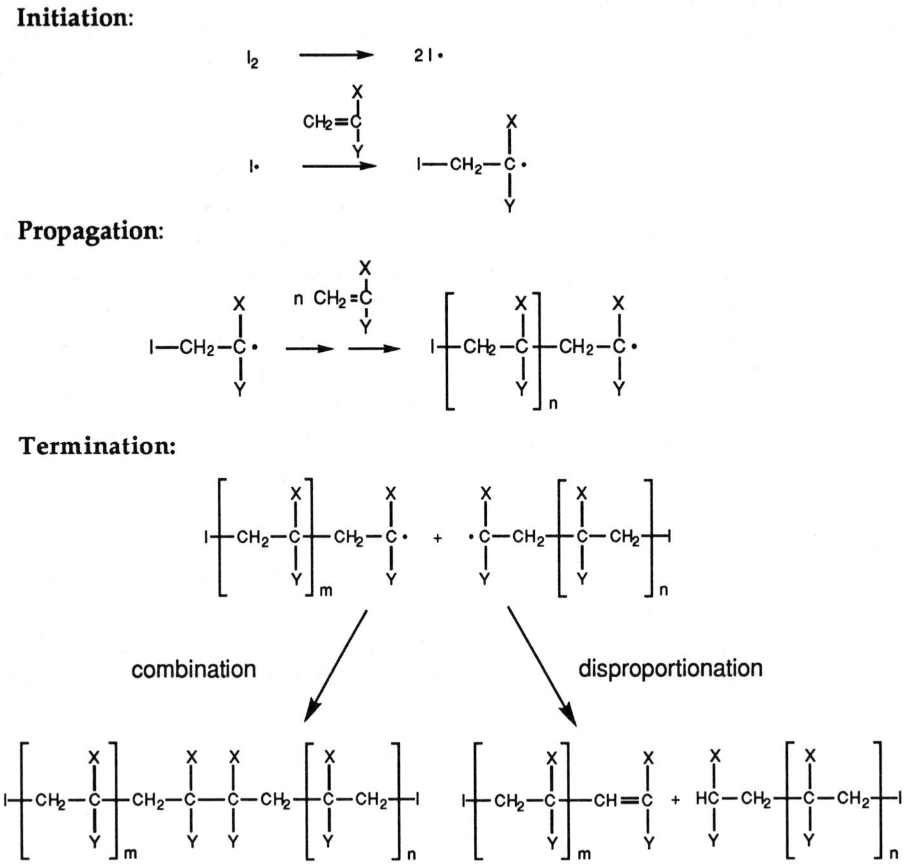

Initiation:

$$I_2 \longrightarrow 2I\cdot$$

$$I\cdot + CH_2=C(X)(Y) \longrightarrow I-CH_2-C(X)(Y)\cdot$$

Propagation:

$$I-CH_2-C(X)(Y)\cdot + n\,CH_2=C(X)(Y) \longrightarrow I-[CH_2-C(X)(Y)]_n-CH_2-C(X)(Y)\cdot$$

Termination:

$$I-[CH_2-C(X)(Y)]_m-CH_2-C(X)(Y)\cdot \;+\; \cdot C(X)(Y)-CH_2-[C(X)(Y)-CH_2]_n-I$$

combination / disproportionation

combination product: $I-[CH_2-C(X)(Y)]_m-CH_2-C(X)(Y)-C(X)(Y)-CH_2-[C(X)(Y)-CH_2]_n-I$

disproportionation products: $I-[CH_2-C(X)(Y)]_m-CH=C(X)(Y) \;+\; HC(X)-CH_2-[C(X)(Y)-CH_2]_n-I$

Scheme 1

In this early work, both initiation and termination were seen to lead to formation of structural units different from those which make up the bulk of the chain. However, the quantity of these groups, when expressed as a weight fraction of the total material, appeared insignificant. In a polymer of molecular weight 100,000 they represent only *ca.* 0.2% of units.* Thus, polymers formed by radical polymerization came to be represented by, and their physical properties and chemistry interpreted in terms of, the simple formula (1) first proposed by Staudinger.[1]

However, it is now quite apparent that the representation (1) while convenient, and in many cases useful as a starting point for discussion, has

---

* Based on a monomer molecular weight of 100 and termination by combination.

serious limitations when it comes to understanding the detailed chemistry of polymeric materials.

For example, how can we rationalize the finding that two polymers with nominally the same chemical and physical composition have markedly different thermal stability? PMMA (**1**, X=CH$_3$, Y=CO$_2$CH$_3$) prepared by anionic polymerization has been reported to be more stable by some 50°C than that prepared by a radical process.[5] The simplified representation, (**1**), also provides no ready explanation for the discrepancy in chemical properties between low molecular weight model compounds and polymers even though both can be represented ostensibly by the same structure (**1**). Consideration of the properties of simple models indicates that the onset of thermal degradation of PVC (**1**, X=H, Y=Cl) should occur at a temperature 100°C higher than is actually found.[6]

Such problems have led to a recognition of the importance of defect groups (structural units different from those described by the generalized formula **1**).[7-11] If we are to achieve an understanding of radical polymerization, and the ability to produce polymers with optimal, or at least predictable, properties, a much more detailed knowledge of the mechanism of the polymerization and of the chemical microstructure of the polymers formed is required.[11] Defect groups may be introduced into the chain during any stage of the polymerization and we must always question whether it is appropriate to use the generalized formula (**1**) for representing the polymer structure.

Obvious examples of defect structures are the groups formed by chain initiation and termination. Initiating species are not only formed directly from initiator decomposition (Scheme 1) but also indirectly by transfer to monomer, solvent, transfer agent, or impurities (see Scheme 2).

Chain Transfer:

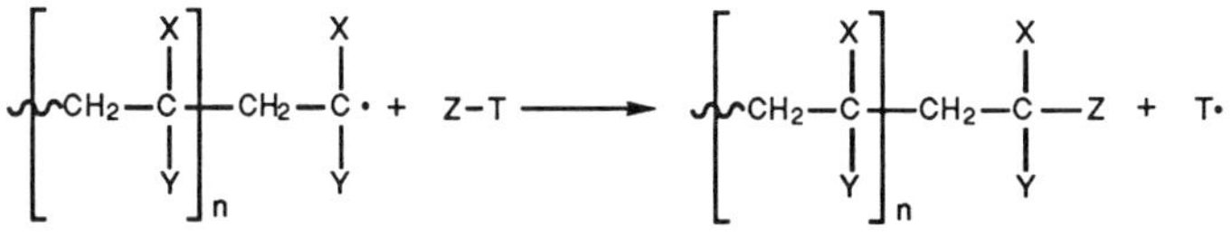

$$\left[\text{\textasciitilde{}CH}_2-\underset{Y}{\overset{X}{C}}\right]_n-\text{CH}_2-\underset{Y}{\overset{X}{C}}\cdot\ +\ Z-T\ \longrightarrow\ \left[\text{\textasciitilde{}CH}_2-\underset{Y}{\overset{X}{C}}\right]_n-\text{CH}_2-\underset{Y}{\overset{X}{C}}-Z\ +\ T\cdot$$

Reinitiation:

$$\text{T}\cdot\ +\ \text{CH}_2=\underset{Y}{\overset{X}{C}}\ \longrightarrow\ \text{T}-\text{CH}_2-\underset{Y}{\overset{X}{C}}\cdot$$

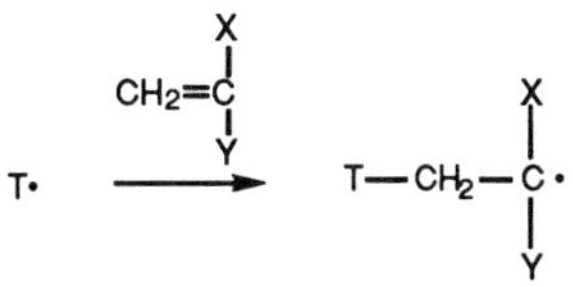

Scheme 2

In termination, unsaturated and saturated ends are formed when the propagating species undergo disproportionation, head-to-head linkages when they combine, and other functional groups may be introduced by reactions with inhibitors or transfer agents (see Scheme 2). In-chain defect structures (within the polymer molecule) can also arise by copolymerization of the unsaturated by-products of initiation or termination.

The generalized structure (1) also overestimates the homogeneity of the repeat units (the specificity of propagation). The traditional explanation offered to rationalize structure (1), which implies exclusive formation of head to tail linkages in the propagation step of radical polymerization, is that the reaction is under thermodynamic control. This explanation was based on the observation that additions of simple radicals to mono- or 1,1-disubstituted olefins typically proceed by tail addition to give secondary or tertiary radicals respectively rather than the less stable primary radical (see Scheme 3) and by analogy with findings for ionic reactions where such thermodynamic considerations are of demonstrable importance.

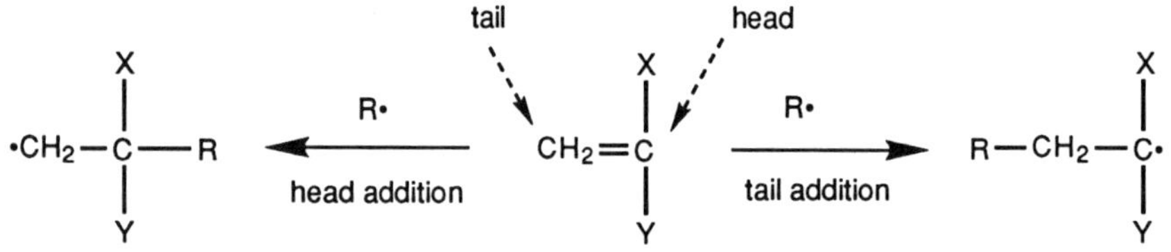

Scheme 3

Until the early 1970s, the absence of suitable techniques for probing the detailed microstructure of polymers or for examining the selectivity and rates of radical reactions prevented the traditional view from being seriously questioned. In more recent times, it has been established that radical reactions, more often than not, are under kinetic rather than thermodynamic control and the predominance of head-to-tail linkages in polymers is determined largely by steric and polar influences (see 2.2).[12]

It is now known that a proportion of "head" addition occurs during the initiation and propagation stages of many polymerizations (see 4.3). For example, poly(vinyl fluoride) chains contain in excess of 10% head-to-head linkages.[13] Benzoyloxy radicals give ca. 5% head addition with styrene (see 3.4.2.2).[14,15] However, one of the first clear-cut examples demonstrating that thermodynamic control is not of overriding importance in determining the outcome of radical reactions is the cyclopolymerization of diallyl compounds (see 4.4.1).[16-19]

The pioneering studies of Butler and coworkers[18,19] established that diallyl compounds, of general structure (3), undergo radical polymerization to give linear saturated polymers. They proposed that the propagation involved a series of inter- and intramolecular addition reactions. The presence of cyclic units in the polymer structure was rigorously established by chemical analysis.[20] Addition of a radical to the diallyl monomer (3) could conceivably lead to the formation of 5-, 6- or even 7-membered rings (Scheme 4). However, application of the then generally accepted hypothesis, that product radical stability was the most important factor determining the course of radical addition, indicated that the intermolecular step should proceed by tail addition (to give 4) and that the intramolecular step should afford a 6-membered ring and a secondary radical (6). On the basis of this theory, it was proposed that the cyclopolymer was composed of 6-membered rings (8).

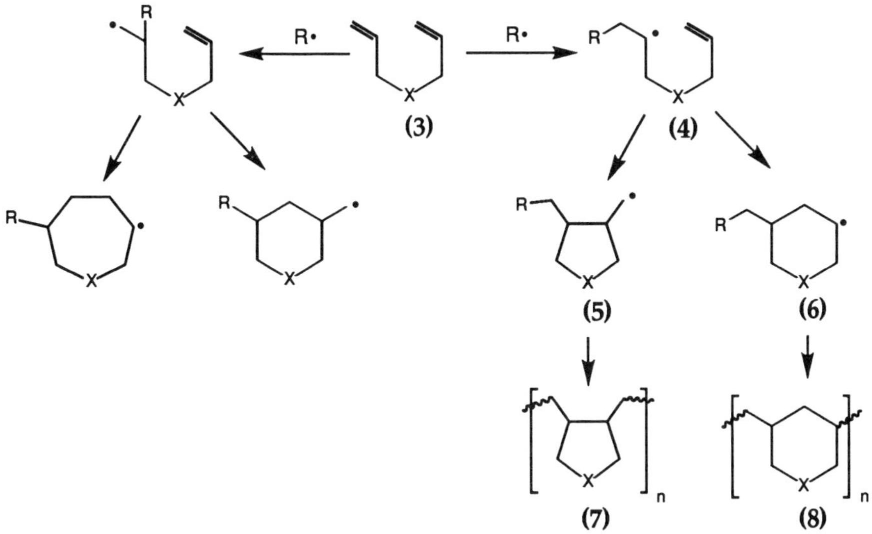

Scheme 4

It was established in the early 1960s that hexenyl radicals and simple derivatives gave 1,5- rather than 1,6-ring closure under conditions of kinetic control.[21] However, it was not until 1976 that Hawthorne et al.[22] proved that in the cyclopolymerization of monomers (3), the intramolecular cyclization step gives preferentially the less stable radical (5) (5-vs. 6-membered ring, primary vs. secondary radical) - i.e. ≥99% head addition. Over the last decade, many other examples of radical reactions which preferentially afford the thermodynamically less stable product have come to light. A discussion of various factors important in determining the course and rate of radical additions will be found in chapter 2.

The examples described in this chapter serve to illustrate two well-recognized, though often overlooked, principles which lie at the heart of polymer and, indeed, all forms of chemistry. These are:

(a) That the dependence of a reaction (polymerization, polymer degradation, *etc.*) on experimental variables cannot be understood until the reaction mechanism is established.
(b) A reaction mechanism cannot be fully defined, when the reaction products are unknown.

In the succeeding chapters we detail the current state of knowledge of the chemistry of each stage of polymerization. We consider the details of the mechanisms, the specificity of the reactions, the nature of the group or groups incorporated in the polymer chain, and any byproducts. The intention is to create an awareness of the factors which must be borne in mind in selecting the conditions for a given polymerization and provide the background necessary for a more thorough understanding of polymerizations and polymer properties.

References

1. Staudinger, H., *Chem. Ber.*, 1920, **53**, 1073.
2. Bevington, J.C., in 'Comprehensive Polymer Science' (Eds. Eastmond, G.C., Ledwith, A., Russo, S., and Sigwalt, P.), Vol. 3, p. 65 (Pergamon: London 1989).
3. Flory, P.J., Principles of Polymer Chemistry, p. 106 (Cornell University Press: Ithaca, N.Y. 1953).
4. Walling, C., Free Radicals in Solution, p. 592 (Wiley: New York 1957).
5. McNeill, I.C., *Eur. Polym. J.*, 1968, **4**, 21.
6. Mayer, Z., *J. Macromol. Sci., Rev. Macromol. Chem.*, 1974, **C10**, 263.
7. Solomon, D.H., Cacioli, P., and Moad, G., *Pure Appl. Chem.*, 1985, **57**, 985.
8. Hwang, E.F.J., and Pearce, E.M., *Polym. Eng. Rev.*, 1983, **2**, 319.
9. Mita, I., in 'Aspects of Degradation and Stabilization of Polymers' (Ed. Jellineck, H.H.G.), p. 247 (Elsevier: Amsterdam 1978).
10. Solomon, D.H., *J. Macromol. Sci., Chem.*, 1982, **A17**, 337.
11. Moad, G., and Solomon, D.H., *Aust. J. Chem.*, 1990, **43**, 215.
12. Tedder, J.M., *Angew. Chem., Int. Ed. Engl.*, 1982, **21**, 401.
13. Cais, R.E., and Kometani, J.M., *ACS Symp. Ser.*, 1984, **247**, 153.
14. Moad, G., Rizzardo, E., and Solomon, D.H., *Macromolecules*, 1982, **15**, 909.
15. Moad, G., Rizzardo, E., Solomon, D.H., Johns, S.R., and Willing, R.I., *Makromol. Chem., Rapid Commun.*, 1984, **5**, 793.
16. Butler, G.B., *Acc. Chem. Res.*, 1982, **15**, 370.
17. Solomon, D.H., and Hawthorne, D.G., *J. Macromol. Sci., Rev. Macromol. Chem.*, 1976, **C15**, 143.
18. Butler, G.B., in 'Encyclopaedia of Polymer Science and Engineering, 2nd Edition' (Eds. Mark, H.F., Bikales, N.M., Overberger, C.G., and Menges, G.), Vol. 4, p. 543 (Wiley: New York 1986).
19. Butler, G.B., in 'Comprehensive Polymer Science' (Eds. Eastmond, G.C., Ledwith, A., Russo, S., and Sigwalt, P.), Vol. 4, p. 423 (Pergamon: London 1989).
20. Butler, G.B., Crawshaw, A., and Miller, W.L., *J. Am. Chem. Soc.*, 1958, **80**, 3615.
21. Beckwith, A.L.J., and Ingold, K.U., in 'Rearrangements in Ground and Excited States' (Ed. de Mayo, P.), Vol. 1, p. 162 (Academic Press: New York 1980).
22. Hawthorne, D.G., Johns, S.R., Solomon, D.H., and Willing, R.I., *Aust. J. Chem.*, 1976, **29**, 1955.

2
Free Radical Reactions

2.1 Introduction

The intention of this chapter is to discuss in some detail the factors which determine the rate and course of radical reactions. Emphasis is placed on those reactions most frequently encountered in radical polymerization:

(a) Addition to carbon-carbon double bonds (*e.g.* initiation - see 3.4, propagation - see 4.2, 4.3).

$$X\cdot \; + \; \;\mathrm{C{=}C} \; \xrightarrow{k_T + k_H} \; X{-}C{-}C\cdot$$

(b) Hydrogen atom transfer (*e.g.* chain transfer - see 5.3).

$$X\cdot \; + \; H{-}C{-} \; \xrightarrow{k_{tr}} \; X{-}H \; + \; \cdot C{-}$$

(c) The self-reaction of carbon-centered radicals (*e.g.* termination - see 5.2).

$$-C{-}C\cdot \; + \; \cdot C{-} \; \xrightarrow{k_{td}} \; \mathrm{C{=}C} \; + \; H{-}C{-}$$
$$\xrightarrow{k_{tc}} \; -C{-}C{-}C{-}$$

Other reactions not covered in this chapter are dealt with in the chapters which follow. These include additions to systems other than carbon-carbon double bonds (*e.g.* additions to aromatic systems - see 3.4.2.2.1 - and strained ring systems - see 4.4.2), heteroatom or group transfer (*e.g.* chain transfer to halocarbons, peroxides, and disulfides - see 5.3.1.1), radical-radical reactions involving heteroatom-centered radicals, and metal ion-radical and metal complex-radical reactions (*e.g.* inhibition by stable radicals and metal complexes - see 3.5.1 and 5.4).

Until the early 1970s, views of radical reactions were dominated by two seemingly contradictory beliefs: (a) that radical reactions, in that they involve highly reactive species, should not be expected to show any particular selectivity, and (b) that (as is often possible with ionic reactions) the outcome could be predicted purely on the basis of the relative thermochemical stability of the product radicals. For condition (a) to apply, a reaction should have an early reactant-like transition state and a near-zero activation energy. For condition (b) to apply, the transition state should be late (or product-like) or the reaction leading to products must be under thermodynamic control by virtue of being rapidly reversible. While either of the above conditions may apply in specific cases, for radical reactions in general, neither applies.

It is now recognized that radical reactions are, more often than not, under kinetic rather than thermodynamic control. The reactions can nonetheless show a high degree of specificity which is imposed by steric (non-bonded interactions), polar (relative electronegativities), stereoelectronic (requirement for overlap of frontier orbitals), bond-strength (relative strengths of bonds formed and broken) and perhaps other constraints.[1-3] In the following sections we discuss these factors, consider their relative importance in specific reactions and suggest guidelines for predicting the outcome of radical reactions.

2.2 Addition to Carbon-Carbon Double Bonds

With few exceptions, radicals are observed to add preferentially to the less highly substituted end of unsymmetrically substituted olefins (*i.e.* give predominantly tail addition* - see Scheme 1). For a long time, this finding was correlated with the observation that any substituent at a radical center tends to enhance its stability. This in turn led to the belief that the degree of stabilization conferred on the product radical by the substituents was the prime factor determining the orientation and rate of radical addition to olefins. That steric, polar, or other factors might favor this same outcome was either considered to be of secondary importance or simply ignored.†

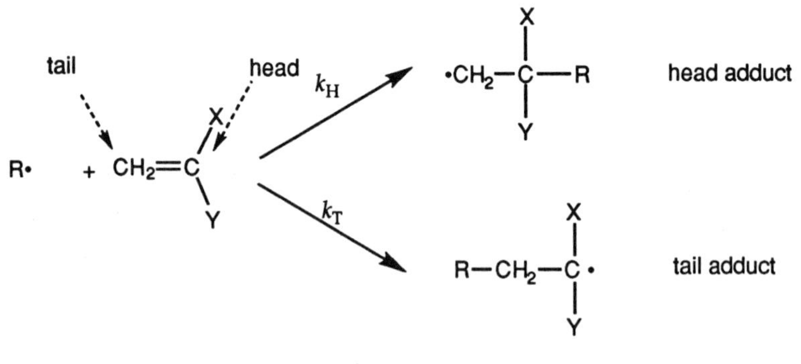

$$R\cdot + CH_2=C\begin{matrix}X\\Y\end{matrix} \xrightarrow{k_H} \cdot CH_2-\underset{Y}{\overset{X}{C}}-R \quad \text{head adduct}$$

$$\xrightarrow{k_T} R-CH_2-\underset{Y}{\overset{X}{C}}\cdot \quad \text{tail adduct}$$

Scheme 1

---

* The term tail addition is used to refer to addition to the less highly substituted end of the double bond.
† To this day some texts put forward product stability as the sole explanation for preferential tail addition.

Indeed, while alternative hypotheses were entertained by some,[4] there was no serious questioning of the dominant role of thermochemistry in the wider community until the 1970s. Many factors were important in bringing about this change in thinking. Three of the more significant were:

(a) A few isolated examples appeared where "wrong way" addition (formation of the less thermodynamically stable radical) was a significant, or even the major, pathway. Notable examples are predominantly head addition in the intramolecular step of cyclopolymerization of 1,6-dienes (Scheme 2),[5]

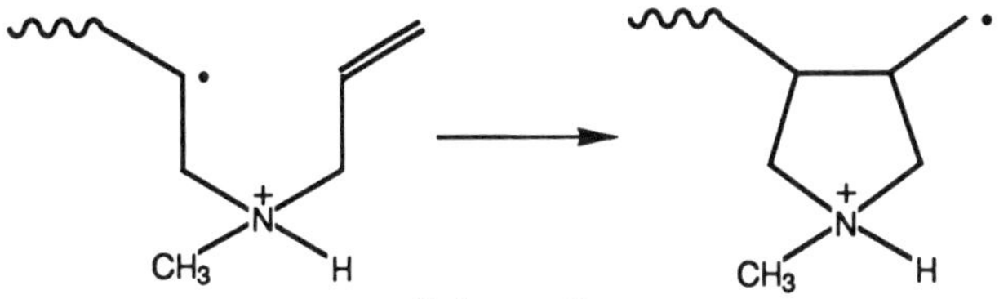

Scheme 2

and in the reaction of t-butoxy radicals with difluoroethylene (Scheme 3).[6]

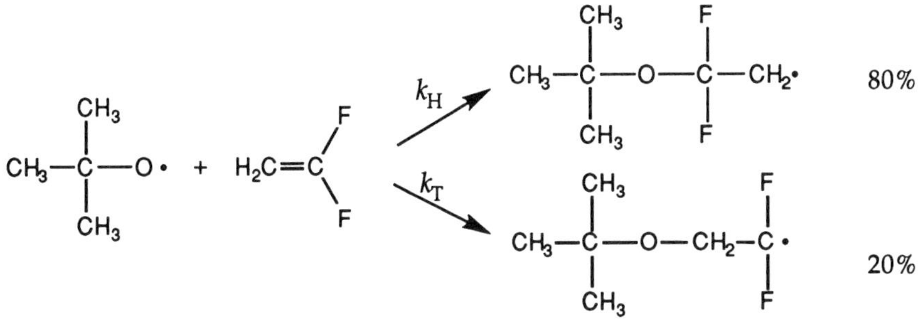

Scheme 3

(b) Dependable measurements of rate constants for radical reactions became available which allowed structure-reactivity relationships to be reliably assessed.[7]
(c) Data on bond dissociation energies were evaluated to demonstrate that the amount of stabilization provided to a radical center by adjacent alkyl substituents is small. The relative stability of primary vs. secondary vs. tertiary radicals, even if fully reflected in the transition state, is not sufficient to account for the degree of regioselectivity observed in additions to alkenes.[8]

It is now established that product radical stability is a consideration in determining the outcome of radical addition reactions only where a substituent provides substantial delocalization of the free spin into a π-system. Even then, because these reactions are generally irreversible and exothermic (and consequently have early transition states), resonance stabilization of the incipient radical center may play only a minor role in determining reaction rate and specificity.[2,9-12] Thermodynamic factors will be the dominant influence only when polar and steric effects are more or less evenly balanced.[13,14]

The relative importance of the various factors in determining the rate and regiospecificity of addition is illustrated by the data shown in Tables 2.1 and 2.2.

Table 2.1 Relative Rate Constants and Regiospecificities for Addition of Radicals to Halo-Olefins[a]

olefin	$(CH_3)_3CO\cdot$[b]		$CH_3\cdot$[c]		$CF_3\cdot$[c]		$CCl_3\cdot$[c]	
	k_{rel}	k_H/k_T	k_{rel}	k_H/k_T	k_{rel}	k_H/k_T	k_{rel}	k_H/k_T
$CH_2=CH_2$	1.0	-	1.0	-	1.0	-	1.0	-
$CH_2=CHF$	0.7	0.35	1.1	0.2	0.5	0.12	0.62	0.11
$CH_2=CF_2$	1.1	4.0	-	1	0.2	0.04	0.25	0.016
$CHF=CF_2$	6.6	4.5	5.8	2.1	0.05	0.55	0.29	0.32

[a] k_{rel} is overall rate constant for addition (k_H+k_T) relative to that for addition to ethylene (=1.0). All data have been rounded to 2 significant figures.
[b] At 60°C.[15]
[c] At 164°C.[7]

Table 2.2 Relative Rate Constants for Reactions of Radicals with Alkyl-Substituted Acrylate Esters, $CHR^1=CR^2CO_2CH_3$[a]

Monomer	R^1	R^2	$PhCO_2\cdot$[b]		$Ph\cdot$[b]		$(CH_3)_3CO\cdot$[b]		$c\text{-}C_6H_{11}\cdot$[c]	
			k_H	k_T	k_H	k_T	k_H	k_T	k_H	k_T
MA	H	H	0.2	1.0	0.03	1.0	0.02	1.0	0.002	1.0
MMA	H	CH_3	0.35	4.5	≤0.01	1.6	0	2.9	≤0.001	0.71
MC[d]	CH_3	H	1.6	1.3	0.07	0.12	≤0.03	0.3	0.001	0.011

[a] Rate constants relative to that for tail addition to MA (=1.0).
[b] At 60°C.[16]
[c] At 20°C.[17]
[d] Methyl *trans*-2-butenoate (methyl crotonate).

Relative rate constants for reaction of methyl, trifluoromethyl, trichloromethyl,[7] and *t*-butoxy radicals[15,18] with the fluoro-olefins are summarized in Table 2.1. Note the following points:

(a) Overall rates of addition for methyl and *t*-butoxy radicals are accelerated by fluorine substitution. In contrast, rates for trifluoromethyl and trichloromethyl radicals are reduced by fluorine substitution.

(b) Trifluoromethyl and trichloromethyl radicals preferentially add to the less substituted end of trifluoroethylene. Methyl and *t*-butoxy radicals add preferentially to the more substituted end.
(c) With vinylidene fluoride, trifluoromethyl and trichloromethyl radicals give predominantly tail addition, methyl radicals give both tail and head addition, *t*-butoxy radicals give predominantly head addition.
(d) The overall trend of reactivities for *t*-butoxy radicals with the fluoro-olefins more closely parallels that for methyl radicals than that for the electrophilic trifluoromethyl or trichloromethyl radicals.

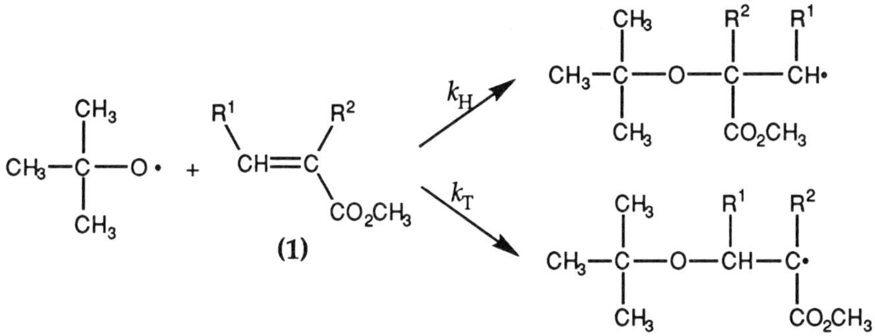

Scheme 4

Different outcomes are obtained from the reactions of radicals with substituted acrylate esters (1) according to the nature of the attacking radical (refer Table 2.2 and Scheme 4). The results may be summarized as follows (note that the methyl substituent is usually considered to be electron donating):

(a) Irrespective of the attacking radical, there is preferential addition to the tail of the double bond (to the end remote from the carbomethoxy group).
(b) For the nucleophilic cyclohexyl radicals (c-C_6H_{11}•), the rate of addition to the unsubstituted end of the double bond is slightly retarded by alkyl substitution (ca. 30% for MMA vs. MA). The rate of addition to the substituted end of the double bond is dramatically retarded by alkyl substitution (ca. 90-fold for MC vs. MA).[17]
(c) For the slightly electrophilic phenyl and *t*-butoxy radicals [Ph•, $(CH_3)_3CO$•], the rate of addition to the unsubstituted end of the double bond is enhanced (2-3-fold) by alkyl substitution, the rate of addition to the substituted end of the double bond is retarded (>3 fold for MC vs. MA) by alkyl substitution.[16,19]
(d) For the electrophilic benzoyloxy radicals ($PhCO_2$•), the rate of addition to the unsubstituted (tail) end of the double bond is enhanced (4.5-fold for MMA vs. MA) by alkyl substitution, the rate of addition to the substituted (head) end of the double bond is slightly enhanced (75% for MMA vs. MA) by alkyl substitution.[16]

The data of Tables 2.1 and 2.2 clearly cannot be rationalized purely in terms of the relative stabilities of the product radicals. Rather, "a complex interplay of polar, steric, and bond strength terms" must be invoked.[7] In the following sections, each of these factors will be examined separately to illustrate their role in determining the outcome of radical addition.

2.2.1 Steric Factors

A clear demonstration of the relative importance of steric and resonance factors in radical additions to carbon-carbon double bonds can be found by considering the effect of (non-polar) substituents on the rate of attack of (non-polar) radicals. Substituents on the double bond strongly retard addition at the substituted carbon while leaving the rate of addition to the other end essentially unaffected (see, for example, Table 2.2). This is in keeping with expectation if steric factors determine the regiospecificity of addition but contrary to expectation if resonance factors are dominant.

It is possible to resolve steric factors into several terms:

(a) B-strain engendered by the change from sp^2 towards sp^3 hybridization at the site of attack.[2,8] B-strain is a consequence of the substituents on the (planar) α-carbon of the double bond being brought closer together on moving towards a tetrahedral disposition (see Figure 1). This term is important in all radical additions and is thought to be the main factor responsible for preferential attack at the less substituted end of the double bond.

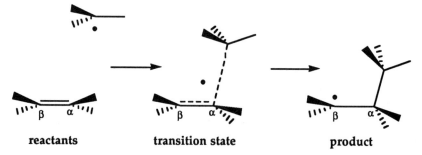

Figure 1 Transition state for radical addition.

(b) Steric hindrance to approach of the attacking radical to the site of attack on the olefin. This term is usually only a minor factor except where substituents on the radical or on the olefin are very bulky.[8,20]
(c) Steric hindrance to adoption of the required transition state geometry. This is not usually a determining factor in intermolecular addition, but is extremely important in intramolecular addition where the approach of the reacting centers is constrained by the molecular architecture (see 2.2.4).[21]

Radical additions are typically highly exothermic and activation energies are small for carbon-[22] and oxygen-centered[23] radicals of the types most often encountered in radical polymerization. Thus, according to the Hammond postulate,* these reactions are expected to have early reactant-like transition states in which there is little localization of the free spin on C_β. However, for steric factors to be important at all, there must be significant bond deformation and movement towards sp^3 hybridization at C_α.

Various *ab initio* and semi-empirical molecular orbital calculations have been carried out on the reaction of radicals with simple alkenes with the aim of defining the nature of the transition state (see 2.2.6).[25-30] These calculations all predict an unsymmetrical transition state for radical addition (*i.e.* Figure 1) though they differ in other aspects. Most calculations also indicate a degree of charge development in the transition state.

The rate of radical addition is most dramatically affected by substituents either at the site of attack or at the radical center. Remote substituents generally have only a small influence on the stereochemistry and regiospecificity of addition unless these groups are very bulky[9,31] or the geometry of the molecules is constrained (*e.g.* intramolecular addition - see 2.2.4).

It is a common assumption that the influence of steric factors will be manifested mainly as a higher activation energy. In fact, there is good evidence[32] to show that steric factors are also reflected in a less favorable entropy of activation or Arrhenius frequency factor. This is due to the degrees of freedom that are lost as the radical center approaches the terminus of the double bond and the α-substituents on the double bond are brought closer together on rehybridization.

2.2.2 Polar Factors

The rates of addition to the unsubstituted terminus of 1- and 1,1-disubstituted olefins (this includes most monomers) are thought to be determined largely by polar factors.[2,10] Polymer chemists were amongst the first to realize that polar factors were an important influence in determining the rate of addition. Polar factors can account for the well known tendency for monomer alternation in many radical copolymerizations and provide the basis for the Q-e,[33] the Patterns of Reactivity,[34,35] and other schemes for estimating monomer reactivity ratios (see 6.2.3).

The traditional means of assessment of the sensitivity of radical reactions to polar factors and establishing the electrophilicity or nucleophilicity of radicals is by way of a Hammett σρ correlation. Thus, the reactions of radicals with substituted styrene derivatives have been examined to demonstrate that simple

---

* A highly exothermic (low activation energy) reaction will generally have a transition state that resembles the reactants.[24]

alkyl radicals have nucleophilic character[36,37] while haloalkyl radicals[38] and oxygen-centered radicals[18] have electrophilic character (see Table 2.3). It is anticipated that electron-withdrawing substituents (Cl, F, CO_2R, CN) will enhance overall reactivity towards nucleophilic radicals and reduce reactivity towards electrophilic radicals. Electron-donating substituents (alkyl) will have the opposite effect.

Table 2.3 Hammett ρ and ρ^+ Parameters for Reactions of Radicals

radical	addition to styrene		abstraction from toluene	
	ρ^+	ρ	ρ^+	ρ^a
$(CH_3)_3C\cdot$	1.1[a,36]	-	-	0.49[b,39]
c-$C_6H_{11}\cdot$	0.68[a,36]	-	-	-
n-$C_6H_{13}\cdot$	0.45[a,36]	-	-	-
n-$C_{11}H_{23}\cdot$	-	-	-	0.45[a,40]
$CH_3\cdot$	-	-	-0.1[c,41]	-0.12[c,41] (-0.21)[d]
$(CH_3)_3CO\cdot$	-0.27[e,18]	-0.31[e,18]	-0.32[f,42]	-0.36[f,42] (-0.36)[d]
$(CH_3)_3COO\cdot$	-	-	-0.56[g,43]	-0.78[g,43] (-0.73)[d]
$(CH_3)_2N\cdot$	-	-	-1.08[h,44]	-1.66[h,44] (-0.96)[d]
$CCl_3\cdot$	-0.42[i,38]	-0.43[i,38]	-1.46[j,45]	-1.46[j,45] (-1.67)[d]

(arrow on left labeled "nucleophilicity" pointing upward)

a 42°C.
b 80°C.
c 100°C.
d ρ values recalculated by Pryor et al.[46] based on m-substituted derivatives only.
e 60°C, benzene.
f 45°C, chlorobenzene. Value shows solvent dependence.
g 40°C.
h 136°C.
i 70°C.
j 50°C.

While steric terms may be the most significant factor in determining that tail addition is the predominant pathway in radical addition, polar terms control the overall reactivity and the degree of regiospecificity. In the reaction of benzoyloxy radicals with MMA, even though there is still a marked preference for tail addition, the methyl substituent enhances the rate constants for attack at both head and tail positions over those seen for MA (see Table 2.2). With cyclohexyl radicals the opposite behavior is seen. Relative rate constants are reduced and the preference for tail addition is reinforced. For olefins substituted with electron-donor substituents, nucleophilic radicals give the greatest tail $vs.$ head specificity. The converse generally also applies.

In the reactions of the fluoro-olefins, steric factors are of lesser importance because of the relatively small size of the fluoro-substituent. Fluorine and hydrogen are of similar bulk. In these circumstances, it should be expected that polar factors could play a role in determining regiospecificity. Application of the usual rules to vinylidene fluoride leads to a prediction that, for nucleophilic radicals, the rate of head addition will be enhanced. Similarly, for electrophilic radicals, the rate of tail addition will be enhanced.

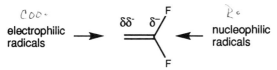

The behavior of methyl and halomethyl radicals in their reactions with the fluoro-olefins (Table 2.1), can thus be rationalized in terms of a more dominant role of polar factors and the nucleophilic or electrophilic character of the radicals involved.[2] Methyl radicals are usually considered to be slightly nucleophilic, trifluoromethyl and trichloromethyl radicals are electrophilic (see Table 2.3).

However, consideration of polar factors in the traditional sense does not provide a ready explanation for the regiospecificity shown by the *t*-butoxy radicals (which are electrophilic, see Table 2.3) in their reactions with the fluoro-olefins (see Table 2.1).[15,18] Apparent ambiphilicity has been reported[14] for other "not very electrophilic radicals" in their reactions with olefins and has been attributed to the polarizability of the radical.

2.2.3 Bond Strengths

Tedder[7] has stated: "If an experimentalist requires a simple qualitative theory, he should seek to estimate the strength of the new bond formed during the initial addition step...". It should be pointed out that the bond strength term cannot be separated rigorously from the polar and steric factors discussed above since the latter both play an important role in determining the strength of the new bond.

Just as steric factors may retard addition, factors which favor bond formation should be anticipated to facilitate addition. A pertinent example is the influence of α-fluorine substitution on C-X bond strength.[47] The C-C bond in CH_3-CF_3 is 46 kJ mol^{-1} stronger than that in CH_3-CH_3. Further fluorine substitution leads to a progressive strengthening of the bond. The effect is even greater for C-O bonds. The C-O bond dissociation energies in CF_3-O-CF_3 and CF_3-OH are greater by 92 and 75 kJ mole^{-1}, respectively, than those in CH_3-O-CH_3 and CH_3-OH. This effect offers an explanation for the differing specificity shown by oxygen- and carbon-centered radicals in their reactions with the fluoro-olefins (Table 2.1).[18,27,48,49] The finding, that *t*-butoxy radicals give predominantly head addition with vinylidene fluoride (Scheme 3), can therefore be attributed to the relative strengths of the CF_2-O and CH_2-O bonds.[18]

2.2.4 Stereoelectronic Factors

A stereoelectronic requirement in radical addition to carbon-carbon double bonds first became apparent from studies on radical cyclization and the reverse (fragmentation) reactions. It provides a rationalization for the preferential formation of the less thermodynamically stable *exo*-product from the cyclization of ω-alkenyl radicals (**2** - see Scheme 5).[12,50-56]

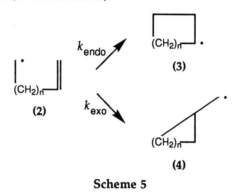

Scheme 5

It was proposed that the transition state requires approach of the radical directly above the site of attack and perpendicular to the plane containing the carbon-carbon double bond. An examination of molecular models shows that for the 3-butenyl and 4-pentenyl radicals (**2**, n=1,2) such a transition state can only be reasonably achieved in *exo*-cyclization (*i.e.* **2**→**4**). With the 5-hexenyl and 6-heptenyl radicals (**2**, n=3,4), the transition state for *exo*-cyclization (**2**→**4**) is more easily achieved than that for *endo*-cyclization (*i.e.* **2**→**3**).

The mode and rate of cyclization can be modified substantially by the presence of substituents at the radical center, on the double bond, and at positions on the connecting chain. As with intermolecular addition, substituents at the site of attack on the double bond strongly retard addition. For the 5-hexenyl system the magnitude of the effect is such that methyl substitution at the 5-position causes *endo*-cyclization to be favored. For the 5,6-disubstituted radical the rates for both *exo*- and *endo*-addition are slowed and *exo*-cyclization again dominates. A full discussion of substituent effects on intramolecular addition can be found in the reviews cited above.

2.2.5 Reaction Conditions

There is ample evidence to show that the outcome of radical reactions is dependent on reaction conditions (temperature, solvent, *etc.*).

2.2.5.1 Temperature

Radical additions to double bonds are, in general, exothermic processes and rates increase with increasing temperature. The regiospecificity of addition to double bonds and the relative reactivity of various olefins towards radicals is also temperature dependent. Typically, specificity decreases with increasing temperature (the reactivity-selectivity principle applies). However, a number of exceptions to this general rule have been reported.[36,57]

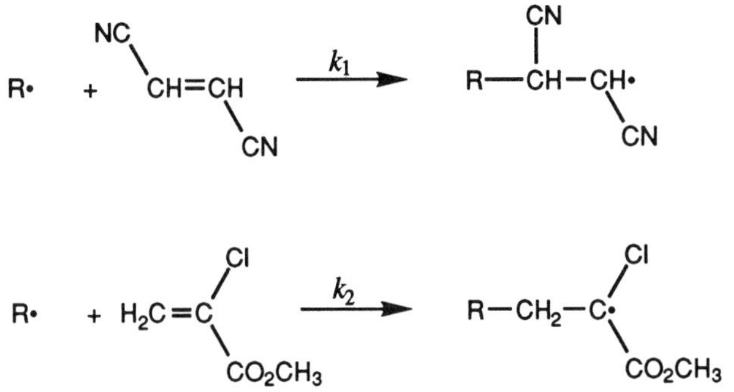

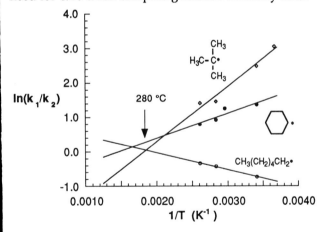

Scheme 6

Giese and Feix[57] examined the temperature dependence of relative reactivity of fumarodinitrile and methyl α-chloroacrylate towards a series of alkyl radicals (Scheme 6). The temperature dependence was such that they predicted that the order of reactivity of the radicals would be reversed for temperatures above 280°C (the isoselective temperature - see Figure 2). This finding clearly indicates the need for care when comparing relative reactivity data.[58]

Figure 2 Relative rate constants for addition of alkyl radicals to fumarodinitrile (k_1) and methyl α-chloroacrylate (k_2) as a function of temperature.[57]

2.2.5.2 Solvent

It is established that rates of propagation in radical polymerization and reactivity ratios in copolymerization can show marked variation according to the solvent employed.[59-63] For polymerizations of ethylene and vinyl acetate, effects on low conversion values of k_p in excess of an order of magnitude have been reported.[60,64] However, conventional wisdom has it that, except for those reactions involving charged intermediates, solvent effects on the rate and regioselectivity of radical addition to olefins are small and, consequently, they have not been widely studied. Nonetheless, reports of measurable solvent effects continue to appear.

For example, Giese and Kretzschmar[65] found the rate of addition of hexenyl radicals to methyl acrylate varied two-fold between aqueous tetrahydrofuran and aqueous ethanol. Bednarek et al.[66] found that the relative reactivity of styrene vs. MMA towards phenyl radicals was ca. 20% greater in ketone solvents than it was in aromatic solvents.

More pronounced solvent effects have been observed in special cases where substrates or products possess ionic character. Ito and Matsuda[67] found a 35-fold variation in the rate of addition of (5) to α-methylstyrene in going from dimethylsulfoxide to cyclohexane. Rates for addition of other arenethiyl radicals do not show such a marked solvent dependence. The different behavior was attributed to the radical (5) existing partly in a zwitterionic quinonoid form.[68]

(5)

2.2.6 Theoretical Treatments

There have been many theoretical studies on radical addition reactions using *ab initio* methods,[28,49,69-72] semi-empirical calculations,[73,74] molecular mechanics[75,76] and other procedures. Sosa et al.[28,70] demonstrated that while geometries do not vary substantially with the level of theory, to obtain meaningful activation parameters with *ab initio* methods, the highest level of theory is required. Such calculations are at this stage only practicable for small systems. Heuts et al.[77] have indicated that reliable Arrhenius A factors may be available using lower levels of theory.

The calculations using semi-empirical and low level *ab initio* methods do not always give good values of absolute activation parameters. However, they have been shown to be useful in predicting relative energies for structurally similar systems and can give valuable insights into mechanism. Methods for obtaining good estimates of relative activation energies by molecular mechanics have also been devised.

Various empirical schemes have also been proposed as predictive tools with respect to the outcome of radical addition reactions. These include Frontier Molecular Orbital (FMO) theory[78] and the State Correlation Diagram (SCD) approach.[49]

The frontier orbital of the radical is that bearing the free spin (the SOMO) and during radical addition this will interact with both the π^* antibonding orbital (the LUMO) and the π-orbital (the HOMO) of the olefin. Both the SOMO-HOMO and the SOMO-LUMO interactions lead to a nett drop in energy [*i.e.* 2(E_2)-E_3 or E_1 respectively - see Figure 3]. The dominant interaction and the reaction rates depend on the relative energies of these orbitals. Most radicals have high energy SOMO's and the SOMO-LUMO interaction is likely to be the most important. However, with highly electron deficient radicals, the SOMO may be of sufficiently low energy for the SOMO-HOMO interaction to be dominant.

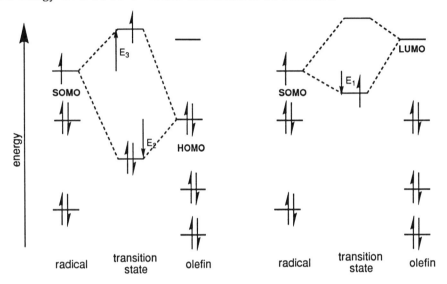

Figure 3 SOMO-HOMO and SOMO-LUMO orbital interaction diagrams.

For olefins with π-substituents, whether electron-withdrawing or electron-donating, both the HOMO and LUMO have the higher coefficient on the carbon atom remote from the substituent. A predominance of tail addition is expected as a consequence. However, for non-conjugated substituents, or those with lone pairs (*e.g.* the halo-olefins), the HOMO and LUMO are polarized in opposite directions. This may result in head addition being preferred in the case of a nucleophilic radical interacting with such an olefin. Thus, the data for attack of alkyl and fluoroalkyl radicals on the fluoro-olefins (see Table 2.1) have been rationalized in terms of FMO theory.[10]

Where the radical and olefin both have near "neutral" philicity, the situation is less clear.[14] Specific knowledge of the relative magnitude of the appropriate orbital coefficients is required in such cases for the method to have any predictive power. This information is usually not readily obtained, although it is, in principle, available from molecular orbital calculations. It should be stressed that FMO theory is not quantitative. It provides no specific value for the differences in relative reactivity of two substrates or the precise degree of regiospecificity observed with a particular olefin. Problems also arise in rationalizing the effects of reaction conditions on reaction specificity.

In spite of their limitations, FMO theory and related schemes have become widespread as a means of rationalizing the outcome and the relative rates of radical additions. They are generally successful in circumstances where steric factors are a minor influence.[10,79]

2.2.7 Summary

No single factor can be identified as determining the outcome of radical addition. Nonetheless, there is a requirement for a set of simple guide-lines to allow qualitative prediction. This need was recognized by Tedder,[2,11] Beckwith et al.[53] and Giese.[10] With the current state of knowledge, any such rules must be largely empirical and, therefore, it is to be expected that they may have to be revised from time to time as more results become available and further theoretical studies are carried out. However, this does not diminish their usefulness.

The following set of guide-lines is based on that suggested by Tedder:[2]

(a) For mono- or 1,1-disubstituted olefins, there will usually be preferential addition to the unsubstituted (tail) end of the double bond. This selectivity can be largely attributed to the degree of steric compression associated with the formation of the new bond which usually overrides other influences on the regioselectivity.

(b) Substituents with π-orbitals (e.g. -CH=CH$_2$, -Ph) which can overlap with the half-filled atomic orbital of the incipient radical center, may enhance the rate of addition at the remote end of the double bond, but substituents with non-bonding pairs of electrons (e.g. -F, -Cl, -OR) have only a very small resonance effect. Most radical additions are exothermic and have early transition states and delocalization of the unpaired electron in the adduct radical is of small importance.

(c) Polarity can have a major effect on the overall rate of addition. Electron withdrawing substituents will facilitate the addition of nucleophilic radicals while electron donating substituents will enhance the addition of electrophilic radicals.

(d) The regioselectivity of addition to polysubstituted olefins is primarily controlled by the degree of steric compression associated with forming the new bond. However, if steric effects are small or mutually opposed, polarity can be the deciding factor.

(e) Even though the regioselectivity of addition to polysubstituted olefins is governed mainly by steric compression, polarity can influence the magnitude of the regioselectivity, making it larger or smaller depending on the relative electronegativity of the radical and the substituents on the olefin. The nett result may be that the more reactive radical is the more selective.

2.3 Hydrogen Atom Transfer

Atom or radical transfer reactions generally proceed by what is known as a S_H2 mechanism (substitution, homolytic, bimolecular) which can be depicted schematically as follows:

$$R\cdot + X\text{-}Y \rightarrow [R\cdots X\cdots Y]^\ddagger \rightarrow R\text{-}X + Y\cdot$$

This area has been the subject of a number of significant reviews.[1-3,20,80-82] The present discussion is limited, in the main, to hydrogen atom abstraction from aliphatic substrates and the factors which influence rate and specificity of this reaction.

2.3.1 Bond Dissociation Energies

Simple thermochemical criteria can often be used to predict the relative facility of hydrogen atom transfer reactions. Evans and Polanyi[83] recognized this and suggested the following relationship (the Evans-Polanyi equation) between the activation energy for hydrogen atom abstraction (E_a) and the difference between the bond dissociation energies for the bonds being formed and broken ($\Delta H°$).

$$E_a = \alpha \, \Delta H° + \beta$$

It follows that for abstraction by a given radical, since the strength of the bond being formed is a constant, there should be a straight line relationship between the activation energy and the strength of the bond being broken [$D(R\text{-}H)$].

$$E_a = \alpha' \, [D(R\text{-}H)] + \beta'$$

Examples of the application of the Evans-Polanyi equation can be found in reviews by Russell[80] and Tedder.[2,3] In the absence of severe steric constraints, straight line correlations between the relative reactivity of substrates towards a given radical can be found for systems: (a) where there is little polarity in the transition state, or (b) when the transition states are of like polarity. Tedder[2,3] has also stressed that, in these reactions, it is important to take note of the strength of the bond being formed. If there is no polarity in the transition state, the more exothermic reaction will generally be the less selective.

2.3.2 Steric Factors

Steric factors fall into four main categories:[20]

(a) The release or occurrence of steric compression due to rehybridization in the transition state where the attacking radical and site of attack are each undergoing rehybridization (from $sp^2 \rightarrow sp^3$ and $sp^3 \rightarrow sp^2$ respectively for aliphatic carbons - refer Figure 4). As a consequence, substituents on the attacking radical are brought closer together while those at the site of attack move apart. Thus, depending on the nature of the substituents at these centers, steric retardation or acceleration may accompany rehybridization.

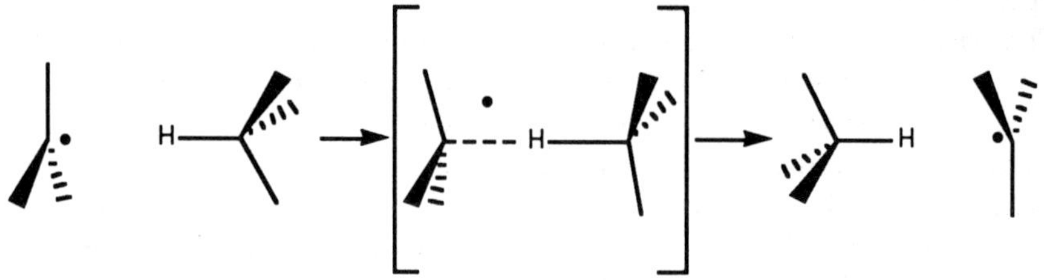

reactants   transition state   products

Figure 4 Transition state for hydrogen atom abstraction.

(b) Steric hindrance of the approach of the attacking radical to the point of reaction in the substrate. This is important for the attack of very bulky radicals on hindered substrates.
(c) Steric inhibition of resonance - important in conformationally constrained molecules (see 2.3.4).
(d) Steric hindrance to adoption of the required co-linear arrangement of atoms in the transition state. This is important in intramolecular reactions (see 2.3.4).

The first term is of importance in all atom abstraction reactions, however, since the reactions are often highly exothermic with consequent early transition states, the effect may be small.

2.3.3 *Polar Factors*

Polar factors can play an extremely important role in determining the overall reactivity and specificity of homolytic substitution.[80] Theoretical studies on atom abstraction reactions support this view by showing that the transition state has a degree of charge separation.[84,85]

The traditional method of assessing the polarity of reactive intermediates is to examine the effect of substituents on rates and establish a linear free energy relationship (*e.g.* the Hammett relationship). The reactions of numerous radicals with substituted toluenes have been examined in this context. The value of the Hammett ρ parameter provides an indication of the sensitivity of the reaction to polar factors and gives a measure of the electrophilic or nucleophilic character of the attacking radical. For example (see Table 2.3), with respect to abstraction of benzylic hydrogens, methyl radicals, usually considered as slightly nucleophilic, have a slightly negative ρ value.[41] Other simple alkyl radicals typically have positive ρ values.[39,40,86,87] Heteroatom-centered radicals (*e.g.* $R_2N\bullet$, $RO\bullet$, $Cl\bullet$) generally have negative ρ values.[42,44,88,89] However, care must be taken in interpreting the results purely in terms of polar effects since electron withdrawing substituents typically also increase bond dissociation energies.[39,46,85,88]

The basic Hammett scheme often does not offer a perfect correlation and a number of variants on this scheme have been proposed to better explain reactivities in radical reactions.[18] However, none of these has achieved widespread acceptance. It should also be noted that linear free energy relationships are the basis of the Q-e and patterns of reactivity schemes for understanding reactivities of propagating species in chain transfer and copolymerization (see 5.3.1 and 6.2.3).

A striking illustration of the influence of polar factors in hydrogen abstraction reactions can be seen in the following examples (Figure 5) where different sites on the molecule are attacked preferentially according to the nature of the attacking radical.[80]

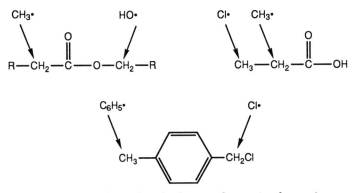

Figure 5 Preferred site of attack in hydrogen abstraction by various radicals.

2.3.4 Stereoelectronic Factors

There is a demonstrated requirement for a near co-linear arrangement of the orbital bearing the unpaired electron and the breaking C-H bond in the transition state for hydrogen atom transfer.[21,56,90,91] This becomes of particular importance for intramolecular atom transfer and accounts for the well-known preference for these reactions to occur by way of a six-membered transition state. The adoption of the chair conformation in the transition state for 1,5-atom transfer allows the requisite arrangement of atoms to be adopted readily. Such a transition state cannot be as readily achieved in smaller rings without significant strain being incurred, or in larger rings due to the severe non-bonded interactions and/or a less favorable entropy of activation.[90,91]

Thus, for the radicals (6), there is a strong preference for 1,5-hydrogen atom transfer (Table 2.4).[92] Although 1,6-transfer is also observed, the preference for 1,5-hydrogen atom transfer over 1,6-transfer is substantial even where the latter pathway would afford a resonance stabilized benzylic radical.[92,93] No sign of 1,2-,

1,3-, 1,4-, or 1,7-transfer is seen in these cases. Similar requirements for a co-linear transition state for homolytic substitution on sulfur and oxygen have been postulated.[12,56]

Table 2.4 Specificity of Intramolecular Hydrogen Abstraction[92]

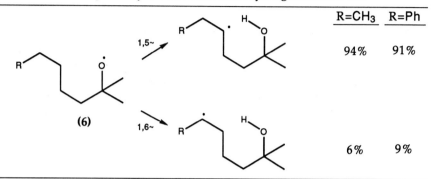

It is expected from simple thermochemical considerations that adjacent π-, σ- or lone pair orbitals should have a significant influence over the facility of atom transfer reactions. Thus, the finding that t-butoxy radicals show a marked preference for abstracting hydrogens α to ether oxygens (Figure 6) is not surprising. The reduced reactivity of the hydrogens β to oxygen in these compounds is attributed to polar influences.[94,95]

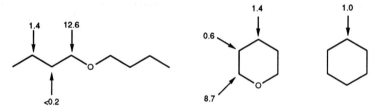

Figure 6 Relative reactivity per hydrogen atom of indicated site towards t-butoxy radicals.[94,95]

The most direct evidence that stereoelectronic effects are also important in these reactions follows from the specificity observed in hydrogen atom abstraction from conformationally constrained compounds.[12,56] Thus, C-H bonds adjacent to oxygen[94-99] or nitrogen[100] and which subtend a small dihedral angle with a lone pair orbital (<30°) are considerably activated in relation to those where the dihedral angle is or approaches 90°. Thus, the equatorial H in (7) is reported to be 12 times more reactive towards t-butoxy radicals than the axial H in (8).[96]

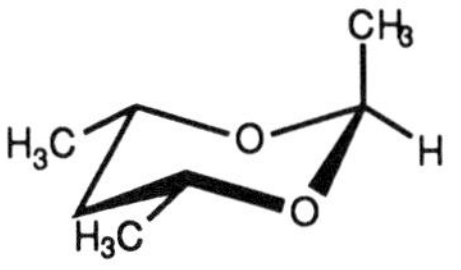

(7) 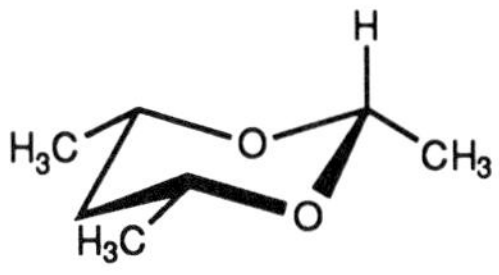   (8)

A further example of the importance of this type of stereoelectronic effect is seen in the reactions of *t*-butoxy radicals with spiro[2,n]alkanes (9) where it is found that hydrogens from the position α- to the cyclopropyl ring are specifically abstracted. This can be attributed to the favorable overlap of the breaking C-H bond with the cyclopropyl σ bonds.[101,102] No such specificity is seen with bicyclo[n,1,0] alkanes (10) where geometric constraints prevent overlap.

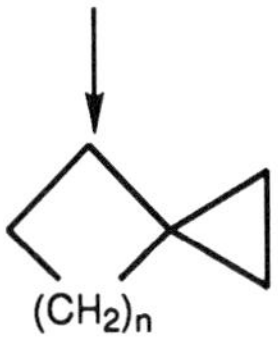

(9) 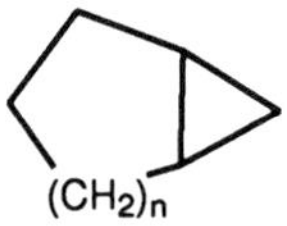   (10)

2.3.5 Reaction Conditions

Even though dissociation energies for X-H bonds appear insensitive to solvent changes,[103,104] the nature of the reaction medium[62,63,105] and the reaction temperature[58] can significantly affect the specificity and rate of atom abstraction reactions. One of the more controversial cases concerns the effect of aromatic solvents on hydrogen abstraction by atomic chlorine.

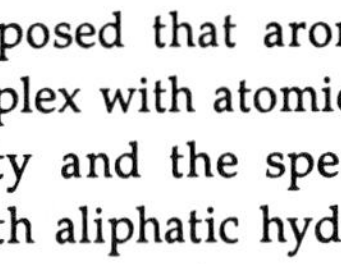

(11) 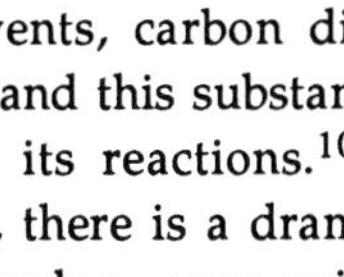   (12)

It has been proposed that aromatic solvents, carbon disulfide, and sulfur dioxide form a complex with atomic chlorine and this substantially modifies both its overall reactivity and the specificity of its reactions.[106] For example, in reactions of Cl• with aliphatic hydrocarbons, there is a dramatic increase in the specificity for abstraction of tertiary or secondary over primary hydrogens in benzene as opposed to aliphatic solvents. At the same time, the overall rate constant for abstraction is reduced by up to two orders of magnitude in the aromatic solvent.[107] The exact nature of the complex responsible for this effect, whether a π-complex (11) or a chlorocyclohexadienyl radical (12), is not yet resolved.[107-113]

Significant, though smaller, solvent effects have also been reported for alkoxy radical reactions (see 3.4.2.1).[114-118]

2.3.6 Abstraction vs. Addition

The relative propensity of radicals to abstract hydrogen or add to double bonds is extremely important. In radical polymerization, this factor determines the significance of transfer to monomer, solvent, *etc.* and hence the molecular weight and end group functionality (see 5.3). It also provides one basis for initiator selection (see 3.1).

The hydrogen abstraction:addition ratio is generally greater in reactions of heteroatom-centered radicals than it is with carbon-centered radicals. One factor is the relative strengths of the bonds being formed and broken in the two reactions (Table 2.5). The difference in exothermicity (Δ) between abstraction and addition reactions is much greater for heteroatom-centered radicals than it is for carbon-centered radicals. For example, for an alkoxy as opposed to an alkyl radical, abstraction is favored over addition by *ca.* 30 kJ mol^{-1}. The extent to which this is reflected in the rates of addition and abstraction will, however, depend on the particular substrate.

Table 2.5 Bond Dissociation Energies (kJ mol^{-1})[a,119]

bond	D	bond	D	bond	D	bond	D
(a) C-R bonds							
C_2H_5-C_2H_5	343	iC_3H_7-C_2H_5	335	tC_4H_9-C_2H_5	326	allyl-C_2H_5	299
C_2H_5-H	410	iC_3H_7-H	395	tC_4H_9-H	384	allyl-H	364
Δ	67		60		58		65
(b) X-R bonds							
H_2N-C_2H_5	351	HO-C_2H_5	381	C_2H_5O-C_2H_5	339	Cl-C_2H_5	339
H_2N-H	460	HO-H	498	C_2H_5O-H	435	Cl-H	431
Δ	109		117		96		92

[a] Values rounded to nearest kJ mol^{-1}.

A number of studies have found that increasing nucleophilicity of the attacking radical favors abstraction over addition to an unsaturated system (benzene ring or double bond).[39,120,121] Bertrand and Surzur[121] surveyed the literature on the reactions of oxygen-centered radicals and observed that the ratio of abstraction to addition increased in the series:

HO•<PhCO$_2$•<CH$_3$O•<nC$_4$H$_9$O•<sC$_4$H$_9$O•<tC$_4$H$_9$O<O$^-$

$$\xrightarrow{\text{nucleophilicity}}$$
$$\xrightarrow{\text{abstraction: addition}}$$

They, and later Houk,[122] attempted to establish a theoretical basis for this trend in terms of FMO theory. Pryor et al.[39] have found a similar trend for a series of aryl and alkyl radicals:

pNO$_2$Ph•<pBrPh•<CH$_3$•~Ph•<iC$_3$H$_7$•<tC$_4$H$_9$~Ph$_3$C•

$$\xrightarrow{\text{nucleophilicity}}$$
$$\xrightarrow{\text{abstraction: addition}}$$

However, the situation is not as clear cut as it might at first seem since a variety of other factors may also contribute to the above-mentioned trend. Abuin et al.[123] pointed out that the transition state for addition is more sterically demanding than that for hydrogen-atom abstraction. Within a given series (alkyl or alkoxy), the more nucleophilic radicals are generally the more bulky (*i.e.* steric factors favor the same trends). It can also be seen from Table 2.5 that, for alkyl radicals, the values of Δ decrease in the series primary>secondary>tertiary (*i.e.* relative bond strengths favor the same trend).

2.3.7 *Summary*

A simple unifying theory to explain rate and specificity in atom abstraction reactions has yet to be developed. However, as with addition reactions, it is possible to devise a set of guide-lines to predict qualitatively the rate and outcome of radical transfer processes. The following are based on those suggested by Tedder:[2]

(a) When there is little polarity in the transition state (or where the polarity is constant in a reaction series), the relative rates of atom transfer by a particular radical (selectivity) will correlate with the strengths of the bonds being broken.
(b) The strength of the bond being formed will be important in determining the absolute rate and the degree of selectivity.
(c) Steric strain relieved or incurred with formation of the new radical center may be important particularly for endothermic or near thermoneutral reactions.
(d) Nucleophilic radicals will prefer to attack electron rich sites. Electrophilic radicals will prefer to attack electron poor sites. If ΔH is small, polar factors may override thermodynamic considerations.

2.4 Radical-Radical Reactions

The last comprehensive review of reactions between carbon-centered radicals appeared in 1973.[124] Rate constants for radical-radical reactions in the liquid phase have been tabulated by Griller.[125] The area has also been reviewed by Alfassi[126] and Moad and Solomon.[127] Radical-radical reactions are, in general, very exothermic and activation barriers are extremely small even for highly resonance stabilized radicals. As a consequence, reaction rate constants often approach the diffusion controlled limit (typically ~10^9 M^{-1} s^{-1}).

The reaction may take several pathways:

(a) Combination, which usually but not invariably (see 2.4.1), takes place by a simple head to head coupling of radicals.

(b) Disproportionation, which involves the transfer of a β-hydrogen from one radical of the pair to the other.

(c) Electron transfer, in which the product is an ion pair.

The latter pathway is rare for reactions involving only carbon-centered radicals and will not be considered further in this chapter.

2.4.1 Pathways for Combination

The combination of carbon-centered radicals usually involves head-to-head (α,α-) coupling. Exceptions to this general rule occur where the free spin can be delocalized into a π-system. The classic example involves the triphenylmethyl radical (13) which combines to give exclusively the α-*para* coupling product (14).[20] This chemistry is also seen in cross reactions of (13) with other tertiary radicals.[128]

Other benzyl radicals, including the parent benzyl radical, give reversible formation of quinonemethide derivatives (typically a mixture of α,*p*- and α,*o*-coupling products) in competition with α,α-coupling (see also 5.2.3.1).[129-133] The kinetic product distribution appears to be determined by steric factors: α-substitution favors quinone methide formation; ring substitution favors α,α-coupling. However, since quinone methide formation is reversible, the only isolable product is often that of α,α-coupling.

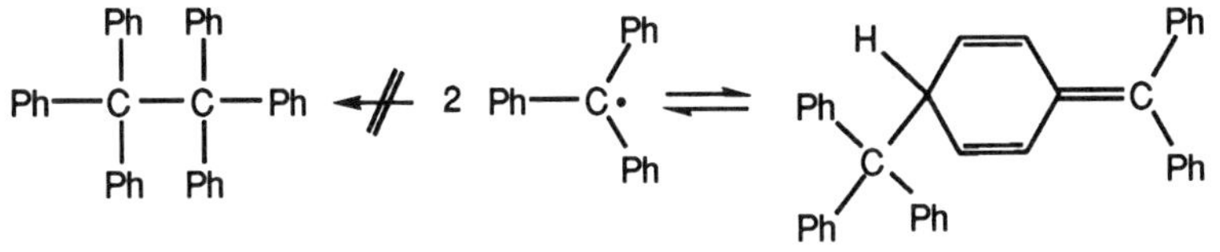

(13) (14)

For combination processes involving cyanoalkyl radicals, reversible C,N-coupling occurs in competition with C,C-coupling. Steric factors appear to be important in determining the relative amounts of C,C- and C,N-coupling[134] and exclusive C,N-coupling is observed when two bulky radicals combine.[135] For cyanoisopropyl radicals, C,N-coupling is the kinetically preferred pathway (Scheme 7).[105-107] However, since the formation of the ketenimine is thermally reversible, the C,C-coupling product is usually the major isolated product (see 5.2.3.3).

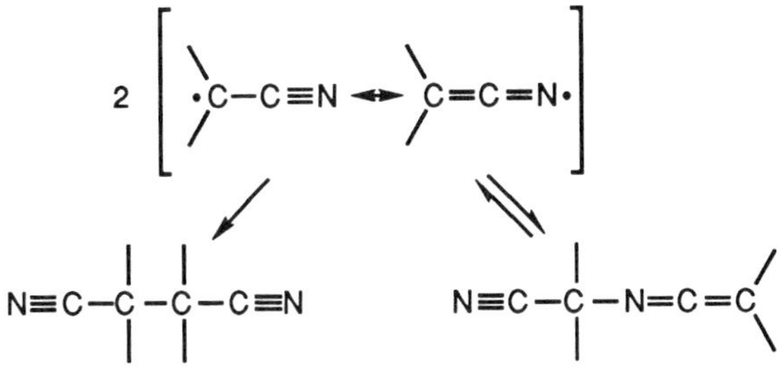

Scheme 7

An example of C,O-coupling of α-ketoalkyl radicals with reversible formation of an enol ether has also been reported for a system where C,C-coupling is very hindered (Scheme 8).[136] However, this pathway is not observed for simpler species (see 5.2.3.2).

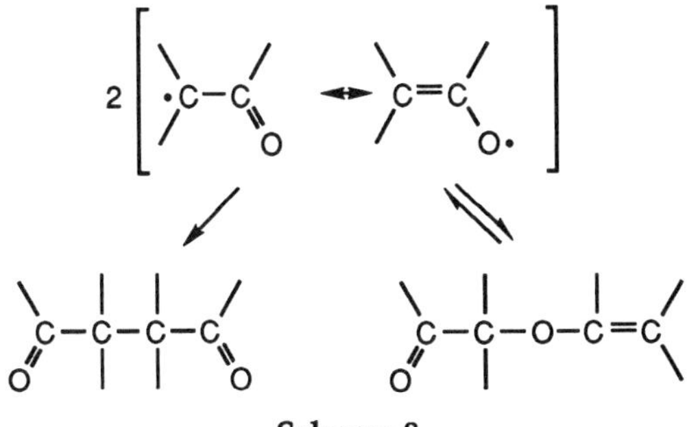

Scheme 8

2.4.2 Pathways for Disproportionation

For simple alkyl radicals, the product distribution appears to be predictable using statistical arguments.

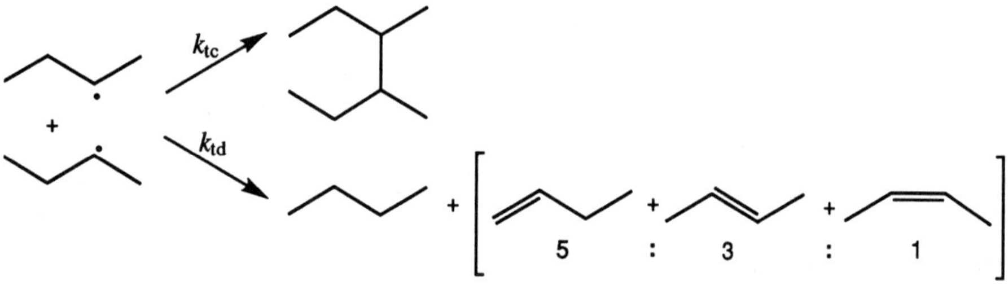

Scheme 9

For example, disproportionation of but-2-yl radicals produces a mixture of butenes as shown (Scheme 9).[120] Thermodynamic considerations suggest that but-1-ene and but-2-enes should be formed in a ratio of ca. 2:98. However, the observed 5:4 ratio of but-1-ene:but-2-enes is little different from the 3:2 ratio that is expected on statistical grounds (i.e. ratio of β-hydrogens in the 1- and 3-positions).

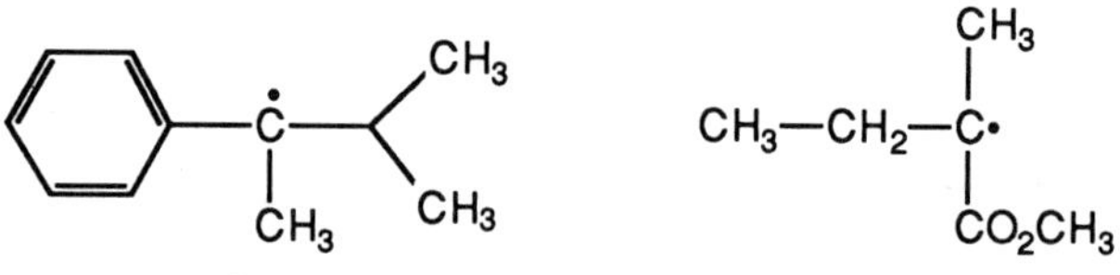

(15) (16)

For more highly substituted examples, it is clear that other factors are also important. Substitution at the radical center has a profound effect. For example, in disproportionation, radicals (15)[137] and (16)[138] show a marked preference for loss of a methyl hydrogen.

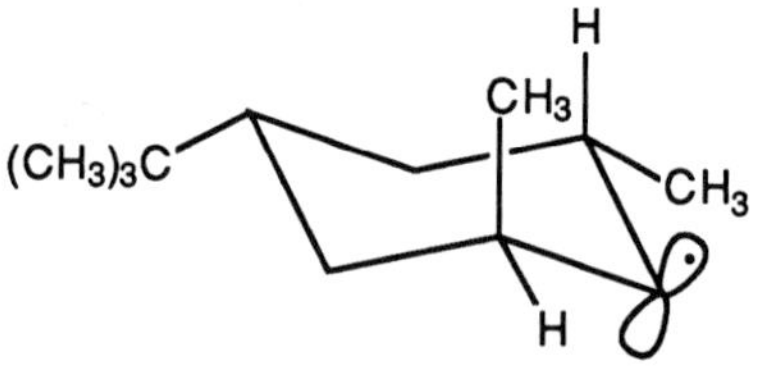

(17)

With the radical (17), even though loss of an equatorial hydrogen should be sterically less hindered and is favored thermodynamically (relief of 1,3 interactions of the axial methyl), there is an 8-fold preference for loss of the axial hydrogen (at 100°C). The selectivity observed in the disproportionation of this and other substituted cyclohexyl radicals led Beckwith[12] to propose that disproportionation is subject to stereoelectronic control which results in preferential breaking of the C–H bond which has best overlap with the orbital bearing the unpaired spin.

2.4.3 Combination vs. Disproportionation

Reactions between carbon-centered radicals generally give a mixture of disproportionation and combination. Much effort has been put into establishing the relative importance of these processes. The ratio of disproportionation to combination (k_{td}/k_{tc}) is dependent on the structural features of the radicals involved and generally shows only minor variation with solvent, pressure, temperature, etc.

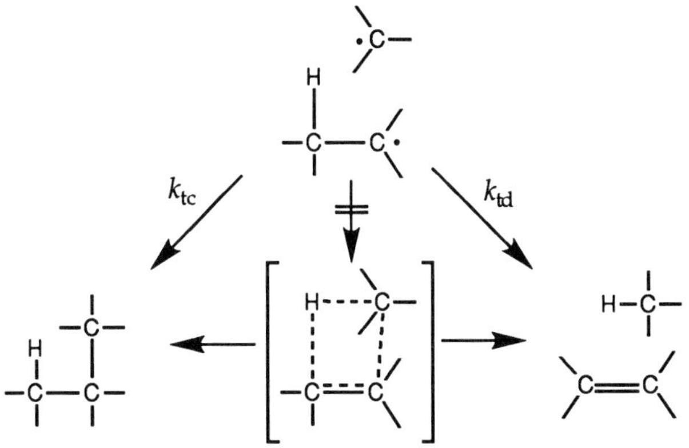

Scheme 10

Early workers in the area[139,140] suggested the involvement of a single 4-center transition state or intermediate which could lead to either disproportionation or combination (Scheme 10). The hypothesis fell from favor when it was established that k_{td}/k_{tc} showed a small though measurable dependence on temperature and pressure.[124] It is now generally recognized that combination and disproportionation should be considered as two separate reactions with distinct transition states. This view is supported by theoretical studies.[141-144]

2.4.3.1 Statistical factors

For a given series of radicals, the ratio k_{td}/k_{tc} increases with the number of β-hydrogen atoms. However, in general, there is no straight-forward relationship between k_{td}/k_{tc} and the number of β-hydrogens and it is clear that other factors are involved.[20,124] It is usually observed that even after allowing for the different number of β-hydrogens, the importance of disproportionation increases with increasing substitution at the radical center. For example, in the self-reaction of simple primary, secondary, and tertiary alkyl radicals, the values of $k_{td}/k_{tc}n$ are ca. 0.06, 0.2, and 0.8 respectively, where n is the number of β-hydrogens.[20,124]

2.4.3.2 Steric factors

It has been suggested that the discrepancies between the value of k_{td}/k_{tc} observed and that predicted on the basis of simple statistics may reflect the greater sensitivity of combination to steric factors. Beckhaus and Rüchardt[145] reported a correlation between $\log(k_{td}/k_{tc})$ (after statistical correction) and Taft steric parameters for a series of alkyl radicals.

A graphic demonstration of the importance of steric factors on k_{td}/k_{tc} is provided by the contrasting behavior of radicals (18) and (19). The self-reaction of cumyl radicals (18) affords predominantly combination while the radical (19), in which the α-methyl is replaced by a *t*-butyl group, give predominantly disproportionation.[146]

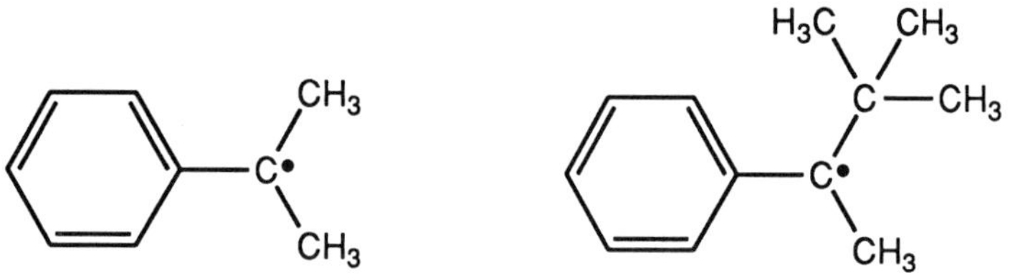

(18)   (19)

In extreme cases, suitably bulky substituents at the radical center can render a radical persistent [*e.g.* di-*t*-butyl methyl radical (20)].[147,148] This radical (20) possesses no β-hydrogens and therefore cannot decay by the normal disproportionation mechanism.

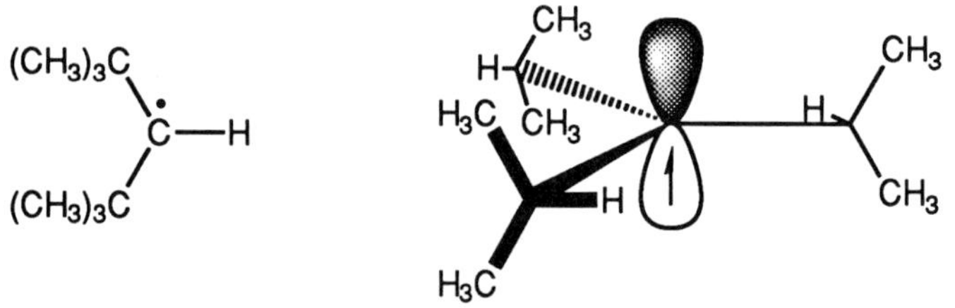

(20)   (21)

The triisopropylmethyl radical (21) is another example of a persistent radical. In this case, both disproportionation and combination are substantially retarded by steric factors.[149,150]

Free Radical Reactions

The examples considered in this section lead to three conclusions:

(a) Disproportionation and combination can both be dramatically slowed by large β- or γ-substituents.
(b) Combination is more sensitive to the presence of bulky β-substituents than disproportionation (i.e. k_{td}/k_{tc} is enhanced).
(c) Steric factors can outweigh simple statistical factors [e.g. even though (19) has less β-hydrogens, it gives more disproportionation than (18)].

Two quite separate influences are important in determining the rate of disproportionation:

(a) Steric hindrance to approach of the attacking radical (important for combination and disproportionation).
(b) Steric hindrance to rotation about the α,β-bond (important for disproportionation).

This latter term is considered in more detail under stereoelectronic factors (see 2.4.3.4).

2.4.3.3 Polar factors

Minato et al.[144] proposed that the transition state for disproportionation has polar character while that for combination is neutral. The finding that polar solvents enhance k_{td}/k_{tc} for ethyl[151] and t-butyl radicals (see 2.4.3.5), the very high k_{td}/k_{tc} seen for alkoxy radicals with α-hydrogens,[152] and the trend in k_{td}/k_{tc} observed for reactions of a series of fluoroalkyl radicals (see Scheme 11, Table 2.6) have been explained in these terms.[126,144]

$$CH_{3-x}F_x-C_2H_5 \xleftarrow{k_{tc}} CH_{3-x}F_x\bullet + C_2H_5\bullet \xrightarrow{k_{td}} CH_{4-x}F_x + C_2H_4$$

Scheme 11

Table 2.6 Values of k_{td}/k_{tc} for the Cross-Reaction between Fluoromethyl and Ethyl Radicals (25 °C)[153-155]

radical	k_{td}/k_{tc}	radical	k_{td}/k_{tc}	radical	k_{td}/k_{tc}
$CH_3\bullet$	0.039	$CHF_2\bullet$	0.068	$C_2F_5\bullet$	0.24
$CH_2F\bullet$	0.038	$CF_3\bullet$	0.11		

2.4.3.4 Stereoelectronic and other factors

The transition state for disproportionation requires overlap of the β-C—H bond undergoing scission and the p-orbital containing the unpaired electron.[12]

This requirement rationalizes the specificity observed in disproportionation of radicals (17) (see 2.4.3) and provides an explanation for the persistency of the triisopropylmethyl radical (21) and related species (see 2.4.3.3).[147] In the case of (21), the β-hydrogens are constrained to lie in the nodal plane of the p-orbital due to steric buttressing between the methyls of the adjacent isopropyls.

It has been noted by a number of workers that the presence of α-substituents which delocalize the free spin favors combination over disproportionation.[108,130,156] For radicals of structure $(CH_3)_2\overset{\bullet}{C}$-X, k_{td}/k_{tc} increases in the series:

X=alkynyl~alkenyl<aryl~nitrile<keto<ester<<alkyl.

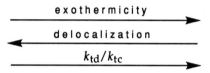

A correlation between the degree of exothermicity and the value of k_{td} has also been found but only for the case of resonance stabilized radicals.[126,157]

It has been suggested that benzylic radicals may form a dimeric association complex which may easily collapse to the combination product but be geometrically unfavorable for disproportionation.[158,159] Even if this applies for the arylalkyl radicals, it cannot account for the behavior of systems with other π-substituents.

Another explanation follows from the above discussion on stereoelectronic factors.[127] If overlap between the semi-occupied orbital and the breaking C-H bond favors disproportionation, then substituents which delocalize the free spin will serve to reduce this interaction and disfavor disproportionation. A proposal along these lines was made originally by Nelson and Bartlett[130] who also noted that diminishment of the spin density at C_α could retard combination. However, it is not necessary that the two effects should cancel one another.

2.4.3.5 Reaction conditions

Values of k_{td}/k_{tc} for simple alkyl radicals are sensitive to reaction conditions (solvent, temperature, pressure). However, the effects appear to be generally small (<2 fold).[124,126] Values of k_{td}/k_{tc} for t-butyl radicals decrease with increasing temperature (the magnitude of the dependence increases with increasing solvent polarity - see Figure 7). For a given solvent type (alkane or alcohol), a very small dependence on the viscosity of the medium is also observed (see Table 2.7). The temperature dependence has been related to the rate of molecular reorientation.[160] This dependence on temperature and medium for small radicals (k_{td}/k_{tc} decreases with temperature) appears in marked contrast with that reported for polymeric species (see 5.2.2) where k_{td}/k_{tc} is observed to increase with increasing temperature.

Table 2.7 Values of k_{td}/k_{tc} for t-Butyl Radicals (25 °C)[160]

solvent	temperature °C	η cP	k_{td}/k_{tc}
n-C_8H_{18}	25	0.51	5.4
n-$C_{10}H_{22}$	25	0.86	5.7
n-$C_{12}H_{26}$	25	1.37	5.9
n-$C_{14}H_{30}$	25	2.10	6.4
n-$C_{16}H_{34}$	25	3.09	6.9
CH_3CN	16.5	7.3	7.5
t-BuOH-pinacol(1:2)	25	-	10.1
3-methyl-3-pentanol	24.5	-	7.5

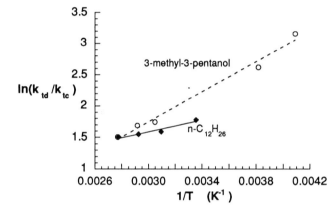

Figure 7 Temperature dependence of k_{td}/k_{tc} values for t-butyl radicals with (a) dodecane (—) or (b) 3-methyl-3-pentanol (---) as solvent.

In studies of radical-radical reactions, radicals are typically generated pairwise and the products come from both cage and encounter (non-cage) reactions. Several studies have indicated that cage $vs.$ encounter product distributions are the same.[138] However, it has been suggested that influences of pressure and viscosity on k_{td}/k_{tc} are more substantial for radicals which undergo self-reaction within the solvent cage.[131]

2.4.4 Summary

The relative importance of combination and disproportionation may be predicted by the following guide-lines:[127]

(a) Radical-radical reactions involving carbon-centered radicals give products from combination and disproportionation.

(b) Simple primary and secondary radicals give predominantly combination. Tertiary radicals give some disproportionation.

(c) The importance of combination is increased by π–substituents at the radical center and decreased by bulky groups at or near the radical center.

References

1. Bamford, C.H., in 'Comprehensive Polymer Science' (Eds. Eastmond, G.C., Ledwith, A., Russo, S., and Sigwalt, P.), Vol. 3, p. 219 (Pergamon: London 1989).
2. Tedder, J.M., *Angew. Chem., Int. Ed. Engl.*, 1982, **21**, 401.
3. Tedder, J.M., *Tetrahedron*, 1982, **38**, 313.
4. Walling, C., Free Radicals in Solution, p. 592 (Wiley: New York 1957).
5. Hawthorne, D.G., Johns, S.R., Solomon, D.H., and Willing, R.I., *Aust. J. Chem.*, 1976, **29**, 1955.
6. Elson, I.H., Mao, S.W., and Kochi, J.K., *J. Am. Chem. Soc.*, 1975, **97**, 335.
7. Tedder, J.M., and Walton, J.C., *Acc. Chem. Res.*, 1976, **9**, 183.
8. Rüchardt, C., *Angew. Chem., Int. Ed. Engl.*, 1970, **9**, 830.
9. Giese, B., *Angew. Chem., Int. Ed. Engl.*, 1989, **28**, 969.
10. Giese, B., *Angew. Chem., Int. Ed. Engl.*, 1983, **22**, 753.
11. Tedder, J.M., and Walton, J.C., *Tetrahedron*, 1980, **36**, 701.
12. Beckwith, A.L.J., *Tetrahedron*, 1981, **37**, 3073.
13. Giese, B., He, J., and Mehl, W., *Chem. Ber.*, 1988, **121**, 2063.
14. Beranek, I., and Fischer, H., in 'Free Radicals in Synthesis and Biology' (Ed. Minisci, F.), p. 303 (Kluwer: Dordrecht 1989).
15. Cuthbertson, M.J., Rizzardo, E., and Solomon, D.H., *Aust. J. Chem.*, 1985, **38**, 315.
16. Moad, G., Rizzardo, E., and Solomon, D.H., *Aust. J. Chem.*, 1983, **36**, 1573.
17. Giese, B., and Lachhein, S., *Angew. Chem., Int. Ed. Engl.*, 1981, **20**, 967.
18. Jones, M.J., Moad, G., Rizzardo, E., and Solomon, D.H., *J. Org. Chem.*, 1989, **54**, 1607.
19. Citterio, A., Minisci, F., and Vismara, E., *J. Org. Chem.*, 1982, **47**, 81.
20. Rüchardt, C., *Top. Curr. Chem.*, 1980, **88**, 1.
21. Beckwith, A.L.J., and Ingold, K.U., in 'Rearrangements in Ground and Excited States' (Ed. de Mayo, P.), Vol. 1, p. 162 (Academic Press: New York 1980).
22. Lorand, J.P., in 'Landoldt-Bornstein, New Series, Radical Reaction Rates in Solution' (Ed. Fischer, H.), Vol. II/13a, p. 135 (Springer-Verlag: Berlin 1984).
23. Howard, J.A., and Scaiano, J.C., in 'Landoldt-Bornstein, New Series, Radical Reaction Rates in Solution' (Ed. Fischer, H.), Vol. II/13d, p. 5 (Springer-Verlag: Berlin 1984).
24. Hammond, G.S., *J. Am. Chem. Soc.*, 1955, **77**, 334.
25. Fujimoto, H., Yamabe, S., Minato, T., and Fukui, K., *J. Am. Chem. Soc.*, 1972, **94**, 9205.

26. Ponec, R., and Malek, J., *Collect. Czech. Chem. Commun.*, 1982, **47**, 802.
27. Canadell, E., Eisenstein, O., Ohanessian, G., and Poblet, J.M., *J. Phys. Chem.*, 1985, **89**, 4856.
28. Gonzalez, C., Sosa, C., and Schlegel, H.B., *J. Phys. Chem.*, 1989, **93**, 2435.
29. Zipse, H., He, J., Houk, K.N., and Giese, B., *J. Am. Chem. Soc.*, 1991, **113**, 4324.
30. Bonacic-Koutecky, V., Koutecky, J., and Salem, L., *J. Am. Chem. Soc.*, 1977, **99**, 842.
31. Nakano, T., Mori, M., and Okamoto, Y., *Macromolecules*, 1993, **26**, 867.
32. Pearson, J.M., and Szwarc, M., *Trans. Faraday Soc.*, 1965, **61**, 1722.
33. Alfrey, T., and Price, C.C., *J. Polym. Sci.*, 1947, **2**, 101.
34. Jenkins, A.D., *Adv. Free Radical Chem.*, 1967, **2**, 139.
35. Jenkins, A.D., in 'Reactivity, Mechanism and Structure in Polymer Chemistry' (Eds. Jenkins, A.D., and Ledwith, A.), p. 117 (Wiley: London 1974).
36. Giese, B., and Meister, J., *Angew. Chem., Int. Ed. Engl.*, 1977, **16**, 178.
37. Giese, B., and Meixner, J., *Chem. Ber.*, 1981, **114**, 2138.
38. Sakurai, H., Hayashi, S., and Hosomi, A., *Bull. Chem. Soc. Japan*, 1971, **44**, 1945.
39. Pryor, W.A., Tang, F.Y., Tang, R.H., and Church, D.F., *J. Am. Chem. Soc.*, 1982, **104**, 2885.
40. Henderson, R.W., and Ward, R.D., Jr., *J. Am. Chem. Soc.*, 1974, **96**, 7556.
41. Pryor, W.A., Tonellato, U., Fuller, D.L., and Jumonville, S., *J. Org. Chem.*, 1969, **34**, 2018.
42. Sakurai, H., and Hosomi, A., *J. Am. Chem. Soc.*, 1967, **89**, 458.
43. Howard, J.A., and Chenier, J.H.B., *J. Am. Chem. Soc.*, 1973, **95**, 3054.
44. Michejda, C.J., and Hoss, W.P., *J. Am. Chem. Soc.*, 1970, **92**, 6298.
45. Huyser, E.S., *J. Am. Chem. Soc.*, 1960, **82**, 394.
46. Pryor, W.A., Lin, T.H., Stanley, J.P., and Henderson, R.W., *J. Am. Chem. Soc.*, 1973, **95**, 6993.
47. Smart, B.E., in 'Molecular Structure and Energetics' (Eds. Liebman, J.F., and Greenberg, A.), Vol. 3, p. 141 (VCH: Deerfield Beach, Florida 1976).
48. Arnaud, R., Subra, R., Barone, V., Lelj, F., Olivella, S., Solé, A., and Russo, N., *J. Chem. Soc., Perkin Trans. 2*, 1986, 1517.
49. Shaik, S.S., and Canadell, E., *J. Am. Chem. Soc.*, 1990, **112**, 1446.
50. Wilt, J.W., in 'Free Radicals' (Ed. Kochi, J.K.), Vol. 1, p. 333 (Wiley: New York 1973).
51. Julia, M., *Acc. Chem. Res.*, 1971, **4**, 386.
52. Julia, M., *Pure Appl. Chem.*, 1974, **40**, 553.
53. Beckwith, A.L.J., Easton, C.J., and Serelis, A.K., *J. Chem. Soc., Chem. Commun.*, 1980, 482.
54. Beckwith, A.L.J., in 'Chem. Soc. Spec. Publ. - Essays on Free Radical Chemistry', Vol. 24, p. 239 (Chem. Soc.: London 1970).

55. Beckwith, A.L.J., and Moad, G., *J. Chem. Soc., Perkin Trans. 2*, 1980, **2**, 1083.
56. Beckwith, A.L.J., *Chem. Soc. Rev.*, 1993, 143.
57. Giese, B., and Feix, C., *Isr. J. Chem.*, 1985, **26**, 387.
58. Giese, B., *Acc. Chem. Res.*, 1984, **17**, 438.
59. Spirin, Y., L., *Russ. Chem. Rev. (Engl. Transl.)*, 1969, **38**, 529.
60. Kamachi, M., *Adv. Polym. Sci.*, 1981, **38**, 55.
61. Gromov, V.F., and Khomiskovskii, P.M., *Russ. Chem. Rev. (Engl. Transl.)*, 1979, **48**, 1040.
62. Huyser, E.S., *Adv. Free Radical Chem.*, 1965, **1**, 77.
63. Martin, J.C., in 'Free Radicals' (Ed. Kochi, J.K.), Vol. 2, p. 493 (Wiley: New York 1973).
64. Shostenko, A.G., and Myshkin, V.E., *Dokl. Phys. Chem. (Engl. Transl.)*, 1979, **246**, 569.
65. Giese, B., and Kretzschmar, G., *Chem. Ber.*, 1984, **117**, 3160.
66. Bednarek, D., Moad, G., Rizzardo, E., and Solomon, D.H., *Macromolecules*, 1988, **21**, 1522.
67. Ito, O., and Matsuda, M., *J. Phys. Chem.*, 1984, **88**, 1002.
68. Fong, C.W., Kamlet, M.J., and Taft, R.W., *J. Org. Chem.*, 1983, **48**, 832.
69. Houk, K.N., Padden-Row, M.N., Spellmeyer, D.C., Rondan, N.G., and Nagase, S., *J. Org. Chem.*, 1986, **51**, 2874.
70. Sosa, C., and Schlegel, H.B., *J. Am. Chem. Soc.*, 1987, **109**, 4193.
71. Arnaud, R., Barone, V., Olivella, S., Russo, N., and Solé, A., *J. Chem. Soc., Chem. Commun.*, 1985, 1331.
72. Delbecq, F., Ilavsky, D., Anh, N.T., and Lefour, J.M., *J. Am. Chem. Soc.*, 1985, **107**, 1623.
73. Dewar, M.J.S., and Olivella, S., *J. Am. Chem. Soc.*, 1978, **17**, 5290.
74. Arnaud, R., Douady, J., and Subra, R., *Nouv. J. Chim*, 1981, **5**, 181.
75. Spellmeyer, D.C., and Houk, K.N., *J. Org. Chem.*, 1987, **52**, 959.
76. Beckwith, A.L.J., and Schiesser, C.H., *Tetrahedron*, 1985, **41**, 3925.
77. Heuts, J.P.A., Clay, P.A., Christie, D.I., Piton, M.C., Hutovic, J., Kable, S.H., and Gilbert, R.G., *Macromol. Symp.*, 1994, in press.
78. Flemming, I., Frontier Orbitals and Organic Chemical Reactions, p. 182 (Wiley: Chichester 1976).
79. Lai, Y., and Butler, G.B., *J. Macromol. Sci., Chem.*, 1984, **A21**, 1547.
80. Russell, G.A., in 'Free Radicals' (Ed. Kochi, J.K.), Vol. 1, p. 275 (Wiley: New York 1973).
81. Poutsma, M.L., in 'Free Radicals' (Ed. Kochi, J.K.), Vol. 2, p. 113 (Wiley: New York 1973).
82. Hendry, D.G., Mill, T., Piszkiewicz, L., Howard, J.A., and Eigenmann, H.K., *J. Phys. Chem. Ref., Data*, 1974, **3**, 937.
83. Evans, M.G., and Polanyi, M., *Trans. Faraday Soc.*, 1938, **34**, 11.
84. Pross, A., Yamataka, H., and Nagase, S., *J. Phys. Org. Chem.*, 1991, **5**, 135.
85. Gilliom, R.D., *J. Mol. Struct.*, 1986, **138**, 157.

86. Pryor, W.A., and Davis, W.H., Jr., *J. Am. Chem. Soc.*, 1974, **96**, 7557.
87. Zavitsas, A.A., and Hanna, G.M., *J. Org. Chem.*, 1975, **40**, 3782.
88. Zavitsas, A.A., and Pinto, J.A., *J. Am. Chem. Soc.*, 1972, **94**, 7390.
89. Walling, C., and McGuinness, J.A., *J. Am. Chem. Soc.*, 1969, **91**, 2053.
90. Huang, X.L., and Dannenberg, J.J., *J. Org. Chem.*, 1991, **56**, 5421.
91. Houk, K.N., Tucker, J.A., and Dorigo, A.E., *Acc. Chem. Res.*, 1990, **23**, 107.
92. Walling, C., and Padwa, A., *J. Am. Chem. Soc.*, 1963, **85**, 1597.
93. Neale, R.S., Walsh, M.R., and Marcus, N.L., *J. Org. Chem.*, 1965, **30**, 3683.
94. Busfield, W.K., Grice, D.I., and Jenkins, I.D., *J. Chem. Soc., Perkin Trans. 2*, 1994, **2**, 1079.
95. Busfield, W.K., Grice, D.I., Jenkins, I.D., and Monteiro, M.J., *J. Chem. Soc., Perkin Trans. 1*, 1994, **2**, 1071.
96. Beckwith, A.L.J., and Easton, C.J., *J. Chem. Soc., Perkin Trans. 2*, 1983, 661.
97. Malatesta, V., and Scaiano, J.C., *J. Org. Chem.*, 1982, **47**, 1455.
98. Easton, C.J., and Beckwith, A.L.J., *J. Am. Chem. Soc.*, 1981, **103**, 615.
99. Malatesta, V., and Ingold, K.U., *J. Am. Chem. Soc.*, 1981, **103**, 609.
100. Griller, D., Howard, J.A., Marriott, P.R., and Scaiano, J.C., *J. Am. Chem. Soc.*, 1981, **103**, 619.
101. Roberts, C., and Walton, J.C., *J. Chem. Soc., Perkin Trans. 2*, 1985, 841.
102. Roberts, C., and Walton, J.C., *J. Chem. Soc., Chem. Commun.*, 1984, 1109.
103. Bausch, M.J., Gostowski, R., Guadalupe-Fasano, C., Selmarten, D., Vaughn, A., and Wang, L.-H., *J. Org. Chem.*, 1991, **26**, 7191.
104. Kanabus-Kaminske, J.M., Gilbert, B.C., and Griller, D., *J. Am. Chem. Soc.*, 1989, **111**, 3311.
105. Reichardt, C., Solvent Effects in Organic Chemistry, p. 110 (Verlag Chemie: Weinheim 1978).
106. Russell, G.A., *J. Am. Chem. Soc.*, 1958, **80**, 4987.
107. Bunce, N.J., Ingold, K.U., Landers, J.P., Lusztyk, J., and Scaiano, J.C., *J. Am. Chem. Soc.*, 1985, **107**, 5464.
108. Walling, *J. Org. Chem.*, 1988, **53**, 305.
109. Tanko, J.M., and Anderson, F.E., III, *J. Am. Chem. Soc.*, 1988, **110**, 3525.
110. Skell, P.S., Baxter, H.N.III., and Taylor, C.K., *J. Am. Chem. Soc.*, 1983, **105**, 120.
111. Skell, P.S., Baxter, H.N.III., Tanko, J.M., and Venkatasuryanarayana, C., *J. Am. Chem. Soc.*, 1986, **108**, 6300.
112. Ponec, R., and Hajeck, J., *Z. Phys. Chem. (Leipzig)*, 1987, **268**, 1233.
113. Ingold, K.U., Lusztyk, J., and Raner, K.D., *Acc. Chem. Res.*, 1990, **23**, 219.
114. Walling, C., and Wagner, P.J., *J. Am. Chem. Soc.*, 1964, **86**, 3368.
115. Mendenhall, G.D., Stewart, L.C., and Scaiano, J.C., *J. Am. Chem. Soc.*, 1982, **104**, 5109.
116. Grant, R.D., Griffiths, P.G., Moad, G., Rizzardo, E., and Solomon, D.H., *Aust. J. Chem.*, 1983, **36**, 2447.

117. Grant, R.D., Rizzardo, E., and Solomon, D.H., *Makromol. Chem.*, 1984, **185**, 1809.
118. Avila, D.V., Brown, C.E., Ingold, K.U., and Lusztyk, J., *J. Am. Chem. Soc*, 1993, **115**, 466.
119. Benson, S.W., Thermochemical Kinetics (Wiley: New York 1976).
120. Sheldon, R.A., and Kochi, J.K., *J. Am. Chem. Soc.*, 1970, **92**, 4395.
121. Bertrand, M.P., and Surzur, J.-M., *Tetrahedron Lett.*, 1976, **38**, 3451.
122. Houk, K.N., in 'Frontiers in Free Radical Chemistry' (Ed. Pryor, W.A.), p. 43 (Academic Press: New York 1980).
123. Abuin, E., Mujica, C., and Lissi, E., *Rev. Latinoamer. Quim.*, 1980, **11**, 78.
124. Gibian, M.J., and Corley, R.C., *Chem. Rev.*, 1973, **73**, 441.
125. Griller, D., in 'Landoldt-Bornstein, New Series, Radical Reaction Rates in Solution' (Ed. Fischer, H.), Vol. II/13a, p. 5 (Springer-Verlag: Berlin 1984).
126. Alfassi, Z.B., in 'Chemical Kinetics of Small Organic Radicals' (Ed. Alfassi, Z.B.), Vol. 1, p. 129 (CRC Press: Boca Raton, Fla. 1988).
127. Moad, G., and Solomon, D.H., in 'Comprehensive Polymer Science' (Eds. Eastmond, G.C., Ledwith, A., Russo, S., and Sigwalt, P.), Vol. 3, p. 147 (Pergamon: London 1989).
128. Engel, P.S., Chen, Y., and Wang, C., *J. Org. Chem.*, 1991, **56**, 3073.
129. Gleixner, G., Olaj, O.F., and Breitenbach, J.W., *Makromol. Chem.*, 1979, **180**, 2581.
130. Nelsen, S.F., and Bartlett, P.D., *J. Am. Chem. Soc.*, 1966, **88**, 137.
131. Neuman, R.C., Jr., and Amrich, M.J., Jr., *J. Am. Chem. Soc.*, 1980, **45**, 4629.
132. Skinner, K.J., Hochster, H.S., and McBride, J.M., *J. Am. Chem. Soc.*, 1974, **96**, 4301.
133. Langhals, H., and Fischer, H., *Chem. Ber.*, 1978, **111**, 543.
134. Barbe, W., and Rüchardt, C., *Makromol. Chem.*, 1983, **184**, 1235.
135. Zarkadis, A.K., Neumann, W.P., Dünnebacke, D., Penenory, A., Stapel, R., and Stewen, U., *Chem. Ber.*, 1993, **126**, 1179.
136. Neumann, W.P., and Stapel, R., *Chem. Ber.*, 1986, **119**, 3422.
137. Bartlett, P.D., and McBride, J.M., *Pure Appl. Chem.*, 1967, **15**, 89.
138. Bizilj, S., Kelly, D.P., Serelis, A.K., Solomon, D.H., and White, K.E., *Aust. J. Chem.*, 1985, **38**, 1657.
139. Bradley, J.N., and Rabinovitch, B.S., *J. Chem. Phys.*, 1962, **36**, 3498.
140. Kerr, J.A., and Trotman-Dickenson, A.F., *Prog. React. Kinet.*, 1961, **1**, 107.
141. Benson, S.W., *Acc. Chem. Res.*, 1986, **19**, 335.
142. Dannenberg, J.J., and Baer, B., *J. Am. Chem. Soc.*, 1987, **109**, 292.
143. Imoto, M., Sakai, S., and Ouchi, T., *J. Chem. Soc. Japan*, 1985, 97.
144. Minato, T., Yamabe, S., Fujimoto, H., and Fukui, K., *Bull. Chem. Soc. Japan*, 1978, **51**, 1.
145. Beckhaus, H.D., and Rüchardt, C., *Chem. Ber.*, 1977, **110**, 878.
146. Fraenkel, G., and Geckle, M.J., *J. Chem. Soc., Chem. Commun.*, 1980, 55.
147. Griller, D., and Ingold, K.U., *Acc. Chem. Res.*, 1976, **9**, 13.

148. Griller, D., and Marriott, P.R., *Int. J. Chem. Kinet.*, 1979, **11**, 1163.
149. Schlüter, K., and Berndt, A., *Tetrahedron Lett.*, 1979, 929.
150. Griller, D., Icli, S., Thankachan, C., and Tidwell, T., *J. Chem. Soc., Chem. Commun.*, 1974, 913.
151. Stefani, A.P., *J. Am. Chem. Soc.*, 1968, **90**, 1694.
152. Druliner, J.D., Krusic, P.D., Lehr, G.F., and Tolman, C.A., *J. Org. Chem.*, 1985, **50**, 5838.
153. Pritchard, G.O., Johnson, K.A., and Nilsson, W.B., *Int. J. Chem. Kinet.*, 1985, **17**, 327.
154. Pritchard, G.O., Nilsson, W.B., and Kirtman, B., *Int. J. Chem. Kinet.*, 1984, **16**, 1637.
155. Pritchard, G.O., Kennedy, V.H., Heldoorn, G.M., Piasecki, M.L., Johnson, K.A., and Golan, D.R., *Int. J. Chem. Kinet.*, 1987, **19**, 963.
156. Ingold, K.U., in 'Free Radicals' (Ed. Kochi, J.K.), Vol. 1, p. 37 (Wiley: New York 1973).
157. Manka, M.J., and Stein, S.E., *J. Phys. Chem.*, 1984, **88**, 5914.
158. Neuman, R.C.Jr., and Alhadeff, E.S., *J. Org. Chem.*, 1970, **35**, 3401.
159. Kopecky, K.R., and Yeung, M.-Y., *Can. J. Chem.*, 1988, **66**, 374.
160. Schuh, H., and Fischer, H., *Helv. Chim. Acta*, 1978, **61**, 2463.

3
Initiation

3.1 Introduction

Initiation is defined as the series of reactions commencing with generation of *primary* radicals* and culminating in addition to the carbon-carbon double bond of the monomer so as to form *initiating* radicals (Scheme 1).[1]

$$I_2 \rightarrow I\cdot \rightarrow \rightarrow \rightarrow X\text{-}M\cdot \rightarrow X\text{-}M\text{-}M\cdot$$

initiator → primary radical → → → initiating radicals → propagating radicals

Scheme 1

Classically, initiation was only considered as the first step in the chain reaction that constitutes radical polymerization. Although the rate and efficiency of initiation were known to be extremely important in determining the kinetics of polymerization, it was generally thought that the detailed mechanism of the process could be safely ignored when interpreting polymer properties. Furthermore, while it was recognized that initiation would lead to formation of structural units different from those which make up the bulk of the chain, the proportion of initiator-derived groups seemed insignificant when compared with total material.† This led to the belief that the physical properties and chemistry of polymers could be interpreted purely in terms of the generalized formula (*i.e.* $CH_2\text{-}CXY)_n$ (see Chapter 1).

This view prevailed until the early 1970s and can still be found in some current-day texts. It is only in recent times that we have begun to understand the complexities of the initiation process and can appreciate the full role of initiation in influencing polymer structure and properties. Three factors may be seen as instrumental in bringing about a revision of the traditional view:

(a) The realization that polymer properties (*e.g.* resistance to weathering, thermal or photochemical degradation) are often not predictable on the basis of the

---

* The term primary radical used in this context should be distinguished from that used when describing the substitution pattern of alkyl radicals.

† For example, in PS, the initiator-derived end groups will account for *ca.* 0.2% of units in a sample of molecular weight 100,000 (termination is mainly by combination).

repeat unit structure but are in many cases determined by the presence of "defect groups".[2-5]
(b) The development of techniques whereby details of the initiation and other stages of polymerization can be studied in depth (see 3.5).
(c) The finding that free radical reactions are typically under kinetic rather than thermodynamic control (see 2.1). Many instances can be cited where the less thermodynamically favored pathway is a significant, or even the major, pathway.

It is the aim of this chapter to describe the nature, selectivity, and efficiency of initiation. The first section summarizes the various reactions associated with initiation and defines the terminology used in describing the process. The second section details the types of initiators, indicating the radicals generated, the by-products formed (initiator efficiency), and any side reactions (*e.g.* transfer to initiator). Emphasis is placed on those initiators which see widespread usage. The third section examines the properties and reactions of the radicals generated, paying particular attention to the specificity of their interaction with monomers and other components of a polymerization system. The final section describes some of the techniques used in the study of initiation.

The intention is to create a greater awareness of the factors which must be borne in mind by the polymer scientist when selecting an initiator for a given polymerization.

3.2 The Initiation Process

The simple initiation process depicted in many standard texts is the exception rather than the rule. The yield of primary radicals produced on thermolysis or photolysis of the initiator is usually not 100%. The conversion of primary radicals to initiating radicals is dependent on many factors and typically is not quantitative. The primary radicals may undergo rearrangement or fragmentation to afford new radical species (secondary radicals) or they may interact with solvent or other species rather than monomer.

The reactions of the radicals (whether primary, secondary, solvent derived, *etc.*) with monomer may not be entirely regio- or chemo-selective. Reactions, such as head addition, abstraction or aromatic substitution, often compete with tail addition. In the sections that follow, the complexities of the initiation process will be illustrated by examining the initiation of polymerization of two commercially important monomers, styrene (S) and methyl methacrylate (MMA), with each of three commonly used initiators, azobisisobutyronitrile (AIBN), dibenzoyl peroxide (BPO), and di-*t*-butyl peroxalate (DBPOX). The primary radicals formed from these three initiators are cyanoisopropyl, benzoyloxy, and *t*-butoxy radicals respectively (Scheme 2). BPO and DBPOX may also afford phenyl and methyl radicals respectively as secondary radicals (see 3.2.2).

3.2.1 Reaction with Monomer

First consider the interaction of radicals with monomers. Some behave as described in the classic texts and give only tail addition in their reaction with monomer (Scheme 3).

However, tail addition to the double bond is only one of the pathways whereby a radical may react with a monomer. The outcome of the reaction is critically dependent on the structure of both radical and monomer.

For reactions with S, specificity is found to decrease in the series cyanoisopropyl~methyl~t-butoxy>phenyl>benzoyloxy. Cyanoisopropyl (Scheme 3),[6] t-butoxy and methyl radicals give exclusively tail addition.[7] Phenyl radicals afford tail addition and *ca.* 1% aromatic substitution.[7] Benzoyloxy radicals give tail addition, head addition, and aromatic substitution (see Scheme 4).[7,8]

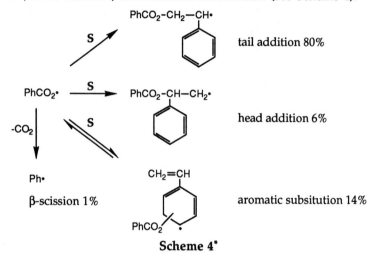

Scheme 4*

With MMA, these radicals show a quite different order of specificity; regiospecificity decreases in the series cyanoisopropyl~methyl>phenyl >benzoyloxy>*t*-butoxy. Cyanoisopropyl and methyl radicals give exclusively tail addition. Benzoyloxy and phenyl radicals also react almost exclusively with the double bond (though benzoyloxy radicals give a mixture of head and tail addition[9]) and abstraction (2), while detectable, is a very minor (<1%) pathway.[9,10] On the other hand, only 63% of *t*-butoxy radicals react with MMA by tail addition to give 1 (see Scheme 5).[11] The remainder abstract hydrogen, either from the α-methyl (predominantly) to give 2 or the ester methyl to give 3.[11,12] The radicals 1-3 and methyl (formed by β-scission) may initiate polymerization.

These examples clearly show that the initiation pathways depend on the structures of the radical and the monomer. Some of the reasons for the differences in behavior have already been discussed in chapter 2. The high degree of specificity shown by a radical (*e.g. t*-butoxy) in its reactions with one monomer (*e.g.* S) must not be taken as a sign that a similarly high degree of specificity will be shown in reactions with all monomers (*e.g.* MMA).

Radicals can be classified according to their tendency to give aromatic substitution, abstraction, double bond addition, or β-scission and further classified in terms of the specificity of these reactions (see 3.4). With this knowledge, it should be possible to choose an initiator according to its suitability for use with a given monomer or monomer system so as to avoid the formation of undesirable end groups or, alternatively, to achieve a desired functionality.

---

* Data are for initiation of bulk polymerization at 60°C and low conversion.

Initiation

The importance of these considerations can be demonstrated by examining some of the possible consequences for radical-monomer systems. For the case of MMA polymerization initiated by a *t*-butoxy radical source, chains may be initiated by the radicals (1), (2) or (3) (see Scheme 5). A significant proportion of chains will therefore have an olefinic end group rather than an initiator-derived end group. These chain ends may be reactive, either during polymerization, leading to chain branching, or afterwards, possibly leading to an impairment in polymer properties (see 7.1.3).

Polystyrene (PS) formed with BPO as initiator will have a proportion of relatively unstable benzoate end groups formed by benzoyloxy radical reacting by head addition and aromatic substitution (see scheme 4).[7,8] There is evidence that PS prepared with BPO as initiator is less thermally stable[13,14] and less resistant to weathering and yellowing[15,16] than that prepared using other initiators (see 7.1.2).

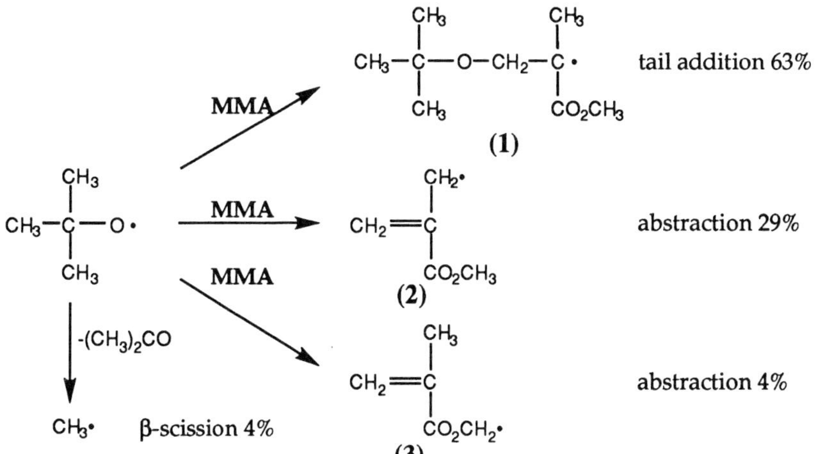

Scheme 5*

3.2.2 Fragmentation

Many radicals undergo fragmentation. For example, *t*-butoxy radicals undergo β-scission to form methyl radicals and acetone (Scheme 6).

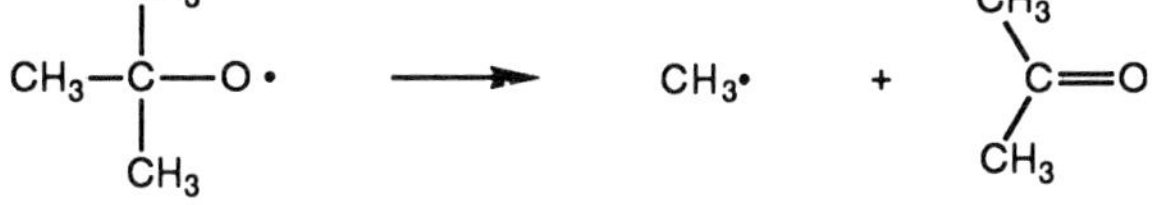

Scheme 6

Benzoyloxy radicals decompose to phenyl radicals and carbon dioxide (Scheme 7).

Scheme 7

The reactivity of the monomer and the reaction conditions determine the relative importance of β-scission. Fragmentation reactions are generally favored by low monomer concentrations, high temperatures, and low pressures. They may also be influenced by the nature of the reaction medium.

Other radicals undergo rearrangement in competition with bimolecular processes. An example is the 5-hexenyl radical (5). The 6-heptenoyloxy radical (4) undergoes sequential fragmentation and rearrangement (Scheme 8).[17]

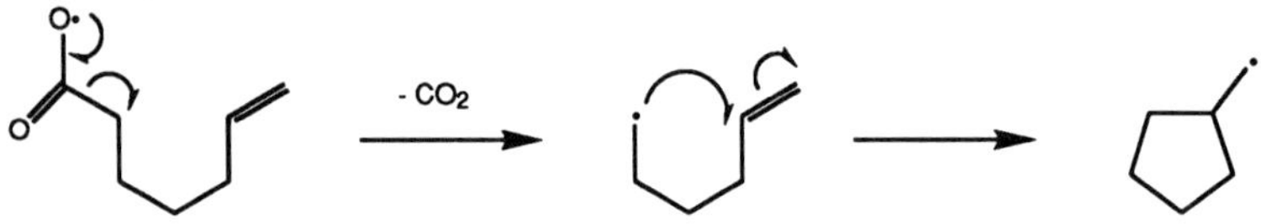

Scheme 8

The radicals formed by unimolecular rearrangement or fragmentation of the primary radicals are commonly termed secondary radicals. Often the absolute rate constants for secondary radical formation are known or can be accurately determined. These reactions may then be used as "radical clocks",[18,19] to calibrate the absolute rate constants for the bimolecular reactions of the primary radicals (*e.g.* addition to monomers - see 3.4). However, care must be taken since the rate constants of some clock reactions (*e.g.* t-butoxy β-scission[20]) are medium dependent (see 3.4.2.1.1).

3.2.3 Reaction with Solvents, Additives, or Impurities

A typical polymerization system comprises many components besides the initiators and the monomers. There will be solvents, additives (*e.g.* transfer agents, inhibitors) as well as a variety of adventitious impurities which may also be reactive towards the initiator-derived radicals.

For the case of MMA polymerization with a source of *t*-butoxy radicals (DBPOX) as initiator and toluene as solvent, most initiation may be by way of solvent-derived radicals[20,21] (see Scheme 9). Thus, a high proportion of chains (>70% for 10% w/v monomers at 60°C[21]) will be initiated by benzyl rather than *t*-butoxy radicals. Other entities with abstractable hydrogens may also be incorporated as polymer end groups. The significance of these processes increases with the degree of conversion and with the (solvent or impurity):monomer ratio.

There is potential for this behavior to be utilized in devising methods for the control of the types of initiating radicals formed and hence the polymer end groups (see 3.4.2.1.1).

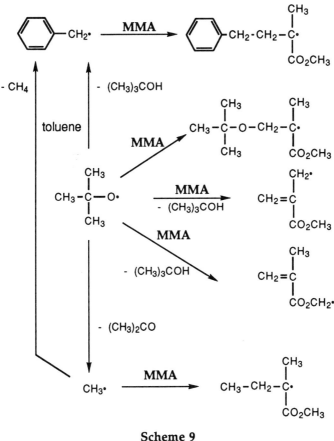

Scheme 9

3.2.4 Effects of the Reaction Medium on Radical Reactivity

The reaction medium may also modify the reactivity of the primary, or other radicals without directly reacting with them. For example, when *t*-butoxy reacts with α-methylvinyl monomers, the ratio of addition:abstraction:β-scission varies according to the nature of the solvent[20] and the reaction temperature[22,23] (see 2.3.7 and 3.4.2.1.1).

3.2.5 Reaction with Oxygen

Radicals, in particular carbon-centered radicals, react with oxygen at near diffusion-controlled rates.[24] Thus, for polymerizations carried out either in air or in incompletely degassed media, oxygen is likely to become involved in, and further complicate, the initiation process.

The reaction of oxygen with carbon-centered radicals (*e.g.* cyanoisopropyl, Scheme 10) affords a alkylperoxy radical (6).[25,26] This species may initiate polymerization so forming a relatively unstable peroxidic end group (7). With respect to most carbon-centered radicals, the hydroperoxy radicals (6) show an enhanced tendency to abstract hydrogen. The hydroperoxy radicals may abstract hydrogen from polymer, monomer, or other components in the system[27] forming a potentially reactive hydroperoxide (8) and a new radical species (R•) which may initiate polymerization. The process is further complicated if (7) or (8) undergo homolysis under the polymerization conditions. The peroxides (that derived from 7) and (8) may also be active as chain transfer agents.

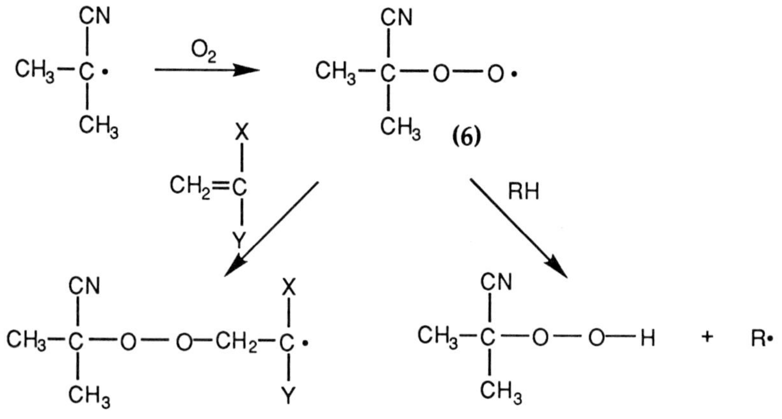

Scheme 10

3.2.6 Initiator Efficiency

The proportion of radicals which escape the solvent cage to form initiating radicals is termed the initiator efficiency (f) which is formally defined as follows:

$$f = \frac{[\text{Rate of initiation of propagating chains}]}{n \cdot [\text{Rate of initiator disappearance}]}$$

where n is the number of moles of radicals generated per mole of initiator. The effective rate of initiation (R_i) in the case of thermal decomposition of an initiator ($I_2 \rightarrow 2I\bullet$) is then:

$$R_i = 2k_d f [I_2]$$

Reactions which lead to loss of initiator or initiator-derived radicals include the cage reaction of the initiator-derived radicals, primary radical termination, transfer to initiator, and various side reactions. It is important to note that the initiator efficiency is typically not a constant. The importance of the above-mentioned processes increases as monomer is depleted and the viscosity of the polymerization medium increases.

3.2.7 Cage Reaction and Initiator-Derived By-products

The decomposition of an initiator seldom produces a quantitative yield of initiating radicals. Most thermal and photochemical initiators generate radicals in pairs. The self-reaction of these radicals is often the major pathway for the direct conversion of primary radicals to non-radical products in solution, bulk or suspension polymerization. This cage reaction is substantial even at low conversion when the medium is essentially monomer. The importance of the process depends on the rate of diffusion of these species away from one another.

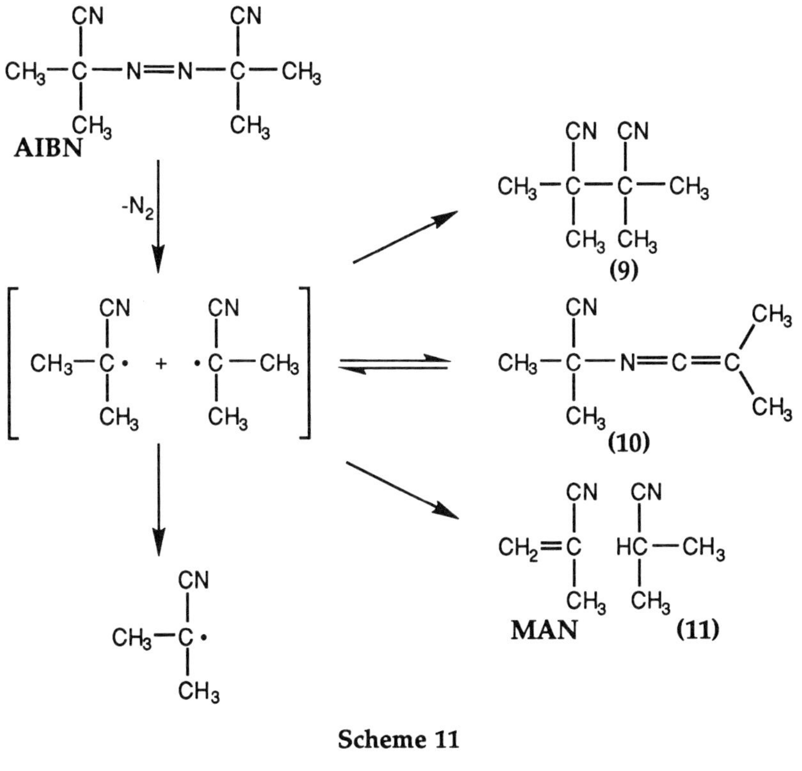

Scheme 11

Thus the size and the reactivity of the initiator-derived radicals and the medium viscosity (or microviscosity) are important factors in determining the initiator efficiency. One implication of the viscosity dependence of f is that the importance of the cage reaction is likely to increase significantly with increasing conversion.[28,29] The cage reaction, as well as lowering the initiation efficiency, can produce a range of by-products. These materials may be reactive under the polymerization conditions or they may themselves have a deleterious influence on polymer properties. For example, the cage reaction of cyanoisopropyl radicals formed from the decomposition of AIBN produces, amongst other products (see Scheme 11), MAN, which readily undergoes copolymerization to be incorporated into the final polymer,[6,30] and tetramethylsuccinonitrile (9), which is toxic and should not be present in polymers used for food contact applications.[31]

In other cases the cage reaction may simply lead to reformation of the initiator. The process is known as cage return and is important during the decomposition of BPO (see 3.3.2.1). Cage return affects the rate but not the efficiency of radical generation.

A variety of methods may be envisioned to decrease the importance of the cage reaction. One method, given the viscosity dependence of the cage reaction, would be to conduct polymerizations in solution rather than in bulk. Another involves carrying out the polymerization in a magnetic field.[32] The latter is thought to reduce the rate of triplet-singlet intersystem crossing for the geminate pair.[33]

3.2.8 Primary Radical Termination

The primary radicals may also interact with other radicals present in the system after they escape the solvent cage. When this involves a propagating radical, the process is known as primary radical termination. Most monomers are efficient scavengers of the initiator-derived radicals and the steady state concentration of propagating radicals is very low (typically $\leq 10^{-7}$M). The concentrations of the primary and other initiator-derived radicals are even lower. Thus, with most initiators, primary radical termination has a very low likelihood during the early stages of polymerization.

Primary radical termination is of demonstrable significance when very high rates of initiation or very low monomer concentrations are employed. It should be noted that these conditions pertain in all polymerizations at high conversion and in starved feed processes.

3.2.9 Transfer to Initiator

Many of the initiators used in radical polymerization are susceptible to induced decomposition by various radical species. When the reaction involves the propagating species, the process is termed transfer to initiator. The importance of this reaction depends on both the initiator and the propagating radical.

Diacyl peroxides are particularly prone to induced decomposition (Scheme 12). Transfer to initiator is of greatest importance for polymerizations taken to high conversion or when the ratio of initiator to monomer is high. It has been shown that, during the polymerization of S initiated by BPO, transfer to initiator can be the major pathway for the termination of chains.[6,34]

Initiation

BPO → **Scheme 12**

Transfer to initiator introduces a new end group into the polymer, lowers the molecular weight of the polymer, reduces the initiator efficiency, and increases the rate of initiator disappearance.

3.2.10 Initiation in Heterogeneous Polymerization

Many polymerizations are carried out in heterogeneous media, usually water-monomer mixtures, where suspending agents or surfactants ensure proper dispersion of the monomer and control the particle size of the product.

Suspension polymerizations are often regarded as "mini-bulk" polymerizations since ideally all reaction occurs within individual monomer droplets. Initiators with high monomer and low water solubility are generally used in this application. The general chemistry, initiator efficiencies, and importance of side reactions are similar to that seen in homogeneous media.

Emulsion polymerizations involve the use of water soluble initiators and polymer chains are initiated in the aqueous phase. A number of mechanisms for particle formation and entry have been described, however, a full discussion of these is beyond the scope of this section. Radicals are generated in the aqueous phase and it is now generally believed that oligomer formation occurs in the aqueous phase prior to particle entry.[35]

In emulsion polymerizations, the concentration of monomers in the aqueous phase is usually very low. This means that there is a greater chance that the initiator-derived radicals will undergo side reactions. Processes such as self-reaction, primary radical termination with oligomeric species, and transfer to initiator are much more significant than in bulk, solution, or suspension polymerization and initiator efficiencies in emulsion polymerization are often very low.

3.3 The Initiators

Certain polymerizations (*e.g.* styrene, see 3.3.6.1) can be initiated simply by applying heat; the initiating radicals are derived from reactions involving only the monomer. More commonly, the initiators are azo-compounds or peroxides that are decomposed to radicals through the application of heat, light, or a redox process.

When initiators are decomposed thermally, the rates of initiator disappearance (k_d) show marked temperature dependence. Since most conventional polymerization processes require that k_d should lie in the range 10^{-6}-10^{-5} s^{-1} (half-life *ca.* 10 h), individual initiators typically have acceptable k_d only within a relatively narrow temperature range (*ca.* 20-30°). For this reason initiators are often categorized purely according to their half-life at a given temperature or *vice versa*.[36]

The initiator in radical polymerization is often regarded simply as a source of radicals. Little attention is paid to the various pathways available for radical generation or to the side reactions that may accompany initiation. The preceding discussion (see 3.2) demonstrated that in selecting initiators (whether thermal, photochemical, redox, *etc.*) for polymerization, they must be considered in terms of the types of radicals formed, their suitability for use with the particular monomers, solvent, and the other agents present in the polymerization medium, and for the properties they convey to the polymer produced.

Many reviews detailing aspects of the chemistry of initiators and initiation have appeared.[36-38] A summary of decomposition rates has been provided by Masson.[39] The subject also receives coverage in most general texts and reviews dealing with radical polymerization. References to reviews that detail the reactions of specific classes of initiator are given under the appropriate subheading below.

Some characteristics of initiators used for thermal initiation are summarized in Table 3.1. These provide some general guidelines for initiator selection. In general, initiators which afford carbon-centered radicals (*e.g.* dialkyl diazenes, aliphatic diacyl peroxides) have lower efficiencies for initiation of polymerization than those which produce oxygen-centered radicals. Exact values of efficiency depend on the particular initiators, monomers, and reaction conditions. Further details of initiator chemistry are summarized in sections 3.3.1 (azo-compounds) and 3.3.2 (peroxides) as indicated in Table 3.1. In these sections, we detail the factors which influence the rate of decomposition (*i.e.* initiator structure, solvent, complexing agents), the nature of the radicals formed, the susceptibility of the initiator to induced decomposition, and the importance of transfer to initiator and other side reactions of the initiator or initiation system. The reactions of radicals produced from the initiator are given detailed treatment in section 3.4.

3.3.1 Azo-Compounds

Two general classes of azo-compound will be considered in this section, the dialkyldiazenes (12) (3.3.1.1) and the dialkyl hyponitrites (13) (3.3.1.2).

$$R\text{---}N\text{=}N\text{---}R' \qquad R\text{---}O\text{---}N\text{=}N\text{---}O\text{---}R'$$
$$(12) \qquad\qquad\qquad (13)$$

Initiator Class	Example	section	radicals generated[a]	efficiency[b]	transfer[c]
dialkyl diazenes	H₃C-C(CN)(CH₃)-N=N-C(CN)(CH₃)-CH₃	3.3.1.1	1° alkyl	low	low
dialkyl hyponitrites	CH₃-C(CH₃)₂-O-N=N-O-C(CH₃)₃	3.3.1.2	1° alkoxy 2° alkyl	high	low
diacyl peroxides	CH₃(CH₂)₁₀CH₂-C(=O)-O-O-C(=O)-CH₂(CH₂)₁₀CH₃	3.3.2.1	(1° acyloxy) 2° alkyl	low	high
diaroyl peroxides	Ph-C(=O)-O-O-C(=O)-Ph	3.3.2.1	1° aroyloxy 2° aryl	high	high
dialkyl peroxydicarbonates	(CH₃)₂CH-O-C(=O)-O-O-C(=O)-O-CH(CH₃)₂	3.3.2.2	1° alkoxycarbonyloxy (2° alkoxy)	high	high
peresters	(CH₃)₃C-O-O-C(=O)-C(CH₃)₃	3.3.2.3	1° alkoxy, acyloxy 2° alkyl	med.	med.
dialkyl peroxalates	(CH₃)₃C-O-O-C(=O)-C(=O)-O-O-C(CH₃)₃	3.3.2.3	1° alkoxy 2° alkyl	high	med.
dialkyl peroxides	(CH₃)₃C-O-O-C(CH₃)₃	3.3.2.4	1° alkoxy 2° alkyl	high	low
dialkyl ketone peroxides	cyclohexane-1,1-diyl bis(tert-butyl peroxide)	3.3.2.5	1° alkoxy 2° alkyl	med.	low

Table 3.1 (continued)

Initiator Class	Example	section	radicals generated[a]	efficiency[b]	transfer[c]
alkyl hydroperoxides	CH₃–C(CH₃)(CH₃)–O–O–H	3.3.2.5	1° hydroxy, alkoxy 2° alkyl	high	high
persulfate	⁻O–S(=O)(=O)–O–O–S(=O)(=O)–O⁻	3.3.2.6	1° sulfate radical anion	low	low
disulfides	(C₂H₅)₂N–C(=S)–S–S–C(=S)–N(C₂H₅)₂	3.3.5	1° thiyl	high	high

a 1° = primary radical from initiator decomposition, 2° = secondary radical derived by fragmentation of 1° radical. Species shown in parentheses may be formed under some conditions but are seldom observed in polymerizations of common monomers.
b Efficiency decreases as the importance of cage reactions increases.
c Susceptibility to radical-induced decomposition.

Initiation

Polymeric azo-compounds and multifunctional initiators with azo-linkages are discussed elsewhere (see 3.3.3 and 6.3.2.1) as are azo compounds which find use as iniferters (see 7.5.2).

3.3.1.1 Dialkyldiazenes

The kinetics and mechanism of the thermal and photochemical decomposition of dialkyldiazenes (12) have been comprehensively reviewed by Engel.[40] The use of these compounds as initiators of radical polymerization has been covered by Moad and Solomon[37] and Sheppard.[41] The general chemistry of azo-compounds has also been reviewed by Koga et al.,[42] Koenig,[43] and Smith.[44]

Dialkyldiazenes (12, R=alkyl) are sources of alkyl radicals. While there is clear evidence for the transient existence of diazenyl radicals (14; see Scheme 13) during the decomposition of certain unsymmetrical diazenes[40,42] and of cis-diazenes,[45] all isolable products formed in thermolysis or photolysis of dialkyldiazenes (12) are attributable to the reactions of alkyl radicals.

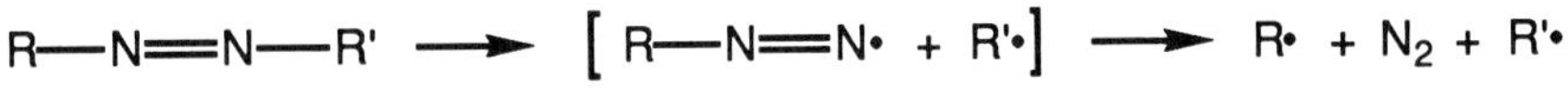

R—N=N—R' → [R—N=N• + R'•] → R• + N$_2$ + R'•

(12)   (14)

Scheme 13

In the decomposition of symmetrical azo compounds the intermediacy of diazenyl radicals remains a subject of controversy. However, it is clear that diazenyl radicals, if they are intermediates, do not have sufficient lifetime to be trapped or to initiate polymerization.

Ayscough et al.[46] photolysed AIBN in a matrix at -196°C and observed EPR signals which were attributed to the diazenyl radical, $(CH_3)_2(CN)C-N=N$• [this assignment has been questioned[42]]. However for AIBN decomposition in solution, at temperatures normally encountered in polymerizations, the finding, that the rate of decomposition is independent of solvent viscosity (i.e. no cage return) is evidence for concerted 2-bond cleavage.[27]

Most available dialkyldiazenes (12) are symmetrical and the R groups are generally tertiary with functionality to stabilize the incipient radical [e.g. cyano AIBN, (15-17), ester (AIBMe), or phenyl (18)]. Those most commonly encountered are the azonitriles, these include 2,2'-azobis(2-methylpropanenitrile) [better known as azobis(isobutyronitrile) or AIBN], 1,1'-azobis(1-cyclohexanenitrile) (15), 2,2'-azobis(2-methylbutanenitrile) (16), and the water soluble initiator 4,4'-azobis(4-cyanovaleric acid) (17).

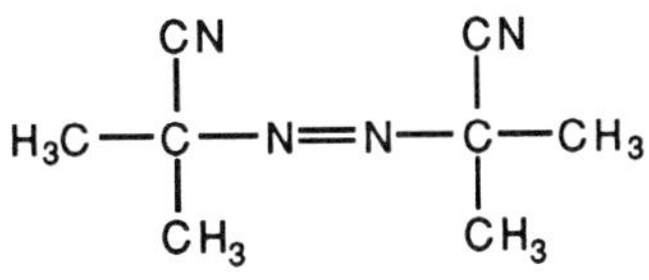

AIBN 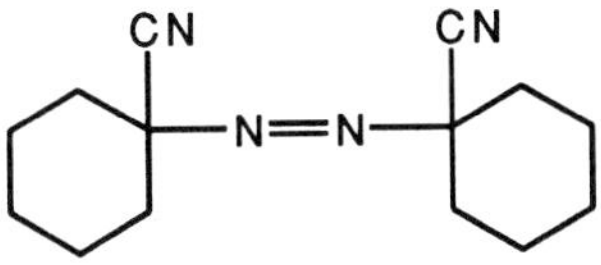   (15)

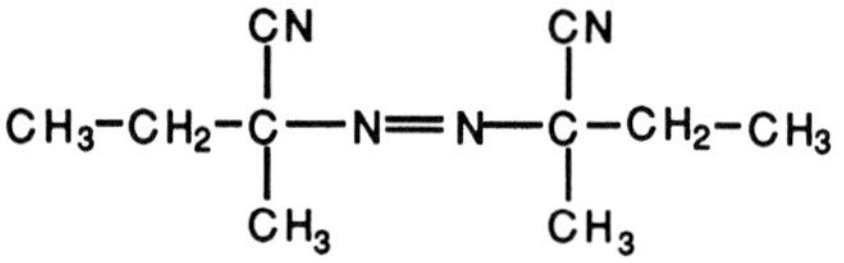

Structures (16), (17), AIBMe, (18)

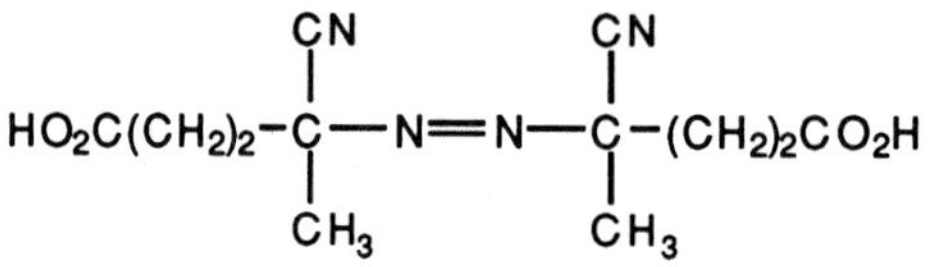

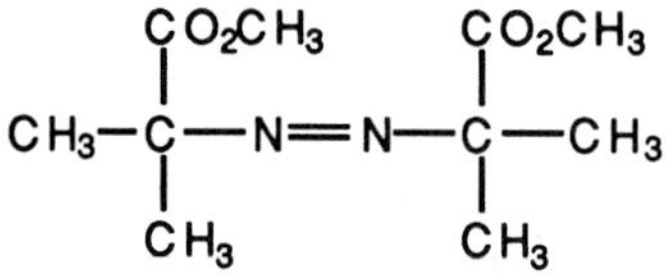

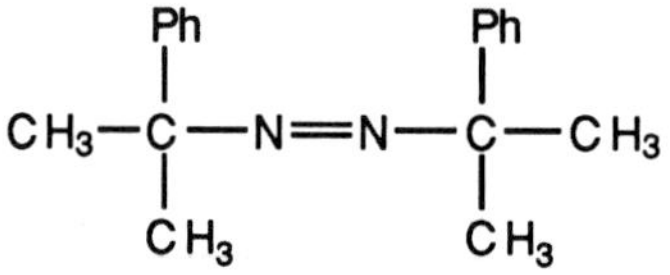

(16) CH₃-CH₂-C(CN)(CH₃)-N=N-C(CH₃)(CH₂-CH₃)-CN

(17) HO₂C(CH₂)₂-C(CN)(CH₃)-N=N-C(CH₃)(CN)-(CH₂)₂CO₂H

AIBMe: CH₃-C(CO₂CH₃)(CH₃)-N=N-C(CH₃)(CO₂CH₃)-CH₃

(18) CH₃-C(Ph)(CH₃)-N=N-C(CH₃)(Ph)-CH₃

Unsymmetrical azo-compounds find application as initiators of polymerization in special circumstances, for example, as initiators of quasi-living polymerization [*e.g.* triphenylmethylazobenzene (19) (see 7.2.2)], as hydroxy radical sources [α-hydroperoxydiazenes (20) (see 3.3.3.1)], for enhanced solubility in organic solvents [*e.g.* t-butylazocyclohexanecarbonitrile (21)], or as high temperature initiators [*e.g.* t-butylazoformamide (22)]. They have also been used as radical precursors in model studies of cross-termination in copolymerization (see 5.2.3.5).

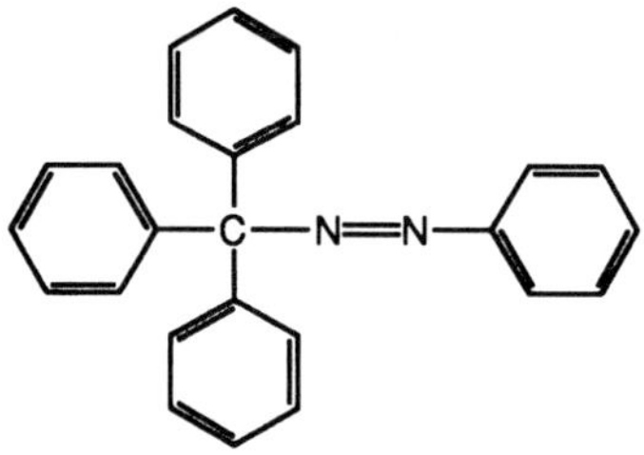

(19) Ph₃C-N=N-Ph

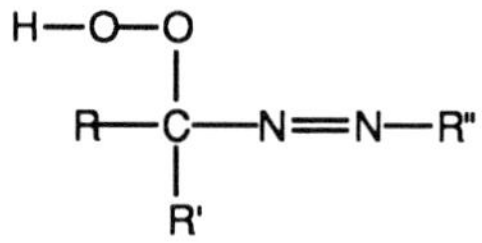

(20) R-C(OOH)(R')-N=N-R''

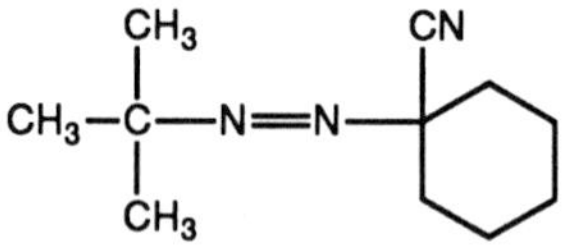

(21) CH₃-C(CH₃)₂-N=N-C(CN)(cyclohexyl)

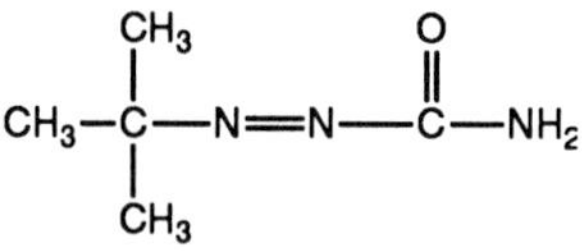

(22) CH₃-C(CH₃)₂-N=N-C(=O)-NH₂

3.3.1.1.1 Thermal decomposition

While some details of the kinetics of radical production from dialkyldiazenes remain to be unraveled, their decomposition mechanism and behavior as polymerization initiators are largely understood. Kinetic parameters for some common azo-initiators are presented in Table 3.2.

Table 3.2 Selected Kinetic Data for Decomposition of Azo-Compounds[a]

R (, R')		solvent	temp. range[b] °C	$k_d \times 10^6$(60°C)[c] $M^{-1} s^{-1}$	E_a kJ mol^{-1}	$A \times 10^{15}$	10 h $t_{1/2}$[d] °C	ref.
(a) diazenes (12)								
(CH$_3$)$_2$(CN)C	**AIBN**	benzene/toluene	37-105(13)	9.7	131.7	4.31	65	28,48-51
(CH$_3$)$_2$(CO$_2$CH$_3$)C	**AIBMe**	benzene	50-70(4)	9.0	124.0	0.248	66	52,53
(CH$_3$)$_2$(Ph)C	**18**	toluene	40-70(17)	170	126.7	12.2	45	54
(CH$_3$)$_3$C		diphenyl ether	165-200(6)	-	180.4	91.7	161	55
(c-C$_6$H$_{10}$)(CN)C	**15**	toluene	80-100(3)	0.30	149.1	71.0	88	56
(CH$_3$)(C$_2$H$_5$)(CN)C	**16**	ethylbenzene	80-100(3)	5.0	137.8	20.3	69	57
(Ph)$_3$C, Ph	**19**	benzene/toluene	25-75(9)	524	114.6	0.486	35	58-60
(b) hyponitrites (13)								
(CH$_3$)$_3$C	**23**	isooctane	45-75(4)	215	119.5	1.17	42	61
(CH$_3$)$_2$(Ph)C	**24**	cyclohexane	40-70(12)	1370	113.9	0.99	29	62

[a] Arrhenius parameters recalculated from original data taken from the indicated references. Values of E_a and A rounded to 4 and 3 significant figures respectively.
[b] Number of data points given in parentheses.
[c] Calculated from Arrhenius parameters shown and rounded to 2 significant figures.
[d] Temperature for ten hour half life rounded to 2 significant figures.

Thermolysis rates (k_d) of dialkyldiazenes (12) show a marked dependence on the nature of R (and R'). The values of k_d increase in the series where R (=R') is aryl, primary, secondary, tertiary, allyl. In general, k_d is dramatically accelerated by α-substituents capable of delocalizing the free spin of the incipient radical.[40] For example, Timberlake[47] has found that for dialkyldiazenes, X-C(CH$_3$)$_2$-N=N-C(CH$_3$)$_2$-X, k_d increase in the series where X is CH$_3$<-OCH$_3$<-SCH$_3$<-CO$_2$R~-CN<-Ph<-CH=CH$_2$ (see also Table 3.2). These results can be rationalized in terms of the relative stability of the generated radicals (R•, R'•).

However, steric factors are also important.[63] Rüchardt et al. showed, for a series of acyclic alkyl derivatives, that a good correlation exists between k_d and ground state strain.[64,65] Additional factors are important for bicyclic and other conformationally constrained azo-compounds.[40,42,66] Wolf[67] has described a scheme for calculating k_d based on radical stability (HOMO pi-delocalization energies) and ground state strain (steric parameters).

There have been numerous studies on the kinetics of decomposition of AIBN and other dialkyldiazenes.[38] Solvent effects on k_d are small by conventional standards but, nonetheless, significant. Data for AIBMe is presented in Table 3.3. There is a factor of two difference between k_d in methanol and k_d in ethyl acetate. For a series of aromatic solvents the value of k_d for AIBN is reported to be slightly higher in aromatic than in hydrocarbon solvents and to increase with the dielectric constant of the medium.[27,68,69] The rate of decomposition of AIBN shows no direct correlation with solvent viscosity (see also 3.3.1.1.3) which is consistent with the irreversible nature of the reaction (*i.e.* no cage return).

Table 3.3 Solvent Dependence of Rate Constants for AIBMe Decomposition

$k_d \times 10^5$ s^{-1}	Solvent	Temperature (°C)	ref.
0.58	cyclohexane	60.0	52
0.72	ethyl acetate	60.0	53
0.74	methyl isobutyrate	60.0	53
0.83	1:1 MMA/S	60.0	70
1.18[a]	aliph. esters	60.0	71
0.88	benzene	60.0	52
0.91	benzene	60.0	53
1.01	acetonitrile	60.0	52
1.13	S	60.0	72
1.20	methanol	60.0	53
1.44	methanol	60.0	52

[a] Calculated from the expression given: ln(k_d)=33.1-(14800/T); said to be valid for a range of aliphatic ester solvents including MMA.

Thermolysis rates are enhanced substantially by the presence of certain Lewis acids (*e.g.* boron and aluminum halides), and transition metal salts (*e.g.* Cu^{2+},

Ag+).[38] There is also evidence that complexes formed between azo-compounds and Lewis acids (*e.g.* ethyl aluminum sesquichloride) undergo thermolysis or photolysis to give complexed radicals which have different specificity to uncomplexed radicals.[73-75]

3.3.1.1.2 Photochemical decomposition

The *trans*-dialkyldiazenes have λ_{max} 350-370 nm and ϵ 2-50 M^{-1} cm^{-1} and are photolabile. They are, therefore, potential photoinitiators.[40,76] The efficiency and rate of radical generation depends markedly on structure.[40] Dialkyldiazenes are often depicted without indicating the stereochemistry about the nitrogen-nitrogen double bond. However, except when constrained in a ring system the dialkyldiazenes can be presumed to have the *trans*-configuration.

Alicyclic *cis*-dialkyldiazenes are very thermolabile when compared to the corresponding *trans*-isomers, often having only transient existence under typical reaction conditions. It has been proposed[40] that the main light-induced reaction of the dialkyldiazenes is *trans-cis* isomerization. Dissociation to radicals and nitrogen is then a thermal reaction of the *cis*-isomer (Scheme 14).

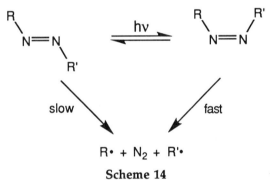

Scheme 14

Therefore, the rate and quantum yield for photoisomerization approximate those for nitrogen formation and are typically *ca.* 0.5. Where the *cis* isomer is thermally stable, quantum yields for initiator disappearance are low (ϕ<0.1).[40]

An important ramification of the photolability of azo-compounds is that, when using dialkyl diazenes as thermal initiators, care must be taken to ensure that the polymerization mixture is not exposed to excessive light during its preparation.

3.3.1.1.3 Initiator efficiency

The proportion of 'useful' radicals generated from common dialkyldiazene initiators is never quantitative; typically it is 50-70% in media of low viscosity (*i.e.* in low conversion polymerizations).[28,77,78] The main cause of this inefficiency is loss of radicals through self-reaction within the solvent cage.

For dialkyldiazenes where the α-positions are not fully substituted, tautomerization to the corresponding hydrazone may also reduce the initiator efficiency[79] (Scheme 15). This rearrangement is catalyzed by light and by acid.

Scheme 15

There is also evidence for a radical-induced mechanism involving initial abstraction of an α-hydrogen (Scheme 16).

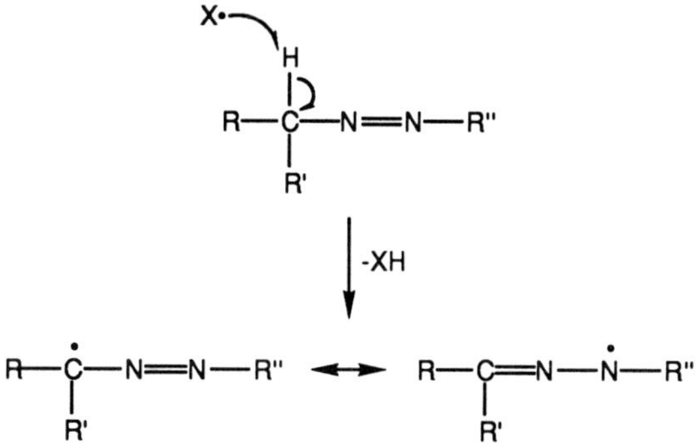

Scheme 16

Conflicting statements have appeared on the sensitivity of f to the nature of the monomer involved. Braun and Czerwinski[80] have reported that for low conversion polymerizations, f is essentially the same in MMA, S, and N-vinylpyrrolidone. Fukuda et al.[81] have reported that f varies between MMA and S. The solvent dependence of k_d may account for this apparent conflict (see Table 3.3).

While the rate of azo-compound decomposition shows only a small dependence on solvent viscosity, the amount of cage reaction (and hence f) varies dramatically with the viscosity of the reaction medium and hence with factors that determine the viscosity (conversion, temperature, solvent, etc.).[27]

Most values of f have been measured at zero or low conversions. During polymerization the viscosity of the medium increases and the concentration of monomer decreases dramatically as conversion increases (i.e. as the volume fraction of polymer increases). The value of f is anticipated to drop accordingly.[28,29,82-85] For example, with S polymerization in 50% (v/v) benzene at 70°C initiated by 0.1M AIBN the 'instantaneous' f varies from 76% at low conversion to <20% at 90-95% conversion (see Figure 1).[28] The assumption that the rate of initiation ($k_d f$) is invariant with conversion (common to most pre 1980s and many recent kinetic studies of radical polymerization) cannot be supported.

Initiation

The viscosity dependence of f may lead to the initiator efficiency being dependent on the molecular weight of the polymer being produced. This, in turn, is a function of the initiator and monomer concentration. For example, initiator efficiencies are expected to be higher during oligomer synthesis than in preparation of high molecular weight polymer. Initiator efficiency has also been shown to depend on the size of the initiator-derived radicals.[29] There is an inverse relationship between the rate of escape from the solvent cage and radical size.

Table 3.4 Zero-Conversion Initiator Efficiency for AIBMe under Various Reaction Conditions

f	scavenger	temperature (°C), solvent	reference
0.81[a]	none	98, S	72
0.72[a]	none	90, S	72
0.77	galvinoxyl	90, chlorobenzene	86
0.76	nitroxide	80, chlorobenzene	87
0.70[b]	triphenylverdazyl	80, MMA	71
0.63[a]	none	80, S	72
0.68-0.60[c]	none	60, MMA/S in benzene	70
0.56[b]	triphenylverdazyl	60, MMA	71
0.48	DPPH	60, not specified	88
0.45	none	60, benzene	53
0.40[a]	none	60, S	72

[a] Estimated by analysis of polymerization kinetics.
[b] Calculated using the expression $\ln f = 0.58-(330/T)$.[71]
[c] [Polymer end groups]/[total products] with AIBMe-α-^{13}C as initiator. Overall efficiency reduces from 0.68 at <16% conversion to 0.60 at 95% conversion.

Initiator efficiency increases with reaction temperature (see Table 3.4). It is also worth noting that apparent zero-conversion initiator efficiencies depend on the method of measurement. Better scavengers trap more radicals. The data in Table 3.4 suggest that monomers (MMA, S) are not as effective at scavenging radicals as the inhibitors used to measure initiator efficiencies. The finding suggests that in polymerization the initiator-derived radicals have a finite probability of undergoing self-reaction after they escape the solvent cage and numbers obtained by the inhibitor method should be considered as upper limits.

The by-products of decomposition of certain dialkyldiazenes can be a concern. Consider the case of AIBN decomposition (Scheme 11). The major by-product is the ketenimine (**10**).[51,89-91] This compound is itself thermally labile and reverts to cyanoisopropyl radicals at a rate constant similar to that for AIBN thermolysis.[49,50,91] This complicates any analysis of the kinetics of initiation.[28,50] A more important concern, is the potential reactivity of (**10**) as a transfer agent under polymerization conditions.[92] Tetramethyl-succinonitrile (**9**) appears essentially inert under polymerization conditions.* However, the compound is reported to be toxic and must be removed from polymers used in food contact applications.[31] Methacrylonitrile (MAN) formed by disproportionation readily copolymerizes.[6,30] The copolymerized MAN may affect the thermal stability of polymers. A suggestion[94] that copolymerized MAN may be a "weak link" in PS initiated with AIBN has been disputed.[13]

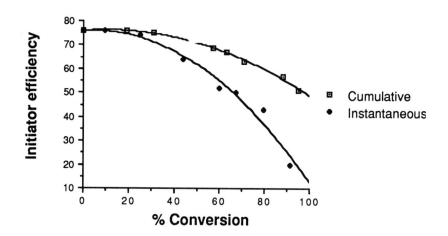

Figure 1 Efficiency of AIBN as initiator in S polymerization (50% v/v toluene, 70°C) as a function of conversion.[1,28]

Some of the complications associated with the use of AIBN may be avoided by use of alternative azo-initiators. Azobis(methyl isobutyrate) (AIBMe) has a decomposition rate only slightly less than AIBN and has been promoted for use in laboratory studies of polymerization[71] because the kinetics and mechanism of its decomposition kinetics are not complicated by ketenimine formation.

The azo-derivative (16) also has similar decomposition kinetics to AIBN (see Table 3.2). The initiators (16) and AIBMe also have greater solubility in organic solvents than AIBN.

3.3.1.1.4 Transfer to initiator

Dialkyldiazenes are often preferred over other (peroxide) initiators because of their lower susceptibility to induced decomposition. The importance of transfer to initiator during polymerizations initiated by AIBN has been the subject of some controversy. While the early work of Baysal and Tobolsky,[95] Bevington

---

* Pryor and Fiske[93] found $C_T = 3.7 \times 10^{-5}$ for tetramethylsuccinonitrile at 60°C in styrene polymerization.

and Lewis,[96] and others suggested that transfer to initiator was insignificant during polymerizations of MMA or S, a number of subsequent studies on polymerization kinetics report a significant transfer constant ($C_I \sim 0.1$).[93,97-101] Studies of S polymerization initiated by ^{13}C-labeled AIBN demonstrate that transfer to initiator has little importance in that system.[6] Thus, other explanations for those irregularities in polymerization kinetics previously attributed to transfer to initiator have to be considered: for example, failure to allow for the variation of initiator efficiency with conversion (see 3.3.1.1.3). There is some evidence that transfer to initiator may be of importance during AIBN-initiated vinyl acetate polymerization.[102]

Even though AIBN has a low transfer constant, the ketenimine formed by combination of cyanoisopropyl radicals (see Scheme 11) is anticipated to be more susceptible to induced decomposition (Scheme 17).[103]

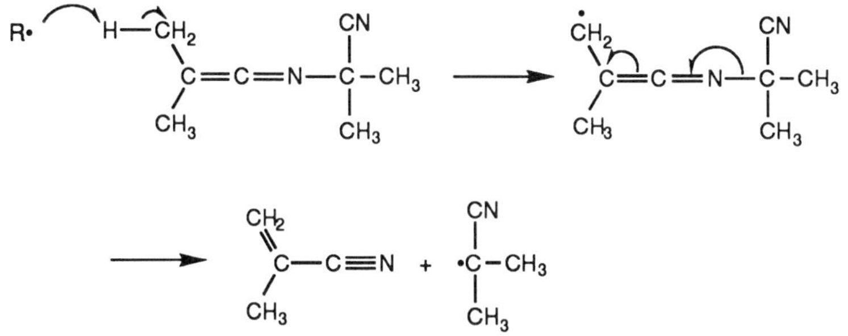

Scheme 17

3.3.1.2 Hyponitrites

The hyponitrites, derivatives of hyponitrous acid (HO–N=N–OH), are low temperature sources of alkoxy or acyloxy radicals. A detailed study of the effect of substituents on k_d for the hyponitrite esters has been reported by Quinga and Mendenhall.[104]

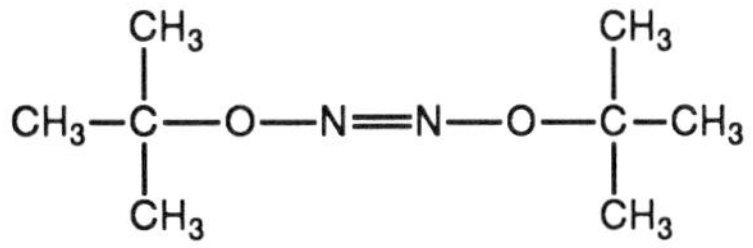

(23) 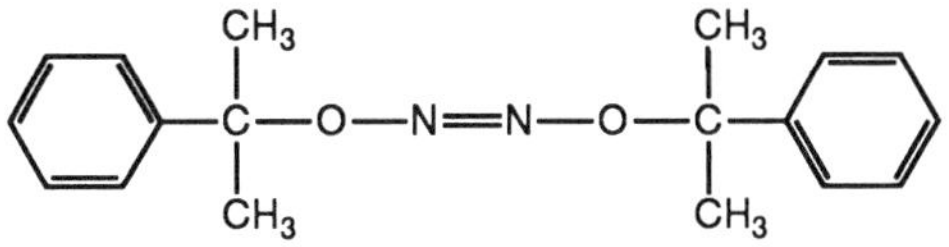   (24)

While di-t-butyl (23) and dicumyl hyponitrites (24) have proved convenient sources of t-butoxy and cumyloxy radicals respectively in the laboratory,[61,62,105,106] the utilization of hyponitrites as initiators of polymerization has been limited by difficulties in synthesis and commercial availability. Dialkyl hyponitrites (13) show only weak absorption at λ>290 nm and their photochemistry is largely a neglected area. The triplet sensitized decomposition of these materials has been investigated by Mendenhall et al.[107]

The hyponitrites generally appear somewhat more efficient with respect to radical generation than the dialkyldiazenes (see 3.3.1.1). However, a proportion of radicals is lost through cage reaction with formation of the corresponding dialkyl peroxides or ketone plus alcohol (Scheme 18).[108,109] The disproportionation pathway is open only to hyponitrites with α-hydrogens. Kiefer and Traylor[110] showed that the extent of cage reaction was strongly dependent on the medium viscosity.

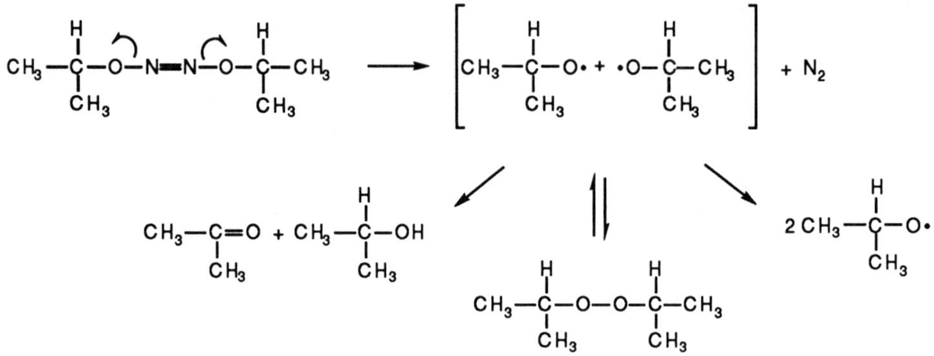

Scheme 18

Approximately 5% of radicals undergo cage recombination when dicumyl hyponitrite (24) is decomposed in bulk MMA or S at 60°C.[62] Dicumyl peroxide, the product of cage recombination is likely to be stable under the conditions where hyponitrites are usually employed. Nonetheless, its formation is a concern since contamination of a product polymer with peroxide may well impair its longer term durability.

Tertiary hyponitrites are not particularly susceptible to induced decomposition. However, the same is not true of primary and secondary hyponitrites.[111] Isopropyl hyponitrite is reported[112] to undergo induced decomposition by a mechanism involving initial abstraction of a α-hydrogen (Scheme 19).

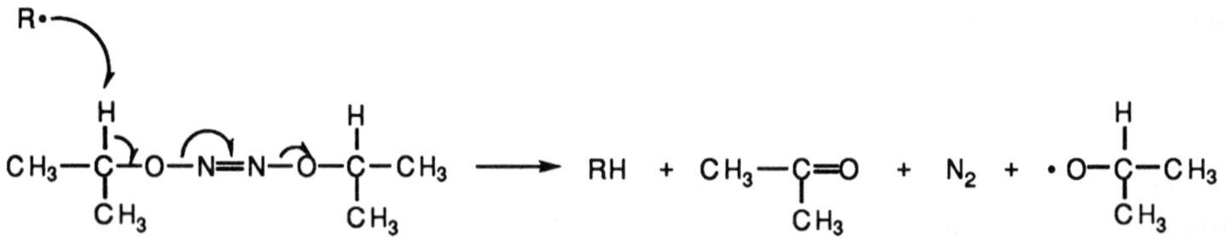

Scheme 19

3.3.2 Peroxides

The general chemistry of the peroxides has been covered in many books and reviews.[37,38,43,113-120] Readers are referred in particular to Swern's Trilogy[116-118] for a comprehensive coverage of the literature through 1970. The chemistry associated with their use as initiators of polymerization was recently described by Moad and Solomon.[37]

Many types of peroxides (R-O-O-R) are known. Those in common use as initiators include: diacyl peroxides (**25**), peroxydicarbonates (**26**), peresters (**27**), dialkyl peroxides (**28**), hydroperoxides (**29**), and inorganic peroxides [*e.g.* persulfate (**30**)]. Multifunctional and polymeric initiators with peroxide linkages are discussed in sections 3.3.3 and 6.3.2.1.

$$\underset{(25)}{R-\overset{O}{\underset{\parallel}{C}}-O-O-\overset{O}{\underset{\parallel}{C}}-R'} \quad \underset{(26)}{R-O-\overset{O}{\underset{\parallel}{C}}-O-O-\overset{O}{\underset{\parallel}{C}}-O-R'}$$

$$\underset{(27)}{R-\overset{O}{\underset{\parallel}{C}}-O-O-R'} \quad \underset{(28)}{R-O-O-R'} \quad \underset{(29)}{R-O-O-H}$$

$$\underset{(30)}{-O-\overset{O}{\underset{\underset{O}{\parallel}}{\overset{\parallel}{S}}}-O-O-\overset{O}{\underset{\underset{O}{\parallel}}{\overset{\parallel}{S}}}-O-}$$

Peroxides are used most commonly either as thermal initiators or as a component in a redox system. While peroxides are photochemically labile, they seldom find use as photoinitiators other than in laboratory studies because of their poor light absorption characteristics. They generally have low extinction coefficients and absorb in the same region as monomer. Kinetic parameters for decomposition of some important peroxides are given in Table 3.5.

3.3.2.1 Diacyl peroxides

Diacyl peroxides (**25**, R= alkyl or aryl) are given specific coverage in reviews by Fujimori,[121] Bouillion *et al.*,[122] and Hiatt.[123] They are sources of acyloxy radicals which in turn are sources of aryl or alkyl radicals.

3.3.2.1.1 Thermal decomposition

The rates of thermal decomposition of diacyl peroxides (**25**) are dependent on the substituents R. The rates of decomposition increase in the series where R is: aryl~primary alkyl<secondary alkyl<tertiary alkyl. This order has been variously proposed to reflect the stability of the radical (R•) formed on β-scission of the acyloxy radical, the nucleophilicity of R, or the steric bulk of R. For peroxides with non-concerted decomposition mechanisms it seems unlikely that the stability of R• should in itself be an important factor.

Table 3.5 Selected Kinetic Data for Decomposition of Peroxides[a]

R	R'		solvent	temp. range[b] °C	$k_d \times 10^6$(60°C)[c] s^{-1}	E_a kJ mol^{-1}	$A \times 10^{15}$	10 h $t_{1/2}$[c,d] °C	ref.
(a) diacyl peroxides (25)									
Ph-	Ph-	BPO	benzene[e]	38-80(17)	1.5	139.0	9.34	78	124
n-C$_{11}$H$_{23}$-	n-C$_{11}$H$_{23}$-	LPO	benzene	35-70(8)	8.9	125.3	0.393	66	125
(b) peroxydicarbonates (26)									
(CH$_3$)$_2$CH-	(CH$_3$)$_2$CH-	34	benzene[e]	35-60(10)	130	126.7	9.75	46	126
(c) peresters (27)									
Ph-	(CH$_3$)$_3$C-	BPB	benzene	110-130(3)	0.04	144.0	1.53	105	127
		DBPOX	benzene	35-55(3)	1800	110.0	0.310	26	128
(d) dialkyl peroxides (28)									
(CH$_3$)$_3$C-	(CH$_3$)$_3$C-	37	benzene	100-135(4)	0.0025	152.7	2.16	125	129,130
(e) alkyl hydroperoxides (29)									
(CH$_3$)$_3$C-	-	40	benzene	155-175(4)	-	174.2	7.97	168	131
(f) inorganic peroxides									
K$_2$S$_2$O$_8$	-	30	0.1M NaOH	50-90(5)	4.4	148.0	709	69	132

[a] Kinetic parameters recalculated from original data taken from the indicated references. Values rounded to 3 significant figures.
[b] Number of data points given in parentheses.
[c] Calculated from Arrhenius parameters shown and rounded to two significant figures.
[d] Temperature for ten hour half life.
[e] In the presence of inhibitor added to prevent induced decomposition.

Initiation

For diaroyl peroxides, *m*- and *p*-electron-withdrawing substituents retard the rate of decomposition while *m*- and *p*-electron-donating and all *o*-substituents enhance decomposition rates. The *o*-substituent effect has been attributed to the sensitivity of homolysis to steric factors.

Only a few diacyl peroxides see widespread use as initiators of polymerization. The reactions of the diaroyl peroxides (25, R=aryl) will be discussed in terms of the chemistry of BPO (scheme 20). The rate of β-scission of thermally generated benzoyloxy radicals is slow relative to cage escape, consequently, both benzoyloxy and phenyl radicals are important as initiating species. In solution, the only significant cage process is reformation of BPO (*ca.* 4% at 80°C in isooctane);[133,134] only minute amounts of phenyl benzoate or biphenyl are formed within the cage. Therefore, in the presence of a reactive substrate (*e.g.* monomer), the production of radicals can be almost quantitative (see 3.3.2.1.3).

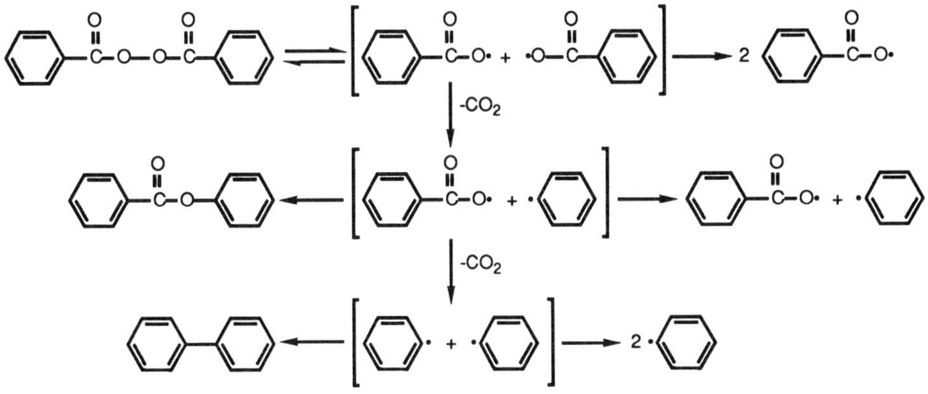

Scheme 20

One of the most commonly encountered aliphatic diacyl peroxides (25, R=alkyl) is didodecanoyl or dilauroyl peroxide (LPO). Lower diacyl peroxides cannot be conveniently handled in a pure state due to their susceptibility to induced decomposition. They are shock sensitive and may decompose explosively.

The aliphatic diacyl peroxide initiators should be considered as sources of alkyl, rather than of acyloxy radicals. Most aliphatic acyloxy radicals have a transient existence at best. For certain diacyl peroxides (25) where R is a secondary or tertiary alkyl group there is controversy as to whether loss of carbon dioxide occurs in concert with O-O bond cleavage. Thus, ester end groups observed in polymers prepared with aliphatic diacyl peroxides are unlikely to arise directly from initiation, but rather from transfer to initiator (see 3.3.2.1.4).

The high rate of decarboxylation of aliphatic acyloxy radicals is also the prime reason behind low initiator efficiencies (see 3.3.2.1.3). Decarboxylation occurs within the solvent cage and recombination gives alkane or ester by-products.

Cage return for LPO is 18-35% at 80°C in *n*-octane as compared to only 4% for BPO under similar conditions.[133]

Observed rates of disappearance for diacyl peroxides show marked dependence on solvent and concentration.[135] In part, this is a reflection of their susceptibility to induced decomposition (see 3.3.2.1.4 and 3.3.2.1.5). However, the rate of disappearance is also a function of the viscosity of the reaction medium. This is evidence for cage return (see 3.3.2.1.3).[134] The observation[133] of slow scrambling of the label in benzoyl-*carbonyl*-^{18}O peroxide between the carbonyl and the peroxidic linkage provides more direct evidence for this process.

3.3.2.1.2 Photochemical decomposition

Diacyl peroxides have continuous weak absorptions in the UV to *ca.* 280 nm ($\varepsilon \sim 50$ M^{-1} cm^{-1} at 234 nm).[136] Although the overall chemistry in thermolysis and photolysis may appear similar, substantially higher yields of phenyl radical products are obtained when BPO is decomposed photochemically. It has been suggested that, during the photodecomposition of BPO, β-scission may occur in concert with O–O bond rupture and give rise to formation of one benzoyloxy radical, one phenyl radical, and a molecule of carbon dioxide (Scheme 21).[137] Time resolved EPR experiments[138] have shown that photochemical decomposition of BPO does produce benzoyloxy radicals with discrete existence. It is, nonetheless, clear that the photochemically generated benzoyloxy radicals have substantially shorter life times in solution than those generated thermally.[139,140] In these circumstances cage products also assume greater importance[140] and initiator efficiencies are anticipated to be lower.

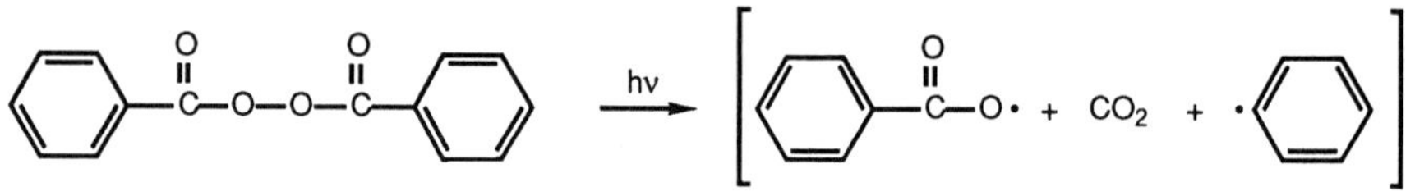

Scheme 21

It has also been suggested that photoexcited benzoyl peroxide is somewhat more susceptible to induced decomposition processes involving electron transfer than the ground state molecule. Rosenthal *et al.*[141] have reported on redox reactions with certain salts (including benzoate ion) and neutral molecules (*e.g.* alcohols).

3.3.2.1.3 Initiator efficiency

Ideally all reactions should result from unimolecular homolysis of the relatively weak O–O bond. However, unimolecular rearrangement and various forms of induced and non-radical decomposition complicate the kinetics of radical generation and reduce the initiator efficiency.[38] Peroxide decomposition

Initiation

induced by radicals and redox chemistry is covered in sections 3.3.2.1.4 and 3.3.2.1.5 respectively.

Cage recombination is also a major factor limiting the efficiency of radical production from aliphatic diacyl peroxides. Initiator efficiency depends on the rate of β-scission of the acyloxy radical formed. If β-scission is slow, the only significant cage reaction involves regeneration of the diacyl peroxide (*e.g.* thermolysis of diaroyl peroxides). Cage return leads to a lowering of the rate of decomposition without reducing the initiator efficiency (see 3.3.2.1.1). However, if β-scission is rapid and decarboxylation occurs within the solvent cage, then combination of the alkyl or aryl radical with another radical to form an ester or alkane will reduce the initiator efficiency (*e.g.* thermolysis or photolysis of aliphatic diacyl peroxides and photolysis of all diacyl peroxides).

The importance of the cage reaction increases according to the viscosity of the reaction medium. This contributes to a decrease in initiator efficiency with conversion.[29,142-144] Stickler[145] determined the initiator efficiency during MMA polymerization at very high conversions (~80%) to be in the range 0.1-0.2 depending on the polymerization temperature. The main initiator-derived by-product was phenyl benzoate.

Diacyl peroxides may also undergo non-radical decomposition *via* the carboxy inversion process to form an acylcarbonate (Scheme 22).[38] The reaction is of greatest importance for diaroyl peroxides with electron withdrawing substituents and for aliphatic diacyl peroxides (25) where R is secondary, tertiary or benzyl.[146] The reaction is thought to involve ionic intermediates and is favored in polar solvents[146] and by Lewis acids.[147] Other heterolytic pathways for peroxide decomposition have been described.[148]

Scheme 22

3.3.2.1.4 *Transfer to initiator and induced decomposition*

Transfer to initiator can be a major complication in polymerizations initiated by diacyl peroxides and the importance of the process typically increases with monomer conversion due to an increase in the [initiator]:[monomer] ratio.[8,95,149-151]

It has been demonstrated that in BPO initiated S polymerization transfer to initiator may be the major chain termination mechanism. For bulk S polymerization with 0.1 M BPO at 60°C up to 75% of chains are terminated by transfer to initiator or primary radical termination (<75% conversion).[6] A further consequence of the high incidence of chain transfer is that high conversion PS formed with BPO initiator tends to have a much narrower molecular weight distribution than that prepared with other initiators (*e.g.* AIBN) under similar conditions.

The mechanism of transfer to BPO involves homolytic attack on one of the oxygen atoms of the peroxidic linkage (Scheme 12) with formation of an ester end group and expulsion of a benzoyloxy radical. The end group formed (a secondary ester) is distinct from that formed in initiation. Such end groups may contribute to the reduced thermal stability of high conversion PS prepared with benzoyl peroxide (see 7.2.1).[13,152] In the case of VAc or VC polymerizations the chain end will be a hydrolytically unstable ketal or α-chloroester group respectively.

The decomposition of BPO and other diacyl peroxides may also be induced by other radicals present in the reaction medium. These include initiator-derived[135] and stable radicals (*e.g.* galvinoxyl,[124] triphenylmethyl[153,154] and nitroxides[155]).

3.3.2.1.5 *Redox reactions*

The decomposition of diacyl peroxides is catalyzed by various transition metal salts,[38,156] for example, Cu^+ (Scheme 23).[157,158] A side reaction is oxidation of alkyl radicals by the oxidized form of the metal salt (*e.g.* Cu^{2+}).

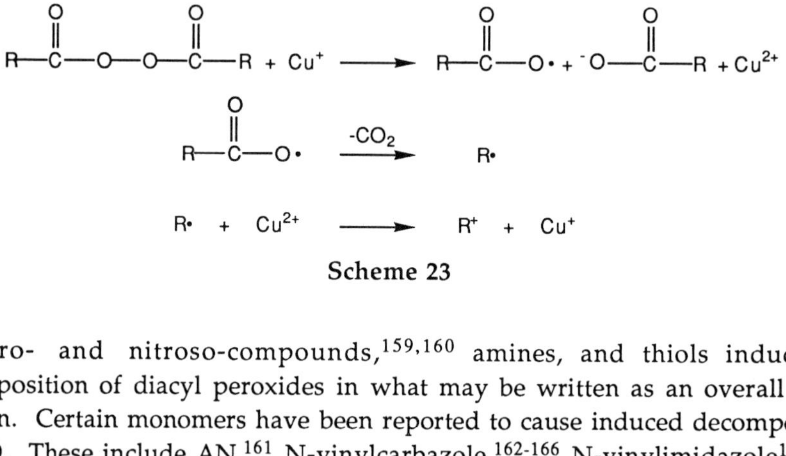

$$R-\underset{\underset{O}{\|}}{C}-O-O-\underset{\underset{O}{\|}}{C}-R + Cu^+ \longrightarrow R-\underset{\underset{O}{\|}}{C}-O\cdot + {}^-O-\underset{\underset{O}{\|}}{C}-R + Cu^{2+}$$

$$R-\underset{\underset{O}{\|}}{C}-O\cdot \xrightarrow{-CO_2} R\cdot$$

$$R\cdot + Cu^{2+} \longrightarrow R^+ + Cu^+$$

Scheme 23

Nitro- and nitroso-compounds,[159,160] amines, and thiols induce the decomposition of diacyl peroxides in what may be written as an overall redox reaction. Certain monomers have been reported to cause induced decomposition of BPO. These include AN,[161] N-vinylcarbazole,[162-166] N-vinylimidazole[167] and N-vinylpyrrolidone.[166]

The mechanism proposed for the production of radicals from the N,N-dimethylaniline/BPO couple[168,169] involves reaction of the aniline with BPO by a S_N2 mechanism to produce an intermediate (31). This thermally decomposes to benzoyloxy radicals and an amine radical cation (32) both of which might in principle initiate polymerization (Scheme 24). Pryor and Hendrikson[170] were able to distinguish this mechanism from a process involving single electron transfer through a study of the kinetic isotope effect.

It has been suggested that the amine radical cation (32) is not directly involved in initiating chains and that most polymerization is initiated by benzoyloxy radicals.[168] However, Sato *et al.*[171] employed spin trapping (3.5.1.1) to

Initiation

demonstrate that anilinomethyl radicals (33) were formed from the radical cation (32) by loss of a proton and proposed that these radicals (33) also initiate polymerization. Overall efficiencies for initiation by amine-peroxide redox couples are very low; Imoto et al.[169] report ca. f=25%; Walling[168] reports f=2-5%.

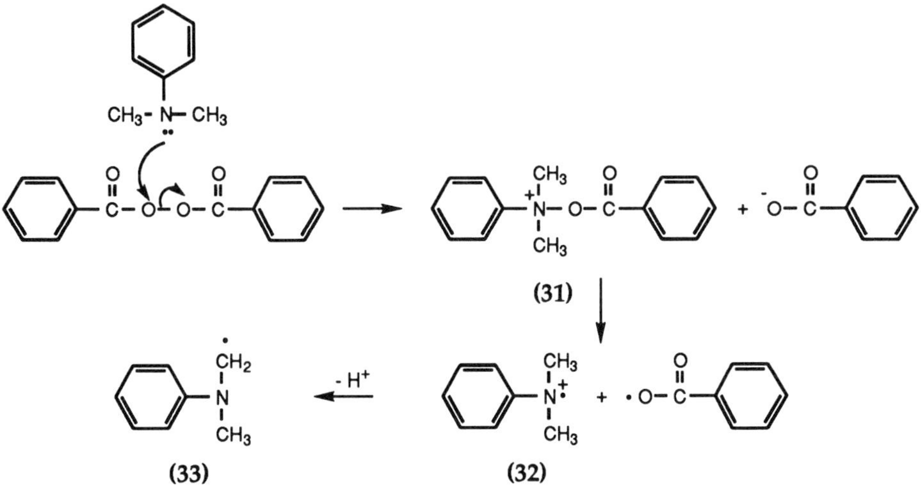

Scheme 24

3.3.2.2 Dialkyl peroxydicarbonates

The chemistry of peroxydicarbonates (26) and their use as initiators of polymerization has been reviewed by Yamada et al.,[126] Hiatt[123] and Strong.[172]

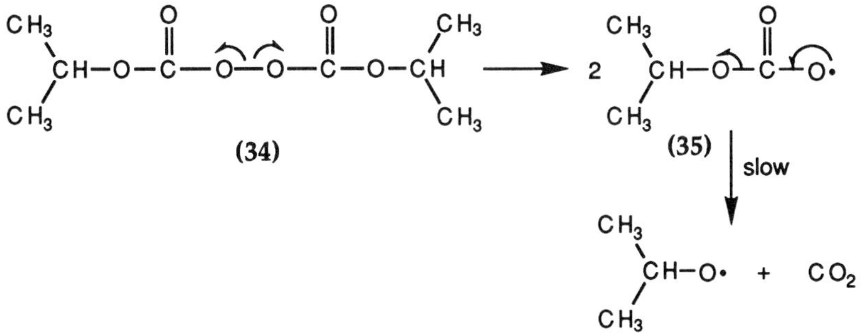

Scheme 25

Dialkyl peroxydicarbonates have been reported as low temperature sources of alkoxy radicals (Scheme 25)[173,174] and these radicals may be formed in relatively inert media. However, it is established, for primary and secondary peroxydicarbonates, that the rate of loss of carbon dioxide is slow compared to the rate of addition to most monomers or reaction with other substrates.[175,176] Thus, in polymerizations carried out with diisopropyl peroxydicarbonate (34), chains will be initiated by isopropoxycarbonyloxy (35) rather than isopropoxy radicals (see 3.4.2.2).[177]

A slow rate of β-scission also means that the main cage recombination process will be to reform peroxydicarbonate. Dialkyl peroxides are typically not found amongst the products of peroxydicarbonate decomposition. In these circumstances, cage recombination is unlikely to be a factor in reducing initiator efficiency.

Laboratory studies have generally focused on the diisopropyl, dicyclohexyl and di-*t*-butyl derivatives. These and the *s*-butyl and 2-ethylhexyl derivatives are commercially available.[178] The rates of decomposition of the peroxydicarbonates show significant dependence on the reaction medium and their concentration. This dependence is, however, less marked than for the diacyl peroxides (see 3.3.2.1.4). Induced decomposition may involve a mechanism analogous to that described for diacyl peroxides. However, a more important mechanism for primary and secondary peroxydicarbonates involves abstraction of an α-hydrogen (Scheme 26).[179]

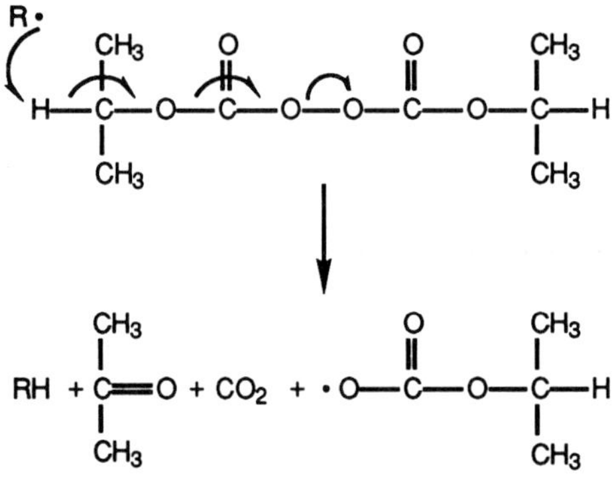

Scheme 26

Crano[180] has investigated the reaction between diisopropyl peroxydicarbonate and tertiary amines. These experiments indicate the formation of radicals by loss of a hydrogen from the α-CH_2 of the amine. It seems likely that the mechanism of radical formation is analogous to that observed for diacyl peroxide-amine systems (see 3.3.2.1.5)

3.3.2.3 Peresters

The chemistry of peresters (27) has been reviewed by Sawaki,[181] Bouillion *et al.*[122] and Singer.[182] The peresters are sources of alkoxy and acyloxy radicals (Scheme 27). Most commonly encountered peresters are derivatives of *t*-alkyl hydroperoxides (*e.g.* cumyl, *t*-butyl, *t*-amyl). Aryl peresters are generally unsuitable as initiators of polymerization owing to the generation of phenoxyl radicals which can inhibit or retard polymerization.

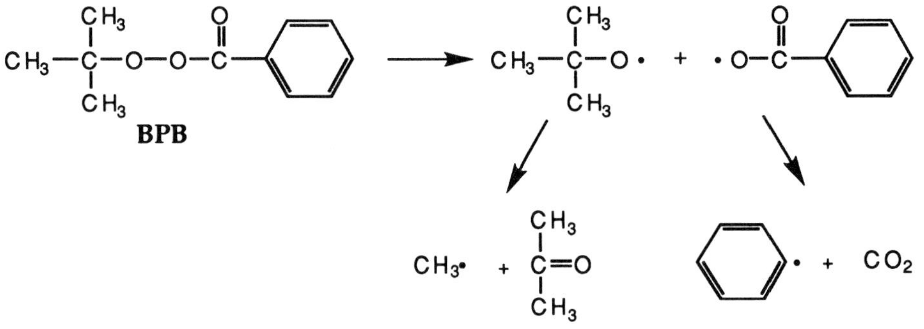

$$\text{Scheme 27}$$

3.3.2.3.1 Thermal decomposition

The rates of decomposition of peresters (27) are very dependent on the nature of the substituents R and R'. The variation in the decomposition rate with R follows the same trends as have been discussed for the corresponding diacyl peroxides (see 3.3.2.1.1).

Peresters derived from secondary (*e.g.* perisobutyrate esters) and tertiary acids (*e.g.* perpivalate esters) are believed to undergo concerted 2-bond cleavage leading to direct production of an alkoxy and an alkyl radical and a molecule of carbon dioxide. On the other hand, primary (*e.g.* peracetate and perpropionate esters) and aromatic peresters (*e.g.* BPB, Scheme 27) are thought to undergo 1-bond scission[134] to generate an acyloxy and an alkoxy radical. Evidence for the transient existence of acyloxy radicals includes the observation of substantial cage return.

Di-*t*-butyl peroxalate (DBPOX) is a clean, low temperature, source of *t*-butoxy radicals (Scheme 28).[128] The decomposition is proposed to take place by concerted 3-bond cleavage to form two alkoxy radicals and two molecules of carbon dioxide.

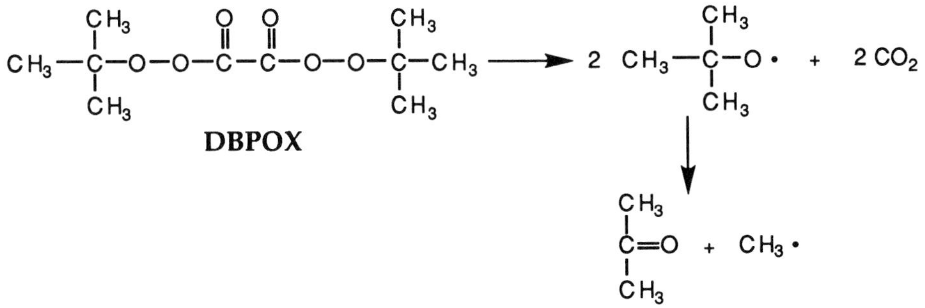

Scheme 28

The initiator efficiency of di-*t*-butyl peroxalate is higher than for other peresters. The dependence of cage recombination on the nature of the reaction medium has been the subject of a number of studies.[110,183,184] The yield of di-*t*-butyl peroxide (the main cage product) depends not only on viscosity but also on the precise nature of the solvent. The affect of solvent is to reduce the yield in the order: aliphatic>aromatic>protic. It has been proposed[183] that this is a consequence of the solvent dependence of β-scission of the *t*-butoxy radical which increases in the same series (see 3.4.2.1.1).

Transfer to initiator is generally of lesser importance than with the corresponding diacyl peroxides. They are, nonetheless, susceptible to the same range of reactions (see 3.3.2.1.4). Radical-induced decomposition usually occurs specifically to give an alkoxy radical and an ester (Scheme 29).

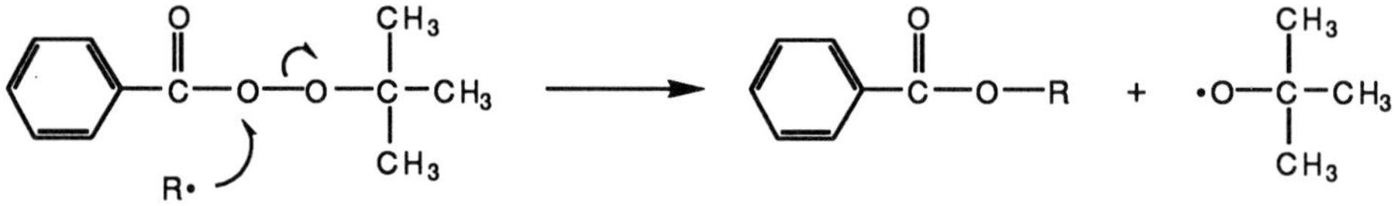

Scheme 29

Peresters may undergo non-radical decomposition *via* the Criegee rearrangement (Scheme 30). This process is analogous to the carboxy inversion process described for diacyl peroxides (see 3.3.2.1.3) and probably involves ionic intermediates.

Scheme 30

The reaction is facilitated when R is electron withdrawing, when R has a high migratory aptitude (ability to stabilize a carbonium ion), and by polar reaction media.

3.3.2.3.2 *Photochemical decomposition*

Peresters seldom find use as photoinitiators since photodecomposition requires light of 250-300 nm, a region where many monomers also absorb. This situation may be improved by the introduction of a suitable chromophore into the molecule or through the use of sensitizers.[185,186] The perester (36) is reported to have λ_{max} 366 nm and φ near unity.[185]

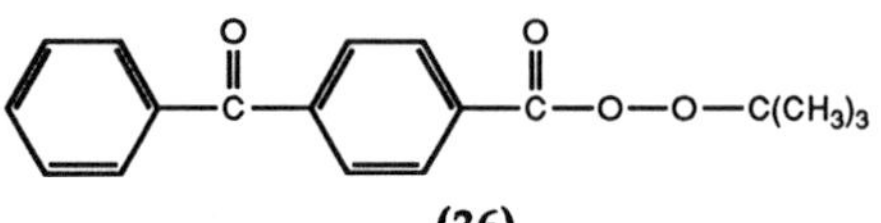

(36)

3.3.2.4 Dialkyl peroxides

The chemistry of the dialkyl peroxides (28) has been reviewed by Matsugo and Saito,[187] Sheldon[188] and Hiatt.[189] Dialkyl peroxides are high temperature sources of alkoxy radicals. Dialkyl peroxides commonly used as initiators have tertiary alkyl substituents. Those available commercially include dicumyl (37) and di-t-butyl peroxides (38), sources of cumyloxy and t-butoxy radicals respectively, and a variety of dialkyl peroxyketals, for example, 1,1-di-t-butylperoxycyclohexane (39).[190,191] These latter initiators (e.g. 39) have decomposition rate constants k_d that are an order of magnitude greater than simple di-t-alkyl peroxides (37, 38)[192] and can be shock sensitive.

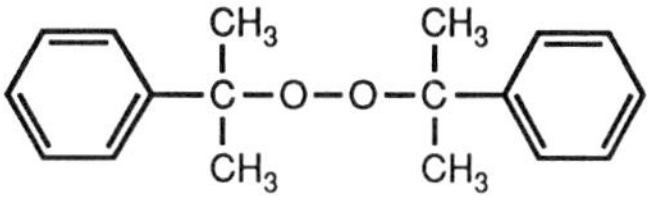

(37) 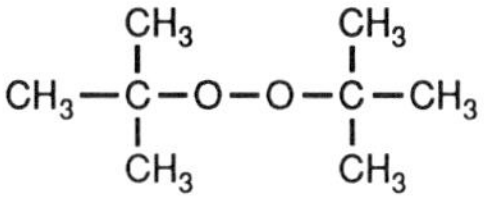   (38) 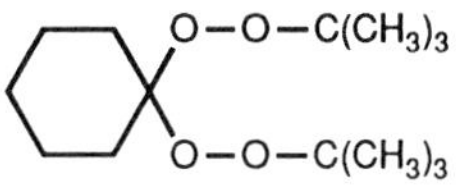   (39)

The decomposition of (39) follows a stepwise, rather than a concerted mechanism. Initial homolysis of one of the O-O bonds gives an alkoxy radical and an α-peroxyalkoxy radical (Scheme 31).[190,192-194] This latter species decomposes by β-scission with loss of either a peroxy radical to form a ketone as by-product or an alkyl radical to form a perester intermediate. The perester formed may also decompose to radicals under the reaction conditions. Thus, four radicals may be derived from the one initiator molecule.

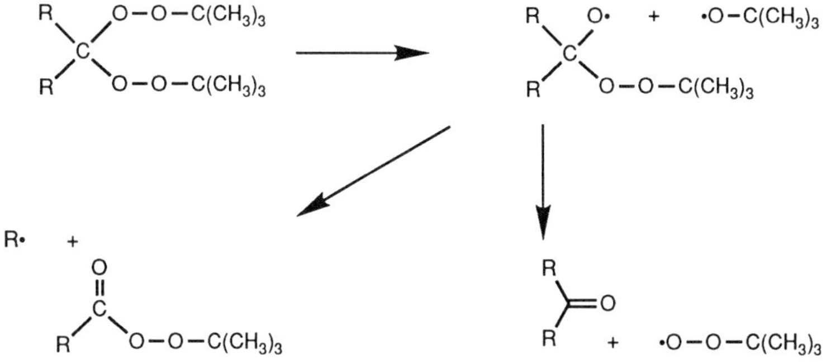

Scheme 31

The relative importance of the various pathways depends on the alkyl groups (R). The rate constants for scission of groups (R•) from t-alkoxy radicals ($R^1R^2R^3C-O•$) increase in the order isopropyl<ethyl<t-butylperoxy<methyl.[194] Thus, the pathway affording perester and an alkyl radical is less important when R is methyl than when R is a higher alkyl group. If the pathway to alkylperoxy radicals is dominant, the resultant polymer is likely to have a proportion of peroxy end groups.[190,195]

Solvent dependence of k_d for di-*t*-alkyl peroxides is small when compared to most other peroxide initiators.[130,196] For di-*t*-butyl peroxide,[130] k_d is slightly greater (up to two-fold at 125°C) in protic (*t*-butanol, acetic acid) or dipolar aprotic solvents than in other media (cyclohexane, triethylamine, tetrahydrofuran).

The chemistry of the di-*t*-butyl and cumyl peroxides is relatively uncomplicated by induced or ionic decomposition mechanisms. However, induced decomposition of di-*t*-butyl peroxide has been observed in primary or secondary alcohols[197,198] (Scheme 32) or primary or secondary amines.[199] The reaction involves oxidation of an α-hydroxyalkyl or α-aminoalkyl radical, to the carbonyl or imino compound respectively, and apparently requires coordination of the hydroxyl or aminyl hydrogen to the peroxidic oxygen.

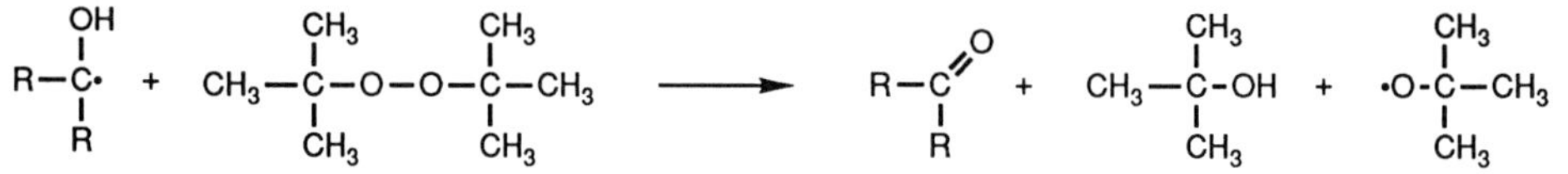

$$\underset{\substack{|\\R}}{\overset{\substack{OH\\|}}{R-C\cdot}} + \underset{\substack{|\\CH_3}}{\overset{\substack{CH_3\\|}}{CH_3-C}}-O-O-\underset{\substack{|\\CH_3}}{\overset{\substack{CH_3\\|}}{C-CH_3}} \longrightarrow \underset{R}{R-C\overset{O}{\diagup\hspace{-0.7em}\diagdown}} + \underset{\substack{|\\CH_3}}{\overset{\substack{CH_3\\|}}{CH_3-C}}-OH + \cdot O-\underset{\substack{|\\CH_3}}{\overset{\substack{CH_3\\|}}{C}}-CH_3$$

Scheme 32

The radical yield from simple di-*t*-alkyl peroxides (*i.e.* dicumyl, di-*t*-butyl) is reported to be almost 100%. The only significant cage reaction is reformation of the peroxide. The efficiencies of dialkyl peroxyketals and primary and secondary peroxides are lower.[191] Lower efficiencies arise when the initially formed radicals undergo β-scission before cage escape or, in the case where primary or secondary alkoxy radicals are formed, by disproportionation within the solvent cage. Primary and secondary peroxides are also susceptible to a variety of induced and non-radical decomposition mechanisms.

3.3.2.5 Alkyl hydroperoxides

The chemistry of alkyl hydroperoxides (**29**) has been reviewed by Porter,[200] Sheldon[188] and Hiatt.[201] Alkyl hydroperoxides are high temperature sources of alkoxy and hydroxy radicals.[202] They are often encountered as components of redox systems.

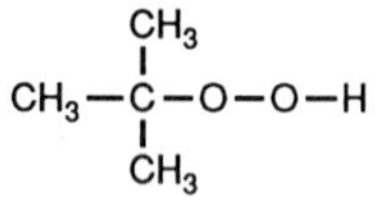

(40) 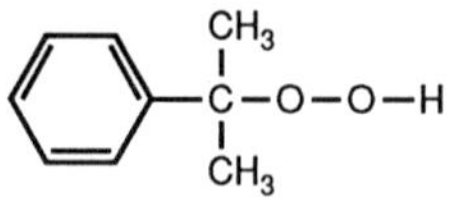 (41) 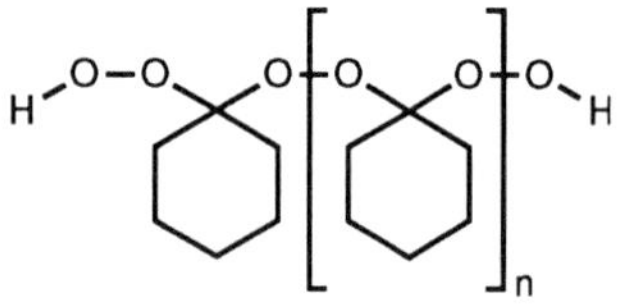 (42)

The common initiators of this class are *t*-alkyl derivatives, for example, *t*-butyl hydroperoxide (**40**), cumene hydroperoxide (**41**), and a range of peroxyketals (**42**). Hydroperoxides formed by hydrocarbon autoxidation have also been used as initiators of polymerization.

Initiation

The ROO-H bond of hydroperoxides is weak compared to most other X-H bonds.* Thus, abstraction of the hydroperoxidic hydrogen by radicals is usually an exothermic process. The hydroperoxides can therefore be extremely efficient transfer agents and radical-induced decomposition may be a major complication in their use as initiators.[205]

Primary and secondary hydroperoxides are also susceptible to induced decomposition through loss of an α-hydrogen. The radical so formed is usually not stable and undergoes β-scission to give a carbonyl compound and hydroxy radical.[206] It is reported that these compounds may also undergo non-radical decomposition with evolution of hydrogen.[129]

Hydroperoxides react with transition metals in lower oxidation states (Ti^{3+}, Fe^{2+}, Cu^+, *etc.*) and a variety of other oxidants to give an alkoxy radical and hydroxide anion (See Scheme 33).[38,207,208]

$$CH_3-\underset{\underset{CH_3}{|}}{\overset{\overset{CH_3}{|}}{C}}-O-O-H \ + \ Fe^{2+} \longrightarrow CH_3-\underset{\underset{CH_3}{|}}{\overset{\overset{CH_3}{|}}{C}}-O\cdot \ + \ ^-OH \ + \ Fe^{3+}$$

Scheme 33

With some systems, the hydroperoxide is reduced to hydroperoxy radical by the metal ion in its higher oxidation state (Scheme 34). Thus, it is possible to set up a catalytic cycle for hydroperoxide decomposition.

$$CH_3-\underset{\underset{CH_3}{|}}{\overset{\overset{CH_3}{|}}{C}}-O-O-H \ + \ Cu^{2+} \longrightarrow CH_3-\underset{\underset{CH_3}{|}}{\overset{\overset{CH_3}{|}}{C}}-O-O\cdot \ + \ H^+ \ + \ Cu^+$$

Scheme 34

With Ti^{4+} and Fe^{3+} this latter pathway is thought not to occur. The formation of ROO•, observed at high hydroperoxide concentrations, is attributed to the occurrence of induced decomposition.[209]

3.3.2.6 *Inorganic peroxides*

Inorganic peroxides [hydrogen peroxide, persulfate (30), peroxydiphosphate, *etc.*] generally have limited usefulness as initiators in bulk or solution polymerization due to their poor solubility in organic media. This means that the main use of these initiators is in aqueous[210] or in part-aqueous heterogeneous media (*e.g.* in emulsion polymerization). They are often encountered as one component in a redox initiation system. The history of these systems has been reviewed by Bacon[211] and Sosnovsky and Rawlinson.[212]

---

* $D_{ROO-H} \sim 375$ kJ mol^{-1}.[203]
$D_{R_3C-H} \sim 385$, $D_{R_2CH-H} \sim 396$, $D_{RCH_2-H} \sim 410$, $D_{RO-H} \sim 435$ kJ mol^{-1}.[204]

The following discussion concentrates on the chemistry of the two most common inorganic peroxides, persulfate and hydrogen peroxide.

3.3.2.6.1 Persulfate

Photolysis or thermolysis of persulfate ion results in homolysis of the O-O bond and formation of two sulfate radical anions. The thermal reaction in aqueous media has been widely studied.[213,214] The rate of decomposition is a complex function of pH, ionic strength, and concentration. Initiator efficiencies for persulfate in emulsion polymerization are low (0.1-0.3) and depend upon reaction conditions (*i.e.* temperature, initiator concentration).[215]

A number of mechanisms for thermal decomposition of persulfate in neutral aqueous solution have been proposed.[214] They include unimolecular decomposition (Scheme 35)

$$^-O-\underset{\underset{O}{\|}}{\overset{\overset{O}{\|}}{S}}-O-O-\underset{\underset{O}{\|}}{\overset{\overset{O}{\|}}{S}}-O^- \longrightarrow 2\ \cdot O-\underset{\underset{O}{\|}}{\overset{\overset{O}{\|}}{S}}-O^-$$

Scheme 35

and various bimolecular pathways for the disappearance of persulfate involving a water molecule and concomitant formation of hydroxy radicals (Scheme 36).

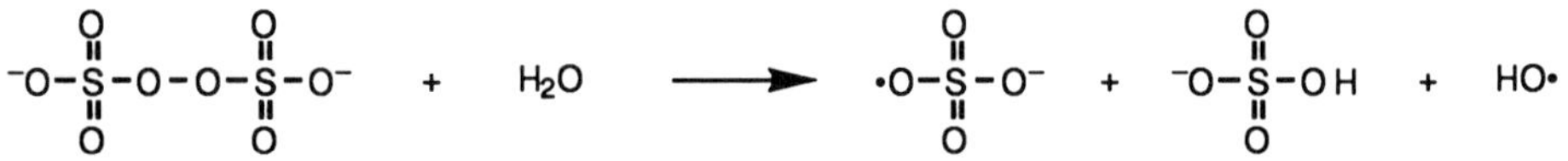

$$^-O-\underset{\underset{O}{\|}}{\overset{\overset{O}{\|}}{S}}-O-O-\underset{\underset{O}{\|}}{\overset{\overset{O}{\|}}{S}}-O^- + H_2O \longrightarrow \cdot O-\underset{\underset{O}{\|}}{\overset{\overset{O}{\|}}{S}}-O^- + {}^-O-\underset{\underset{O}{\|}}{\overset{\overset{O}{\|}}{S}}-OH + HO\cdot$$

Scheme 36

The formation of polymers with negligible hydroxy end groups is evidence that the unimolecular process dominates in neutral solution. Heterolytic pathways for persulfate decomposition can be important in acidic media.

Normally, persulfate can only be used to initiate polymerization in aqueous or part aqueous (emulsion) media because it has poor solubility in most organic solvents and monomers. However, it has been reported that polymerizations in organic solvent may be initiated by crown ether complexes of potassium persulfate.[216-219] Quaternary ammonium persulfates can also serve as useful initiators in organic media.[218,220] The rates of decomposition of both the crown ether complexes and the quaternary ammonium salts appear dramatically greater than those of conventional persulfate salts (K^+, Na^+, NH_4^+) in aqueous solution. The crown ether complex can be used to initiate polymerization at ambient temperature.[216]

In part, the accelerated decomposition might be attributed to the occurrence of induced decomposition and primary radical transfer.[221] Persulfate is also known to be a strong oxidant and, in this context, has been widely applied in synthetic organic chemistry.[222] It is established that the rate of disappearance of persulfate in aqueous media is accelerated by the presence of organic compounds[213] and induced decomposition is an integral step in the oxidation of organic substrates (including ethers) by persulfate.[223]

Persulfate absorbs only weakly in the UV (ε~25 M^{-1} cm^{-1} at 250 nm).[224] Nonetheless, direct photolysis of persulfate ion has been used as a means of generating sulfate radical anion in laboratory studies.[224,225]

Persulfate reacts with transition metal ions (*e.g.* Ag^+, Fe^{2+}, Ti^{3+}) according to Scheme 37.

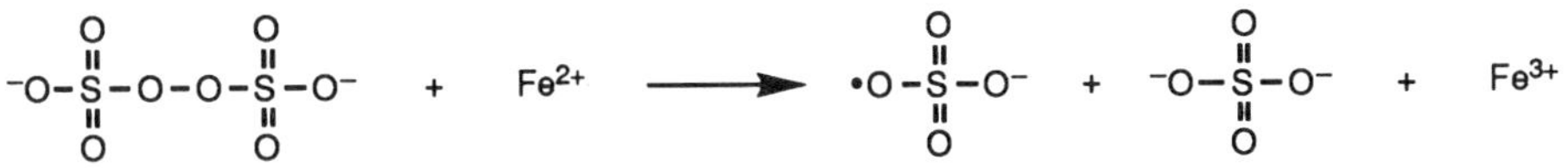

$$\text{}^-\text{O-SO}_2\text{-O-O-SO}_2\text{-O}^- + Fe^{2+} \longrightarrow {}^\bullet\text{O-SO}_2\text{-O}^- + {}^-\text{O-SO}_2\text{-O}^- + Fe^{3+}$$

Scheme 37

Various other reductants have been described. These include halide ions, thiols (*e.g.* 2-mercaptoethanol, thioglycolic acid, cysteine, thiourea), bisulfite, thiosulfate, amines (triethanolamine, hydrazine hydrate), ascorbic acid, and solvated electrons (*e.g.* in radiolysis). The mechanisms and the initiating species produced have not been fully elucidated for many systems.[226]

Various multicomponent systems have also been described. Three component systems in which a second reducing agent (*e.g.* sulfite) acts to recycle the transition metal salt, have the advantage that less metal is used (Scheme 38).

$$SO_3^{=} + Cu^{2+} \longrightarrow SO_3^{-}\bullet + Cu^+$$

Scheme 38

Redox initiation is commonly employed in aqueous emulsion polymerization. Initiator efficiencies obtained with redox initiation systems in aqueous media are generally low. One of the reasons for this is the susceptibility of the initially formed radicals to undergo further redox chemistry. For example, potential propagating radicals may be oxidized to carbonium ions (Scheme 39). The problem is aggravated by the low solubility of the monomers (*e.g.* MMA, S) in the aqueous phase.

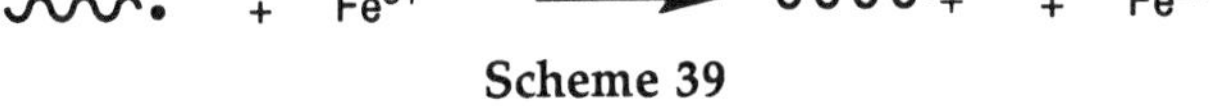

$$\sim\sim\bullet + Fe^{3+} \longrightarrow \sim\sim^+ + Fe^{2+}$$

Scheme 39

3.3.2.6.2 Hydrogen peroxide

Homolytic scission of the O-O bond of hydrogen peroxide may be effected by heat or UV irradiation.[227] The thermal reaction requires relatively high temperatures (>90°C). Photolytic initiation generally employs 254 nm light. Reactions in organic media require a polar cosolvent (*e.g.* an alcohol).

Hydrogen peroxide also reacts with reducing agents (transition metals, metal complexes, solvated electrons, and some organic reagents) to produce hydroxyl radicals. It reacts with oxidizing agents to give hydroperoxy radicals. The reaction between hydrogen peroxide and transition metal ions in their lower oxidation state is usually represented as the simple process first described by Haber and Weiss (Scheme 40).[228] However the mechanism is significantly more complex.

$$H_2O_2 + Fe^{2+} \longrightarrow HO\cdot + {}^-OH + Fe^{3+}$$
Scheme 40

It has been suggested that the reactive species are metal complexed hydroxy radicals rather than "free" hydroxyl radicals.[229-231] The reactions observed show dependence on the nature of the metal ion and quite different product distributions can be obtained from reaction of organic substrates with Fe^{2+}-H_2O_2 (Fenton's Reagent) and Ti^{3+}-H_2O_2. However, it is not clear whether these findings reflect the involvement of a different active species or simply the different rates and/or pathways for destruction of the initially formed intermediates.[232] Metal ions in their higher oxidation states (*e.g.* Fe^{3+}) can bring about the destruction of hydrogen peroxide according to Scheme 41.

$$H_2O_2 + Fe^{3+} \longrightarrow HOO\cdot + H^+ + Fe^{2+}$$
Scheme 41

The Ti^{3+}-H_2O_2 system is preferred over Fenton's reagent because Ti^{4+} is a less powerful oxidizing agent than Fe^{3+} and the above mentioned pathway and other side reactions are therefore of less consequence.[233] Much of the discussion on redox initiation in section 3.3.2.6.1 is also relevant to hydrogen peroxide.

3.3.3 Multifunctional Initiators

Multifunctional initiators contain two or more radical generating functions within the one molecule. They can be considered in two distinct classes according to whether they undergo concerted (see 3.3.3.1) or non-concerted decomposition (see 3.3.3.2).

3.3.3.1 Concerted decomposition

Multifunctional initiators where the radical generating functions are in appropriate proximity may decompose in a concerted manner or in a way such that the intermediate species can neither be observed nor isolated. Examples of such behavior are peroxalate esters (see 3.3.2.3.1) and α-hydroperoxy diazenes [*e.g.* (43)] and derived peresters [*e.g.* (44)].

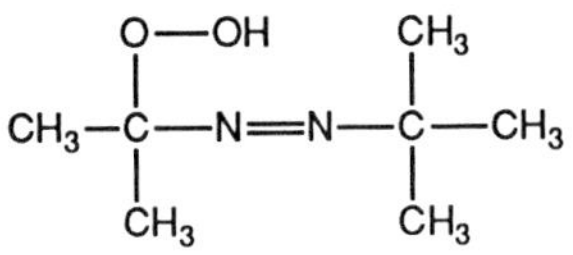

Initiation

(43) CH₃−C(CH₃)(O−OH)−N=N−C(CH₃)₂−CH₃

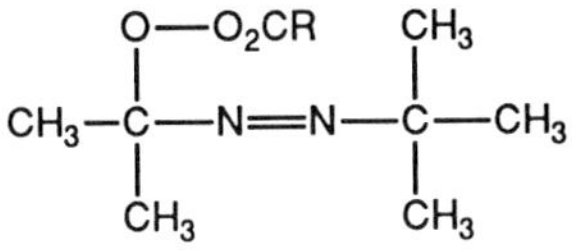

(44) CH₃−C(CH₃)(O−O₂CR)−N=N−C(CH₃)₂−CH₃

These initiators (43) and (44) are low temperature sources of alkyl and hydroxy or acyloxy radicals respectively (Scheme 42).[234-236] The α-hydroperoxy diazenes [*e.g.* (44)] are one of the few convenient sources of hydroxy radicals in organic solution.[236,237]

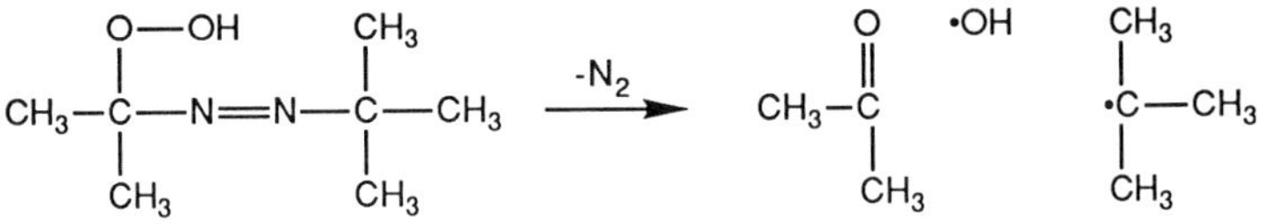

Scheme 42

(43) CH₃−C(CH₃)(O−OH)−N=N−C(CH₃)₂−CH₃ → −N₂ → CH₃−C(=O)−CH₃ + •OH + •C(CH₃)₂−CH₃

It has been reported that the α-hydroperoxy diazenes may undergo induced decomposition either by OH or H transfer.[238]

3.3.3.2 Non-concerted decomposition

Initiators where the radical generating functions are sufficiently remote from each other break-down in a non-concerted fashion. Examples include the azo-peroxide (45). Their chemistry is understandable in terms of the chemistry of analogous monofunctional initiators.[239] This class also includes the dialkyl peroxyketals (see 3.3.2.4) and hydroperoxyketals (see 3.3.2.5).

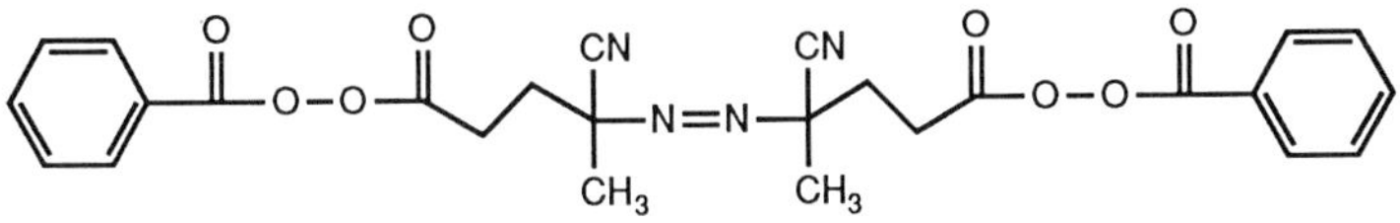

(45)

The use of these initiators has been promoted for achieving higher molecular weights or higher conversions in conventional polymerization and for the production of block and graft copolymers. The use and applications of multifunctional initiators in the synthesis of block and graft copolymers is briefly described in section 6.3.2.

3.3.4 Photochemical Initiators

Photoinitiation is most commonly used in curing or crosslinking processes and in initiating graft copolymerization. It also finds utility in small scale experiments and in kinetic and mechanistic studies.

The applications of azo-compounds and peroxides as photoinitiators have already been considered in the sections on those initiators (see 3.3.1.1.2, 3.3.2.1.2, & 3.3.2.3.2). Recent general reviews on photoinitiation include those by Pappas,[240-242] Bassi,[243] Mishra[244] and Oster and Yang.[245] References to reviews on specific photoinitiators are given in the appropriate section below.

3.3.4.1 Aromatic carbonyl compounds

Many reviews have been written on the photochemistry of aromatic carbonyl compounds[246] and on the use of these compounds as photoinitiators.[247-252] Primary radicals are generated by one of the following processes:

(a) A unimolecular fragmentation involving, most commonly, either α-scission (Scheme 43; e.g. benzoin ethers, acyl phosphine oxides)

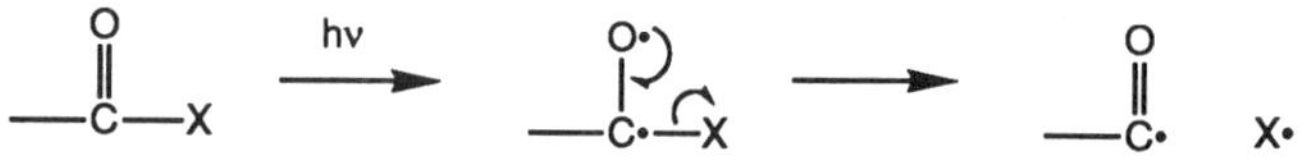

Scheme 43

or β-scission (Scheme 44; e.g. α-haloketones).

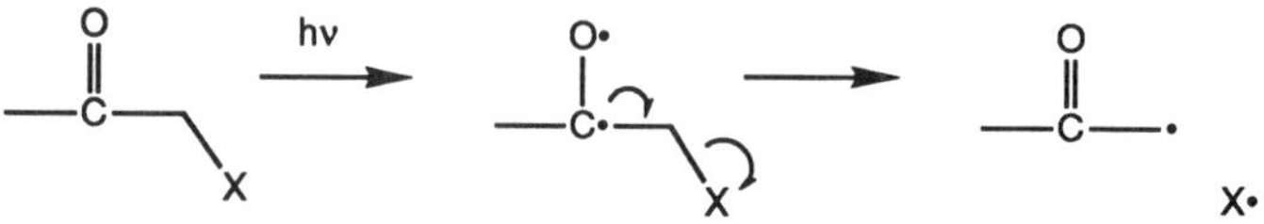

Scheme 44

Examples of scission of bonds separated from the carbonyl group by a double bond or an aromatic ring are also known. For example, the benzil monooxime (46) undergoes γ-scission (Scheme 45) (possibly by consecutive α- then β-scissions).

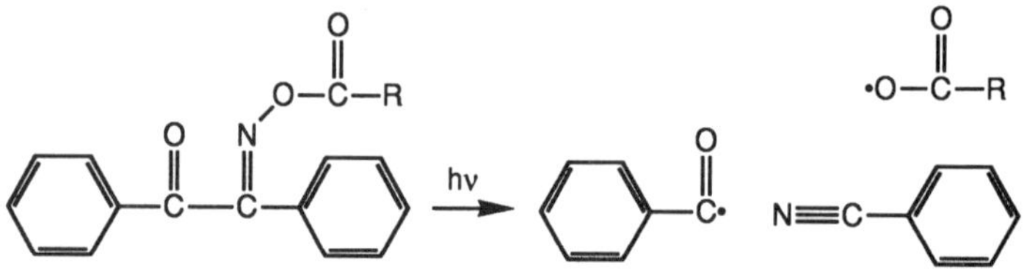

(46)

Scheme 45

(b) A bimolecular process involving direct abstraction of hydrogen from a suitable donor (Scheme 46; e.g. with hydrocarbons, ethers, alcohols),

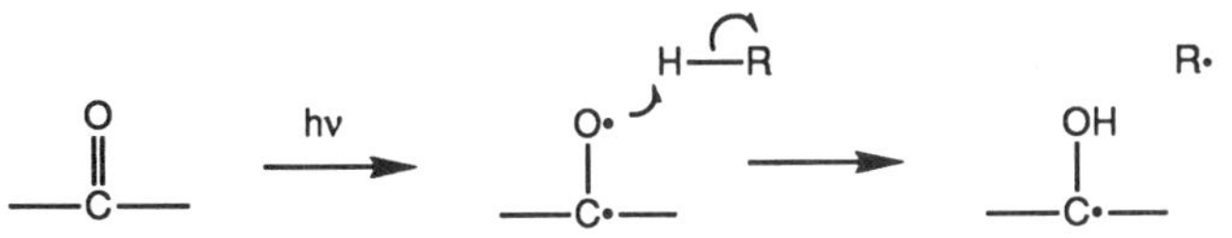

Scheme 46

or sequential electron and proton transfer (Scheme 47; e.g. with amines, thiols).

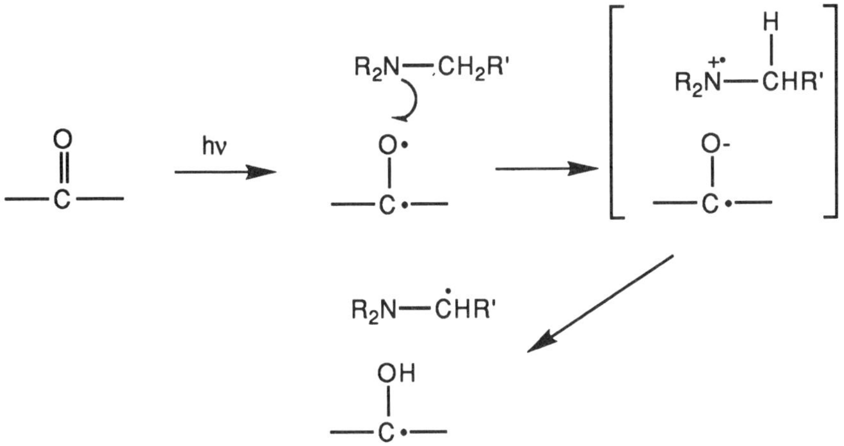

Scheme 47

The reaction pathway followed depends on whether H-donors or electron acceptors are present and the relative strengths of the bonds to the α-and β-substituents.

3.3.4.1.1 Benzoin derivatives

Benzoin and a variety of derivatives (**47**) have been extensively studied both as initiators of polymerization and in terms of their general photochemistry.[248,250] The acetophenone chromophore absorbs in the near UV (300-400 nm). The mechanism of radical generation is usually depicted as excitation to the $S_1(n,\pi^*)$ state followed by intersystem crossing to the $T_1(n,\pi^*)$ state and fragmentation; usually by α-scission (Scheme 48).

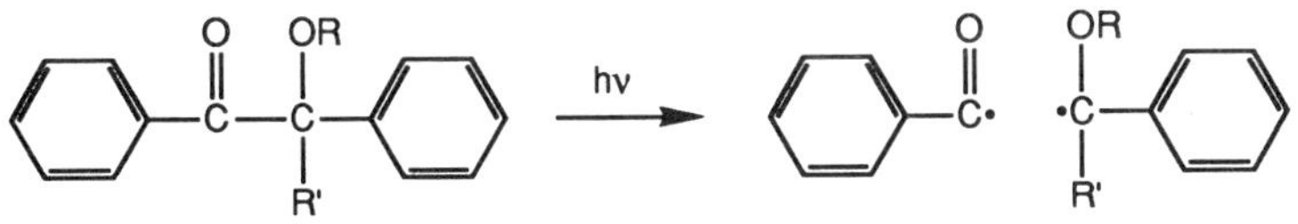

(**47**)   (**48**)   (**49**)

Scheme 48

Those benzoin derivatives most used as initiators are the benzoin ethers (**47**, R=alkyl; R'=H) and the α-alkyl benzoin derivatives (**47**, R=H, alkyl; R'=alkyl). The α-scission process is extremely facile and is not quenched by oxygen or conventional triplet quenchers.[253] This means that the initiators can be used for UV-curing in air. The products of α-scission of benzoin derivatives (see Scheme 48) are a benzoyl radical (**48**) and an α-substituted benzyl radical (**49**) both of which may, in principle, initiate polymerization.[253,254] However, depending on the nature of the substituent R', the radical (**49**) may be slow to add to double bonds and primary radical termination can be a severe complication.[255]

It should be pointed out that not all benzoin derivatives (**47**) are suitable for use as photoinitiators. Benzoin esters (**47**, R=acyl) undergo a side reaction leading to furan derivatives. Aryl ethers (**47**, R=aryl) undergo β-scission to give a phenoxy radical (an inhibitor) in competition with α-scission (Scheme 49). Benzoin derivatives with α-hydrogens (**47**, R'=H) are readily autoxidized and consequently have poor shelf lives.

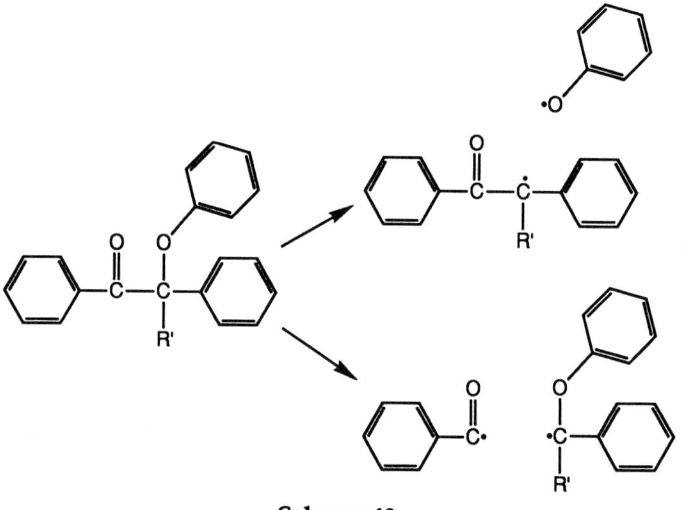

Scheme 49

The problems associated with formation of a relatively stable radical are mitigated with the α-alkoxy (**50**) and α-methylsulfonyl derivatives (**51**).[256] In both cases the substituted benzyl radicals formed by α-scission (**52** and **53** respectively) themselves undergo an extremely facile fragmentation to form a more reactive radical which is less likely to be involved in primary radical termination (Scheme 50).

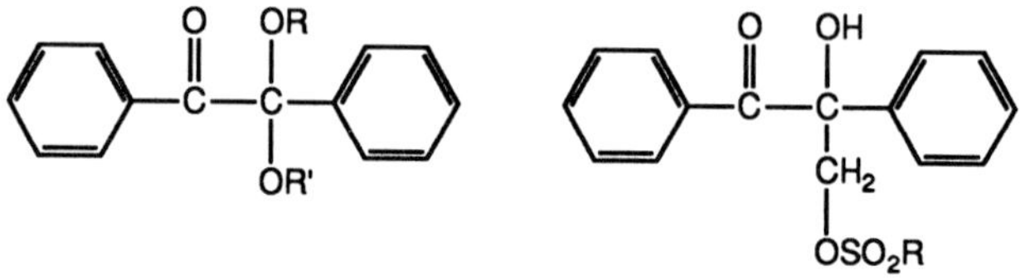

(**50**)   (**51**)

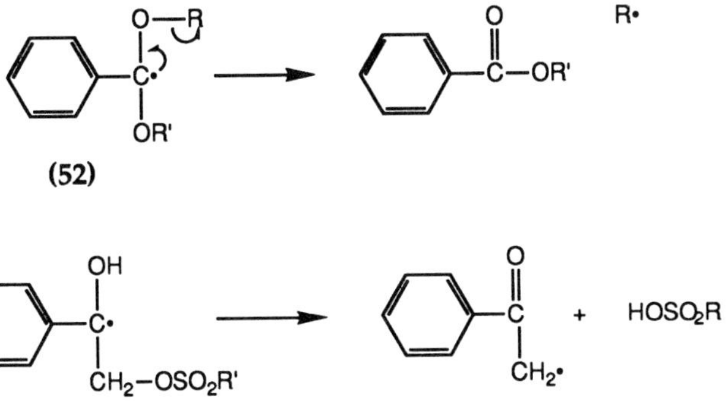

(52)

(53)

Scheme 50

The acyl phosphonates and acyl phosphine oxides (*e.g.* **54**) absorb strongly in the near UV (350-400 nm) and generally decompose by α-scission in a manner analogous to the benzoin derivatives.[257-259] Quantum yields vary from 0.3 to 1.0 depending on structure. The phosphinyl radicals are highly reactive towards unsaturated substrates and appear to have a high specificity for addition *vs.* abstraction (see 3.4.3.2).

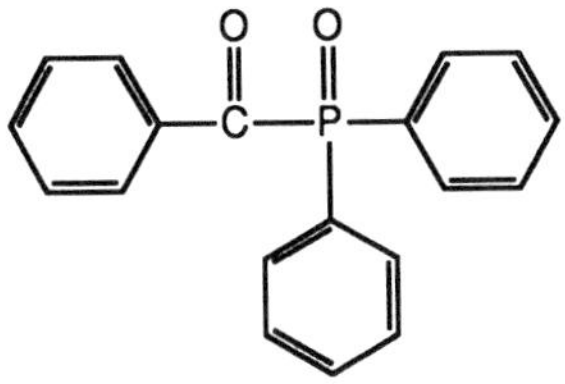

(54)

Klos *et al.*[260] have described a range of polymerizable photoinitiators (*e.g.* **55, 56**). The polymers derived from these monomers have advantages over low molecular weight analogous when migration stability is a problem.

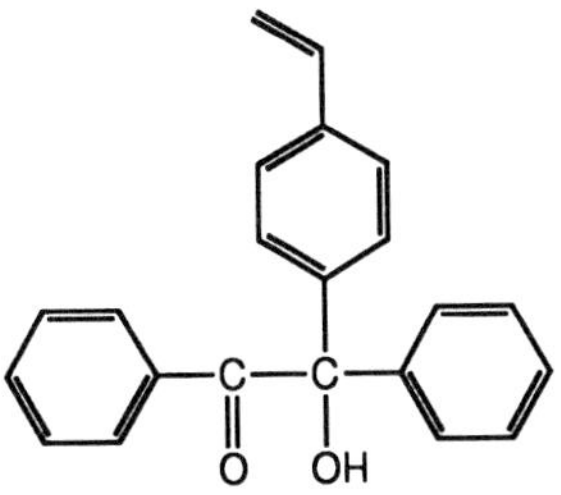

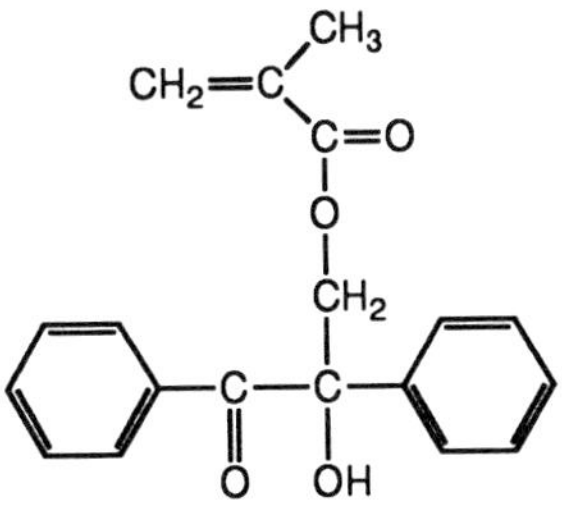

(55) (56)

3.3.4.1.2 *Carbonyl compound-tertiary amine systems*

Photoredox systems involving carbonyl compounds and amines are used in many applications. Carbonyl compounds employed include benzophenone and derivatives, α-diketones (*e.g.* benzil, camphoroquinone (57),[261] 9,10-phenanthrene quinone), and xanthone and coumarin derivatives. The amines are tertiary and must have α-hydrogens (*e.g.* N,N-dimethylaniline, Michler's ketone (58)). The radicals formed are an α-aminoalkyl radical and a ketyl radical.

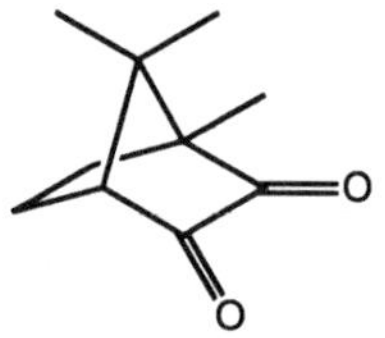

(57) 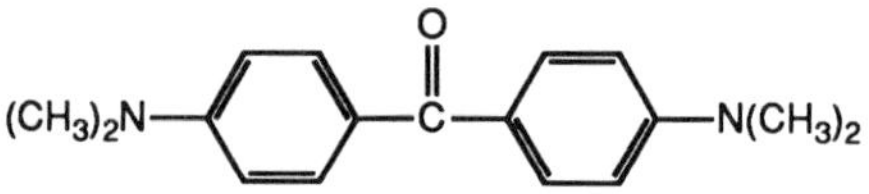 (58)

The reaction between the photoexcited carbonyl compound and an amine occurs with substantially greater facility than that with most other hydrogen donors. The rate constants for triplet quenching by amines show little dependence on the amine α–C–H bond strength. However, the ability of the amine to release an electron is important.[262] This is in keeping with a mechanism of radical generation which involves initial electron (or charge) transfer from the amine to the photoexcited carbonyl compound. Loss of a proton from the resultant complex (exciplex) results in an α-aminoalkyl radical which initiates polymerization.

The concurrently formed ketyl radicals are generally slow to initiate polymerization and consequently primary radical termination is a common complication with these initiator systems.

The electron transfer step is typically fast and efficient. Griller *et al.*[262] have measured absolute rate constants for decay of benzophenone triplet in the presence of aliphatic tertiary amines in benzene as solvent. Values lie in the range $3-4 \times 10^9$ M^{-1} s^{-1} and quantum yields are close to unity.

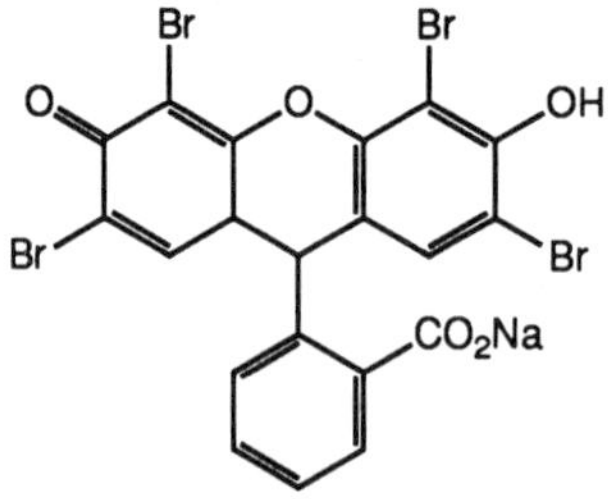

(59)

Visible light systems comprising a photoreducible dye molecule (*e.g.* 59)[263] or an α-diketone (*e.g.* 57)[261] and an amine have also been described.[263] The mechanism of radical production is probably similar to that described for the ketone amine systems described above (*i.e.* electron transfer from the amine to the photoexcited dye molecule and subsequent proton transfer). Ideally, the dye molecule is reduced to a colorless by-product.

3.3.4.2 Sulfur compounds

The S-S linkage of disulfides and the C-S linkage of certain sulfides can undergo photoinduced homolysis. The low reactivity of the sulfur-centered radicals in addition or abstraction processes means that primary radical termination can be a complication. The disulfides may also be extremely susceptible to transfer to initiator (C_I for (**60**) is ca. 0.5, see 5.3.2.2 and 7.3.1). However, these features are used to advantage when the disulfides are used as initiators in the synthesis of telechelics[264] or in living radical polymerizations (see 7.3.1). The most common initiators in this context are the dithiuram disulfides (**60**) which are both thermal and photochemical initiators. The corresponding monosulfides [*e.g.* (**61**)] are thermally stable but can be used as photoinitiators. The chemistry of these initiators is discussed in more detail in section 7.3.1.

(**60**)   (**61**)

3.3.5 Redox Initiators

The early history of redox initiation has been described by Bacon.[211] More recently the subject has been reviewed by Misra and Bajpai,[265] and Bamford.[266] The mechanism of redox initiation is usually bimolecular and involves a single electron transfer as an essential feature of the mechanism which distinguishes it from other initiation processes. Redox initiation systems are in common use when initiation is required at or below ambient temperature and they are frequently used for initiation of emulsion polymerization.

Common components of many redox systems are a peroxide and a transition metal ion or complex. The redox reactions of peroxides are covered in the sections on those compounds. Discussion on specific redox systems can be found in sections on diacyl peroxides (3.3.2.1.5), hydroperoxides (3.3.2.5), persulfate (3.3.2.6.1) and hydrogen peroxide (3.3.2.6.2).

Numerous redox systems have been described which do not involve peroxides including many metal ion free systems. These include carbonyl compound-tertiary amine (3.3.4.1.2), metal complex-organic halide (3.3.5.1), and ceric ion-organic substrate systems (3.3.5.2). The following two sections describe redox systems based on the use of metal complexes and simple organic molecules.

3.3.5.1 Metal complex-organic halide systems

The kinetics and mechanism of redox and photoredox systems involving transition metal complexes has been reviewed by Bamford.[266] One system which has seen extensive use comprises a transition metal in a low, typically zero, oxidation state (*e.g.* $Mo(CO)_6$, $Re(CO)_6$) and an organic halide. Radical production involves single electron transfer from the metal to the halogen substituent of the alkyl halide which then fragments to form a halide ion and an alkyl radical.[267] Accordingly, the organic fragment of the alkyl halide should be a good electron acceptor, for example, CCl_4, $CHCl_3$, α-haloketones, α-haloesters, etc. The use of polymeric halo compounds allows this chemistry to be used in the preparation of block and graft copolymers (see 6.3).[268,269]

The metal complexes most commonly used in these photoredox systems are manganese and rhenium carbonyls. The proposed mechanism of the photoredox reaction involving $Mn_2(CO)_{10}$ is represented schematically as follows (Scheme 51). Quantum yields for photoinitiation are high.[266] Redox couples involving similar metal complexes and an electron deficient monomer (typically a fluoro-olefin) have also been described.[266]

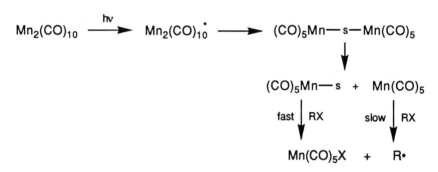

Scheme 51

s = solvent, monomer or coordinating additive (*e.g.* acetylacetone)

3.3.5.2 Ceric ion systems

Ceric ions oxidize various organic substrates and the mechanisms typically involve free radical intermediates.[270] When conducted in the presence of a monomer these radicals may initiate polymerization.

The reaction of ceric ion with alcohols,[271] amides and urethanes[272] is thought to involve single electron transfer to the ceric ion and loss of a proton to give the corresponding oxygen- or nitrogen-centered radical (Scheme 52). The reaction may involve ligation of cerium. Mechanisms for ceric ion oxidation of alcohols which yield α-hydroxyalkyl radicals as initiating species have also been proposed.

Initiation

$$XH + Ce^{4+} \longrightarrow X\cdot + Ce^{3+} \quad H^+$$

Scheme 52

Ceric ions react rapidly with 1,2-diols. There is evidence for chelation of cerium and these complexes are likely intermediates in radical generation.[273,274] The overall chemistry may be understood in terms of an intermediate alkoxy radical which undergoes β-scission to give a carbonyl compound and a hydroxyalkyl radical (Scheme 53). However, it is also possible that there is concerted electron transfer and bond-cleavage. There is little direct data on the chemical nature of the radical intermediates.

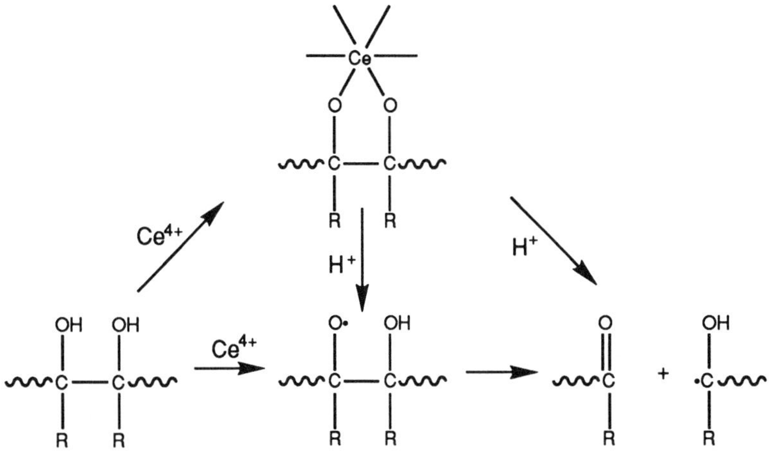

Scheme 53

The specificity for reaction with 1,2-diols over mono-ols and 1,3-diols accounts for the finding that oxidation of PVA gives specific cleavage of the 1,2-diol groups present as a consequence of head addition to monomer (see 4.3.1.1). The 1,3-glycol units in PVA also complex ceric ion and, while these complexes decompose only slowly under normal conditions, they undergo a facile photoinduced decomposition to generate initiating species.[275]

The reaction of ceric ions with polymer-bound functionalities gives polymer-bound radicals. Thus, one of the major applications of ceric ion initiation chemistry has been in grafting onto starch, cellulose,[273,274,276] polyurethanes and other polymers.[272] The advantage of this over conventional initiating systems is that, ideally, no low molecular weight radicals which might give homopolymer contaminant are formed.

The ceric ion also traps carbon-centered radicals (initiator-derived species, propagating chains) by single electron transfer (Scheme 54).

$$\sim\sim\cdot + Ce^{4+} \longrightarrow \sim\sim^+ + Ce^{3+}$$

Scheme 54

3.3.6 Thermal Initiation

This section describes polymerizations of monomer(s) where the initiating radicals are formed from the monomer(s) by a purely thermal reaction (*i.e.* no other reagents are involved). The adjectives, thermal, self-initiated and spontaneous, are used interchangeably to describe these polymerizations which have been reported for many monomers and monomer combinations. While homopolymerizations of this class typically require above ambient temperatures, copolymerizations involving certain electron-acceptor-electron-donor monomer pairs can occur at or below ambient temperature.

Aspects of thermal initiation have been reviewed by Moad *et al.*,[277] Pryor and Laswell,[278] Kurbatov,[279] and Hall.[280] It is often difficult to establish whether initiation is actually a process involving only the monomer. Trace impurities in the monomers or the reaction vessel may prove to be the actual initiators. Purely thermal homopolymerizations to high molecular weight polymers have only been demonstrated unequivocally for S and its derivatives and MMA. For these and other systems, the identity of the initiating radicals and the mechanisms by which they are formed remain subjects of controversy.

3.3.6.1 Styrene homopolymerization

The thermal polymerization of S has a long history.[278] The process was first reported in 1839, though the involvement of radicals was only proved in the 1930s. Carefully purified S undergoes spontaneous polymerization at a rate of 0.1% per hour at 60° and 2% per hour at 100°C. At 180°C, 80% conversion of monomer to polymer is achieved in approximately 40 minutes. Polymer production is accompanied by the formation of styrene dimers and trimers which comprise *ca.* 2% by weight of total products. The dimer fraction consists largely of *cis*- and *trans*-1,2-diphenylcyclobutanes (62) while the stereoisomeric tetrahydronaphthalenes (63) are the main constituents of the trimer fraction.[281]

(62)   (63)

Initiation

The two most widely accepted mechanisms for the spontaneous generation of radicals from S are the biradical mechanism (Scheme 55) first proposed by Flory[282] and the Mayo[283] or MAH (molecule assisted homolysis) mechanism (Scheme 56).

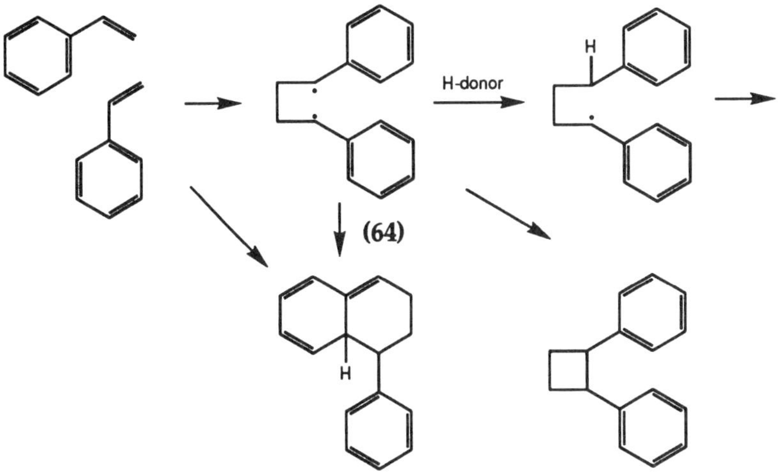

Scheme 55

The Mayo mechanism involves a thermal Diels-Alder reaction between two molecules of S to generate the adduct (65) (Scheme 55) which donates a hydrogen atom to another molecule of S to give the initiating radicals (66) and (67) (Scheme 56). The driving force for the molecule assisted homolysis is provided by formation of an aromatic ring. The Diels-Alder intermediate (65) has never been isolated. However, related compounds have been synthesized and shown to initiate styrene polymerization.[278]

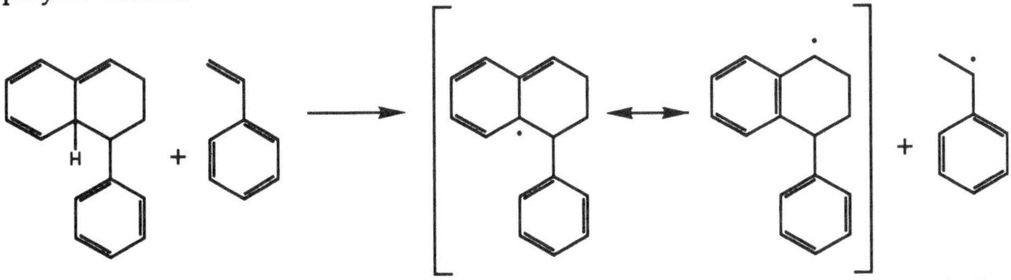

Scheme 56

The identification of phenylethyl and 1-phenyl-1,2,3,4-tetrahydronaphthalenyl end groups in polymerizations retarded by FeCl$_3$/DMF provides the most compelling evidence for the Mayo mechanism.[284] The Diels-Alder intermediate is rapidly trapped by aromatization in the presence of acids. Thus, the observation by

Buzanowski et al.,[285] of dramatically lower rates for S polymerizations carried out in the presence of various acid catalysts, is further evidence for the Mayo mechanism.

To account for certain features of the polymerization kinetics, it has been postulated that radical production proceeds mainly through the isomer of (65) in which the phenyl group is axial.[281,286] Both isomers of (65) can give rise to trimers (63), possibly by an ene reaction between (65) and S. However, trimers (63) could also form by cage combination of radicals (66) and (67).

Despite the body of evidence in favor of the Mayo mechanism, the formation of diphenylcyclobutanes (62) must still be accounted for. It is possible that they arise via 1,4-diradicals (64) and it is also conceivable that diradicals (64) are intermediates in the formation of the Diels-Alder adduct (65) (see Scheme 55) and could provide a second (minor) source of initiation. Direct initiation by diradicals is suggested in the thermal polymerization of 2,3,4,5,6-pentafluorostyrene where transfer of a fluorine atom from Diels-Alder dimer to monomer seems highly unlikely (high C-F bond strength) and for derivatives which cannot form a Diels-Alder adduct.

3.3.6.2 Acrylate homopolymerization

Various acrylates, methacrylates and related compounds have been reported to undergo spontaneous polymerization.[278] A complication in studying thermal polymerization of MMA is the difficulty in eliminating impurity initiated polymerization. The monomer is extremely difficult to purify or retain in a "pure" state. These problems have lead some to question whether there is any true spontaneous initiation.[287] It is, in any event, clear that the rate of thermal polymerization of MMA is substantially less than that of S at the same temperature (at least 70-fold less at 90°C).[278,288]

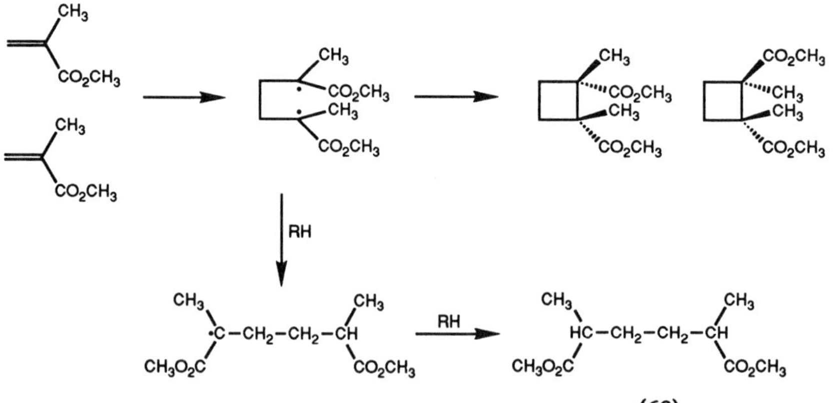

Scheme 57

Dimer and trimer by-products have been isolated from MMA polymerizations and these are suggestive of 1,4-diradical intermediates.[289-292] Lingnau and Meyerhoff[289] found that rates of spontaneous polymerization of MMA were

Initiation

substantially higher in the presence of transfer agents (RH). They were able to isolate the compound (**68**) which might come from trapping of the biradical intermediate (Scheme 57).

3.3.6.3 Copolymerization

Monomers which are strong electron donors may undergo spontaneous copolymerization with strong electron acceptor monomers by a free radical mechanism. In certain cases homopolymers formed by an ionic mechanism accompany copolymer formation.[280,293]

Examples where free radical initiation is believed to be dominant include:

(a) S with MAH,[294,295] AA,[296] AN,[297,298] vinylidene cyanide,[299] or dimethyl 1,1-dicyanoethane-2,2-dicarboxylate.[280]
(b) *p*-Methoxystyrene with trimethyl ethylenetricarboxylate[280] or dimethyl cyanofumarate.[300]
(c) 1,2-Dimethoxyethylene with MAH.[301]
(d) Vinyl sulfides with a range of electrophilic monomers.[302]

Various mechanisms have been proposed to explain the initiation processes. The self-initiated copolymerizations of the monomer pairs S-MMA and S-AN proceed at substantially faster rates than pure S polymerization. For S-AN[298] and S-MAH[295] the mechanism of initiation was proposed to be analogous to that of S homopolymerization (Scheme 57) but with acrylonitrile acting as the dienophile in the formation of the Diels-Alder adduct (Scheme 58).

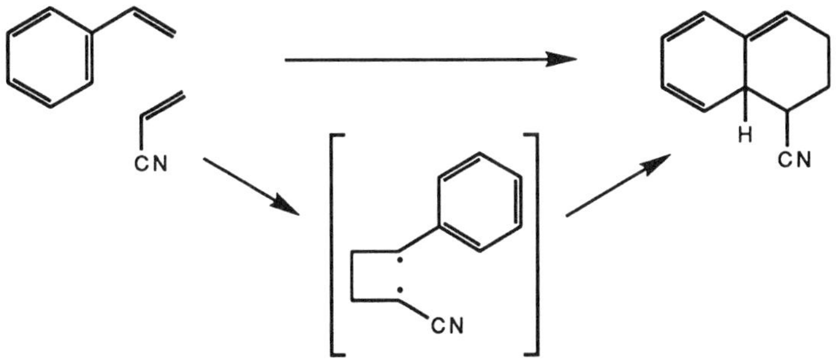

Scheme 58

Various oligomers formed by Diels-Alder/ene reactions are observed.[297,298] For S-MAH polymerization Sato *et al.*[295] used spin trapping to identify the initiating species. On the other hand, in the case of S-AN copolymerization, the finding that acid catalysts do not affect the rate of polymerization argues against the involvement of this species in the initiation mechanism.[297] Acid catalysts, which effectively trap the Diels-Alder intermediate (**65**) by aromatization, have been found to lower the rate of thermal styrene homopolymerization dramatically.[285]

Other postulated mechanisms for spontaneous initiation include electron transfer followed by proton transfer to give two monoradicals,[302] hydrogen atom transfer between a charge-transfer complex and solvent,[294] and formation of a diradical from a charge-transfer complex.[303]

Hall[280,293] has proposed a unifying concept based on tetramethylenes (resonance hybrids of 1,4-diradical and zwitterionic limiting structures - see Scheme 59) to rationalize all donor-acceptor polymerizations. The predominant character of the tetramethylenes (zwitterionic or diradical) depends on the nature of the substituents.[280,304] However, more evidence is required to prove the global application of the mechanism.

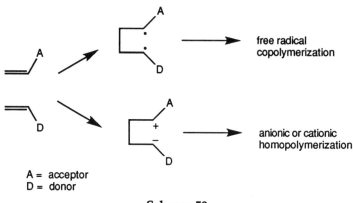

Scheme 59

3.4 The Radicals

In this section, the reactions undergone by radicals generated in the initiation or chain transfer processes are detailed. Emphasis is placed on the specificity of radical-monomer reactions and other processes likely to take place in polymerization media under typical polymerization conditions. The various factors important in determining the rate and selectivity of radicals in addition and substitution processes have been discussed in general terms in section 2.2.

3.4.1 Carbon-Centered Radicals

Carbon-centered radicals are produced as primary radicals in the decomposition of azo-compounds (*e.g.* Scheme 60),

Scheme 60

as secondary radicals from peroxides by β-scission of the initially formed acyloxy or alkoxy radicals (*e.g.* Scheme 61),

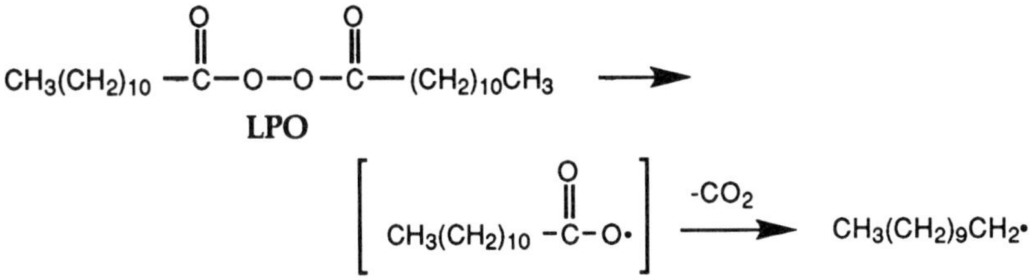

$$CH_3(CH_2)_{10}-\overset{O}{\overset{\|}{C}}-O-O-\overset{O}{\overset{\|}{C}}-(CH_2)_{10}CH_3 \longrightarrow$$
LPO

$$\left[CH_3(CH_2)_{10}-\overset{O}{\overset{\|}{C}}-O\cdot \right] \xrightarrow{-CO_2} CH_3(CH_2)_9CH_2\cdot$$

Scheme 61

and by transfer reactions (*e.g.* Scheme 62).

$$R\cdot \;+\; PhCH_3 \xrightarrow{-RH} PhCH_2\cdot$$

Scheme 62

Primary radical termination involving carbon-centered radicals is described in sections 2.4 and 5.2.1.

3.4.1.1 Alkyl radicals

Alkyl radicals, when considered in relation to heteroatom-centered radicals (*e.g.* t-butoxy, benzoyloxy), show a high degree of regiospecificity in their reactions. Their chemistry has been reviewed by Giese,[305] Tedder,[306] Beckwith,[307] Rüchardt,[65] and Tedder and Walton.[308,309] Rate constants for reactions of carbon-centered radicals for the period through 1982 have been compiled by Lorand[310] and Asmus and Bonifacic.[311]

The relative reactivity of various monomers, solvents, inhibitors, and other materials towards methyl radicals and representative primary, secondary, and tertiary alkyl radicals has been studied (see Table 3.6). In the absence of heteroatom containing substituents (*e.g.* halo-, cyano-) at or conjugated with the radical center, carbon-centered radicals have nucleophilic character and absolute rate constants for reaction with monomers generally lie in the range 10^5-10^6 M^{-1} s^{-1}. Thus, simple alkyl radicals generally show higher reactivity toward electron-deficient monomers (*e.g.* acrylic monomers) than towards electron-rich monomers (*e.g.* vinyl esters, S) - see Table 3.6.

Alkyl radicals show a high degree of regiospecificity for tail *vs.* head addition. Significant amounts of head addition are observed only when addition to the tail-position is sterically inhibited as it is in α,β-disubstituted monomers. For example, with β-alkylacrylates, head addition is observed and the proportion shows a good correlation with the steric size of the β-substituent.[312]

Table 3.6 Rate Data for Reactions of Carbon-Centered Radicals

radical	Temp °C	k_S M^{-1}s^{-1}	AMS	MA	MMA	AA	MAA	AN	MAN	VAc	PAc	VC	PhCH$_3$	refs.
PhCH$_2\dot{C}$(CO$_2$Et)$_2$	60 [a]	-	-	.0071	-	-	-	0.0088	-	0.0074	-	-	-	313
(CH$_3$)$_3$CO$_2$CH$_2\cdot$	23 [b]	1.95 × 10^6	1.8	0.23	0.64	-	-	0.26	-	0.031	0.042	-	-	314
(CH$_3$)$_2\dot{C}$CN	30 [d]	-	1.06[c]	-	0.56	-	-	-	-	0.02	-	-	-	315,316
(CH$_3$)$_2\dot{C}$CN	60 [d,f]	3 × 10^3	0.95	0.3	0.56	-	-	0.44	0.34	0.03	-	0.04[e]	-	50,315-321
(CH$_3$)$_2\dot{C}$CN	100 [f]	-	0.87	-	0.56	-	-	-	0.49	0.05	-	-	-	316,320,322
(CH$_3$)$_2\dot{C}$CN	58 [f]	2.4 × 10^3	0.96	0.15	0.66	-	-	0.84	0.44	0.017	0.033	-	-	323
(CH$_3$)$_2\dot{C}$CO$_2$CH$_3$	60 [f]	-	-	-	0.7	-	-	-	-	0.038[g]	-	-	-	315,324
CH$_3\cdot$	65	-	1.16	1.3	1.8	-	-	2.2	2.7	0.038	-	-	0.000015	325,326
PhCH$_2\cdot$	23 [f]	1.3 × 10^4	-	-	-	-	-	1.7	-	0.011	-	-	-	327
(CH$_3$)$\dot{C}$HPh	100 [f]	-	1.1	1.5	1.9	-	-	5.0	-	-	-	-	-	328,329
	69 [b]	3.2 × 10^5 [h]	0.6	3.5	-	4.7	4.4	7.5	-	-	-	-	-	17
	20 [i]	-	0.93	6.7	5.0	-	-	24	13	0.12	-	0.016	-	305
(CH$_3$)$_3$C$\cdot$	27 [f]	1.3 × 10^5	-	6.7	-	-	-	25.3	-	0.032	-	0.125	-	330,331
Ph$\cdot$	25 [j]	1.1 × 10^8	-	-	1.6	-	-	-	-	-	-	-	0.015	332
Ph$\cdot$	60 [k]	-	1.31	0.73	1.16	-	-	1.14	1.30	0.14	0.14	0.18	-	333
ClPh$\cdot$	25 [l]	-	-	0.66	1.03	-	-	0.68	-	-	-	-	-	341

[a] In acetic acid. [b] In acetonitrile. [c] 40 °C in toluene.
[d] In benzene. Value based on the reported rate constant for addition to MAN^{50} and value of k_{MAN}/k_S shown.
[e] 45 °C. [f] In toluene. [g] 30 °C, in ethyl acetate.
[h] Reported values corrected using a more recent rate constant for the 5-hexenyl clock (Chatgilialoglu, C., Ingold, K.U., and Scaiano, J.C., *J. Am. Chem. Soc.*, 1981, 103, 7739).
[i] In methylene chloride. [j] In Freon 113. [k] In carbon tetrachloride. [l] In aqueous acetone.

Initiation

Alkyl radicals also show a high specificity for addition *vs.* abstraction. Rate constants for hydrogen abstraction from monomers and solvents (*e.g.* toluene) are generally much smaller (*ca.* 100-fold less) than those for addition to double bonds.

3.4.1.2 α-Cyanoalkyl radicals

Thermal or photochemical decomposition of azonitriles (*e.g.* AIBN) affords α-cyanoalkyl radicals (Scheme 63).[25]

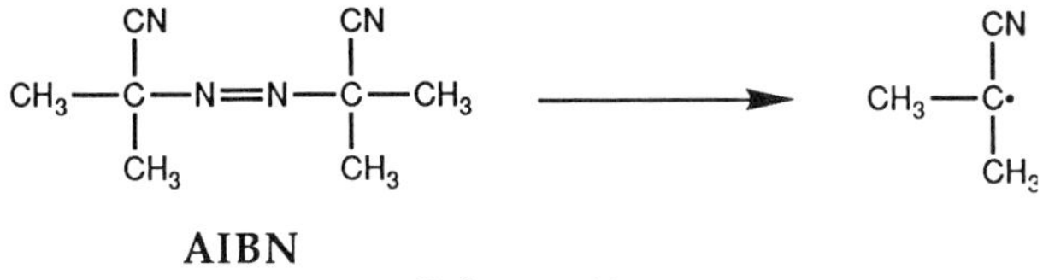

$$CH_3-\underset{\underset{CH_3}{|}}{\overset{\overset{CN}{|}}{C}}-N=N-\underset{\underset{CH_3}{|}}{\overset{\overset{CN}{|}}{C}}-CH_3 \longrightarrow CH_3-\underset{\underset{CH_2}{|}}{\overset{\overset{CN}{|}}{C}}\cdot$$

AIBN

Scheme 63

The reactions of cyanoisopropyl radicals with monomers have been widely studied. Relative reactivity data (see Table 3.6) demonstrates that they possess some electrophilic character. However, this is not sufficient to overcome the bond strength term (*e.g.* the more electron-rich VAc is very much less reactive than the electron-deficient AN or MA). Heberger and Fisher[323] have undertaken a detailed analysis of factors controlling rate of addition.

Cyanoisopropyl radicals generally show a high degree of specificity in reactions with unsaturated substrates. They react with most monomers (*e.g.* S, MMA) exclusively by tail addition (Scheme 3). However, Bevington *et al.*[102,320] have provided evidence that cyanoisopropyl radicals give ~ 10% head addition with VAc at 60°C and that the proportion of head addition increases with increasing temperature.

α-Cyanoalkyl radicals show relatively little tendency to abstract hydrogen from monomer, solvent, or polymer even in relation to other alkyl radicals.[335] However, these radicals, like other carbon-centered radicals,[24] react with oxygen at diffusion controlled rates (see 3.2.5). For polymerizations carried out in poorly degassed media, it has been proposed[25,26] that abstraction products, peroxide linkages, and other defect structures may arise through the intermediacy of a alkylperoxy radical (Scheme 10).

The α-cyanoalkyl radicals can, in principle, react with substrates either at carbon or at nitrogen (Scheme 64). However, reaction at nitrogen to give a ketenimine has only been observed in cases of reactions with other radicals (see 5.2.1.2) or organometallic reagents.[336] There is as yet no evidence for ketenimine being produced in reactions with monomers or spin traps[6,337] despite several studies aimed specifically at detecting such processes. It is anticipated that reaction through nitrogen would be favored by steric hindrance at the site of attack and by electron donating substituents on the substrate. It is also likely that addition via the nitrogen will be readily reversible (*i.e.* rapid trapping of the initial adduct will be required to observe this pathway).

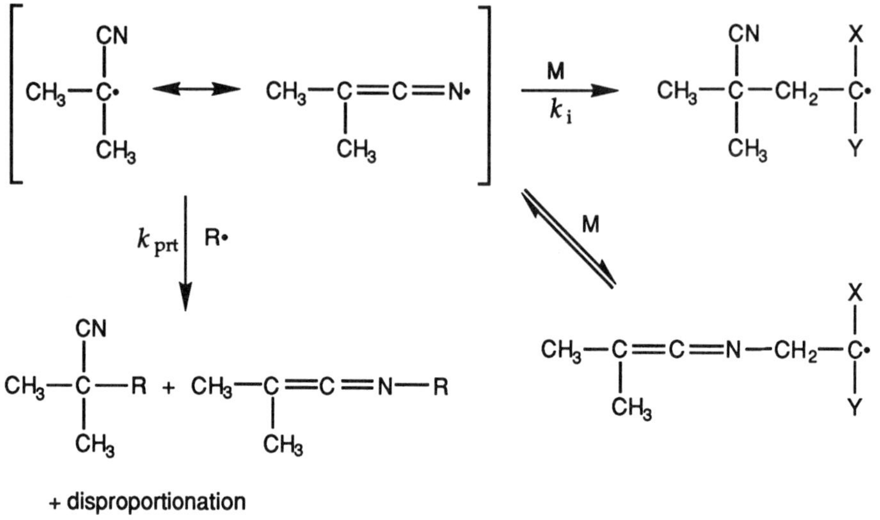

Scheme 64

Herberger and Fischer[323] have measured absolute rate constants for the reactions of cyanoisopropyl with monomers by time resolved EPR spectroscopy (see Table 3.6). The numbers are in general agreement with values estimated by analysis of polymerization data. Krstina *et al.*[50] determined the rate constant for addition of cyanoisopropyl to MAN to be in the range $10^3 \leq k_i \leq 5 \times 10^3$ M^{-1} s^{-1}. Mahabadi and O'Driscoll[338] estimated the ratio $k_{prt}/k_i.k_p$ for AIBN initiated S polymerization on the basis of an analysis of polymerization kinetics. Their data would suggest that if $k_{prt} \sim 10^9$ M^{-1} s^{-1}, then $k_i \sim 10^3$ M^{-1} s^{-1} at 60°C.

A number of reports[93,97,338] have indicated that primary radical termination can be important during polymerizations initiated by azonitriles. However, for the case of S polymerization initiated by AIBN, NMR end group determination[6] shows that primary radical termination is of little importance except when very high rates of initiation are employed (*e.g.* with high initiator concentrations at high temperatures). Cyanoalkyl radicals give a mixture of combination and disproportionation in their reactions with other radicals (see also 2.4, 5.2.2, 5.2.3.3 and 5.2.4.3). This finding is significant for those who use azonitriles as initiators in producing telechelics (see 6.3.1.1).

3.4.1.3 Aryl radicals

Aryl radicals are produced in decomposition of alkylazobenzenes and diazonium salts, and by β-scission of aroyloxy radicals (Scheme 65).

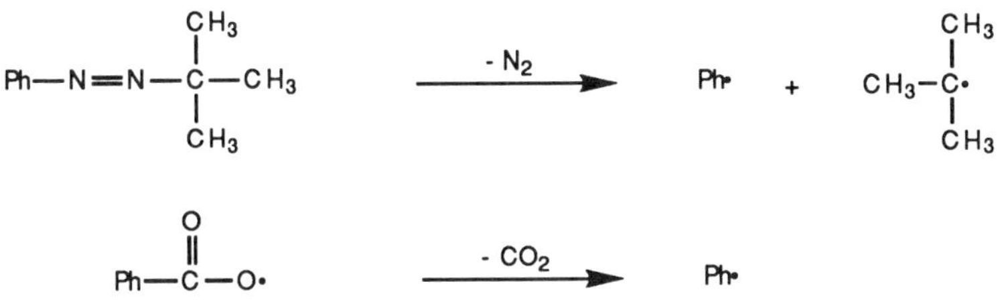

$$Ph-N=N-\underset{\underset{CH_3}{|}}{\overset{\overset{CH_3}{|}}{C}}-CH_3 \xrightarrow{-N_2} Ph\bullet + CH_3-\underset{\underset{CH_3}{|}}{\overset{\overset{CH_3}{|}}{C}}\bullet$$

$$Ph-\overset{\overset{O}{\|}}{C}-O\bullet \xrightarrow{-CO_2} Ph\bullet$$

Scheme 65

Aryl radicals have been reported to attack aromatic rings (*e.g.* S[7]) or abstract hydrogen (*e.g.* MMA[9]) in competition with adding to a monomer double bond. However, these processes typically account for ≤1% of the total. The degree of specificity for tail *vs.* head addition is also very high. Significant head addition has been observed only where tail addition is retarded by steric factors (*e.g.* methyl crotonate[9] and β-substituted methyl vinyl ketones[334,339]).

Absolute rate constants for the attack of aryl radicals on a variety of substrates have been reported by Scaiano and Stewart (Ph•)[332] and Citterio *et al.* (*p*-ClPh•).[334,339] The reactions are extremely facile in comparison with additions of other carbon-centered radicals [*e.g.* $k(S) = 1.1 \times 10^8$ M^{-1} s^{-1} at 25°C].[332] Relative reactivities are available for a wider range of monomers and other substrates (see Table 3.6).[332,333,340-342] Phenyl radicals do not show clear cut electrophilic or nucleophilic behavior.

3.4.2 Oxygen-Centered Radicals

Oxygen-centered radicals are arguably the most common of initiator-derived species generated during initiation of polymerization and many studies have dealt with these species. The class includes alkoxy, hydroxy and acyloxy radicals and the sulfate radical anion (formed as primary radicals by homolysis of peroxides or hyponitrites) and alkylperoxy radicals (produced by the interaction of carbon-centered radicals with molecular oxygen or by the induced decomposition of hydroperoxides).

Howard and Scaiano[343] have provided an excellent compilation of absolute and relative rate data for reactions of oxygen-centered radicals covering the literature through 1982. Some of these data and other more recent information are summarized in Tables 3.7 and 3.8.

The pathways whereby oxygen-centered radicals interact with monomers show marked dependence on the structure of the radical (see Table 3.8). For example, with MMA the proportion of tail addition varies from 66% for *t*-butoxy to 99% for isopropoxycarbonyloxy radical. The reactions of oxygen-centered radicals are discussed in detail in the following sections.

Table 3.7 Selected Rate Data for Reactions of Oxygen-Centered Radicals[a]

radical	Temp. °C	k_S $M^{-1}s^{-1}$	AMS	MA	MMA	AN	MAN	VAc	PhCH$_3$	refs.
(CH$_3$)$_3$CO•	60	~9 × 10^5 [b]	1.3	0.06	0.28	0.05	0.12	0.06	0.19	7,11,21,344
(CH$_3$)$_2$(Ph)CO•	60	~3 × 10^6 [c]	–	–	0.1	–	–	–	–	62
HO•	60	–	1.2	0.34	0.63	–	–	–	–	345
HO•	25	1.97 × 10^{10}	–	–	1.0	0.27	0.96	–	–	317
PhCO$_2$•	24	5.1 × 10^7	–	–	–	–	–	–	–	346
PhCO$_2$•	60	–	–	0.05	0.12	<0.05	–	0.36	–	347
PhCO$_2$•	60	–	–	0.02	0.11	0.02	–	0.26	–	9,10,21,348,349

[a] Overall reactivity. For pathways observed see Table 3.8.
[b] Based on rate constant for β-scission as clock reaction[9] and the yield of methyl radical-derived products observed in bulk styrene polymerization.[7]
[c] Based on the analysis of Rizzardo et al.[62] but assuming a rate constant for β-scission for cumyloxy radical of 1.5 × 10^6 at 60°C.

Table 3.8 Specificity Observed in the Reactions of Oxygen-Centered Radicals with Various Monomers at 60°C

monomer	radical	pathway[a]			
		A	B	C	D
CH$_2$=C(H)(Ph) (A↑, B←, D↘)	$(CH_3)_3CO \cdot$[b,c,7]	100	-	-	-
	$(CH_3)_2CH_2O \cdot$[b,c,350]	100	-	-	-
	$CH_3CH_2O \cdot$[b,c,350]	100	-	-	-
	$(CH_3)_2(Ph)CO \cdot$[b,c,62]	100	-	-	-
	$HO \cdot$[b,345]	87	6	-	7
	$PhCO_2 \cdot$[c,d,7]	80	6	-	14
	$(CH_3)_2CHOCO_2 \cdot$[b,c,351]	95	-	-	5
CH$_2$=C(CH$_3$)(Ph) (A↑, B←, C↖, D↘)	$(CH_3)_3CO \cdot$[b,c,f,344]	85	-	15	-
	$HO \cdot$[b,345]	83	3	5	9
CH$_2$=C(H)(CO-O-CH$_3$) (A↑, B←, D↘)	$(CH_3)_3CO \cdot$[b,c,9]	83	2	-	15
	$HO \cdot$[b,345]	80	17	-	3
	$PhCO_2 \cdot$[c,d,9]	84	16	-	-
CH$_2$=C(CH$_3$)(CO-O-CH$_3$) (A↑, B←, C↖, D↘)	$(CH_3)_3CO \cdot$[b,c,f,11]	66	-	30	4
	$(CH_3)_2(Ph)CO \cdot$[b,c,62]	70	-	26	3
	$(CH_3)_2CHO \cdot$[b,c,112]	88	-	12	-
	$CH_3CH_2O \cdot$[b,c,350]	92	-	8	-
	$HO \cdot$[b,345]	87	6	5	2
	$PhCO_2 \cdot$[c,d,9]	93	7	<1[e]	-
	$(CH_3)_2CHOCO_2 \cdot$[b,177]	>99	-	-	-

Table 3.8 (continued)

monomer	radical	pathway[a]			
		A	B	C	D
CH₂=C(H)(CN) (A, B)	(CH₃)₃CO•[b,c,352]	100	-	-	-
	PhCO₂•[c,d,349]	98	2	-	-
CH₂=C(CH₃)(CN) (A, B, C)	(CH₃)₃CO•[b,c,353]	74	-	26	-
CH₂=C(H)(O-C(=O)-CH₃) (A, B, D)	(CH₃)₃CO•[b,c,352]	79	15	-	6
	PhCO₂•[c,d,349]	76	24	-	-
CH₂=C(CH₃)(O-C(=O)-CH₃) (A, B, C, D)	(CH₃)₃CO•[b,c,f,344]	48	-	48	4

a Relative yields of products formed by pathway indicated. All data rounded to nearest 1%. A dash indicates that the product was not detected.
b In bulk monomer.
c Yields have been normalized to exclude β-scission products.
d In 50% v/v acetone/monomer.
e Total abstraction by benzoyloxy and phenyl radicals.
f Addition:abstraction ratio shows solvent dependence.[20,344]

3.4.2.1 Alkoxy radicals

Alkoxy radicals are frequently encountered as initiating species in polymerizations and have been the subject of numerous laboratory studies. Most work has concentrated on the chemistry of *t*-butoxy radical and relatively little attention has been paid to the chemistry of other alkoxy radicals. The chemistry of alkoxy radicals has been the subject of several reviews.[354-357]

3.4.2.1.1 *t*-Butoxy radicals

The reactions of *t*-butoxy radicals are amongst the most studied of all radical processes. These radicals are generated by thermal or photochemical decomposition of peroxides or hyponitrites (Scheme 66).

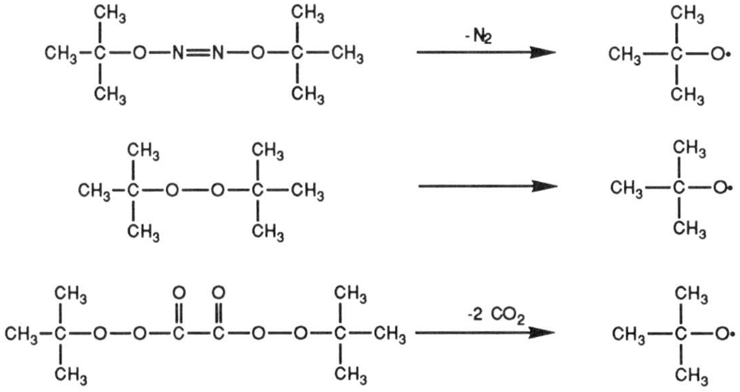

Scheme 66

In a polymerization reaction they may: (a) initiate a chain by adding to the double bond of a monomer; (b) abstract a hydrogen atom from the monomer, solvent, or another component of the reaction mixture to afford a new radical species and *t*-butanol (primary radical transfer); (c) undergo β-scission to give methyl radicals and acetone (*e.g.* Scheme 5). The relative importance of these processes depends strongly on the particular monomer(s) and the reaction conditions.

In contrast to most other oxygen-centered radicals [*e.g.* benzoyloxy (3.4.2.2.1), hydroxy (3.4.2.3)], *t*-butoxy radicals show relatively high regiospecificity in reactions with carbon-carbon double bonds (see Table 3.8). Nonetheless, significant amounts of head addition are observed with the halo-olefins,[23,358] simple alkenes,[359] vinyl acetate and methyl acrylate (see Table 3.8).[352] Head addition is generally not observed with 1,1-disubstituted monomers. The exception is vinylidene fluoride[23,358] where head addition predominates (see 2.2). With allyl methacrylate (**69**)[360] and acrylate (**70**),[361] *t*-butoxy radicals give substantially more addition to the acrylate double bond than to the allyl double bond (see Figure 2).

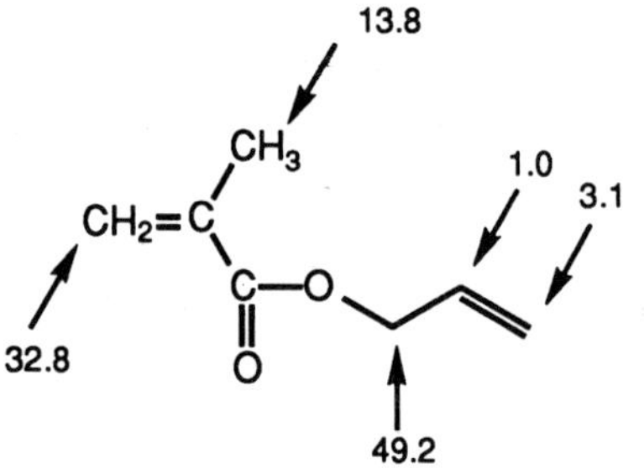

[Structures of (69) allyl methacrylate and (70) allyl acrylate with relative reactivity numbers:

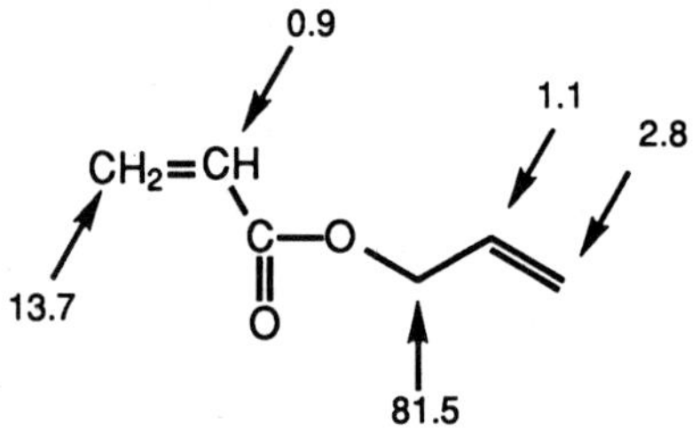

(69): 32.8 (CH₂=C), 13.8 (CH₃), 1.0, 3.1, 49.2 (C=O arrow)

(70): 13.7 (CH₂=CH), 0.9, 1.1, 2.8, 81.5 (C=O arrow)]

Figure 2 Relative reactivity of indicated site towards t-butoxy radicals for allyl methacrylate (69) and allyl acrylate (70).

Studies of the relative reactivity of *t*-butoxy radicals with substituted styrenes,[362] toluenes[363,364] and other substrates (see 2.3.3) indicate that they are slightly electrophilic in character. However, Sato and Otsu [12] found that the order of reactivity of *t*-butoxy radicals towards a series of monomers was different from that of the more electrophilic benzoyloxy radicals. They concluded that product radical stability was important in determining reactivity. Cuthbertson et al.[358] examined the reactions of *t*-butoxy radicals toward fluoro-olefins and found a pattern of reactivities more characteristic of a nucleophilic species. The strength of the bond being formed plays an important role in determining regiospecificity. The factors influencing the specificity and rate of addition are discussed in greater detail in section 2.2.2.

Recent investigations[11,12,20,344,365,366] have shown that the reaction of *t*-butoxy radicals with monomers bearing sp^3 hydrogens invariably produces a mixture of initiating radicals arising from hydrogen abstraction and addition (see Table 3.8). Simple alkenes (*e.g.* butenes),[359] vinyl ethers[367] and higher acrylates [*e.g.* BMA - see Figure 3][11] may give predominantly abstraction.

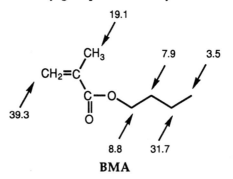

BMA

Figure 3 Relative reactivity of indicated site towards *t*-butoxy radicals for BMA.[368,369]

t-Butoxy radicals also undergo unimolecular fragmentation to produce acetone and methyl radicals (Scheme 5). Significant amounts of the β-scission products are obtained in the presence of even the most reactive monomers (*e.g.* S[7]). The reactions of methyl radicals have been discussed above (see 3.4.1.1).

The relative amounts of double bond addition, hydrogen abstraction and β-scission observed are dependent on the reactivity and concentration of the particular monomer(s) employed and the reaction conditions. Higher reaction temperatures are reported to favor abstraction over addition in the reaction of *t*-butoxy radicals with AMS[365] and cyclopentadiene.[370] However, the opposite trend is seen with isobutylene.[22,23]

Pioneering work by Walling *et al.*[356] established that the specificity shown by *t*-butoxy radical is solvent dependent. Recent work[20,21,344] on the reactions of *t*-butoxy radicals with a series of α-methylvinyl monomers has shown that polar and aromatic solvents favor abstraction over addition, and β-scission over either addition or abstraction. Data for cumyloxy radicals indicate that the rate constant for β-scission process is strongly solvent dependent under conditions where that for hydrogen atom abstraction is solvent independent (see 3.4.2.1.2).[371] Although there is as yet insufficient absolute rate data on the solvent dependence of *t*-butoxy radical reactions to prove similar behavior, it seems likely that a similar situation applies.

3.4.2.1.2 *Other t-alkoxy radicals*

Various *t*-alkoxy radicals may be formed by processes analogous to those described for *t*-butoxy radicals. The limited data available suggest that their propensities for addition *vs.* abstraction are similar.[62] However, rate constants for β-scission of *t*-alkoxy radicals show marked dependence on the nature of substituents α to oxygen.[194,372] For *t*-alkoxy radicals of general structure [$RC(CH_3)_2O\bullet$], the rate constant for β-scission increases in the series R = methyl<<ethyl<isopropyl<*t*-butyl. Polar, steric and thermodynamic factors are all thought to play a part in favoring this trend.[355]

$$C_2H_5-\underset{\underset{CH_3}{|}}{\overset{\overset{CH_3}{|}}{C}}-O\bullet \qquad Ph-\underset{\underset{CH_3}{|}}{\overset{\overset{CH_3}{|}}{C}}-O\bullet$$

(71) (72)

Thus, even if *t*-amyloxy (71) shows similar specificity for addition *vs.* abstraction to *t*-butoxy radicals, abstraction will be of lesser importance.[373,374] The reason is that most *t*-amyloxy radicals do not react directly with monomer. They undergo β-scission and initiation is mainly by ethyl radicals. Ethyl radicals are much more selective and give addition rather than abstraction. This behavior has led to *t*-amyl peroxides and peresters being promoted as superior to the corresponding *t*-butyl derivatives as polymerization initiators.[374]

The rate constant of β-scission of cumyloxy radicals (**72**) is also significantly greater than that for *t*-butoxy radicals.[372,375] β-Scission gives exclusively acetophenone and methyl radicals. For the case of S or MMA polymerization initiated by cumyloxy radicals at 60°C, the proportion of methyl radical initiation is six-fold greater than is seen with *t*-butoxy radicals.[62]

The absolute rate constant for β-scission of cumyloxy radicals has been shown to be solvent dependent. The absolute rate constant (2.6×10^5 s^{-1} at 30°C in CCl$_4$) increases ca. seven-fold over the series CCl$_4$, C$_6$H$_6$, C$_6$H$_5$Cl, (CH$_3$)$_3$COH, CH$_3$CN, CH$_3$COOH.[371] The rate constant for abstraction from cyclohexane remains at $1.2 \pm 0.1 \times 10^6$ M^{-1} s^{-1} in all solvents.

3.4.2.1.3 *Primary and secondary alkoxy radicals*

Relatively few studies have dealt with the reactions of primary and secondary alkoxy radicals (isopropoxy, methoxy, *etc.*) with monomers. These radicals are conveniently generated from the corresponding hyponitrites (Scheme 67).[112,350]

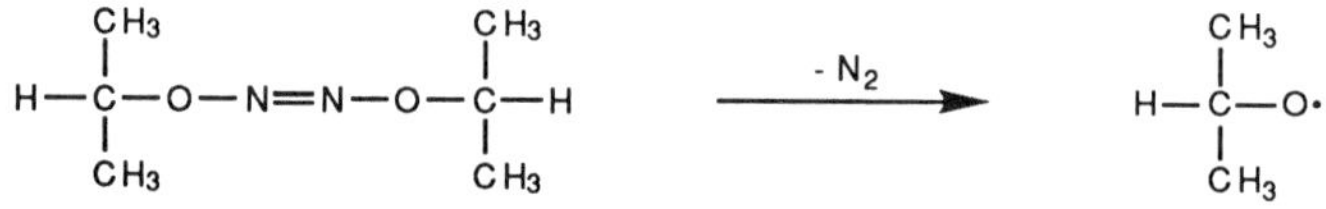

Scheme 67

Primary and secondary alkoxy radicals generally show a reduced tendency to abstract hydrogen or to undergo β-scission when compared to the corresponding *t*-alkoxy radical.[112,350] This has been correlated with the lesser nucleophilicity of these radicals.[376]

It has been suggested[112,350] that primary and secondary alkoxy radicals may react with S by donation of a hydrogen atom to the monomer and production of an aldehyde.

3.4.2.2 *Acyloxy and alkoxycarbonyloxy radicals*

Aroyloxy radicals are formed by thermal or photochemical decomposition of diaroyl peroxides and aromatic peresters; alkoxycarbonyloxy radicals are similarly produced from peroxydicarbonates (Scheme 68).

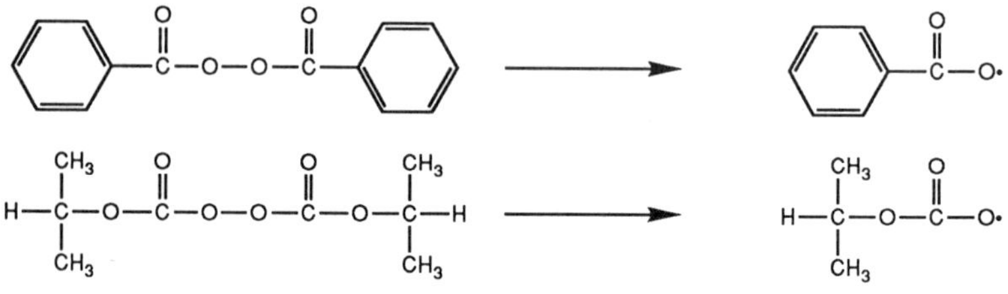

Scheme 68

Initiation

Aliphatic acyloxy radicals undergo facile fragmentation with loss of carbon dioxide (see Scheme 61) and, with few exceptions,[377] do not have sufficient lifetime to enable direct reaction with monomers or other substrates. The rate constants for decarboxylation of aliphatic acyloxy radicals are in the range $1\text{-}10 \times 10^9$ M^{-1} s^{-1} at 20°C.[378] Ester end groups in polymers produced with aliphatic diacyl peroxides as initiators most likely arise by transfer to initiator (see 3.3.2.1.3). The chemistry of the carbon-centered radicals formed by β-scission of acyloxy radicals is discussed above (see 3.4.1.1).

3.4.2.2.1 Benzoyloxy radicals

Benzoyloxy radicals are electrophilic and show higher reactivity towards electron-rich (*e.g.* S, VAc) than electron-deficient (*e.g.* MMA, AN) monomers (see Table 3.7).[349,379] Product studies on the reactions of benzoyloxy radicals with simple olefins and monomers[7,9,157,347,349,379-382] show that they have remarkably poor regiospecificity when adding to carbon-carbon double bonds. Their reactions invariably give a mixture of products from head addition and tail addition (see Scheme 4 and Table 3.8).[7,9,349,382] They also display a marked propensity for aromatic substitution.[7,34,346] On the other hand, compared with alkoxy radicals, they show little tendency to abstract hydrogen.[9]

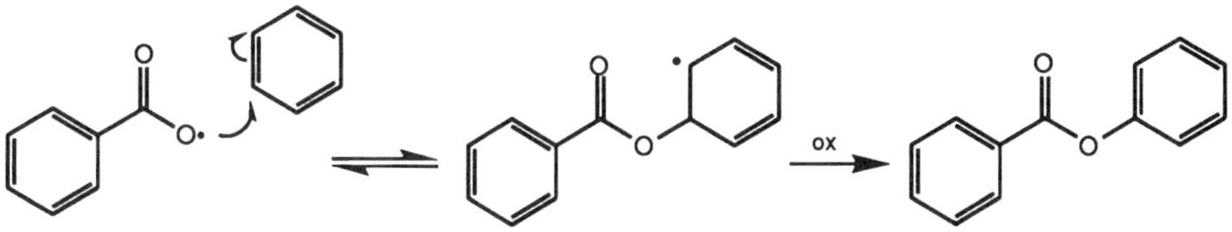

Scheme 69

Additions of benzoyloxy radicals to double bonds[383,384] and aromatic rings (Scheme 69)[137] are potentially reversible. For double bond addition, the rate constant for the reverse fragmentation step is slow ($k=10^2\text{-}10^3$ s^{-1} 25°C) with respect to the rate of propagation during polymerizations. Thus, double bond addition is effectively irreversible. However, for aromatic substrates, the rate of the reverse process is extremely fast. While the aromatic substitution products may be trapped with efficient scavenging agents (*e.g.* a nitroxide[7,34] or a transition metal[158]), they are generally not observed under polymerization conditions.[8] A different situation may pertain when redox initiation is used, as the oxidants employed may be effective radical traps. A small proportion of aromatic benzoate residues can be detected in high conversion PS prepared with benzoyl peroxide. However, it is likely that these arise through attack on PS rather than S.[8,143]

The rate of β-scission of benzoyloxy radicals is such that in most polymerizations initiated by these radicals both phenyl and benzoyloxy end groups will be formed (Scheme 4). A reliable value for the rate constant for β-scission would enable the absolute rates of initiation by benzoyloxy radical to be estimated. Various values for the rate constant for β-scission have appeared. Many of the early estimates are low. The activation parameters (CCl_4) determined by Chateauneuf et al.[346] are $\log_{10} A = 12.6$ and $E_a = -35.97$ kJ mol^{-1} which corresponds to a rate constant of 9×10^6 s^{-1} at 60°C.

The rate constant for β-scission is dependent on ring substituents. Rate constants for radicals R-$C_6H_4CO_2\bullet$ are reported to increase in the series where R is p-F≤p-CH_3O<p-CH_3~p-Cl<H<m-Cl.[346] There is qualitative evidence that the relative rates for β-scission and addition are insensitive to solvent changes. For benzoyloxy radicals, similar relative reactivities are obtained from direct competition experiments[9] as from studies on individual monomers when β-scission is used as a clock reaction.[347,349]

The rate constants for benzoyloxy and phenyl radicals adding to monomer are high (> 10^7 M^{-1} s^1 for S at 60°C - see Table 3.7). In these circumstances primary radical termination should have little importance under normal polymerization conditions. Some kinetic studies indicating substantial primary radical termination during S polymerization may need to be re-evaluated in this light.[150] Secondary benzoate end groups in PS with BPO initiator most likely arise by transfer to initiator.

3.4.2.2.2 Alkoxycarbonyloxy radicals

The chemistry of alkoxycarbonyloxy radicals in many ways parallels that of the aroyloxy radicals (*e.g.* benzoyloxy, see 3.4.2.2.1). Products attributable to the reactions of alkoxy radicals generally are not observed. This indicates that the rate of β-scission is slow relative to the rate of addition to monomers or other substrates.[177,380]

The alkoxycarbonyloxy radicals show little tendency to abstract hydrogen.[177,380] For example, in the reaction of isopropoxycarbonyloxy radicals with MMA, hydrogen abstraction, while observed, is a minor pathway (≤1%). When isopropoxycarbonyloxy radicals abstract hydrogen, isopropanol is the expected by-product since the intermediate acid undergoes facile decarboxylation. Formation of isopropanol is not evidence for the involvement of isopropoxy radicals (Scheme 70).

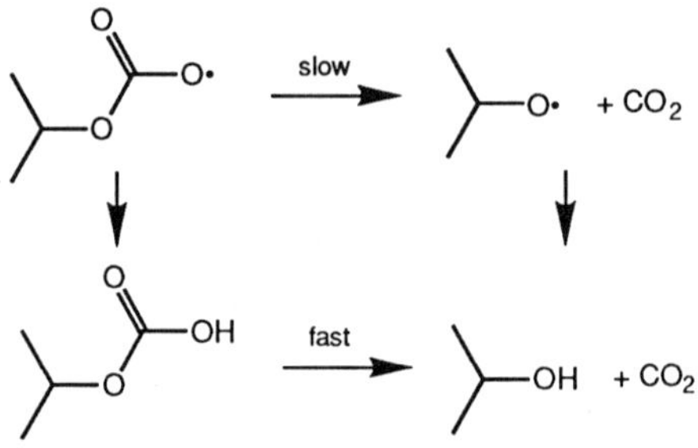

Scheme 70

Isopropoxycarbonyloxy radicals undergo facile reaction with aromatic substrates (*e.g.* toluene) by reversible aromatic substitution.[158,385] Isopropoxycarbonyloxy radicals react with S to give ring substitution (*ca.* 1%) as well as the expected double bond addition.[351]

3.4.2.3 Hydroxy radicals

The transient radicals produced in reactions of hydroxy radicals with vinyl monomers in aqueous solution have been detected directly by EPR[386-388] or UV spectroscopy.[389,390] These studies indicate that hydroxy radicals react with monomers and other species at or near the diffusion-controlled rates (see Table 3.7). However, high reactivity does not mean a complete lack of specificity. Hydroxy radicals are electrophilic and trends in the relative reactivity of the hydroxy radicals toward monomers can be explained on this basis.[345]

Grant *et al.*[345] have examined the reactions of hydroxy radicals with a range of vinyl and α-methylvinyl monomers in organic media. Hydroxy radicals on reaction with AMS give significant yields of products from head addition, abstraction and aromatic substitution (see Table 3.8) even though resonance and steric factors combine to favor "normal" tail addition. However, it is notable that the extents of abstraction (with AMS and MMA) are less than obtained with *t*-butoxy radicals and the amounts of head addition (with MMA and S) are no greater than those seen with benzoyloxy radicals under similar conditions. It is clear that there is no direct correlation between reaction rate and low specificity.

Yields of aromatic substitution on S and AMS obtained by Grant *et al.*[345] should be regarded as minimum yields until the efficiency of trapping of the cyclohexadienyl radicals under their reaction conditions is known. This may help reconcile the finding that, in aqueous media, aromatic substitution is reported to be the main reaction pathway.[390] Grant *et al.*[345] also found that aromatic substitution on S proceeded by preferential *para* attack. This preference agrees with the calculated relative reactivity of the ring carbons based on frontier electron densities, but is otherwise unprecedented.[391]

3.4.2.4 Sulfate radical anion

The sulfate radical anion is formed by thermal, photochemical or redox decomposition of persulfate salts. Consequently, it is usually used in aqueous solution. However, crown ether complexes or alkylammonium salts may be used to generate sulfate radical anion in organic solution (see 3.3.2.6.1).

Two pathways for the reaction of sulfate radical anion with monomers have been described (scheme 71).[233] These are: (A) direct addition to the double bond or (B) electron transfer to generate a radical cation. The radical cation may also be formed by an addition-elimination sequence. It has been postulated that the radical cation can propagate by either cationic or a radical mechanism (both mechanisms may occur simultaneously). However, in aqueous media the cation is likely to hydrate rapidly to give a hydroxyethyl chain end.

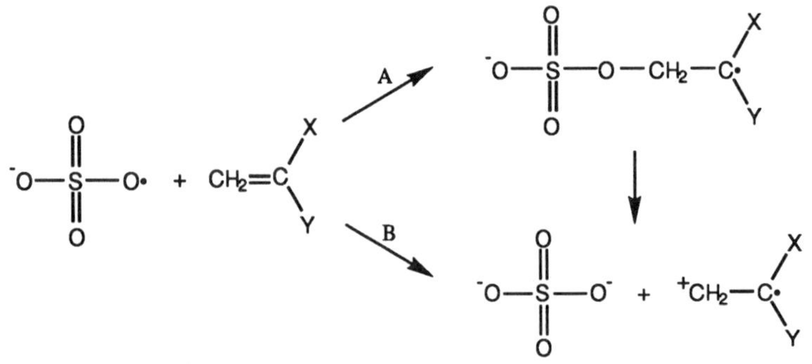

$$\text{Scheme 71}$$

The preferred initiation pathway is dependent on the particular monomer involved and the reaction conditions. Generally radical cation formation (by either mechanism) is facilitated by low pH. The failure to detect an intermediate sulfate adduct led workers to propose that reactions of the sulfate radical anion with electron-rich alkenes and S derivatives proceeded by pathway (B) over a wide range of pH and reaction conditions.[392-394] However, other workers have rationalized similar data by allowing the initial formation of a sulfate adduct (pathway A).[395] Detection of an intermediate in the reaction of sulfate radical anion with S[396] or with cyclohexene[224] clearly points to addition being a major pathway in those cases. Moreover, PS formed with persulfate initiation is known to possess a high proportion of sulfate end groups.[397-400] Thus, the bulk of available evidence suggests that in initiation of S polymerization there is initial formation of a sulfate adduct (pathway A) and that, radical cations, if formed, are produced by subsequent elimination (see Scheme 71).

In the case of electron-deficient monomers (e.g. acrylics) it is accepted that reaction occurs by initial addition of the sulfate radical anion to the monomer. Reactions of sulfate radical anion with acrylic acid derivatives have been shown to give rise to the sulfate adduct under neutral or basic conditions but under acidic conditions give the radical cation probably by an addition-elimination process.

Hydroxy radical and sulfate radical anion, though they may sometimes give rise to similar products, show quite different selectivity in their reactions with unsaturated substrates. In particular, the sulfate radical anion has a somewhat lower propensity for hydrogen abstraction than the hydroxyl radical. For example, the sulfate radical anion shows little tendency to abstract hydrogen from methacrylic acid.[233]

Sulfate radical anion may be converted to the hydroxyl radical in aqueous solution. Evidence for this pathway under polymerization conditions is the formation of a proportion of hydroxy end groups in some polymerizations.

However, the hydrolysis of sulfate radical anion at neutral pH is slow ($k=10^7$ M^{-1} s^{-1}) compared with the rate of reaction with most monomers ($k=10^8$-10^9 M^{-1} s^{-1}, see Table 3.8)[389] under typical reaction conditions. Thus, hydrolysis should only be competitive with addition when the monomer concentration is very low. The formation of hydroxy end groups in polymerizations initiated by sulfate radical anion can also be accounted for by the hydration of an intermediate radical cation or by the hydrolysis of an initially formed sulfate adduct either during the polymerization or subsequently.

3.4.2.5 Alkylperoxy radicals

Alkylperoxy radicals are generated by the reactions of carbon-centered radicals with oxygen and in the induced decomposition of hydroperoxides (Scheme 72). Their reactions have been reviewed by Howard[401] and rate constants for their self reaction and for their reaction with a variety of substrates including various inhibitors have been tabulated.[402]

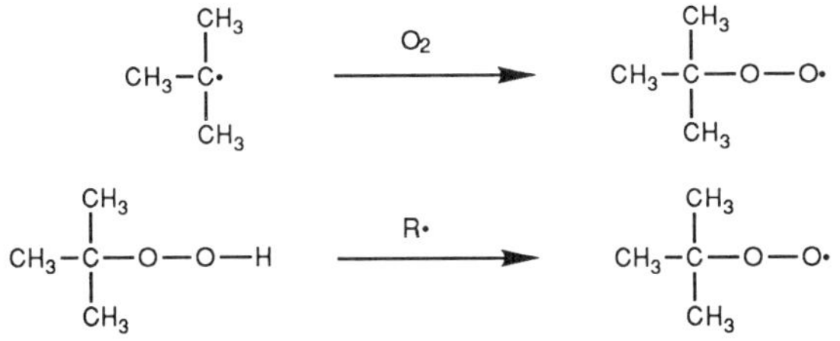

Scheme 72

Because of the importance of hydroperoxy radicals in autoxidation processes, their reactions with hydrocarbons are well known. However, reactions with monomers have not been widely studied. Absolute rate constants for addition to common monomers are in the range 0.09 - 3 M^{-1} s^{-1} at 40°C. These are substantially lower than k_i for other oxygen-centered radicals (see Table 3.7).[403]

Epoxide formation may be a side reaction occurring during initiation by *t*-butylperoxy radicals. The mechanism proposed for this process is as follows (Scheme 73).[195]

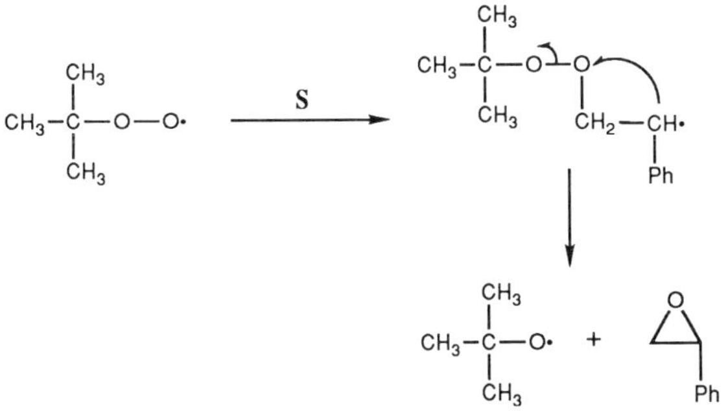

Scheme 73

3.4.3 Other Heteroatom-Centered Radicals

Various other heteroatom-centered radicals have been generated as initiating species. These include sulfur-, selenium- (see 3.4.3.1), nitrogen- and phosphorous-centered species (see 3.4.3.2). Kinetic data for reactions of these radicals with monomers is summarized in Table 3.9.

Table 3.9 Selected Rate Data for Reactions of Heteroatom-Centered Radicals

radical	Temp. °C	k_S M^{-1}s^{-1}	k/k_S[a]						refs.
			AMS	MA	MMA	AN	MAN	VAc	
C$_2$H$_5$S•	60	-	-	0.036	-	-	-	-	404
CH$_3$(CH$_2$)$_2$CH$_2$S•	60	-	-	-	0.17	-	-	-	405
PhS•	23	2.0 x 10^7	-	-	0.16	-	-	-	406
p-ClPhS•	23	5.1 x 10^7	-	-	0.10	0.0090	0.045	0.0009	407
Ph(CO)S•	22	-	-	0.03	0.12	0.0091	-	0.0025	408
PhSe•	23	2.2 x 10^6	0.76	0.0078	0.019	0.0064	0.012	0.0005	409
Ph$_2$(O)P•	b	6.0 x 10^7	-	0.58	1.33	0.33	0.83	0.027	410
Ph$_2$(O)P•	20	1.1 x 10^7	1.27	-	1.45	-	-	0.25	411
(CH$_3$O)$_2$(O)P•	b,c	2.2 x 10^8	-	0.077	0.26	0.26	0.42	0.013	410

[a] Data rounded to two significant figures.
[b] Room temperature.
[c] Similar relative reactivities for VAc and AN have been reported at 60°C.[412]

3.4.3.1 Sulfur- and selenium-centered radicals

Thiyl radicals are formed by transfer to thiols or by thermal or photochemical decomposition of disulfides (Scheme 74).

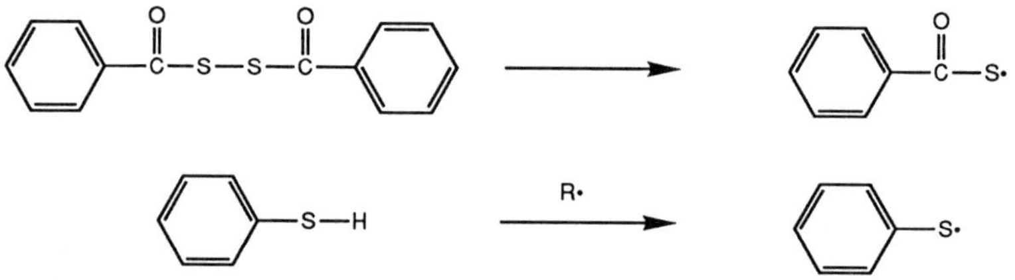

Scheme 74

Most studies have concerned the kinetics of arenethiyl radicals with monomers including S and its derivatives[413-417] and MMA.[414,418] The radicals have electrophilic character and add more rapidly to electron-rich systems (see Table 3.9). Relative reactivities of the monomers towards the benzoylthiyl radical have also been examined.[408]

It is established that the initial reaction involves predominantly tail addition to monomer.[418] There is no evidence that abstraction competes with addition. It should be noted that the addition of arenethiyl radicals to double bonds is readily reversible.

Initiation

A study on the kinetics of the reactions of phenylseleno radicals with vinyl monomers has also been reported.[409]

3.4.3.2 Phosphorous-centered radicals

Phosphinyl radicals (*e.g.* **73-76**) are generated by photodecomposition of acyl phosphinates or acyl phosphine oxides (3.3.4.1)[258,411,419] or by hydrogen abstraction from the appropriate phosphine oxide.[412]

$$\underset{(73)}{\underset{Ph}{\overset{O}{\underset{|}{Ph-\overset{\|}{P}\cdot}}}} \quad \underset{(74)}{\underset{OCH(CH_3)_2}{\overset{O}{\underset{|}{Ph-\overset{\|}{P}\cdot}}}} \quad \underset{(75)}{\underset{O^-}{\overset{O}{\underset{|}{Ph-\overset{\|}{P}\cdot}}}} \quad \underset{(76)}{\underset{OCH_3}{\overset{O}{\underset{|}{CH_3O-\overset{\|}{P}\cdot}}}}$$

The reactivities of the various phosphinyl radicals with monomers have been examined (see Table 3.9).[259,410,412] Absolute rate constants are high, lying in the range 10^6-10^8 M^{-1} s^{-1} and show some solvent dependence. The rate constants are higher in aqueous acetonitrile solvent than in methanol. The high magnitude of the rate constants has been linked to the pyramidal structure of the phosphinyl radicals.[410]

The phosphinyl radicals (**73-76**) all show nucleophilic character (*e.g.* VAc is substantially less reactive than the acrylic monomers). However, the nucleophilicity varies according to the number of oxygen substituents on phosphorous.[410,412]

3.5 Techniques

The low concentration of initiator residues in polymers formed by radical polymerization means that initiator residues can usually only be observed directly in exceptional circumstances or in very low molecular weight polymers (see 3.5.2). Thus, the study of the reactions of initiator-derived radicals with monomers has required the development of a number of new techniques. Two basic approaches have been employed. These involve:

(a) Isolation of the initiator-monomer reaction by employing a reagent designed to trap the first-formed adduct. This involves conducting the polymerization in the presence of an appropriate inhibitor (see 3.5.1).
(b) Labeling the initiator such that the initiator-derived residues in the polymer can be more readily detected and quantified by chemical or spectroscopic analysis (see 3.5.3).

3.5.1 Radical Trapping

Radical traps used for the study of radical monomer reactions should meet a number of criteria:

(a) The trap should ideally show a degree of specificity for reaction with the propagating species as opposed to the initiator-derived radicals.
(b) All products from the reaction with monomer should be trapped with equal efficiency.
(c) The trapped products should be stable under the polymerization conditions.

Various reagents have been employed as radical traps. Those most commonly encountered are summarized in Table 3.10. The advantages, limitations and applications of each are considered in the following sections.

Table 3.10 Radical Trapping Agents for Studying Initiation

Trap	Species Trapped	Section
spin traps:		
nitroso-compounds	most radicals	3.5.1.1
nitrones	most radicals	3.5.1.1
transition metal ions:		
cupric ions	nucleophilic carbon-centered radicals	3.5.1.2
ferric ions	nucleophilic carbon-centered radicals	3.5.1.2
titanous ions	electrophilic carbon-centered radicals	3.5.1.2
metal hydrides:		
mercuric hydride	electrophilic carbon-centered radicals	3.5.1.3
Group VI hydrides	carbon-centered radicals	3.5.1.3
nitroxides	carbon centered radicals	3.5.1.4

3.5.1.1 Spin traps

In spin trapping, radicals are trapped by reaction with a diamagnetic molecule to give a radical product.[420] This feature (i.e. that the free spin is retained in the trapped product) distinguishes it from the other trapping methods. The technique involves EPR detection of the relatively stable nitroxides which result from the trapping of the more transient radicals (Schemes 75, 76). No product isolation or separation is required.

The two most commonly employed spin traps are 2-methyl-2-nitrosopropane (77) (more commonly known as nitroso-*t*-butane) and phenyl *t*-butyl nitrone (78).

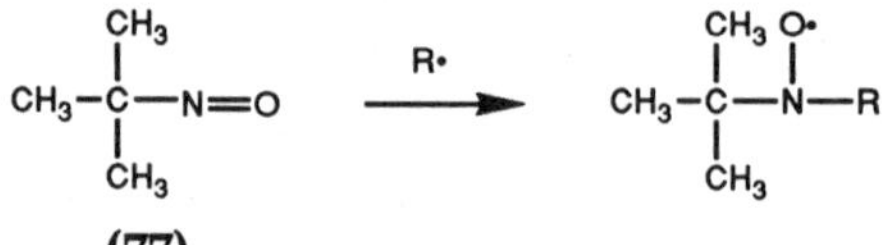

$$CH_3-\underset{\underset{CH_3}{|}}{\overset{\overset{CH_3}{|}}{C}}-N{=}O \quad \xrightarrow{R\cdot} \quad CH_3-\underset{\underset{CH_3}{|}}{\overset{\overset{CH_3}{|}}{C}}-\underset{}{\overset{\overset{O\cdot}{|}}{N}}-R$$

(77)

Scheme 75

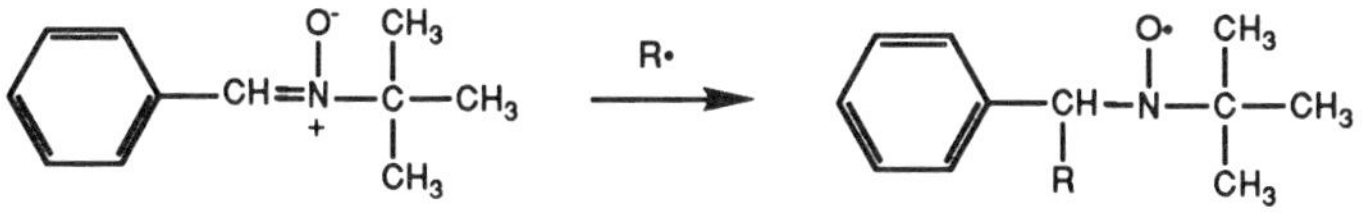

Scheme 76

(78)

Chalfont et al.[421] were the first to apply the spin trapping technique in the study of radical polymerization. They studied radicals produced during S polymerization initiated by *t*-butoxy radicals with (77) as the radical trap. Since that time many other systems have been studied using this trap (77).[12,171,408] The use of 2,4,6-tri-*t*-butylnitrosobenzene (79) in the study of polymerization, has been described by Savedoff and Ranby[422] and by Lane and Tabner.[382] This nitroso-compound is reported to be more thermally and photochemically stable than (77).

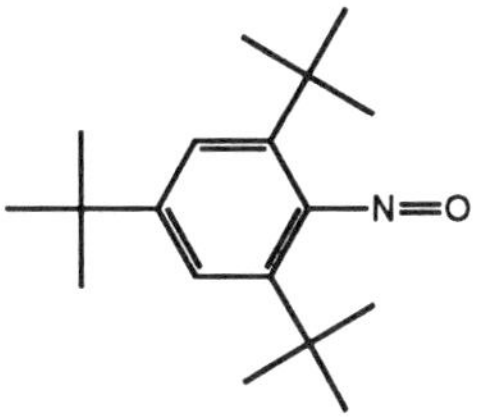

(79)

Nitrones are generally more stable than nitroso-compounds and are therefore easier to handle. However, the nitroxides formed by reaction with nitrones [e.g. phenyl *t*-butyl nitrone (78)][423,424] have the radical center one carbon removed from the trapped radical (see Scheme 76). The EPR spectra are therefore less sensitive to the nature of that radical and there is greater difficulty in resolving signals. Nitrones are generally less efficient traps than nitroso-compounds.[420]

There are several limitations on the use of the spin trapping technique which must be borne in mind, particularly when quantitative results are required. These are:

(a) Not all radicals may be trapped at equal rates or with equal efficiency.[425]
(b) The product nitroxides may not be stable under the reaction conditions. Nitroxide stability is strongly dependent on the nature of the trapped species. Nitroxides react with radicals at or near diffusion controlled rates and they can also undergo β–scission either to regenerate the trapped radical or to form a new radical.
(c) Side reactions involving the trap and the monomer may give rise to products which complicate the interpretation of the EPR spectra. Various side reactions are described in the literature:[420] the nitroso-compound (77) reacts with α–methylvinyl monomers by an ene reaction (see Scheme 77);[177]

t-butyl radicals are produced by thermal or photochemical decomposition of (77) and are trapped as di-t-butylnitroxide.

Many of the above-mentioned complications can be avoided or allowed for by carrying out appropriate control experiments. A further difficulty lies with the sensitivity of the method. Minor initiation pathways (≤5%) may be extremely difficult to detect.

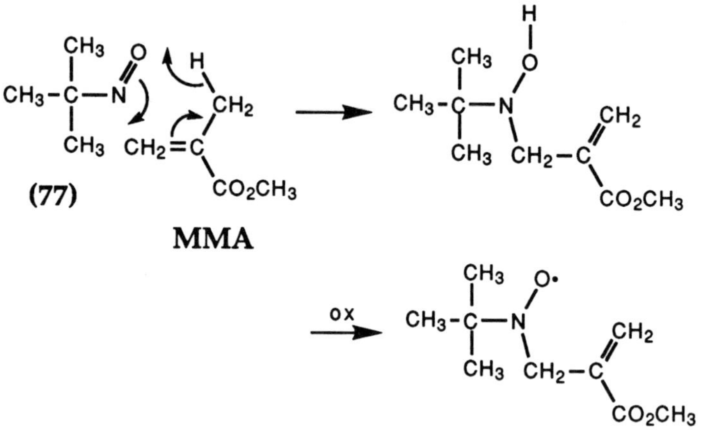

Scheme 77

3.5.1.2 Transition metal salts

Certain transition metal salts can be used as radical traps (Schemes 78, 79).[426] These include various cupric (e.g. Cu(OAc)$_2$, CuCl$_2$, Cu(SCN)$_2$),[17,157,355,381,427] ferric (e.g. FeCl$_3$),[284,428] and titanous salts (e.g. TiCl$_3$).[339] These traps react with radicals by ligand- or electron-transfer to give products which can be determined by conventional analytical techniques.

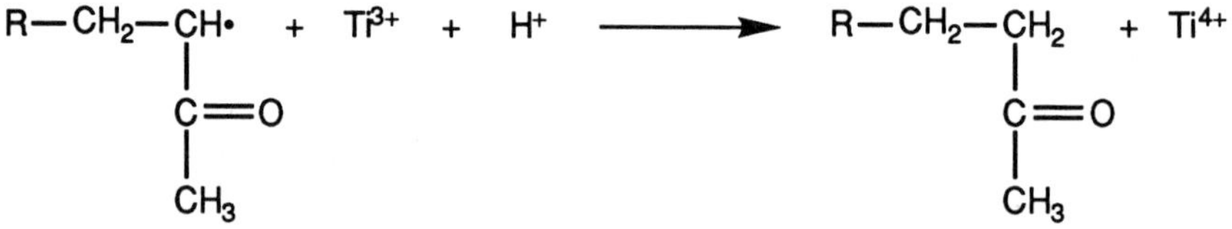

Scheme 78

The rate of oxidation/reduction of radicals is strongly dependent on radical structure. Transition metal reductants (e.g. Ti^{III}) show selectivity for electrophilic radicals (e.g. those derived by tail addition to acrylic monomers or alkyl vinyl ketones - Scheme 78)[339] while oxidants (Cu^{II}, Fe^{III}) show selectivity for nucleophilic radicals (e.g. those derived from addition to S - Scheme 79).[17] A consequence of this specificity is that the various products from the reaction of an initiating radical with monomers will not all be trapped with equal efficiency and complex mixtures can arise.

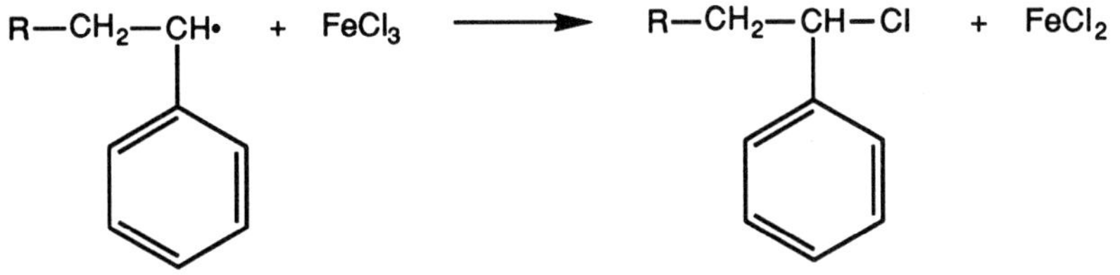

Scheme 79

3.5.1.3 Metal hydrides

Metal hydride trapping agents have been used extensively in studying the reaction of alkyl radicals with monomers.[429,430]

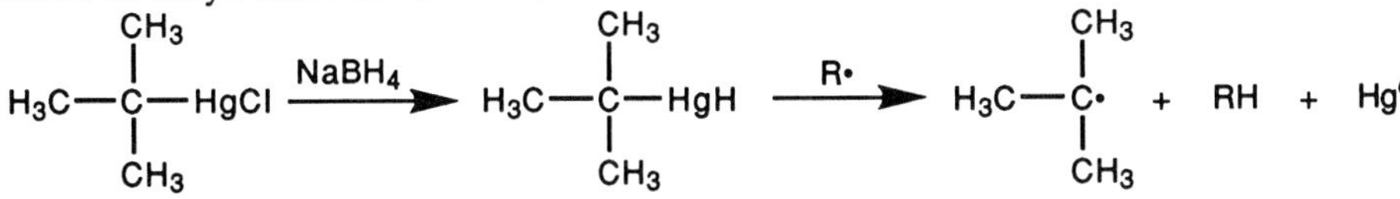

Scheme 80

Alkyl mercuric hydrides are generated *in situ* by reduction of an alkyl mercuric salt with sodium borohydride (Scheme 80). Their use as radical traps was first reported by Hill and Whitesides[431] and developed for the study of radical-olefin reactions by Giese *et al.*[429,430] and Tirrell *et al.*[432] Careful choice of reagents and conditions provides excellent yields of adducts of nucleophilic radicals (*e.g.* n-hexyl, cyclohexyl, *t*-butyl, alkoxyalkyl) to electron-deficient monomers (*e.g.* acrylics).

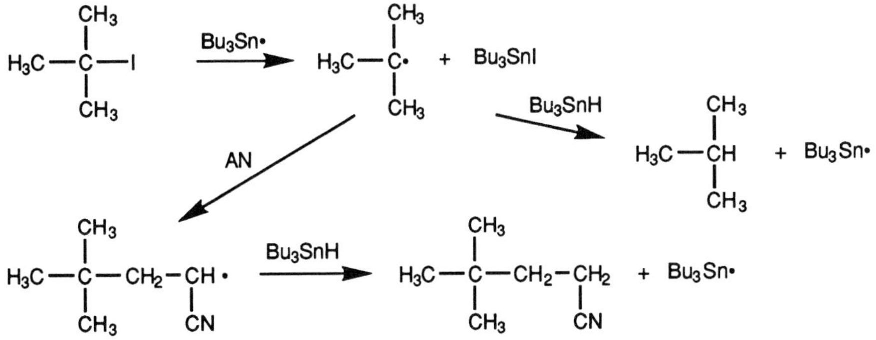

Scheme 81

A consequence of the selectivity for electrophilic radicals is that not all products are trapped with equal efficiency. Furthermore, with electron-rich monomers (*e.g.* S) oligomerization may occur. Other possible complications in the utilization of this method have been discussed by Russell.[433]

Group IV hydrides (R₃SnH, R₃GeH) have also been used as trapping reagents.[434,435] The reduction of alkyl halides by stannyl or germyl radicals affords alkyl radicals. These react with the group IV hydrides to set up a radical chain (Scheme 81).[435] The alkyl radicals may react with a substrate (*e.g.* monomer) in competition with being trapped by the hydride. Absolute rate constants for the reactions of group IV hydrides with radicals are known. Thus the H-atom transfer step may be used as a radical clock to calibrate radical-monomer reactions.[19] This technique has seen widespread use in the study of intramolecular radical reactions.[307]

One limitation of the use of the group IV hydrides as radical traps in the study of polymerization is that the stannyl and germyl radicals may themselves add monomer, *albeit* reversibly.

3.5.1.4 Nitroxides

A well known feature of the chemistry of nitroxides (*e.g.* 80-83) is that they rapidly combine with carbon-centered radicals to give alkoxyamines. This feature led to the use of nitroxides as the reagents of choice in the inhibitor method for the determination of initiator efficiency.[81]

(80) (81) (82) (83)

Rizzardo and Solomon[436] applied nitroxide chemistry to develop one of the most versatile techniques for examining the initiation step of polymerization. The method is reliant on the initiator-derived radicals either not reacting or reacting only slowly with the nitroxide while the propagating radicals are efficiently scavenged to yield stable alkoxyamines (Scheme 81). The technique has been successfully used to study the reactions of heteroatom-centered (*t*-butoxy,[7,9,11,20,21,166,177,344,352,358-362,436,437] cumyloxy,[62] isopropoxy,[112,350] ethoxy,[350] benzoyloxy,[7,9,21,34,155,177,349] isopropoxycarbonyloxy,[177] hydroxy,[236,345] thiyl, phosphinyl[412,419]) and more reactive carbon-centered radicals (methyl, undecyl, *t*-butyl, phenyl[7,9,21,34,166]). The reaction has been employed to detect radical intermediates in organic reactions and also in the identification of primary radicals produced from photoinitiators.[419]

The reaction between nitroxides and carbon-centered radicals occurs at near (but not at) diffusion controlled rates.[438-441] Rate constants and Arrhenius parameters for coupling of nitroxides and various carbon-centered radicals have been determined. The rate constants (20°C) for the reaction of (80) with primary, secondary and tertiary alkyl and benzyl radicals are 1.2, 1.0, 0.8 and 0.5 x 10^9 M^{-1} s^{-1} respectively. The corresponding rate constants for reaction of (83) are slightly higher. If due allowance is made for the afore-mentioned sensitivity to radical structure[440] and some dependence on reaction conditions,[441] the reaction can be applied as a clock reaction to estimate rate constants for reactions between carbon-centered radicals and monomers[442] or other substrates.[19]

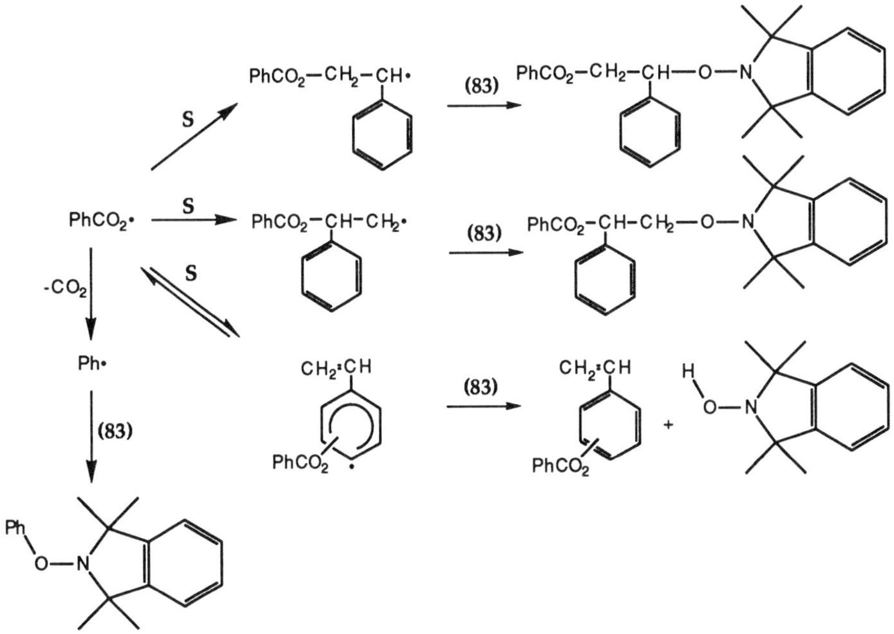

Scheme 82

Major advantages of this method over other trapping techniques are that typical conditions for solution/bulk polymerization can be employed and that a very wide range of initiating systems can be examined. The application of the technique is greatly facilitated by the use of a nitroxide possessing a UV chromophore (e.g. 81-83).

Some limitations of the method arise due to side reactions involving the nitroxide. However, such problems can usually be avoided by the correct choice of nitroxide and reaction conditions. Nitroxides, while stable in the presence of most monomers, may act as oxidants or reductants under suitable reaction conditions.[443] The induced decomposition of certain initiators (e.g. diacyl peroxides) can be a problem (Scheme 83).[155,166] There also is some evidence that nitroxides may disproportionate with alkoxy radicals bearing α-hydrogens.[112]

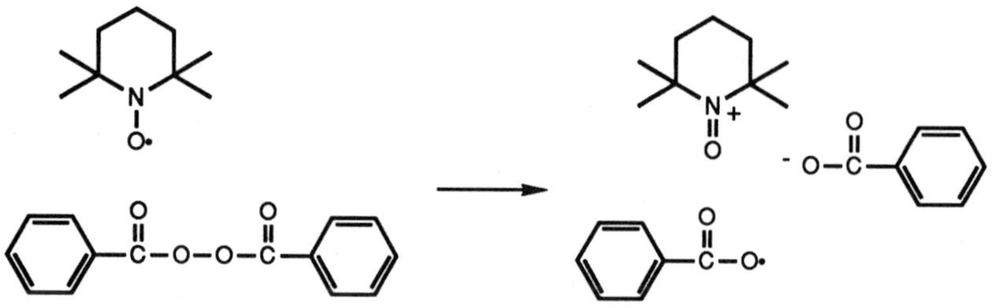

Scheme 83

Various light induced reactions including hydrogen atom abstraction, electron transfer and β-scission occur under the influence of UV light.[444-447] Certain radicals, for example cyclohexadienyl radicals (Scheme 84), are trapped by disproportionation rather than coupling.[7] Nitroxides are also reported to react by hydrogen abstraction with molecules that are extremely good hydrogen donors [*e.g.* S dimer (**65**),[448] the ketenimine (**10**)[103]].

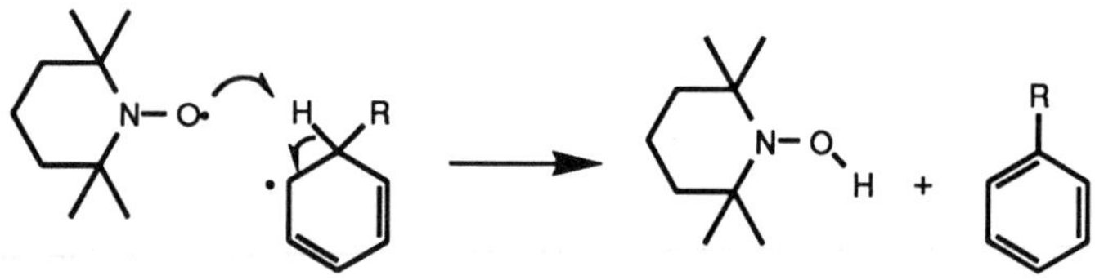

Scheme 84

The reaction of radicals with nitroxides is reversible.[277] This means that the highest temperature that the technique can reasonably be employed at is *ca.* 80°C for tertiary propagating species and *ca.* 120°C for secondary propagating species.[21] These maximum temperatures are only a guide because the stability of alkoxyamines is also dependent on solvent (polar solvents favor decomposition) and the structure of the propagating species. This chemistry has led to certain alkoxyamines being useful as initiators of living polymerization (see 7.5.3).

3.5.2 *Direct Detection of End Groups*

In favorable circumstances initiator-derived end groups may be detected by spectroscopic methods or by chemical or chromatographic analysis. Most of the methods are sensitive only to a given type of end group in a given class of polymers. However, they have the advantage that no special chemistry or isolation steps are required. The main disadvantages associated with these methods are that they require foreknowledge of what the end groups are likely to be and, in general, they can only be applied to low molecular weight polymers.

3.5.2.1 Infra-red and UV-visible spectroscopy

UV[142,449] and IR spectroscopy[83,450] have been used to study the kinetics and efficiency of initiation of polymerization. These techniques are not universally applicable. Ideally, it is required (a) that the chromophores are in a clear region of the spectrum and (b) that the positions of the absorptions are sensitive to the chemical environment of the chromophore such that end groups can be distinguished from residual initiator and initiator-derived by-products.

Buback et al.[83,450,451] applied FTIR to follow the course of the initiation of S polymerization by AIBN and to determine initiator efficiency. Contributions to the IR signal due to cyanoisopropyl end groups, AIBN, and the ketenimine can be separated using curve resolution techniques.

Garcia-Rubio et al.[142,449] have examined S and MMA polymerizations initiated by BPO and have shown that UV can be used to quantitatively determine aliphatic and aromatic benzoate groups in MMA and S polymerizations.

3.5.2.2 Nuclear magnetic resonance spectroscopy

The sensitivity of modern NMR allows initiator residues to be determined directly in polymers of moderate molecular weight where the desired signals are discrete from those of the backbone carbons.[30,452-455] The molecular weight limit is imposed both by sensitivity and the dynamic range of the spectrometer.

In some cases it is possible to suppress NMR signals due to backbone carbons or hydrogens thus allowing obscured end group resonances to be observed. Four basic methods have been described in the literature. These are:

(a) Subtraction of the spectrum of an exactly similar polymer but without the defect structure being sought.[328,456] The procedure has the disadvantages that noise is added to the spectrum and that it requires preparation of a reference polymer. The method does not alleviate the dynamic range problems discussed above.

(b) Use of a Hahn spin echo experiment to suppress signals from backbone atoms. Johns et al.[6,457] demonstrated that end group signals usually persist longer than backbone signals because of longer T_2 relaxation times. The method was applied to detect obscured cyanoisopropyl end groups in PMMA.

(c) Use of pulse sequences that selected for the number of attached hydrogens. For example, for PS prepared with AIBN a 'quaternary only' pulse sequence can be used to better visualize signals due to the quaternary carbons of the AIBN-derived residues.[6]

(d) Analysis of polymers prepared from NMR-inactive monomers. Hatada et al. have used ^{1}H NMR to determine end groups in perdeuterated PS[458] and

PMMA.[459,460] Similarly, the use of ^{12}C-enriched monomers has been envisaged as an aid in detecting end groups in ^{13}C NMR experiments.[461]

These four techniques rely on suppressing signals due to the backbone carbons. The end group signals are not enhanced. Therefore, the sensitivity problems associated with detecting end groups in high molecular weight polymers are not entirely solved. However, the latter three methods (b-d) do allow acquisition at higher spectrometer gain settings and assist in overcoming spectrometer dynamic range problems. A drawback of the pulse sequence methods is that quantification may not be a straightforward exercise.

Selective labeling of the initiator with ^{13}C allows substantial enhancement of the signals of the initiator residues relative to signals due to the backbone in ^{13}C NMR spectra. This technique is described in section 3.5.3.2.

3.5.2.3 *Electron paramagnetic resonance spectroscopy*

The application of EPR in the detection and quantification of species formed by spin-trapping the products of radical-monomer reactions is described in section 3.5.1.1. The direct detection of the products of radical-monomer reactions can also yield valuable information.[386,388,462,463] Fischer *et al.*[323,331,464,465] have applied time-resolved EPR spectroscopy to study intermolecular radical-alkene reactions in solution. Absolute rate constants for the reactions of *t*-butyl[331] and cyanoisopropyl[323] radicals with monomers have been reported (see Table 3.6).

3.5.2.4 *Mass spectrometry*

New mass spectrometric techniques allow analysis of polymers of relatively high molecular weight. Meisters *et al.*[466] have reported that fast atom bombardment mass spectrometry (FAB-MS) can be applied in the analysis of MMA oligomers to at least hexadecamer. Laser desorption mass spectrometry offers new possibilities for analysis of polymers with molecular weights in the tens of thousands.

For polymers that degrade by unzipping, pyrolysis GCMS can provide extremely useful data. Farina *et al.*[405,467] and Ohtani *et al.*[468] have described the application of pyrolysis GCMS to determine end groups in PMMA.

3.5.2.5 *Chemical methods*

Chemical analysis often allows end groups to be determined with high precision. Techniques have been developed for the chemical derivatization of polymer end groups so they can be more readily measured by spectrophotometric methods. One of the most used is the dye-partition method introduced by Palit.[400,469-471] Variants of this method have now been applied to detect hydroxy,[471,472] amino and sulfate end groups.[399,400] A two step dealkylation-derivatization method[473] has also been successfully used for determining *t*-

butoxy end groups in PS. In that case the t-butoxy ends were first cleaved with trifluoroacetic acid to give hydroxy chain ends. This method was not applicable to PMMA. It was found the t-butoxy ends could be determined by measuring the release of *t*-butyl chloride formed on treating the polymer with boron trichloride.[473]

3.5.3 Labeling Techniques

Various methods have been described whereby polymers are formed with an initiator which is labeled with chromophores or other functionality to permit ready detection of initiator-derived end groups by chemical or spectroscopic methods.[474,475] A potential disadvantage of this procedure is that the initiator is chemically modified and the specificity shown by the initiator-derived radicals may be different from that of the corresponding unlabeled species.

The best labeling system in this regard is isotopic labeling since it involves the minimum change from the standard initiator. Methods based on radiolabeling (3.5.2.2) and stable isotopes (detectable by NMR) are described in sections 3.5.2.1 and 3.5.3.2 respectively.

3.5.3.1 Radiolabeling

Polymer formed using radiolabeled initiators may be isolated and analyzed to determine the concentration of initiator-derived residues and calculate the initiator efficiency. Radiolabeled initiators have also been used extensively to establish the relative reactivity of monomers towards radicals.[77,96,475-477]

Radiolabeling offers greater sensitivity than most other labeling methods. However, the technique has the disadvantage that end groups formed by initiation cannot be directly distinguished from initiator residues produced by other processes (*e.g.* primary radical termination or copolymerization of initiator by-products) or from residual initiator. In general, the method gives the total initiator residues in the polymer. Analysis of the kinetics of polymerization can help to resolve these problems. A further disadvantage is that polymer isolation and purification is required.

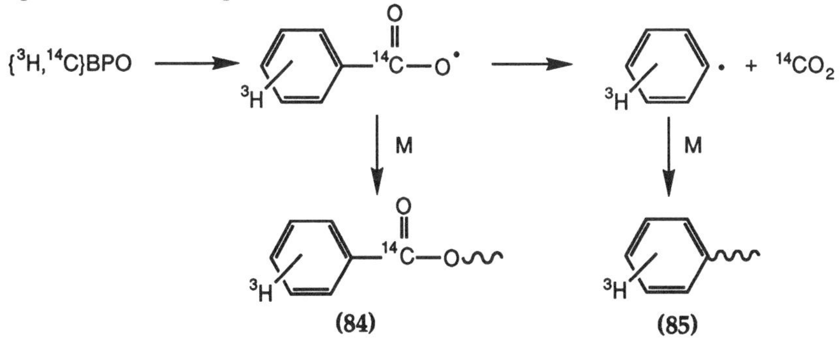

Scheme 85

For the case of initiators which produce both primary and secondary radicals (*e.g.* BPO) use of a doubly labeled initiator allows the different types of end groups to be distinguished [*e.g.* (84) and (85) - see Scheme 85] and the reactivities of monomers towards the primary radicals to be readily established by using the fragmentation step as a clock reaction.[347,478]

3.5.3.2 Stable isotopes and nuclear magnetic resonance

NMR methods can be applied to give quantitative determination of initiator-derived and other end groups and provide a wealth of information on the polymerization process. They provide a chemical probe of the detailed initiation mechanism and a greater understanding of polymer properties. The main advantage of NMR methods over alternative techniques for initiator residue detection is that NMR signals (in particular ^{13}C NMR) are extremely sensitive to the structural environment of the initiator residue. This means that functionality formed by tail addition, head addition, transfer to initiator or primary radical termination, and various initiator-derived by-products can be distinguished.

Selective labeling of the initiator allows substantial enhancement of the signals of the initiator residues relative to the signals due to the backbone. While a number of stable isotopes have been employed in this context (including D, F and ^{15}N), most work has involved the use of ^{13}C-labeling. The method has been reviewed.[455,479,480] The power of the technique is illustrated by the fact that one experiment allows the determination of:

(a) The total fate of the initiator as a function of conversion (initiator efficiency, nature and amount of by-products).
(b) The chain ends (reactivity of primary radicals towards monomers, head *vs.* tail addition, *etc.*).
(c) The rate of polymerization.
(d) The number average molecular weight = ([end groups]/[monomer used]).

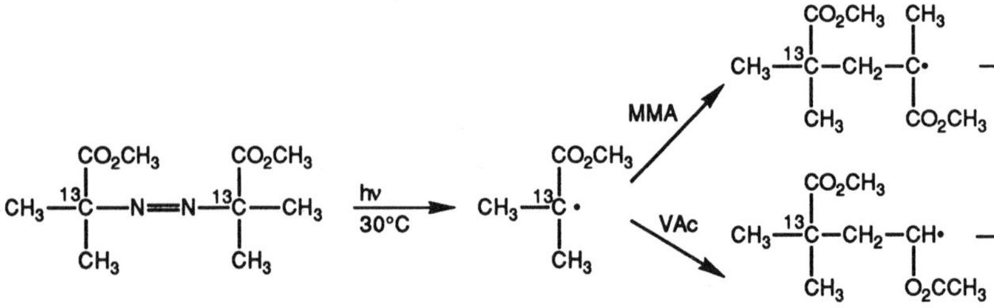

Scheme 86

The use of labeled initiators to evaluate the relative reactivity of monomers towards radicals was pioneered by Bevington *et al.*[317] The method involves determination of the relative concentrations of the end groups formed by addition to two monomers [eg. (86) and (87)] in a binary copolymer formed with use of a labeled initiator. For example, when AIBMe-α-^{13}C is used to initiate copolymerization of MMA and VAc (Scheme 86),[315] the following simple relationship gives the relative rate constants for addition to the two monomers.

$$\frac{k_{MMA}}{k_{VAc}} = \frac{[VAc].[86]}{[MMA].[87]}$$

A further use of ^{13}C-labeled initiators is in assessing the kinetics and efficiency of initiation.[13,28,37,50,70] This application requires that the polymer end groups, residual initiator, and various initiator-derived by-products should each give rise to discrete signals in the NMR spectrum. So far this method has been demonstrated for homo- and copolymerizations of S and MMA prepared with AIBN-α-^{13}C, AIBMe-α-^{13}C or BPO-*carbonyl*-^{13}C/BPO-*ring*-^{13}C (1:1) as initiator.

References

1. Solomon, D.H., and Moad, G., *Makromol. Chem., Macromol. Symp.*, 1987, **10/11**, 109.
2. Solomon, D.H., Cacioli, P., and Moad, G., *Pure Appl. Chem.*, 1985, **57**, 985.
3. Hwang, E.F.J., and Pearce, E.M., *Polym. Eng. Rev.*, 1983, **2**, 319.
4. Mita, I., in 'Aspects of Degradation and Stabilization of Polymers' (Ed. Jellineck, H.H.G.), p. 247 (Elsevier: Amsterdam 1978).
5. Solomon, D.H., *J. Macromol. Sci., Chem.*, 1982, **A17**, 337.
6. Moad, G., Solomon, D.H., Johns, S.R., and Willing, R.I., *Macromolecules*, 1984, **17**, 1094.
7. Moad, G., Rizzardo, E., and Solomon, D.H., *Macromolecules*, 1982, **15**, 909.
8. Moad, G., Solomon, D.H., Johns, S.R., and Willing, R.I., *Macromolecules*, 1982, **15**, 1188.
9. Moad, G., Rizzardo, E., and Solomon, D.H., *Aust. J. Chem.*, 1983, **36**, 1573.
10. Moad, G., Rizzardo, E., and Solomon, D.H., *Polym. Bull. (Berlin)*, 1984, **12**, 471.
11. Griffiths, P.G., Rizzardo, E., and Solomon, D.H., *J. Macromol. Sci., Chem.*, 1982, **A17**, 45.
12. Sato, T., and Otsu, T., *Makromol. Chem.*, 1977, **178**, 1941.
13. Krstina, J., Moad, G., and Solomon, D.H., *Eur. Polym. J.*, 1989, **25**, 767.
14. Singh, M., and Nandi, U.S., *J. Polym. Sci., Polym. Lett. Ed.*, 1979, **17**, 121.
15. Schildknecht, C.E., in 'Polymerization Processes' (Eds. Schildknecht, C.E., and Skeist, I.), p. 88 (Wiley: New York 1977).
16. Boundy, R.H., and Boyer, R.F., Styrene (Reinhold: New York 1952).
17. Citterio, A., Arnoldi, A., and Minisci, F., *J. Org. Chem.*, 1979, **44**, 2674.
18. Griller, D., and Ingold, K.U., *Acc. Chem. Res.*, 1980, **13**, 317.
19. Newcomb, M., *Tetrahedron*, 1993, **49**, 1151.

20. Grant, R.D., Griffiths, P.G., Moad, G., Rizzardo, E., and Solomon, D.H., *Aust. J. Chem.*, 1983, **36**, 2447.
21. Bednarek, D., Moad, G., Rizzardo, E., and Solomon, D.H., *Macromolecules*, 1988, **21**, 1522.
22. Walling, C., and Thaler, W., *J. Am. Chem. Soc.*, 1961, **83**, 3877.
23. Elson, I.H., Mao, S.W., and Kochi, J.K., *J. Am. Chem. Soc.*, 1975, **97**, 335.
24. Maillard, B., Ingold, K.U., and Scaiano, J.C., *J. Am. Chem. Soc.*, 1983, **105**, 5095.
25. Hartzler, H.D., in 'The Chemistry of the Cyano Group' (Ed. Rappoport, Z.), p. 671 (Wiley: London 1970).
26. Bevington, J.C., and Troth, H.G., *Trans. Faraday Soc.*, 1962, **58**, 186.
27. Niki, E., Kamiya, Y., and Ohta, N., *Bull. Chem. Soc. Japan*, 1969, **42**, 3220.
28. Moad, G., Rizzardo, E., Solomon, D.H., Johns, S.R., and Willing, R.I., *Makromol. Chem., Rapid Commun.*, 1984, **5**, 793.
29. Achilias, D.S., and Kiparissides, C., *Macromolecules*, 1992, **25**, 3739.
30. Starnes, W.H., Jr., Plitz, I.M., Schilling, F.C., Villacorta, G.M., Park, G.S., and Saremi, A.H., *Macromolecules*, 1984, **17**, 2507.
31. Ishiwata, H., Inoue, T., and Yoshihira, K., *J. Chromatogr.*, 1986, **370**, 275.
32. Simionescu, C.I., Chiriac, A.P., and Chiriac, M.V., *Polymer*, 1993, **34**, 3917.
33. Turro, N.J., and Kraeutler, B., *Acc. Chem. Res.*, 1980, **13**, 239.
34. Moad, G., Rizzardo, E., and Solomon, D.H., *J. Macromol. Sci., Chem.*, 1982, **A17**, 51.
35. Morrison, B.R., Maxwell, I.A., Gilbert, R.G., and Napper, D.H., *ACS Symp. Ser.*, 1992, **492**, 28.
36. Sheppard, C.S., and Kamath, V.R., *Polym. Eng. Sci.*, 1979, **19**, 597.
37. Moad, G., and Solomon, D.H., in 'Comprehensive Polymer Science' (Eds. Eastmond, G.C., Ledwith, A., Russo, S., and Sigwalt, P.), Vol. 3, p. 97 (Pergamon: London 1989).
38. Barton, J., and Borsig, E., Complexes in Free Radical Polymerization (Elsevier: Amsterdam 1988).
39. Masson, J.C., in 'Polymer Handbook, 3rd Edition' (Eds. Brandup, J., and Immergut, E.H.), p. II/1 (Wiley: New York 1989).
40. Engel, P.S., *Chem. Rev.*, 1980, **80**, 99.
41. Sheppard, C.S., in 'Encyclopaedia of Polymer Science and Engineering', 2nd Edn (Eds. Mark, H.F., Bikales, N.M., Overberger, C.G., and Menges, G.), Vol. 2, p. 143 (Wiley: New York 1985).
42. Koga, G., Koga, N., and Anselme, J.-P., in 'The Chemistry of the Hydrazo, Azo and Azoxy Groups' (Ed. Patai, S.), Vol. 16, part 2, p. 861 (Wiley: London 1975).
43. Koenig, T., in 'Free Radicals' (Ed. Kochi, J.K.), Vol. 1, p. 113 (Wiley-Interscience: New York 1973).
44. Smith, P.A.S., in 'The Chemistry of Open Chain Organic Nitrogen Compounds', Vol. 2, p. 269 (Benjamin: New York 1966).

45. Neuman, R.C., Jr., Grow, R.H., Binegar, G.A., and Gunderson, H.J., *J. Org. Chem.*, 1990, **55**, 2682.
46. Ayscough, P.B., Brooks, B.R., and Evans, H.E., *J. Phys. Chem.*, 1964, **68**, 3889.
47. Timberlake, J.W., in 'Substituent Effects in Radical Chemistry' (Eds. Viehe, H.G., Janousek, Z., and Merenyi, R.), p. 271 (Reidel: Dordecht 1986).
48. Van-Hook, J.P., and Tobolsky, S., *J. Am. Chem. Soc.*, 1958, **80**, 779.
49. Barbe, W., and Rüchardt, C., *Makromol. Chem.*, 1983, **184**, 1235.
50. Krstina, J., Moad, G., Willing, R.I., Danek, S.K., Kelly, D.P., Jones, S.L., and Solomon, D.H., *Eur. Polym. J.*, 1993, **29**, 379.
51. Talat-Erben, M., and Bywater, S., *J. Am. Chem. Soc.*, 1954, **77**, 3712.
52. Otsu, T., and Yamada, B., *J. Macromol. Sci. Chem*, 1969, **A3**, 187.
53. Krstina, J., Moad, G., and Solomon, D.H., to be submitted.
54. Nelsen, S.F., and Bartlett, P.D., *J. Am. Chem. Soc.*, 1966, **88**, 137.
55. Martin, J.C., and Timberlake, J.W., *J. Am. Chem. Soc.*, 1970, **92**, 978.
56. Overberger, C.G., and Berenbaum, M.B., *J. Am. Chem. Soc.*, 1953, **75**, 2078.
57. Duisman, W., and Rüchardt, C., *Chem. Ber.*, 1978, **111**, 596.
58. Adler, M.G., and Leffler, J.E., *J. Am. Chem. Soc.*, 1954, **76**, 1425.
59. Cohen, S.G., Cohen, F., and Wang, C.H., *J. Org. Chem.*, 1963, **28**, 1479.
60. Solomon, S., Wang, C.H., and Cohen, S.G., *J. Am. Chem. Soc.*, 1957, **79**, 4104.
61. Kiefer, H., and Traylor, T.G., *Tetrahedron Lett.*, 1966, 6163.
62. Rizzardo, E., Serelis, A.K., and Solomon, D.H., *Aust. J. Chem.*, 1982, **35**, 2013.
63. Overberger, C.G., Hale, W.F., Berenbaum, M.B., and Finestone, A.B., *J. Am. Chem. Soc.*, 1954, **76**, 6185.
64. Duisman, W., and Rüchardt, C., *Tetrahedron Lett.*, 1974, 4517.
65. Rüchardt, C., *Top. Curr. Chem.*, 1980, **88**, 1.
66. Firestone, R.A., *J. Org. Chem.*, 1980, **45**, 3604.
67. Wolf, R.A., *ACS Symp. Ser.*, 1989, **404**, 416.
68. Henrici-Olivé, G., and Olivé, S., *Makromol. Chem.*, 1962, **58**, 188.
69. Tanaka, H., Fukuoka, K., and Ota, T., *Makromol. Chem., Rapid, Commun.*, 1985, **6**, 563.
70. Spurling, T.H., Deady, M., Krstina, J., and Moad, G., *Makromol. Chem., Macromol. Symp.*, 1991, **51**, 127.
71. Stickler, M., *Makromol. Chem.*, 1986, **187**, 1765.
72. O'Driscoll, K.F., and Huang, J., *Eur. Polym. J.*, 1989, **7/8**, 629.
73. Lyons, R.A., Moad, G., and Senogles, E., *Eur. Polym. J.*, 1993, **29**, 389.
74. Lyons, R.A., Moad, G., and Senogles, E., in 'Pacific Polymer Conference Preprints', Vol. 3, p. 249 (Polymer Division, Royal Australian Chemical Insitute: Brisbane 1993).
75. Krstina, J., Moad, G., and Solomon, D.H., *Polym. Bull. (Berlin)*, 1992, **27**, 425.

76. Drewer, R.J., in 'The Chemistry of the Hydrazo, Azo and Azoxy Groups' (Ed. Patai, S.), Vol. 16, part 2, p. 935 (Wiley: London 1975).
77. Ayrey, G., *Chem. Rev.*, 1963, **63**, 645.
78. Fink, J.K., *J. Polym. Sci., Polym. Chem. Ed.*, 1983, **21**, 1445.
79. Cox, R.A., and Buncel, E., in 'The Chemistry of the Hydrazo, Azo and Azoxy Groups' (Ed. Patai, S.), Vol. 16, part 2, p. 9 (Wiley: London 1975).
80. Braun, D., and Czerwinski, W.K., *Makromol. Chem.*, 1987, **188**, 2371.
81. Fukuda, T., Ma, Y.-D., and Inagaki, H., *Macromolecules*, 1985, **18**, 17.
82. Russell, G.T., Napper, D.H., and Gilbert, R.G., *Macromolecules*, 1988, **21**, 2141.
83. Buback, M., Huckestein, B., Kuchta, F.-D., Russell, G.T., and Schmid, E., *Macromol. Chem. Phys.*, 1994, **195**, 2117.
84. Sack, R., Schulz, G.V., and Meyerhoff, G., *Macromolecules*, 1988, **21**, 3346.
85. Faldi, A., Tirrell, M., Lodge, T.P., and von Meewwall, E., *Macromolecules*, 1994, **27**, 4184.
86. Bizilj, S., Kelly, D.P., Serelis, A.K., Solomon, D.H., and White, K.E., *Aust. J. Chem.*, 1985, **38**, 1657.
87. Trecker, D.J., and Foote, R.S., *J. Org. Chem.*, 1968, **33**, 3527.
88. Kodaira, K., Ito, K., and Iyoda, S., *Polym. Commun.*, 1987, **28**, 86.
89. Talat-Erben, M., and Bywater, S., *J. Am. Chem. Soc.*, 1954, **77**, 3710.
90. Jaffe, A.B., Skinner, K.J., and McBride, J.M., *J. Am. Chem. Soc.*, 1972, **94**, 8510.
91. Hammond, G.S., Trapp, O.D., Keys, R.T., and Neff, D.L., *J. Am. Chem. Soc.*, 1959, **81**, 4878.
92. Chung, R.P.-T., Danek, S.K., Quach, C., and Solomon, D.H., *J. Macromol. Sci., Chem.*, 1994, **A31**, 329.
93. Pryor, W.A., and Fiske, T.R., *Macromolecules*, 1969, **2**, 62.
94. Cascaval, C.N., Straus, S., Brown, D.W., and Florin, R.E., *J. Polym. Sci., Polym. Symp.*, 1976, **57**, 81.
95. Baysal, B., and Tobolsky, A.V., *J. Polym. Sci.*, 1952, **8**, 529.
96. Bevington, J.C., and Lewis, T.D., *Polymer*, 1960, **1**, 1.
97. Pryor, W.A., and Coco, J.H., *Macromolecules*, 1970, **3**, 500.
98. May, J.A., Jr., and Smith, W.B., *J. Phys. Chem.*, 1968, **72**, 2993.
99. Ayrey, G., and Haynes, A.C., *Makromol. Chem.*, 1974, **175**, 1463.
100. Athey, R.D., *J. Polym. Sci., Polym. Chem. Ed.*, 1977, **15**, 1517.
101. Braks, J.G., and Huang, R.Y.M., *J. Appl. Polym. Sci.*, 1978, **22**, 3111.
102. Bevington, J.C., Breuer, S.W., Heseltine, E.N.J., Huckerby, T.N., and Varma, S.C., *J. Polym. Sci., Part A: Polym. Chem.*, 1987, **25**, 1085.
103. Hill, D.J.T., O'Donnell, J.H., and O'Sullivan, P.W., *Prog. Polym. Sci.*, 1982, **8**, 215.
104. Quinga, E.M.Y., and Mendenhall, G.D., *J. Org. Chem.*, 1985, **50**, 2836.
105. Dulog, L., and Klein, P., *Chem. Ber.*, 1971, **104**, 902.
106. Protasiewicz, J., and Mendenhall, G.D., *J. Org. Chem.*, 1985, **50**, 3220.

107. Mendenhall, G.D., Stewart, L.C., and Scaiano, J.C., *J. Am. Chem. Soc.*, 1982, **104**, 5109.
108. Mendenhall, G.D., and Quinga, E.M.Y., *Int. J. Chem. Kinet.*, 1985, **17**, 1187.
109. Druliner, J.D., Krusic, P.D., Lehr, G.F., and Tolman, C.A., *J. Org. Chem.*, 1985, **50**, 5838.
110. Kiefer, H., and Traylor, T.G., *J. Am. Chem. Soc.*, 1967, **89**, 6667.
111. Mendenhall, G.D., and Cary, L.W., *J. Org. Chem.*, 1975, **40**, 1646.
112. Busfield, W.K., Jenkins, I.D., Rizzardo, E., Solomon, D.H., and Thang, S.H., *J. Chem. Soc., Perkin Trans. 1*, 1991, 1351.
113. Ando, W., 'Organic Peroxides' (Wiley: Chichester 1992).
114. Sheppard, C.S., in 'Encyclopaedia of Polymer Science and Engineering', 2nd Edn (Eds. Mark, H.F., Bikales, N.M., Overberger, C.G., and Menges, G.), Vol. 11, p. 1 (Wiley: New York 1987).
115. Patai, S., 'The Chemistry of Peroxides', p 1006 (Wiley: Chichester, UK 1983).
116. Swern, D., 'Organic Peroxides', Vol. 1 (Wiley-Interscience: New York 1970).
117. Swern, D., 'Organic Peroxides', Vol. 2 (Wiley-Interscience: New York 1971).
118. Swern, D., 'Organic Peroxides', Vol. 3 (Wiley-Interscience: New York 1971).
119. Davies, A.G., Organic Peroxides (Butterworths: London 1961).
120. Hawkins, E.G.R., Organic Peroxides - Their Formation and Reactions (Van Nostrand: Princeton 1961).
121. Fujimori, K., in 'Organic Peroxides' (Ed. Ando, W.), p. 319 (Wiley: Chichester 1992).
122. Bouillion, G., Lick, C., and Schank, K., in 'The Chemistry of the Peroxides' (Ed. Patai, S.), p. 287 (Wiley: London 1983).
123. Hiatt, R., in 'Organic Peroxides' (Ed. Swern, D.), Vol. 2, p. 799 (Wiley-Interscience: New York 1971).
124. Janzen, E.G., Evans, C.A., and Nishi, Y., *J. Am. Chem. Soc.*, 1972, **94**, 8236.
125. Bawn, C.E.H., and Halford, R.G., *Trans. Faraday Soc.*, 1955, **51**, 780.
126. Yamada, M., Kitagawa, K., and Komai, T., *Plast. Ind. News*, 1971, **17**, 131.
127. Blomquist, A.T., and Ferris, A., *J. Am. Chem. Soc.*, 1951, **73**, 3412.
128. Bartlett, P.D., Benzing, E.P., and Pincock, R.E., *J. Am. Chem. Soc.*, 1960, **82**, 1762.
129. Hiatt, R., Mill, T., Irwin, K.C., and Castleman, J.K., *J. Org. Chem.*, 1968, **33**, 1421.
130. Huyser, E.S., and VanScoy, R., *J. Org. Chem.*, 1968, **33**, 3524.
131. Hiatt, R., and Strachan, W.M.J., *J. Org. Chem.*, 1963, **28**, 1893.
132. Kolthoff, I.M., and Miller, I.K., *J. Am. Chem. Soc.*, 1951, **73**, 3055.
133. Martin, J.C., and Hargis, J.H., *J. Am. Chem. Soc.*, 1969, **91**, 5399.
134. Pryor, W.A., Morkved, E.H., and Bickley, H.T., *J. Org. Chem.*, 1972, **37**, 1999.
135. Nozaki, K., and Bartlett, P.D., *J. Am. Chem. Soc.*, 1946, **68**, 1686.
136. Sheldon, R.A., and Kochi, J.K., *J. Am. Chem. Soc.*, 1970, **92**, 4395.
137. Saltiel, J., and Curtis, H.C., *J. Am. Chem. Soc.*, 1971, **93**, 2056.

138. Yamauchi, S., Hirota, N., Takahara, S., Sakuragi, H., and Tokumaru, K., *J. Am. Chem. Soc.*, 1985, **107**, 5021.
139. Grossi, L., Lusztyk, J., and Ingold, K.U., *J. Org. Chem.*, 1985, **50**, 5882.
140. Lefort, D., and Nedelec, J.Y., *Tetrahedron*, 1980, **36**, 3199.
141. Rosenthal, I., Mossoba, M.M., and Riesz, P., *J. Magn. Reson.*, 1982, **47**, 200.
142. Garcia-Rubio, L.H., and Mehta, J., *ACS Symp. Ser.*, 1986, 202.
143. Garcia-Rubio, L.H., Ro, N., and Patel, R.D., *Macromolecules*, 1984, **17**, 1998.
144. Navolokina, R.A., Zilberman, E.N., Krasavina, N.B., and Kharitonova, O.A., *Izv. Vyssh. Uchebn. Zaved. Khim. Khim. Tekhnol.*, 1986, **29**, 83.
145. Stickler, M., and Dumont, E., *Makromol. Chem.*, 1986, **187**, 2663.
146. Walling, C., Waits, H.P., Milovanovic, J., and Pappiaonnou, C.G., *J. Am. Chem. Soc.*, 1970, **92**, 4927.
147. Sivaram, S., Singhae, R.K., and Bhardwaj, I.S., *Polym. Bull. (Berlin)*, 1980, **3**, 27.
148. Curci, R., and Edwards, J.O., in 'Organic Peroxides' (Ed. Swern, D.), Vol. 1, p. 200 (Wiley: New York 1971).
149. Mayo, F.R., Gregg, R.A., and Matheson, M.S., *J. Am. Chem. Soc.*, 1951, **73**, 1691.
150. Berger, K.C., Deb, P.C., and Meyerhoff, G., *Macromolecules*, 1977, **10**, 1075.
151. Anisimov, Y.N., Ivanchev, S.S., and Yurzhenko, A.I., *Polym. Sci. USSR (Engl. Transl.)*, 1967, **9**, 692.
152. Moad, G., Solomon, D.H., and Willing, R.I., *Macromolecules*, 1988, **21**, 855.
153. Suehiro, T., Kanoya, A., Yamauchi, T., Komori, T., and Igeta, S.-I., *Tetrahedron*, 1968, **24**, 1551.
154. Suehiro, T., Kanoya, A., Hara, H., Nakahama, T., and Komori, T., *Bull. Chem. Soc. Japan*, 1967, **40**, 668.
155. Moad, G., Rizzardo, E., and Solomon, D.H., *Tetrahedron Lett.*, 1981, **22**, 1165.
156. Sosnovsky, G., and Rawlinson, D.J., 'Metal Ion-Catalyzed Reactions of Symmetric Peroxides', Vol. 1, p 561 (Wiley-Interscience: New York 1970).
157. Kochi, J.K., *J. Am. Chem. Soc.*, 1962, **84**, 1572.
158. Kurz, M.E., and Kovacic, P., *J. Org. Chem.*, 1968, **33**, 1950.
159. Chalfont, G.R., Hey, D.H., Liang, K.S.Y., and Perkins, M.J., *J. Chem. Soc., Chem. Commun.*, 1967, 367.
160. Chalfont, G.R., Hey, D.H., Liang, K.S.Y., and Perkins, M.J., *J. Chem. Soc. (B)*, 1971, 233.
161. Rusakova, A., and Margaritova, M.F., *Vysokomol. Soedin. Ser. B*, 1967, **9**, 515. (Chem. Abstr. 1967, 67, 73912s)
162. Jones, R.G., *J. Chem. Soc., Chem. Commun.*, 1972, 22.
163. Bevington, J.C., Dyball, C.J., and Leech, J., *Makromol. Chem.*, 1977, **178**, 2741.
164. Bevington, J.C., Dyball, C.J., and Leech, J., *Makromol. Chem.*, 1979, **180**, 657.
165. Sato, T., Abe, M., and Otsu, T., *Makromol. Chem.*, 1977, **178**, 1259.

166. Bottle, S., Busfield, W.K., Jenkins, I.D., Thang, S., Rizzardo, E., and Solomon, D.H., *Eur. Polym. J.*, 1989, **25**, 671.
167. Dambatta, B.B., and Ebdon, J.R., *Eur. Polym. J.*, 1986, **22**, 783.
168. Walling, C., Free Radicals in Solution, p. 592 (Wiley: New York 1957).
169. Imoto, M., and Choe, S., *J. Polym. Sci.*, 1955, **15**, 485.
170. Pryor, W.A., and Hendrickson, W.H., Jr, *Tetrahedron Lett.*, 1983, **24**, 1459.
171. Sato, T., Kita, S., and Otsu, T., *Makromol. Chem.*, 1975, **176**, 561.
172. Strong, W.A., *Ind. Eng. Chem.*, 1964, **56(12)**, 33.
173. McBay, H.C., and Tucker, O., *J. Org. Chem.*, 1954, **19**, 869.
174. Razuvaev, G.A., Terman, L.M., and Petukhow, G.G., *Dokl. Akad. Nauk. USSR (Engl. Transl.)*, 1961, **136**, 111.
175. Cohen, S.G., and Sparrow, D.B., *J. Am. Chem. Soc.*, 1950, **72**, 611.
176. Van Sickle, D.E., *J. Org. Chem.*, 1969, **34**, 3446.
177. Cuthbertson, M.J., Moad, G., Rizzardo, E., and Solomon, D.H., *Polym. Bull. (Berlin)*, 1982, **6**, 647.
178. Pastorino, R.L., and Lewis, R.N., in 'Modern Plastics Encyclopaedia', p. 165 (McGraw-Hill: New York 1988).
179. Duynstee, E.F.J., Esser, M.L., and Schellekens, R., *Eur. Polym. J.*, 1980, **16**, 1127.
180. Crano, J., *J. Org. Chem.*, 1966, **31**, 3615.
181. Sawaki, Y., in 'Organic Peroxides' (Ed. Ando, W.), p. 426 (Wiley: Chichester 1992).
182. Singer, L.A., in 'Organic Peroxides' (Ed. Swern, D.), Vol. 1, p. 265 (Wiley-Interscience: New York 1970).
183. Niki, E., and Kamiya, Y., *J. Am. Chem. Soc.*, 1974, **96**, 2129.
184. Hiatt, R., and Traylor, T.G., *J. Am. Chem. Soc.*, 1965, **87**, 3766.
185. Gupta, S.N., Gupta, I., and Neckers, D.C., *J. Polym. Sci., Polym. Chem. Ed.*, 1981, **19**, 103.
186. Allen, N.S., Hardy, S.J., Jacobine, A., Glaser, D.M., Catalina, F., Navaratnam, S., and Parsons, B.J., in 'Radiation Curing of Polymers II' (Ed. Randell, D.R.), p. 182 (Royal Society of Chemistry: Cambridge 1990).
187. Matsugo, S., and Saito, I., in 'Organic Peroxides' (Ed. Ando, W.), p. 157 (Wiley: Chichester 1992).
188. Sheldon, R.A., in 'The Chemistry of the Peroxides' (Ed. Patai, S.), p. 161 (Wiley: London 1983).
189. Hiatt, R., in 'Organic Peroxides' (Ed. Swern, D.), Vol. 3, p. 1 (Wiley-Interscience: New York 1971).
190. Suyama, S., Sugihara, Y., Watanabe, Y., and Nakamura, T., *Polym. J. (Tokyo)*, 1992, **24**, 971.
191. Drumright, R.E., Kastl, P.E., and Priddy, D.B., *Macromolecules*, 1993, **26**, 2246.
192. Matsuyama, K., and Kimura, H., *J. Org. Chem.*, 1993, **58**, 1766.
193. Bischoff, C., and Platz, K.-H., *J. Prakt. Chem*, 1973, **315**, 175.

194. Suyama, S., Wanatabe, Y., and Sawaki, Y., *Bull. Chem. Soc. Japan*, 1990, **63**, 716.
195. Wanatabe, Y., Ishigaki, H., Okada, H., and Suyama, S., *Bull. Chem. Soc. Japan*, 1991, **64**, 1231.
196. Yamamoto, T., Nakashio, Y., Onishi, and Hirota, M., *Nippon Kagaku Kaishi*, 1985, 2296.
197. Huyser, E.S., and Feng, R.H.C., *J. Org. Chem.*, 1969, **34**, 1727.
198. Huyser, E.S., and Bredeweg, C.J., *J. Am. Chem. Soc.*, 1964, **86**, 2401.
199. Huyser, E.S., Bredeweg, C.J., and Vanscoy, R.M., *J. Am. Chem. Soc.*, 1964, **86**, 4148.
200. Porter, N.A., in 'Organic Peroxides' (Ed. Ando, W.), p. 102 (Wiley: Chichester 1992).
201. Hiatt, R., in 'Organic Peroxides' (Ed. Swern, D.), Vol. 2, p. 1 (Wiley-Interscience: New York 1971).
202. Hiatt, R., and Irwin, K.C., *J. Org. Chem.*, 1968, **33**, 1436.
203. Nangia, P.S., and Benson, S.W., *J. Phys. Chem.*, 1979, **83**, 1138.
204. Benson, S.W., Thermochemical Kinetics (Wiley: New York 1976).
205. Hiatt, R., Mill, T., and Mayo, F.R., *J. Org. Chem.*, 1968, **33**, 1416.
206. Hiatt, R., Mill, T., Irwin, K.C., and Castleman, J.K., *J. Org. Chem.*, 1968, **33**, 1428.
207. Hiatt, R., Irwin, K.C., and Gould, C.W., *J. Org. Chem.*, 1968, **33**, 1430.
208. Sosnovsky, G., and Rawlinson, D.J., in 'Organic Peroxides' (Ed. Swern, D.), Vol. 2, p. 153 (Wiley-Interscience: New York 1971).
209. Mulcahy, M.F.R., Steven, J.R., and Ward, J.C., *Aust. J. Chem.*, 1965, **18**, 1177.
210. Hamilton, C.J., and Tighe, B.J., in 'Comprehensive Polymer Science' (Eds. Eastmond, G.C., Ledwith, A., Russo, S., and Sigwalt, P.), Vol. 3, p. 261 (Pergamon: London 1989).
211. Bacon, R.G.R., *Chem. Soc., Quart. Rev.*, 1955, **9**, 287.
212. Sosnovsky, G., and Rawlinson, D.J., 'Metal Ion-Catalyzed Reactions of Hydrogen Peroxide and Peroxydisulfate', Vol. 2, p 153 (Wiley-Interscience: New York 1971).
213. House, D.A., *Chem. Rev.*, 1962, **62**, 185.
214. Behrman, E.J., and Edwards, J.O., *Rev. Inorg. Chem.*, 1980, **2**, 179.
215. Rudin, A., Samanta, M.C., and Van Der Hoff, B.M.E., *J. Polym. Sci., Polym. Chem. Ed.*, 1979, **17**, 493.
216. Rasmussen, J.K., and Smith, H.K., *J. Am. Chem. Soc.*, 1981, **103**, 730.
217. Choi, K.Y., and Lee, C.Y., *Ind. Eng. Chem. Res.*, 1987, **26**, 2079.
218. Rasmussen, J.K., and Smith, H.K., *Polym. Prepr. (Am. Chem. Soc., Div. Polym. Chem)*, 1982, **23(1)**, 152.
219. Takeishi, M., Ohkawa, H., and Hayama, S., *Makromol. Chem., Rapid Commun.*, 1981, **2**, 457.
220. Rasmussen, J.K., Heilmann, S.M., Krepski, L.R., and Smith, H.K., *ACS Symp. Ser.*, 1987, **326**, 116.

221. Rasmussen, J.K., Heilmann, S.M., Toren, P.E., Pocius, A.V., and Kotnour, T.A., *J. Am. Chem. Soc.*, 1983, **105**, 6845.
222. Kim, Y.H., in 'Organic Peroxides' (Ed. Ando, W.), p. 387 (Wiley: Chichester 1992).
223. Curci, R., Delano, G., DiFuria, F., Edwards, J.O., and Gallopo, A.R., *J. Org. Chem.*, 1974, **39**, 3020.
224. Chawla, O.P., and Fessenden, R.W., *J. Phys. Chem.*, 1975, **79**, 2693.
225. Tang, Y., Thorn, R.P., Mauldin, R.L., and Wine, P.H., *J. Photochem. Photobiol.*, *A*, 1988, **44**, 243.
226. Ebdon, J.R., Huckerby, T.N., and Hunter, T.C., *Polymer*, 1994, **35**, 250.
227. Brosse, J.-C., Derouet, D., Epaillard, F., Soutif, J.-C., Legeay, G., and Dusek, K., *Adv. Polym. Sci.*, 1986, **81**, 167.
228. Haber, F., and Weiss, J.J., *Proc. R. Soc., London, Ser. A*, 1934, **147**, 332.
229. Shiga, T., Boukhors, A., and Douzou, P., *J. Phys. Chem.*, 1967, **71**, 3559.
230. Chiang, Y.S., Craddock, J., and Turkevich, J., *J. Phys. Chem.*, 1966, **70**, 3508.
231. Norman, R.O.C., and Dixon, W.T., *J. Chem. Soc.*, 1963, 3119.
232. Walling, C., *Acc. Chem. Res.*, 1975, **8**, 125.
233. Norman, R.O.C., in 'Chem. Soc. Spec. Publ. - Essays on Free Radical Chemistry', Vol. 24, p. 117 (Chem. Soc.: London 1970).
234. Nazran, A.S., and Warkentin, J., *J. Am. Chem. Soc.*, 1982, **104**, 6405.
235. Dixon, D.W., *Adv. Oxygenated Processes*, 1988, **1**, 179.
236. Grant, R.D., Rizzardo, E., and Solomon, D.H., *J. Chem. Soc., Chem. Commun.*, 1984, 867.
237. Tezuka, T., and Narita, N., *J. Am. Chem. Soc.*, 1979, **101**, 7413.
238. Osei-Twum, E.Y., McCallion, D., Nazran, A.S., Pannicucci, R., Risbood, P.A., and Warkentin, J., *J. Org. Chem.*, 1984, **49**, 336.
239. Simionescu, C., Comanita, E., Pastravanu, M., and Dumitriu, S., *Prog. Polym. Sci.*, 1986, **12**, 1.
240. Pappas, S.P., in 'Comprehensive Polymer Science' (Eds. Eastmond, G.C., Ledwith, A., Russo, S., and Sigwalt, P.), Vol. 4, p. 337 (Pergamon: London 1989).
241. Pappas, S.P., in 'Encyclopaedia of Polymer Science and Engineering', 2nd Edn (Eds. Mark, H.F., Bikales, N.M., Overberger, C.G., and Menges, G.), Vol. 11, p. 186 (Wiley: New York 1987).
242. Pappas, S.P., *J. Radiat. Curing*, 1987, **14**, 6.
243. Bassi, G.L., *J. Radiat. Curing*, 1987, **14**, 18.
244. Mishra, M.K., *J. Macromol. Sci., Rev. Macromol. Chem.*, 1982, **C22**, 409.
245. Oster, G., and Yang, N., *Chem. Rev.*, 1968, **68**, 125.
246. Wagner, P.J., *Top. Curr. Chem.*, 1976, **66**, 1.
247. Hageman, H.J., *Prog. Org. Coatings*, 1985, **13**, 123.
248. McGinniss, V.D., *Dev. Polym. Photochem.*, 1982, **3**, 1.
249. Berner, G., Kirchmayr, R., and Rist, G., *J. Oil Colour Chem. Assoc.*, 1978, **61**, 105.

250. Ledwith, A., *J. Oil Colour Chem. Assoc.*, 1976, **59**, 157.
251. Pappas, S.P., *Prog. Org. Coatings*, 1973, **2**, 333.
252. Heine, H.-G., Rosenkranz, H.-J., and Rudolf, H., *Angew. Chem., Int. Ed. Engl.*, 1972, **11**, 974.
253. Pappas, S.P., Chattopadhyay, A.K., and Carlblom, L.H., *ACS Symp. Ser.*, 1976, **25**, 12.
254. Hageman, H.J., and Overeem, T., *Makromol. Chem., Rapid Commun.*, 1981, **2**, 719.
255. Lipscomb, N.T., and Tarshiani, Y., *J. Polym. Sci., Part A: Polym. Chem.*, 1988, **26**, 529.
256. Hageman, H.J., and Jansen, L.G.J., *Makromol. Chem.*, 1988, **189**, 2781.
257. Schnabel, W., and Sumiyoshi, T., in 'New Trends in the Photochemistry of Polymers' (Eds. Allen, N.S., and Rabek, J.F.), p. 133 (Elsevier Applied Science: London 1985).
258. Baxter, J.E., Davidson, R.S., Hageman, H.J., and Overeem, T., *Makromol. Chem.*, 1988, **189**, 2769.
259. Majima, T., and Schnabel, W., *Makromol. Chem.*, 1991, **192**, 2307.
260. Klos, R., Gruber, H., and Greber, G., *J. Macromol. Sci., Chem.*, 1991, **A28**, 925.
261. Cook, W.D., *Polymer*, 1992, **33**, 600.
262. Griller, D., Howard, J.A., Marriott, P.R., and Scaiano, J.C., *J. Am. Chem. Soc.*, 1981, **103**, 619.
263. Alexander, I.J., and Scott, R.J., *Br. Polym. J.*, 1983, **15**, 30.
264. Nair, C.P.R., and Clouet, G., *J. Macromol. Sci., Rev. Macromol. Chem. Phys.*, 1991, **C31**, 311.
265. Misra, N., and Bajpai, U.D.N., *Prog. Polym. Sci.*, 1982, **8**, 61.
266. Bamford, C.H., in 'Comprehensive Polymer Science' (Eds. Eastmond, G.C., Ledwith, A., Russo, S., and Sigwalt, P.), Vol. 3, p. 123 (Pergamon: London 1989).
267. Bamford, C.H., in 'Reactivity, Mechanism, and Structure in Polymer Chemistry' (Eds. Jenkins, A.D., and Ledwith, A.), p. 52 (Wiley-Interscience: London 1974).
268. Eastmond, G.C., and Grigor, J., *Makromol. Chem., Rapid Commun.*, 1986, **7**, 375.
269. Bamford, C.H., Eastmond, G.C., Woo, J., and Richards, D.H., *Polym. Commun.*, 1982, **23**, 643.
270. Ho, T.-L., *Synthesis*, 1973, 347.
271. Mino, G., Kaizerman, S., and Rasmussen, E., *J. Am. Chem. Soc.*, 1959, **81**, 1494.
272. Bamford, C.H., Middleton, I.P., Sataka, Y., and Al-Lamee, K.G., in 'Advances in Polymer Synthesis' (Eds. Culbertson, B.M., and McGrath, J.E.), p. 291 (Plenum: New York 1986).

273. Hebeish, A., and Guthrie, J.T., The Chemistry and Technology of Cellulosic Copolymers, p. 155 (Springer: Berlin 1981).
274. McDowell, D.J., Gupta, B.S., and Stannett, V.T., *Prog. Polym. Sci.*, 1984, **10**, 1.
275. Hill, D.J.T., McMillan, A.M., O'Donnell, J.H., and Pomery, P.J., *Makromol. Chem., Macromol. Symp.*, 1990, **33**, 201.
276. Casinos, I., *Polymer*, 1992, **33**, 1304.
277. Moad, G., Rizzardo, E., and Solomon, D.H., in 'Comprehensive Polymer Science' (Eds. Eastmond, G.C., Ledwith, A., Russo, S., and Sigwalt, P.), Vol. 3, p. 141 (Pergamon: London 1989).
278. Pryor, W.A., and Lasswell, L.D., *Adv. Free Radical Chem.*, 1975, **5**, 27.
279. Kurbatov, V.A., *Russ. Chem. Rev. (Engl. Transl.)*, 1987, **56**, 505.
280. Hall, H.K., Jr., *Agnew. Chem. Int. Ed. Engl.*, 1983, **22**, 440.
281. Kirchner, K., and Riederle, K., *Angew. Makromol. Chem.*, 1983, **111**, 1.
282. Flory, P.J., *J. Am. Chem. Soc.*, 1937, **59**, 241.
283. Mayo, F.R., *Polym. Prepr. (Am. Chem. Soc., Div. Polym. Chem)*, 1961, **2**, 55.
284. Chong, Y.K., Rizzardo, E., and Solomon, D.H., *J. Am. Chem. Soc.*, 1983, **105**, 7761.
285. Buzanowski, W.C., Graham, J.D., Priddy, D.B., and Shero, E., *Polymer*, 1992, **33**, 3055.
286. Kaufmann, H.F., Olaj, O.F., and Breitenbach, J.W., *Makromol. Chem.*, 1977, **178**, 2707.
287. Clouet, G., Chaumont, P., and Corpart, P., *J. Polym. Sci., Part A: Polym. Chem.*, 1993, **31**, 2815.
288. Walling, C., and Briggs, E.R., *J. Am. Chem. Soc.*, 1946, **68**, 1141.
289. Lingnau, J., and Meyerhoff, G., *Polymer*, 1983, **24**, 1473.
290. Lingnau, J., and Meyerhoff, G., *Macromolecules*, 1984, **17**, 941.
291. Lingnau, J., Stickler, M., and Meyerhoff, G., *Eur. Polym. J.*, 1980, **16**, 785.
292. Stickler, M., and Meyerhoff, G., *Makromol. Chem.*, 1978, **179**, 2729.
293. Hall, H.K., Jr., and Padias, A.B., *Acc Chem. Res.*, 1990, **23**, 3.
294. Matsuda, M., and Abe, K., *J. Polym. Sci., Part A-1*, 1968, **6**, 1441.
295. Sato, T., Abe, M., and Otsu, T., *Makromol. Chem.*, 1977, **178**, 1061.
296. Spychaj, T., and Hamielec, A.E., *J. Appl. Polym. Sci.*, 1991, **42**, 2111.
297. Hasha, D.L., Priddy, D.B., Rudolf, P.R., Stark, E.J., de Pooter, M., and Van Damme, F., *Macromolecules*, 1992, **25**, 3046.
298. Kirchner, K., and Schlapkohl, H., *Makromol. Chem.*, 1976, **177**, 2031.
299. Stille, J.K., and Chung, D.C., *Macromolecules*, 1975, **8**, 83.
300. Hall, H.K., Jr., Padias, A.B., Pandya, A., and Tanaka, H., *Macromolecules*, 1987, **20**, 247.
301. Kokubo, T., Iwatsuki, S., and Yamashita, Y., *Makromol. Chem.*, 1969, **123**, 256.
302. Sato, T., Abe, M., and Otsu, T., *J. Macromol. Sci., Chem.*, 1981, **A15**, 367.
303. Gaylord, N.G., and Takahashi, A., *Adv. Chem. Ser.*, 1969, **91**, 94.
304. Jug, K., *J. Am. Chem. Soc.*, 1987, **109**, 3534.

305. Giese, B., *Angew. Chem., Int. Ed. Engl.*, 1983, **22**, 753.
306. Tedder, J.M., *Angew. Chem., Int. Ed. Engl.*, 1982, **21**, 401.
307. Beckwith, A.L.J., *Tetrahedron*, 1981, **37**, 3073.
308. Tedder, J.M., and Walton, J.C., *Acc. Chem. Res.*, 1976, **9**, 183.
309. Tedder, J.M., and Walton, J.C., *Tetrahedron*, 1980, **36**, 701.
310. Lorand, J.P., in 'Landoldt-Bornstein, New Series, Radical Reaction Rates in Solution' (Ed. Fischer, H.), Vol. II/13a, p. 135 (Springer-Verlag: Berlin 1984).
311. Asmus, K.-D., and Bonifacic, M., in 'Landoldt-Börnstein, New Series, Radical Reaction Rates in Solution' (Ed. H., F.), Vol. II/13b (Springer-Verlag: Berlin 1984).
312. Giese, B., and Lachhein, S., *Angew. Chem., Int. Ed. Engl.*, 1981, **20**, 967.
313. Santi, R., Bergamini, F., Citterio, A., Sebastiano, R., and Nicolini, M., *J. Org. Chem.*, 1992, **57**, 4250.
314. Beranek, I., and Fischer, H., in 'Free Radicals in Synthesis and Biology' (Ed. Minisci, F.), p. 303 (Kluwer: Dordrecht 1989).
315. Krstina, J., Moad, G., and Solomon, D.H., *Eur. Polym. J.*, 1992, **28**, 275.
316. Behari, K., Bevington, J.C., and Huckerby, T.N., *Makromol. Chem.*, 1987, **188**, 2441.
317. Bevington, J.C., Huckerby, T.N., and Hutton, N.W.E., *J. Polym. Sci., Polym. Chem. Ed.*, 1982, **20**, 2655.
318. Bevington, J.C., Huckerby, T.N., and Hutton, N.W.E., *Eur. Polym. J.*, 1984, **20**, 525.
319. Ayrey, G., Jumangat, K., Bevington, J.C., and Huckerby, T.N., *Polym. Commun.*, 1983, **24**, 275.
320. Bevington, J.C., Huckerby, T.N., and Varma, S.C., *Eur. Polym. J.*, 1986, **22**, 427.
321. Barson, C.A., Bevington, J.C., and Huckerby, T.N., *Polym. Bull. (Berlin)*, 1986, **16**, 209.
322. Bevington, J.C., Breuer, S.W., and Huckerby, T.N., *Polym. Commun.*, 1984, **25**, 260.
323. Heberger, K., and Fischer, H., *Int. J. Chem. Kinet.*, 1993, **25**, 249.
324. Bevington, J.C., Lyons, R.A., and Senogles, E., *Eur. Polym. J.*, 1992, **28**, 283.
325. Szwarc, M., *J. Polym. Sci.*, 1955, **16**, 367.
326. Herk, L., Stefani, A., and Szwarc, M., *J. Am. Chem. Soc.*, 1961, **83**, 3008.
327. Heberger, K., Walbiner, M., and Fischer, H., *Angew. Chem. Int. Ed. Engl.*, 1992, **31**, 635.
328. Bevington, J.C., Cywar, D.A., Huckerby, T.N., Senogles, E., and Tirrell, D.A., *Eur. Polym. J.*, 1990, **26**, 41.
329. Bevington, J.C., Cywar, D.A., Huckerby, T.N., Senogles, E., and Tirrell, D.A., *Eur. Polym. J.*, 1990, **26**, 871.
330. Russell, G.A., Jiang, W., Hu, S.S., and Khanna, R.K., *J. Org. Chem.*, 1986, **51**, 5498.
331. Münger, K., and Fischer, H., *Int. J. Chem. Kinet.*, 1985, **17**, 809.

332. Scaiano, J.C., and Stewart, L.C., *J. Am. Chem. Soc.*, 1983, **105**, 3609.
333. Levin, Y.A., Abul'khanov, A.G., Nefedov, A.G., Skorobogatova, M.S., and Ivanov, B.E., *Dokl. Phys. Chem. (Engl. Transl.)*, 1977, **235**, 728.
334. Citterio, A., Vismara, E., and Bernardi, R., *J. Chem. Res., Miniprint*, 1983, **4**, 876.
335. Kuwae, Y., and Kamachi, M., *Bull. Chem. Soc. Japan*, 1989, **62**, 2474.
336. Dzhabiyeva, Z.M., Matkovskii, P.Y., Pechatnikov, Y.L., and Byrikhina, N.A., *Polym. Sci. USSR (Engl. Transl.)*, 1985, **27**, 2416.
337. Bevington, J.C., Huckerby, T.N., and Hutton, N.W.E., *Eur. Polym. J.*, 1982, **18**, 963.
338. Mahabadi, H.K., and O'Driscoll, K.F., *Makromol. Chem.*, 1977, **178**, 2629.
339. Citterio, A., Minisci, F., and Vismara, E., *J. Org. Chem.*, 1982, **47**, 81.
340. Pryor, W.A., and Fiske, T.R., *Trans. Faraday Soc.*, 1969, **65**, 1865.
341. Dickerman, S.C., Megna, I.S., and Skoultchi, M.M., *J. Am. Chem. Soc.*, 1959, **81**, 2270.
342. Bevington, J.C., and Ito, T., *Trans. Faraday Soc.*, 1968, **64**, 1329.
343. Howard, J.A., and Scaiano, J.C., in 'Landoldt-Bornstein, New Series, Radical Reaction Rates in Solution' (Ed. Fischer, H.), Vol. II/13d, p. 5 (Springer-Verlag: Berlin 1984).
344. Grant, R.D., Rizzardo, E., and Solomon, D.H., *Makromol. Chem.*, 1984, **185**, 1809.
345. Grant, R.D., Rizzardo, E., and Solomon, D.H., *J. Chem. Soc., Perkin Trans. 2*, 1985, 379.
346. Chateauneuf, J., Lusztyk, J., and Ingold, K.U., *J. Am. Chem. Soc.*, 1988, **110**, 2866.
347. Bevington, J.C., Harris, D.O., and Johnson, M., *Eur. Polym. J.*, 1965, **1**, 235.
348. Bevington, J.C., Breuer, S.W., and Huckerby, T.N., *Macromolecules*, 1989, **22**, 55.
349. Moad, G., Rizzardo, E., and Solomon, D.H., *Makromol. Chem., Rapid Commun.*, 1982, **3**, 533.
350. Busfield, W.K., Jenkins, I.D., Thang, S.H., Rizzardo, E., and Solomon, D.H., *Eur. Polym. J.*, 1993, **29**, 397.
351. Cuthbertson, M.C., and Rizzardo, E., personal communication.
352. Griffiths, P.G., Rizzardo, E., and Solomon, D.H., *Tetrahedron Lett.*, 1982, **23**, 1309.
353. Rizzardo, E., personal communication.
354. Heicklein, J.P., *Adv. Photochem.*, 1988, **14**, 177.
355. Kochi, J.K., in 'Free Radicals' (Ed. Kochi, J.K.), Vol. 2, p. 665 (Wiley: New York 1973).
356. Walling, C., *Pure. Appl. Chem.*, 1967, **15**, 69.
357. Ingold, K.U., *Pure Appl. Chem.*, 1967, **15**, 49.
358. Cuthbertson, M.J., Rizzardo, E., and Solomon, D.H., *Aust. J. Chem.*, 1985, **38**, 315.

359. Cuthbertson, M.J., Rizzardo, E., and Solomon, D.H., *Aust. J. Chem.*, 1983, **36**, 1957.
360. Busfield, W.K., Jenkins, I.D., Thang, S.H., Rizzardo, E., and Solomon, D.H., *Aust. J. Chem.*, 1985, **38**, 689.
361. Busfield, W.K., Jenkins, I.D., Thang, S.H., Rizzardo, E., and Solomon, D.H., *J. Chem. Soc., Perkin Trans. 1*, 1988, 485.
362. Jones, M.J., Moad, G., Rizzardo, E., and Solomon, D.H., *J. Org. Chem.*, 1989, **54**, 1607.
363. Sakurai, H., and Hosomi, A., *J. Am. Chem. Soc.*, 1967, **89**, 458.
364. Walling, C., and McGuinness, J.A., *J. Am. Chem. Soc.*, 1969, **91**, 2053.
365. Encina, M.V., Rivera, M., and Lissi, E.A., *J. Polym. Sci., Polym. Chem. Ed.*, 1978, **16**, 1709.
366. Kunitake, T., and Murakami, S., *J. Polym. Sci., Polym. Chem. Ed.*, 1974, **12**, 67.
367. Korth, H.-G., and Sustmann, R., *Tetrahedron Lett.*, 1985, **26**, 2551.
368. Busfield, W.K., Grice, D.I., and Jenkins, I.D., *J. Chem. Soc., Perkin Trans. 2*, 1994, 1079.
369. Busfield, W.K., Grice, D.I., Jenkins, I.D., and Monteiro, M.J., *J. Chem. Soc., Perkin Trans. 1*, 1994, 1071.
370. Wong, P.C., Griller, D., and Scaiano, J.C., *J. Am. Chem. Soc.*, 1982, **104**, 5106.
371. Avila, D.V., Brown, C.E., Ingold, K.U., and Lusztyk, J., *J. Am. Chem. Soc*, 1993, **115**, 466.
372. Walling, C., and Padwa, A., *J. Am. Chem. Soc.*, 1963, **85**, 1593.
373. Huyser, E.S., and Jankauskas, K.J., *J. Org. Chem.*, 1970, **35**, 3196.
374. Kamath, V.R., and Sargent, J.D., Jr., *J. Coatings Technol.*, 1987, **59**, 51.
375. Baignee, A., Howard, J.A., Scaiano, J.C., and Stewart, L.C., *J. Am. Chem. Soc.*, 1983, **105**, 6120.
376. Bertrand, M.P., and Surzur, J.-M., *Tetrahedron Lett.*, 1976, **38**, 3451.
377. Bertrand, M.P., Halou, O.-M., and Surzur, J.-M., *Bull. Soc. Chim. Fr.*, 1985, 115.
378. Hilborn, J.W., and Pincock, J.A., *J. Am. Chem. Soc.*, 1991, **113**, 2683.
379. Bevington, J.C., *Angew. Makromol. Chem.*, 1991, **185/186**, 1.
380. Edge, D.J., and Kochi, J.K., *J. Am. Chem. Soc.*, 1973, **95**, 2635.
381. Kochi, J.K., *J. Am. Chem. Soc.*, 1962, **84**, 774.
382. Lane, J., and Tabner, B.J., *J. Chem. Soc., Perkin Trans. 2*, 1984, 1823.
383. Barclay, L.R.C., Griller, D., and Ingold, K.U., *J. Am. Chem. Soc.*, 1982, **104**, 4399.
384. Beckwith, A.L.J., and Thomas, C.B., *J. Chem. Soc., Perkin Trans. 2*, 1973, 861.
385. Nakata, T., Tokumaru, K., and Simamura, O., *Tetrahedron Lett.*, 1967, **34**, 3303.
386. Fischer, H., *Z. Naturforsch.*, 1964, **19a**, 866.
387. Fischer, H., and Giacometti, G., *J. Polym. Sci., Polym. Symp.*, 1967, **16**, 2763.
388. Roth, H.K., and Wunsche, P., *Acta Polym.*, 1981, **32**, 491.

389. Maruthamuthu, P., *Makromol. Chem., Rapid, Commun.*, 1980, **1**, 23.
390. McAskill, N.A., and Sangster, D.F., *Aust. J. Chem.*, 1984, **37**, 2137.
391. Sloane, T.M., and Brudzynski, R.J., *J. Am. Chem. Soc.*, 1979, **101**, 1495.
392. Ledwith, A., and Russell, P.J., *J. Polym. Sci., Polym. Lett.*, 1975, **13**, 109.
393. Citterio, A., Arnoldi, C., Giordano, C., and Castaldi, G., *J. Chem. Soc., Perkin Trans. 1*, 1983, 891.
394. Arnoldi, C., Citterio, A., and Minisci, F., *J. Chem. Soc., Perkin Trans. 2*, 1983, 531.
395. Fristad, W.E., and Peterson, J.R., *Tetrahedron*, 1984, **40**, 1469.
396. McAskill, N.A., and Sangster, D.F., *Aust. J. Chem.*, 1979, **32**, 2611.
397. Ghosh, N.N., and Mandal, B.M., *Macromolecules*, 1986, **19**, 19.
398. Misra, N., and Mandal, B.M., *Macromolecules*, 1984, **17**, 495.
399. Misra, N., and Mandal, B.M., *J. Polym. Sci., Polym. Lett. Ed.*, 1985, **23**, 63.
400. Banthia, A.K., Mandal, B.M., and Palit, S.R., *J. Polym. Sci., Polym. Chem. Ed.*, 1977, **15**, 945.
401. Howard, J.A., *Rev. Chem. Intermed.*, 1984, **5**, 1.
402. Neta, P., Huie, R.E., and Ross, A.B., *J. Chem. Phys. Ref., Data*, 1990, **19**, 413.
403. Howard, J.A., in 'Free Radicals' (Ed. Kochi, J.K.), Vol. 2, p. 3 (Wiley: New York 1973).
404. Scott, G.P., Soong, C.C., Allen, J.L., and Reynolds, J.L., *Polym. Prepr. (Am. Chem. Soc., Div. Polym. Chem)*, 1963, **4(1)**, 67.
405. Farina, M., Di Silvestro, G., and Sozzani, P., *Makromol. Chem.*, 1989, **190**, 213.
406. Ito, O., and Matsuda, M., *J. Am. Chem. Soc.*, 1979, **101**, 5732.
407. Ito, O., and Matsuda, M., *J. Am. Chem. Soc.*, 1979, **101**, 1815.
408. Sato, T., Abe, M., and Otsu, T., *Makromol. Chem.*, 1977, **178**, 1951.
409. Ito, O., *J. Am. Chem. Soc*, 1983, **105**, 850.
410. Sumiyoshi, T., and Schnabel, W., *Makromol. Chem.*, 1985, **186**, 1811.
411. Kajiwara, A., Konishi, Y., Morishima, Y., Schnabel, W., Kuwata, K., and Kamachi, M., *Macromolecules*, 1993, **26**, 1656.
412. Bottle, S.E., Busfield, W.K., Grice, I.D., Heiland, K., Meutermans, W., and Monteiro, M., *Prog. Pac. Polym. Sci.*, 1994, **3**, 85.
413. Geers, B.N., Gleicher, G.J., and Church, D.F., *Tetrahedron*, 1980, **36**, 997.
414. Ito, O., and Masuda, M., *J. Am. Chem. Soc.*, 1979, **101**, 5732.
415. Ito, O., and Matsuda, M., *J. Phys. Chem.*, 1984, **88**, 1002.
416. Ito, O., and Matsuda, M., *J. Org. Chem.*, 1983, **48**, 2748.
417. Ito, O., and Matsuda, M., *J. Org. Chem.*, 1983, **48**, 2410.
418. Bessiere, J.-M., Boutevin, B., and Sarraf, L., *Polym. Bull. (Berlin)*, 1987, **18**, 253.
419. Baxter, J.E., Davidson, R.S., Hageman, H.J., and Overeem, T., *Makromol. Chem., Rapid Commun.*, 1987, **8**, 311.
420. Perkins, M.J., *Adv. Phys. Org. Chem*, 1981, **17**, 1.

421. Chalfont, G.R., Perkins, M.J., and Horsfield, A., *J. Am. Chem. Soc.*, 1968, **90**, 7141.
422. Savedoff, L.G., and Ranby, B., *Polym. Prepr. (Am. Chem. Soc., Div. Polym. Chem)*, 1978, **19(1)**, 629.
423. Bevington, J.C., Tabner, B.J., and Fridd, P.F., *Rev. Roum. Chim.*, 1980, **25**, 947. (Chem. Abstr. 1981, 94, 4294n)
424. Bevington, J.C., Fridd, P.F., and Tabner, B.J., *J. Chem. Soc., Perkin Trans. 2*, 1982, 1389.
425. Pichot, C., Spitz, R., and Guyot, A., *J. Macromol. Sci.- Chem.*, 1977, **A11**, 251.
426. Minisci, F., *Acc. Chem. Res.*, 1975, **8**, 165.
427. Caronna, T., Citterio, A., Ghirardini, M., and Minisci, F., *Tetrahedron*, 1977, **33**, 793.
428. Bamford, C.H., Jenkins, A.D., and Johnston, R., *Proc. R. Soc., London, Ser. A*, 1957, **239**, 214.
429. Giese, B., Radicals in Organic Synthesis: Formation of Carbon-Carbon Bonds (Pergamon Press: Oxford 1986).
430. Giese, B., *Rev. Chem. Intermed.*, 1986, **7**, 3.
431. Hill, C.L., and Whitesides, G.M., *J. Am. Chem. Soc.*, 1974, **96**, 870.
432. Jones, S.A., Prementine, G.S., and Tirrell, D.A., *J. Am. Chem. Soc.*, 1985, **107**, 5275.
433. Russell, G.A., *Acc. Chem. Res.*, 1989, **22**, 1.
434. Pike, P., Hershberger, S., and Hershberger, J., *Tetrahedron Lett.*, 1985, **26**, 6289.
435. Giese, B., Gonzalez-Gomez, J.A., and Witzel, T., *Angew. Chem., Int. Ed. Engl.*, 1984, **23**, 69.
436. Rizzardo, E., and Solomon, D.H., *Polym. Bull. (Berlin)*, 1979, **1**, 529.
437. Busfield, W.K., Jenkins, I.D., Thang, S.H., Rizzardo, E., and Solomon, D.H., *Tetrahedron Lett*, 1985, **26**, 5081.
438. Beckwith, A.L.J., Bowry, V.W., and Moad, G., *J. Org. Chem.*, 1988, **53**, 1632.
439. Chateauneuf, J., Lusztyk, J., and Ingold, K.U., *J. Org. Chem.*, 1988, **53**, 1629.
440. Bowry, V.W., and Ingold, K.U., *J. Am. Chem. Soc.*, 1992, **114**, 4992.
441. Beckwith, A.L.J., Bowry, V.W., and Ingold, K.U., *J. Am. Chem. Soc.*, 1992, **114**, 4983.
442. Moad, G., Rizzardo, E., Solomon, D.H., and Beckwith, A.L.J., *Polym. Bull. (Berlin)*, 1992, **29**, 647.
443. Golubev, V.A., Kozlov, Y.N., Petrov, A.N., and Purmal, A.P., *Prog. React. Kinet.*, 1991, **16**, 35.
444. Anderson, D.R., Keute, J., Chapel, H.L., and Koch, T.H., *J. Am. Chem. Soc.*, 1979, **101**, 1904.
445. Keana, J.F.W., Dinerstein, R.J., and Baitis, F., *J. Org. Chem.*, 1971, **36**, 209.
446. Johnston, L.J., Tencer, M., and Scaiano, J.C., *J. Org. Chem.*, 1986, **51**, 2806.
447. Coxon, J.M., and Pattsalides, E., *Aust. J. Chem.*, 1982, **35**, 509.

448. Moad, G., Rizzardo, E., and Solomon, D.H., *Polym. Bull. (Berlin)*, 1982, **6**, 589.
449. Shetty, S., and Garcia-Rubio, L.H., *Polym. Mater. Sci. Eng.*, 1991, **65**, 103.
450. Buback, M., Huckestein, B., and Ludwig, B., *Makromol. Chem., Rapid Commun.*, 1992, **13**, 1.
451. Buback, M., Huckestein, B., and Leinhos, U., *Makromol. Chem., Rapid Commun.*, 1987, **8**, 473.
452. Carduner, K.R., Carter, R.O., Zinbo, M., Gerlock, J.L., and Bauer, D.R., *Macromolecules*, 1988, **21**, 1598.
453. Meijs, G.F., Morton, T.C., Rizzardo, E., and Thang, S.H., *Macromolecules*, 1991, **24**, 3689.
454. Meijs, G.F., Rizzardo, E., and Thang, S.H., *Macromolecules*, 1988, **21**, 3122.
455. Bevington, J.C., Ebdon, J.R., and Huckerby, T.N., in 'NMR Spectroscopy of Polymers' (Ed. Ibbett, R.N.), p. 51 (Blackie: London 1993).
456. Bevington, J.C., Cywar, D.A., Huckerby, T.N., Senogles, E., and Tirrell, D.A., *Eur. Polym. J.*, 1988, **24**, 699.
457. Johns, S.R., Rizzardo, E., Solomon, D.H., and Willing, R.I., *Makromol. Chem., Rapid Commun.*, 1983, **4**, 29.
458. Hatada, K., Kitayama, T., and Masuda, E., *Polym. J. (Tokyo)*, 1985, **17**, 985.
459. Kashiwagi, T., Inaba, A., Brown, J.E., Hatada, K., Kitayama, T., and Masuda, E., *Macromolecules*, 1986, **19**, 2160.
460. Hatada, K., Kitayama, T., and Masuda, E., *Polym. J. (Tokyo)*, 1986, **18**, 395.
461. Bevington, J.C., Ebdon, J.R., and Huckerby, T.N., *Eur. Polym. J.*, 1985, **21**, 685.
462. Shiraishi, H., and Ranby, B., *Chemica Scripta*, 1977, **12**, 118.
463. Smith, P., Gilman, L.B., Stevens, R.D., and de Hargrave, C.V., *J. Magn. Resonance*, 1978, **29**, 545.
464. Fischer, H., in 'Substituent Effects in Radical Chemistry' (Ed. H. G. Viehe, Z.J.a.R.M.), p. 123 (Reidel: Dordecht 1986).
465. Fischer, H., and Paul, H., *Acc. Chem. Res.*, 1987, **20**, 200.
466. Meisters, A., Moad, G., Rizzardo, E., and Solomon, D.H., *Polym. Bull. (Berlin)*, 1988, **20**, 499.
467. Farina, M., *Makromol. Chem., Macromol. Symp.*, 1987, **10/11**, 255.
468. Ohtani, H., Ishiguro, S., Tanaka, M., and Tsuge, S., *Polym. J. (Tokyo)*, 1989, **21**, 41.
469. Palit, S.R., *Makromol. Chem.*, 1959, **36**, 89.
470. Palit, S.R., *Makromol. Chem.*, 1960, **38**, 96.
471. Rizzardo, E., and Solomon, D.H., *J. Macromol. Sci., Chem.*, 1979, **A13**, 997.
472. Ghosh, N.N., Sengputa, P.K., and Pramanik, A., *J. Polym. Sci., Part A*, 1965, **3**, 1725.
473. Rizzardo, E., and Solomon, D.H., *J. Macromol. Sci., Chem.*, 1979, **A13**, 1005.
474. Kern, W., *Chem. Ztg.*, 1976, **100**, 401.
475. Bevington, J.C., Radical Polymerization (Academic Press: London 1961).

476. Bevington, J.C., *Trans. Faraday Soc.*, 1955, **51**, 1392.
477. Bevington, J.C., and Ebdon, J.R., *Developments in Polymerization*, 1979, **2**, 1.
478. Bevington, J.C., *Makromol. Chem., Macromol. Symp.*, 1987, **10/11**, 89.
479. Moad, G., *Chem. Aust.*, 1991, **58**, 122.
480. Moad, G., in 'Annual Reports in NMR Spectroscopy' (Ed. Webb, G.A.), Vol. 29, p. 287 (Academic Press: London 1994).

4
Propagation

4.1 Introduction

The propagation step of radical polymerization comprises a sequence of radical additions to carbon-carbon double bonds. The factors which govern the rate and specificity of radical addition have been dealt with in general terms in chapter 2. In order to produce high molecular weight polymers, a propagating radical must show a high degree of specificity in its reactions with unsaturated systems. It must give addition to the exclusion of side reactions which bring about the cessation of growth of the polymer chain. Despite this limitation, there is considerable scope for structural variation in homopolymers.

The asymmetric substitution pattern of most monomers means that addition gives rise to a chiral center and their polymers will have tacticity (see 4.2).

Addition to double bonds may not be completely regiospecific. The predominant head-to-tail structure may be interrupted by head-to-head and tail-to-tail linkages (see 4.3).

Intramolecular rearrangement of the initially formed radical may occur occasionally (*e.g.* backbiting - see 4.4.3) or even be the dominant pathway (*e.g.* cyclopolymerization - see 4.4.1, ring-opening polymerization - see 4.4.2).

$$\text{\textasciitilde}CH_2\text{-}CH(\text{-}CH_2\text{-}CH_2\text{-}CH_2)\cdot CH_2\text{-}H \xrightarrow{\text{backbiting}} \text{\textasciitilde}CH_2\text{-}\dot{C}(\text{-}CH_2\text{-}CH_2\text{-}CH_2)\text{-}CH_2\text{-}H$$

This chapter is primarily concerned with the chemical microstructure of the products of free radical homopolymerization. Variations on the general $(CH_2\text{-}CXY)_n$ structure are described and the mechanisms for their formation and the associated rate parameters are examined. With this background established, aspects of the kinetics and thermodynamics of propagation are considered (see 4.5).

4.2 Stereosequence Isomerism - Tacticity

The classical representation of a homopolymer chain (1), in which the end groups are disregarded and only one monomer unit is considered, allows no possibility for structural variation. However, possibilities for stereosequence isomerism arise as soon as a monomer unit is considered in relation to its neighbors and the substituents X and Y are different. The chains have tacticity (see 4.2.1). Experimental methods for tacticity determination are summarized in 4.2.2 and the tacticity of some common polymers is considered in 4.2.3.

$$CH_2=C(X)(Y) \longrightarrow \longrightarrow \text{\textasciitilde}[CH_2-C(X)(Y)]_n\text{\textasciitilde}$$

(1)

The following discussion is limited to polymers of mono- or 1,1-disubstituted monomers. Other factors become important in describing the types of stereochemical isomerism possible for polymers formed from other monomers (*e.g.* 1,2-disubstituted monomers).[1]

4.2.1 Terminology and Mechanisms

Detailed discussion of polymer tacticity can be found in texts by Randall,[2] Bovey,[1,3] Koenig,[4] and Tonelli.[5] In order to understand stereoisomerism in polymer chains formed from mono- or 1,1-disubstituted monomers, consider four idealized chain structures:

(a) The isotactic chain where the relative configuration of all the substituted carbons in the chain is the same.

For the usual diagrammatic representation of a polymer chain, this corresponds to the situation where similar substituents lie on the same side of a plane perpendicular to the page and containing the polymer backbone.

$$\sim\sim\sim C(X)(Y)-CH_2-C(X)(Y)-CH_2-C(X)(Y)-CH_2-C(X)(Y)-CH_2-C(X)(Y)-CH_2-C(X)(Y)-CH_2-C(X)(Y)\sim\sim\sim$$

(b) The syndiotactic chain where the relative configuration of centers alternates along the chain.

$$\sim\sim\sim C(X)(Y)-CH_2-C(Y)(X)-CH_2-C(X)(Y)-CH_2-C(Y)(X)-CH_2-C(X)(Y)-CH_2-C(Y)(X)-CH_2-C(X)(Y)\sim\sim\sim$$

(c) The heterotactic chain where the dyad configuration alternates along the chain.

$$\sim\sim\sim C(X)(Y)-CH_2-C(X)(Y)-CH_2-C(Y)(X)-CH_2-C(Y)(X)-CH_2-C(X)(Y)-CH_2-C(X)(Y)-CH_2-C(Y)(X)\sim\sim\sim$$

(d) The atactic chain where there is a random arrangement of centers along the chain.*

$$\sim\sim\sim C(X)(Y)-CH_2-C(X)(Y)-CH_2-C(Y)(X)-CH_2-C(X)(Y)-CH_2-C(Y)(X)-CH_2-C(Y)(X)-CH_2-C(X)(Y)\sim\sim\sim$$

For polymers produced by free radical polymerization, while one of these structures may predominate, the idealized structures do not occur. It is necessary to define parameters to more precisely characterize the tacticity of polymer chains.

It should be stressed that this treatment of polymer stereochemistry only deals with relative configurations; whether a substituent is "up or down" with respect to that on a neighboring unit. Therefore, the smallest structural unit which contains stereochemical information is the dyad. There are two types of dyad; meso (m),

---

* In the literature the term atactic is sometimes used to refer to all polymers that are not entirely isotactic or entirely syndiotactic.

where the two chiral centers have like configuration, and racemic (*r*), where the centers have opposite configuration.

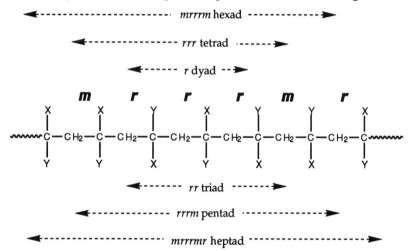

Confusion can arise because of the seemingly contradictory nomenclature established for analogous model compounds with just two asymmetric centers.[6] In such models the diastereomers are named as in the following example:

It is usual to discuss triads, tetrads, pentads, *etc.* in terms of the component dyads. For example, the *mrrrmr* heptad is represented as shown in Figure 1.

Figure 1 Representation of *mrrrmr* heptad.

It is informative to consider how tacticity arises in terms of the mechanism for propagation. The radical center on the propagating species will usually have a planar sp^2 configuration. As such it is achiral and it will only be locked into a specific configuration after the next monomer addition. This situation should be contrasted with that which pertains in anionic or coordination polymerizations where the active center is pyramidal and therefore has chirality.

Propagation

The configuration of a center in radical polymerization is established in the transition state for addition of the next monomer unit when it becomes a tetrahedral sp^3 center. If the stereochemistry of this center is established at random (Scheme 1; $k_m = k_r$) then a pure atactic chain is formed and $P(m)=P(r)=0.5$.

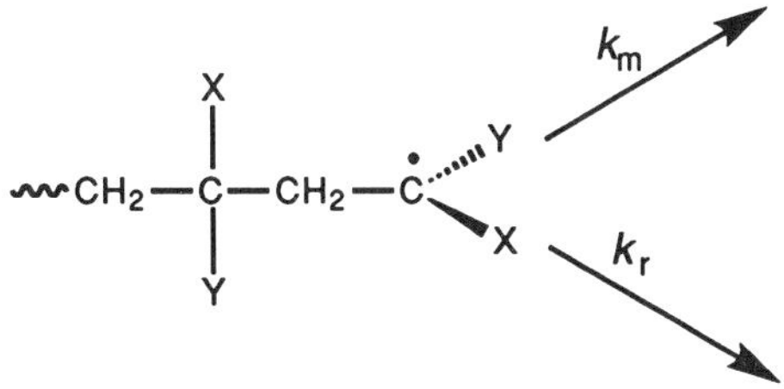

Scheme 1

If the center adopts a preferred configuration with respect to the configuration of the penultimate unit in the chain (Scheme 1; $k_m \ne k_r$) then Bernoullian statistics apply. The stereochemistry of the chain is characterized by a single parameter, $P(m)$ or $P(r)$ [$= 1-P(m)$]; the probabilities that a given dyad will be m or r respectively.

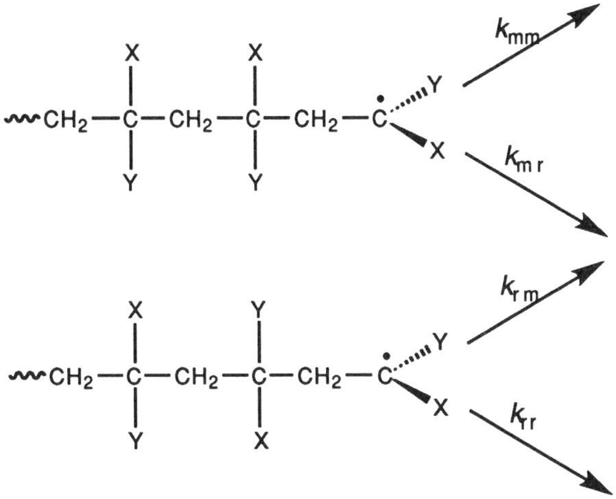

Scheme 2

Where the nature of the preceding dyad is important in determining the configuration of the new chiral center (see Scheme 2), first order Markov statistics apply. Propagation may be subject to a penpenultimate unit effect (also called an antepenultimate effect). Two parameters are required to specify the stereochemistry, $P(m|r)$ [$=1-P(m|m)$] and $P(r|r)$ [$=1-P(r|m)$], where $P(i|j)$ is the conditional probability that given a j dyad, the next unit in the chain will be an i dyad.[*]

---

[*] In texts by Bovey[1,3] and Tonelli[5] $P(i|j)$ is written $P_{j/i}$.

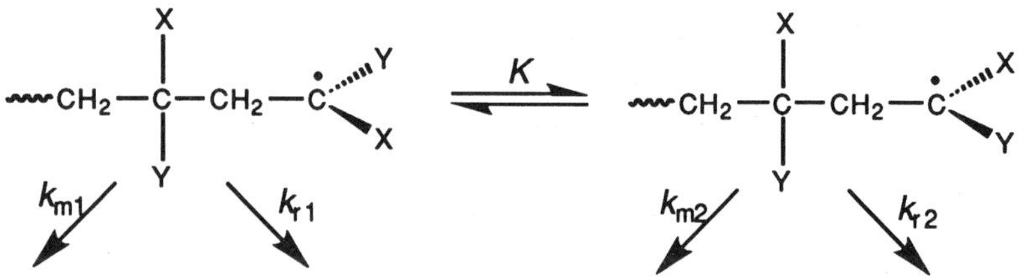

Scheme 3

The Coleman-Fox two state model describes the situation where there is restricted rotation about the bond to the preceding unit (Scheme 3). If this is slow with respect to the time scale for addition, then at least two conformations of the propagating radical need to be considered each of which may react independently with monomer. The rate constants associated with the conformational equilibrium and two values of P(*m*) are required to characterize the process.

More complex situations may also be envisaged and it should always be borne in mind that the fit of experimental data to a simple model provides support for but does not prove that model. The power of the experiment to discriminate between models has to be considered.

4.2.2 *Experimental Methods*

The application of NMR spectroscopy to tacticity determination of synthetic polymers was pioneered by Bovey *et al.*[7] NMR spectroscopy is the most used method and often the only technique available for directly assessing tacticity of polymer chains.[1,2,6,8] The chemical shift of a given nucleus in or attached to the chain may be sensitive to the configuration of centers three or more monomer units removed. Other forms of spectroscopy (*e.g.* IR spectroscopy[9,10]) and various physical properties (*e.g.* the Kerr effect[11]) may also be correlated with tacticity.

The ambiguity of the NMR peak assignments may cause problems in tacticity determination. The usual method of assigning peaks to configurational sequences involves matching expected and measured peak intensities. There are obvious problems inherent in this approach and these are being redressed by the application of 2D NMR methods which in many cases can provide unambiguous assignments.[12,13] These methods have been applied to make absolute tacticity assignments for PAA,[14] PMMA,[15-18] PMAN,[19] PVA,[20,21] PVC[22,23] and PVF.[24]

Attention must also be paid to sample preparation methods.[25] The number average molecular weight of the polymer must be sufficiently high that signals due to sequences near the chain ends make no significant contribution to the spectrum. For PMMA with heptad resolution, this would require that $\bar{M}_n$ is in excess of 30,000.

4.2.3 Tacticities of Common Polymers

Many radical polymerizations have been examined from the point of view of establishing the stereosequence distribution. For most systems it is claimed that the tacticity is predictable within experimental error* by Bernoullian statistics [*i.e.* by the single parameter P(m) - see 4.2.1].

Most polymers formed by radical polymerization have an excess of syndiotactic over isotactic dyads [*i.e.* P(m) ≤ 0.5]. P(m) typically lies in the range 0.4-0.5 for vinyl monomers and 0.2-0.5 for 1,1-disubstituted monomers. It is also generally found that P(m) decreases with decreasing temperature (see Table 4.1).

Exceptions to this general rule are polymers with very bulky substituents. For example, methacrylates with bulky ester substituents can have a preference for syndiotacticity.[28] Data on tacticity for some common polymers are presented in Table 4.1.

An explanation for the preference for syndiotacticity during MMA polymerization was proposed by Tsuruta *et al.*[29] They considered that the propagating radical should exist in one of two conformations and showed, with models, that attack on the less hindered side of the preferred conformation (where steric interactions between the substituent groups are minimized) would lead to formation of a syndiotactic dyad while similar attack on the less stable conformation would lead to an isotactic dyad.

MMA polymerization is one of the most studied systems and until recently was thought to be explicable, within experimental error, in terms of Bernoullian statistics. Moad *et al.*[30] have made precise measurements of the configurational sequence distribution for PMMA. It is clear that Bernoullian statistics do not provide a satisfactory description of the tacticity.[30] This finding is supported by other work.[26,31,32] First order Markov statistics provide an adequate fit of the data. Possible explanations include: a) penpenultimate unit effects are important; and/or b) conformational equilibrium is slow (see 4.2.1). At this stage, the experimental data do not allow these possibilities to be distinguished.

It seems likely that other polymerizations will be found to depart from Bernoullian statistics as the precision of tacticity measurements improves. One study[9] indicated that vinyl chloride polymerizations are also more appropriately described by first order Markov statistics. However, there has been some reassignment of signals since that time.[22,23]

The triad fractions for PVA[20,33] seem to obey Bernoullian statistics. However, the concentrations of higher order n-ads cannot be explained even by first (or second) order Markov statistics suggesting either that ambiguities still remain in the signal assignments at this level or that there are unresolved complexities in the polymerization mechanism.

---

* It should be noted that, in some studies, deviations of 5-10% in expected and measured NMR peak intensities have been ascribed to experimental error. Such error is sufficient to hide significant departures from Bernoullian statistics.[26,27]

Tacticities have been found to vary with reaction temperature[34] and may be influenced by the reaction media, by Lewis acids (see 7.2) and possibly by the presence of preformed polymer (see 7.3).

Table 4.1 Tacticities of Some Common Homopolymers

Polymer	Temp. °C	P(m)[a]	P(m|m)	P(r|m)	Solvent	Conv. %	Ref.
PAN	35	0.52	-	-	H$_2$O	-	35
PMAN	60	0.406	-	-	bulk	15	19
PMA	60	0.49	-	-	toluene	<50	36,37
PMMA	60	(0.202)	0.159	0.212	benzene	5	30
PS	80	0.46	-	-	benzene	92	38-40
PVA	-	0.46 ± .01[b]	-	-	c	-	20,41,42
PVC	90	(0.454)	0.437	0.465	d	-	9
	5	(0.406)	0.391	0.424	d	-	9
	-30	(0.377)	0.337	0.391	c	-	9

a Best fit number for P(m). The polymerization is believed to follow first order Markov statistics.
b See text.
c Commercial samples or conditions of preparation unknown.
d Suspension polymerization.

4.3 Regiosequence Isomerism - Head *vs.* Tail Addition

Most monomers have an asymmetric substitution pattern and the two ends of the double bond are distinct. For mono- and 1,1-disubstituted monomers (see 4.3.1) it is usual to call the less substituted end "the tail" and the more substituted end "the head". Thus the terminology evolved for two modes of addition: head and tail; and for the three types of linkages: head-to-tail, head-to-head and tail-to-tail. For 1,2-di-, tri- and tetrasubstituted monomers the definitions of head and tail are necessarily more arbitrary. The term "head" has been used for that end bearing the most substituents, the largest substituents or the best radical stabilizing substituent.

With 1,3-diene based polymers, greater scope for structural variation is introduced because there are two double bonds to attack and the propagating species is a delocalized radical with several modes of addition possible (see 4.3.2).

4.3.1 Monoene Polymers

Various terminologies for describing regiosequence isomerism have been proposed.[1,4] By analogy with that used to describe stereosequence isomerism (see 4.2), it has been suggested that a polymer chain with the monomer units connected by "normal" head-to-tail linkages should be termed isoregic, that with alternating head-to-head and tail-to-tail linkages, syndioregic, and that with a random arrangement of connections, aregic.[1]

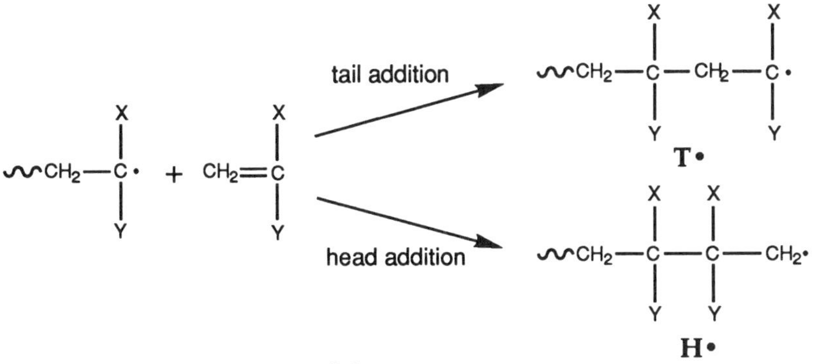

Scheme 4

For mono- and 1,1-disubstituted monomers, steric, polar, resonance, and bond-strength terms (see chapter 2) usually combine to favor a preponderance of tail addition; *i.e.* an almost completely isoregic structure. However, the occurrence of head addition has been unambiguously demonstrated during many polymerizations. During the intramolecular steps of cyclopolymerization, 100% head addition may be obtained (see 4.5.2).

The tendency for radicals to give tail addition means that a head-to-head linkage will, most likely, be followed by a tail-to-tail linkage (Scheme 5).

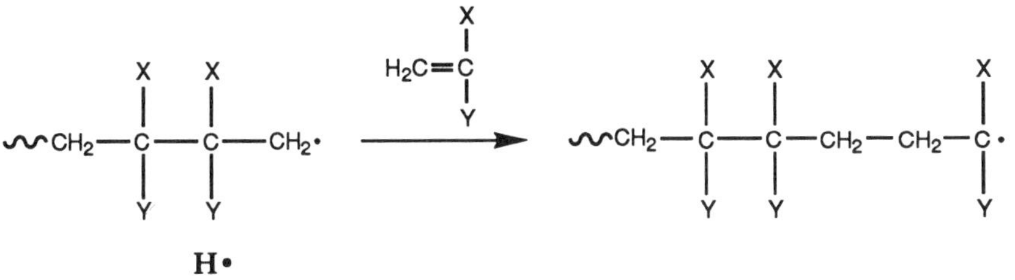

Scheme 5

Thus, head-to-head linkages formed by an "abnormal" addition reaction are chemically distinct from those formed in termination by combination of propagating radicals (Scheme 6).

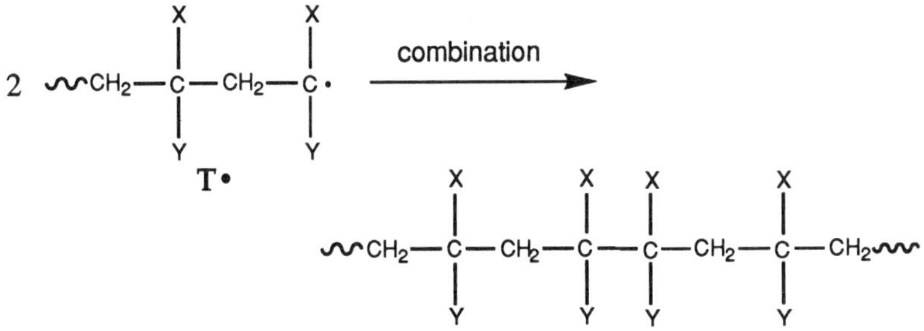

Scheme 6

In view of the potential problems associated with discriminating between the various types of head-to-head linkages, it is perhaps curious that, while much effort has been put into finding head-to-head linkages, relatively little attention has been paid to applying spectroscopic methods to detect tail-to-tail linkages where no such difficulty arises.

Even allowing for the above-mentioned complication, the number of head-to-head linkages is unlikely to equate exactly with the number of tail-to-tail linkages. The radicals formed by tail addition (T•) and those formed by head addition (H•) are likely to have different reactivities.

Consideration of data on the reactions for small radicals (see 2.2) suggest that the primary alkyl radical (H•) is more likely to give head addition than the normal propagating species (T•) for three reasons:

(a) The propensity for head addition, which usually corresponds with attack at the more substituted end of the double bond, should decrease as the steric bulk of the attacking radical increases. Note that H• (a primary alkyl radical in the case of mono- and 1,1-disubstituted monomers) will usually be less sterically bulky than T•.
(b) Most common monomers have some dipolar character. H• and T• will usually be polarized similarly to the head and tail ends of the monomer respectively. This should favor T• adding tail and H• adding head.
(c) The primary alkyl radical (H•) has no α-substituent to delocalize the free spin.

However, head addition is usually a very minor pathway and is difficult to determine experimentally. Analysis of the events which follow head addition presents an even more formidable problem. Therefore, there is little experimental data on polymers with which to test the above-mentioned hypothesis. Data for fluoro-olefins indicate that H• gives less head addition than T• (see 4.3.1.3). No explanation for the observation was proposed.

The primary alkyl radical, H•, is anticipated to be generally more reactive and may show different specificity to the secondary or tertiary radical, T•. In VAc and VC polymerizations the radical H• appears more prone to undertake intermolecular (see 4.3.1.1 and 4.3.1.2) or intramolecular (see 4.3.1 and 4.3.2) atom transfer reactions.

4.3.1.1 Poly(vinyl acetate)

It is generally agreed that *ca.* 1-2% of propagation steps during VAc polymerization involve head addition. There is some evidence that, depending on reaction conditions, a high proportion of the head-to-head linkages may appear at chain ends and that the number of head-to-head linkages may not equate with tail-to-tail linkages. The importance of head addition in VAc polymerization increases with the polymerization temperature.

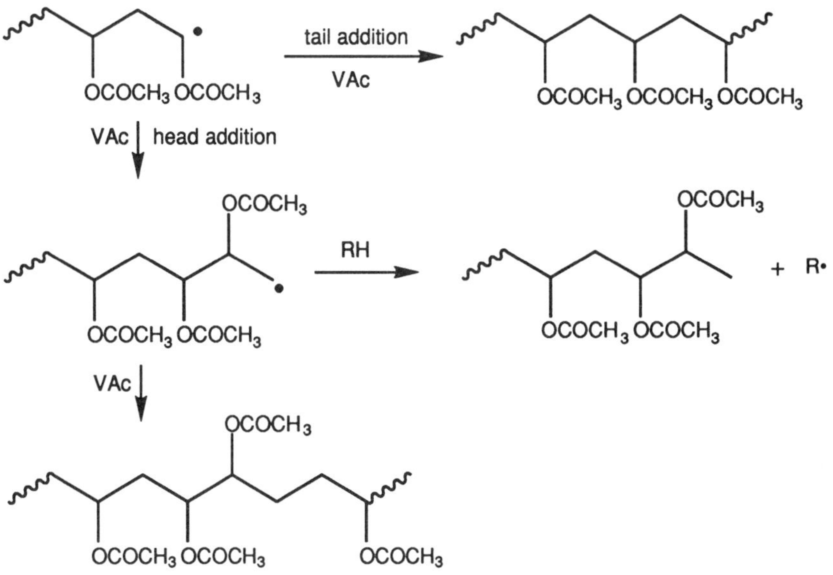

Scheme 7

The classic method for establishing the proportion of head addition occurring in VAc polymerization involves a two step process.[43] The PVAc is converted to PVA by exhaustive hydrolysis and the number of 1,2-glycol units is determined by periodate cleavage.

The reliability of the chemical method has been assessed by Adelman and Ferguson.[44] They showed that, for low molecular weight PVA, a significant proportion of the 1,2-glycol units appear at chain ends as 2,3-dihydroxybutyl groups (ca. 20% for $\bar{M}_n$ = 5,000, PVAc prepared in methanol at 75°C). The inference is that the radical formed by head addition is particularly active in inter- or intramolecular transfer and/or termination reactions. The result suggests that measurements of the molecular weight decrease after periodate cleavage could underestimate the amount of head addition.[43]

Analysis of ^{13}C NMR spectra of PVA provides a direct estimate of the extent of head addition occurring in VAc polymerizations.[33,45,46] Another advantage of the NMR method over chemical methods is that both head-to-head and tail-to-tail linkages can be observed. The polymers examined in these studies[33,45] were of relatively high molecular weight and prepared by emulsion polymerization. They possessed an equal number of head-to-head and tail-to-tail linkages. We have found that NMR can also be used to determine the fraction of head-to-head linkages in PVAc directly.

The reaction conditions (solvent, temperature) may also influence the amount of head addition and determine whether the radical formed undergoes propagation or chain transfer.

4.3.1.2 Poly(vinyl chloride)

Establishment of the detailed microstructure of PVC has attracted considerable interest. This has been spurred by the desire to rationalize the poor thermal stability of the polymer (see Chapter 1). Many reviews have appeared on the chemical microstructure of PVC and the mechanisms of "defect group" formation.[47-50]

Although head addition occurs during PVC polymerization to the extent of ca. 1%, it is now thought that PVC contains few, if any, head-to-head linkages (<0.05%).[51,52] Propagation from the radical formed by head addition is not competitive with a unimolecular pathway for its disappearance, namely, 1,2-chlorine atom transfer (see Scheme 8).

Rigo et al.[53] were the first to propose that head addition does occur but is immediately followed by a 1,2-chlorine atom shift. The viability of 1,2-chlorine atom shifts is well established in model studies and theoretical calculations.[54] Experimental support for this occurring during VC polymerization has been provided by NMR studies on reduced PVC.[55,56] Starnes et al.[51] proposed that head addition may be followed by one or two 1,2-chlorine atom shifts to give chloromethyl or dichloroethyl branch structures respectively (see Scheme 8).

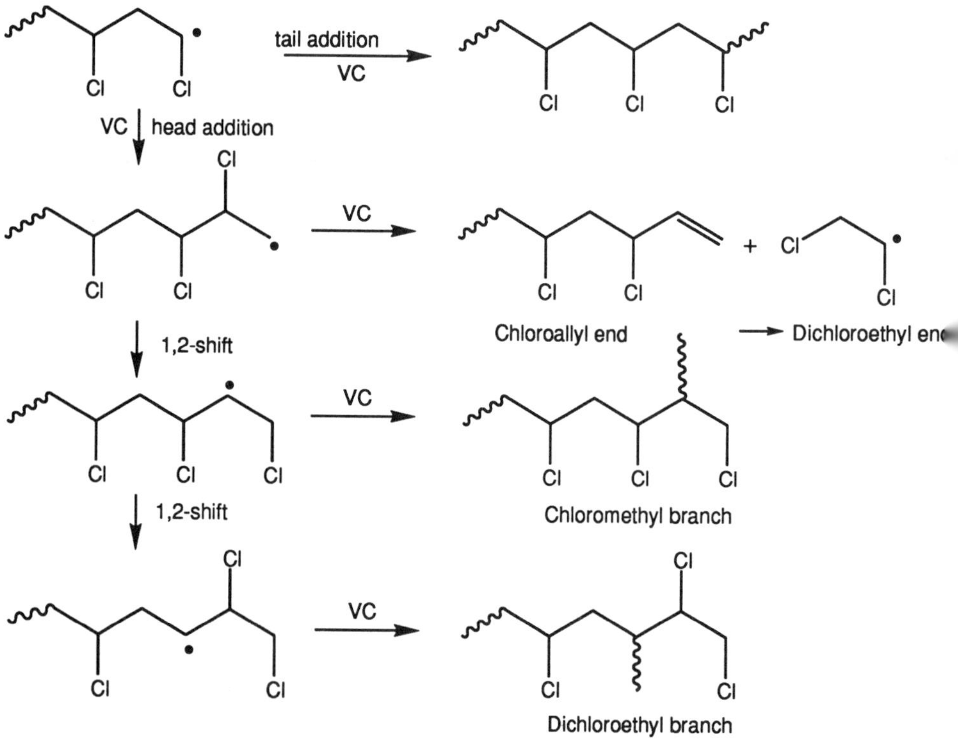

Scheme 8

Starnes et al.[57] have also suggested that the head adduct may undergo β-scission to eliminate a chlorine atom which in turn adds VC to initiate a new polymer chain. Kinetic data suggest that the chlorine atom does not have discrete existence. This addition-elimination process is proposed to be the principal mechanism for transfer to monomer during VC polymerization and it accounts for the reaction being much more important than in other polymerizations. The reaction gives rise to terminal chloroallyl and 1,2-dichloroethyl groups (see Scheme 8).

The presence of 1,2-dichloroethyl end groups and branch structures is likely to confuse attempts to determine head-to-head linkages by chemical methods (e.g. iodometric titration[58]).

4.3.1.3 Fluoro-olefin polymers

Propagation reactions involving the fluoro-olefins, vinyl fluoride (VF),[59-62] vinylidene fluoride (VF2)[59,62-64] and trifluoroethylene (VF3),[65] show relatively poor regiospecificity. This poor specificity is also seen in additions of small radicals to the fluoro-olefins (see 2.1). Since the fluorine atom is small, the major factors affecting the regiospecificity of addition are anticipated to be polarity and bond strength.

The fraction of head-to-head linkages in the poly(fluoro-olefins) increases in the series PVF2 < PVF ~ PVF3 (see Table 4.2). This can be rationalized in terms of the propensity of electrophilic radicals to add preferentially to the more electron rich end of monomers (i.e. that with the lowest number of fluorines). This trend is also seen in the reactions of trifluoromethyl radicals with the fluoro-olefins (see 2.2).

The proportion of head-to-head linkages in fluoro-olefin polymers also depends on the polymerization temperature (see Table 4.2).[59,60,62,63]

Table 4.2 Temperature Dependence of Head vs. Tail Addition for Fluoro-olefin Monomers

Temperature	% head addition		
°C	VF3[65]	VF2[63]	VF[60]
100			13.0
80	13.8	5.7	12.5
70			12.5
60			12.5
0	11.8	3.45	-
-80	10.0	3.0	-

^{19}F NMR studies have allowed regiosequence information to be determined at the pentad (VF) or heptad (VF2) level. Early studies[66] found that polymers formed by radical polymerization could be adequately described by Bernoullian statistics. However, Cais et al.[64] found that it was more appropriate to use first order Markov statistics to interpret regiospecificity. Their analysis suggests that the -CH$_2\bullet$ radical (formed by head addition) is much less likely to add head than the -CFX$\bullet$ radical [by a factor of ~14-18 for VF2 (depending on the polymerization temperature) or ~4 for VF]. No explanation for this selectivity was offered. The findings for fluoro-olefin propagation appear at variance with the considerations discussed above (see 4.3) and observations made for simple models. For example, with VF2, methyl radical is known to give much more head addition than trifluoromethyl radical (see 2.2).

4.3.1.4 Allyl polymers

Matsumoto et al.[67-71] have reported that substantial amounts (5-20%) of head addition occur during polymerization of allyl esters and that the proportion increases with polymerization temperature. The most recent study indicates that the proportion of head-to-head linkages in poly(allyl esters) is also dependent on the molecular weight of the polymer chain. For short chains, the fraction is reported to be ca. 10% irrespective of the nature of the ester group. For longer chains the proportion of head-to-head linkages decreases and the molecular weight dependence of this fraction increases according to the size/polarity of the ester group.

The very high levels of head addition and the substituent effects found in these studies are not consistent with expectations based on knowledge of the reactions of small radicals (see 2.2). If borne out, the results have implications in cyclopolymerization of diallyl monomers (see 4.4.1.1).

4.3.1.5 Acrylic polymers

Before the advent of NMR spectroscopy, a number of reports appeared suggesting the possibility of substantial head addition during polymerization of acrylate ester derivatives. Marvel et al.[72,73] reported chemical degradation experiments which suggested that α-haloacrylate polymers contain halogen substituents in a 1,2-relationship. On this basis they proposed that these monomers polymerize in a head-to-head, tail-to-tail fashion. McCurdy and Laidler[74] suggested that irregularities in the heats of polymerization of methyl and higher acrylates and methacrylates could be rationalized if a fraction of units were arranged in head-to-head, tail-to-tail arrangement.

Since that time, many studies by NMR and other techniques on the microstructure of acrylic and methacrylic polymers formed by radical polymerization have proved their predominant head-to-tail structure.

There is, however, some evidence that a small amount of head addition during propagation occurs in the polymerization of acrylic monomers. On the basis of chemical analysis, Sawant and Morawetz[75] suggested that 4.6% of amide groups in PAAm may be present as head-to-head linkages. Minigawa[76] has indicated the presence of a small percentage of head-to-head linkages in PAN.

4.3.2 Conjugated Diene Polymers

There is greater scope for structural variation in the diene based polymers than for the monoene polymers already discussed. The polymers contain units from overall 1,2- and *cis*- and *trans*-1,4-addition. Two mechanisms for overall 1,2-addition may be proposed. These are illustrated in Schemes 9 and 10:

(a) By analogy with the chemistry seen with monoene monomers the propagating species could, in principle, add to one of the internal (2- or 3-) positions of the diene (Scheme 9).

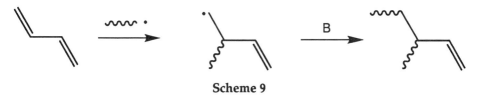

Scheme 9

(b) The delocalized allyl radical produced by addition to the 1- (or 4-) position may react in two ways to give overall 1,2-addition or 1,4-addition (Scheme 10).

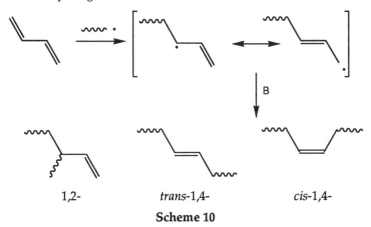

Scheme 10

Analyses of polymer microstructures do not allow these possibilities to be unambiguously distinguished. However, EPR experiments demonstrate that radicals add exclusively to one of the terminal methylenes.[77]

When used in conjunction with unsymmetrical dienes with substituents in the 2-position, the term 'tail addition' has been used to refer to addition to the methylene remote from the substituent. 'Head addition' then refers to addition to the methylene bearing the substituent (*i.e.* head addition ≡ 4,1- or 4,3-addition, tail addition ≡ 1,4- or

1,2-addition) as illustrated below for chloroprene (Scheme 11). Note that 1,2- and 4,3-addition give different structures while 1,4- and 4,1-addition give equivalent structures and a chain of two or more monomer units must be considered to distinguish between head and tail addition.

Tacticity is only a consideration for units formed by 1,2-addition. However, units formed by 1,4-addition may have a *cis*- or a *trans*-configuration.

In anionic or Ziegler-Natta polymerizations, reaction conditions can be chosen to yield polymers of specific microstructure. However, in radical polymerization while some sensitivity to reaction conditions has been reported, the product is typically a mixture of microstructures in which 1,4-addition is favored. Substitution at the 2-position (*e.g.* isoprene or chloroprene - see 4.3.2.2) favors 1,4-addition and is attributed to the influence of steric factors. The reaction temperature does not affect the ratio of 1,2:1,4-addition but does influence the configuration of the double bond formed in 1,4-addition. Lower reaction temperatures favor *trans*-1,4-addition (see 4.3.2.1 and 4.3.2.2).

Early work on the microstructure of the diene polymers has been reviewed.[1] While polymerizations of a large number of 2-substituted and 2,3-disubstituted dienes have been reported,[78] little is known about the microstructure of diene polymers other than PB,[79] polyisoprene,[80] and polychloroprene.[81]

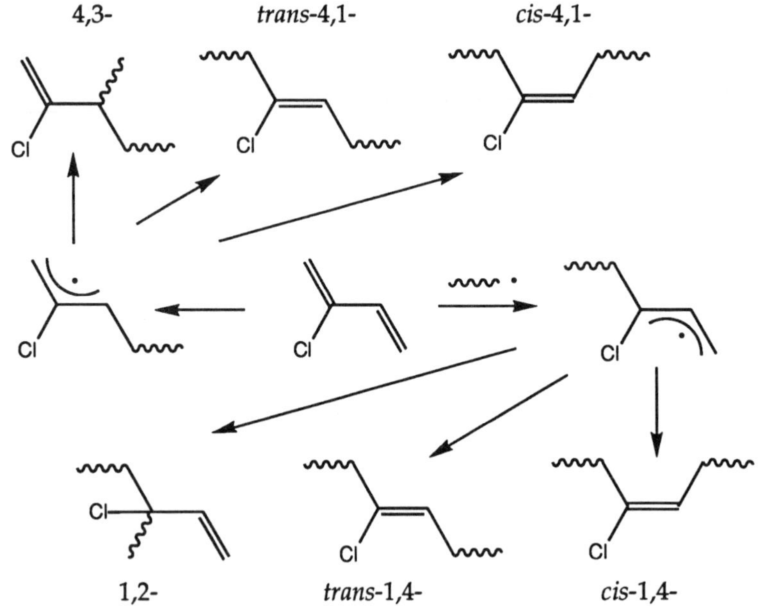

Scheme 11

4.3.2.1 Polybutadiene

The mechanism of B polymerization is summarized in Scheme 10. 1,2-, and *cis*- and *trans*-1,4-butadiene units may be discriminated by infra-red, Raman, or NMR spectroscopy.[1] PB comprises predominantly 1,4-*trans*-units. A typical composition is 57.3:23.7:19.0 for *trans*-1,4-:*cis*-1,4-:1,2-. While the ratio of 1,2- to 1,4-units shows only a small temperature dependence, the effect on the *cis-trans* ratio is substantial. Sato et al.[82] have determined dyad sequences by ^{13}C NMR and found that the distribution of isomeric structures and tacticity is adequately described by Bernoullian statistics.

4.3.2.2 Polychloroprene, polyisoprene

The mechanism of chloroprene polymerization is summarized in Scheme 11. Coleman et al.[83,84] have applied ^{13}C NMR in a detailed investigation of the microstructure of poly(chloroprene) (also known as neoprene). They report a substantial dependence of the microstructure on temperature and perhaps on reaction conditions (Table 4.3). The polymer prepared at -150°C essentially has a homogeneous 1,4-*trans*-microstructure. The polymerization is less specific at higher temperatures. Note that different polymerization conditions were employed as well as different temperatures and the influence of these has not been considered separately.

Table 4.3 Microstructure of Poly(chloroprene) vs. Temperature

Temp. (°C)	unit				
	1,4-*trans*	4,1-*trans*	1,4-*cis*[a]	1,2-[b]	4,3-
90[c]	75.1	10.3	7.8	2.9	4.1
40[d]	81.6	9.2	5.2	2.5	1.4
0[d]	90.4	5.5	1.8	2.1	1.1
40[d]	93.2	4.2	0.7	1.4	0.5
150[e]	98.0	2.0	<0.2	<0.2	<0.2

[a] 1,4- and 4,1-*cis* not distinguished.
[b] 25-50% of 1,2- are isomerized.
[c] Reaction conditions not stated.
[d] Emulsion polymer.
[e] Polymer prepared by irradiation of crystalline monomer.

Poly(isoprene) can also be prepared by radical polymerization.[85] Although the ratio of 1,4-:1,2-:4,3- units is stated to be *ca.* 90:5:5 irrespective of the polymerization temperature (range -20–50°C), the proportion of *cis*-1,4-addition increases from 0 at -20°C to 17.6% at 50°C. EPR studies also indicate that radicals add preferentially to the 1-position.[77]

4.4 Structural Isomerism - Rearrangement

During most radical polymerizations, the basic carbon skeleton of the monomer unit is maintained intact. However, in some cases the initially formed radical may undergo intramolecular rearrangement leading to the incorporation of new structural units into the polymer chain. The rearrangement may take the form of ring closure (see 4.4.1), ring opening (see 4.4.2) or intramolecular atom transfer (see 4.4.3).

The unimolecular rearrangement must compete with normal propagation. As a consequence, for systems where there is <100% rearrangement, the concentration of rearranged units in the polymer chain will be dependent on reaction conditions. The use of low monomer concentrations will favor the unimolecular process and it follows that the rearrangement process will become increasingly favored over normal propagation as polymerization proceeds and monomer is depleted (*i.e.* at high conversion). Higher reaction temperatures generally also favor rearrangement.

4.4.1 Cyclopolymerization

Diene monomers with suitably disposed double bonds may undergo intramolecular ring-closure in competition with propagation (Scheme 12). The term cyclopolymerization was coined to cover such systems. Many systems which give cyclopolymerization to the exclusion of "normal" propagation and crosslinking are now known. The subject is reviewed in a series of works by Butler.[86-90]

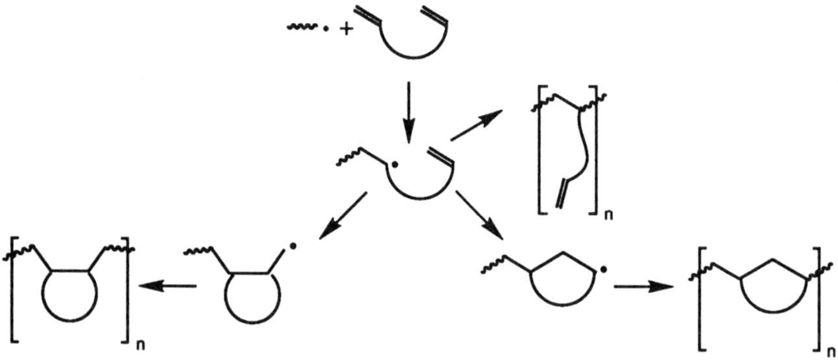

Scheme 12

Intramolecular cyclization is subject to the same factors as intermolecular addition (see 2.2). However, stereoelectronic factors achieve greater significance because the relative positions of the radical and double bond are constrained by being part of the one molecule (see 2.2.4) and can lead to head addition being the preferred pathway for the intramolecular step.

Geometric considerations in cyclopolymerization are optimal for 1,6-dienes (see 4.4.1.1). Instances of cyclopolymerization involving formation of larger rings have also been reported (see 4.4.1.4), as have examples where sequential intramolecular additions lead to bicyclic structures within the chain (see 4.4.1.2). Various 1,4- and 1,5-dienes are proposed to undergo cyclopolymerization by a mechanism involving two sequential intramolecular additions (see 4.4.1.3).

4.4.1.1 1,6-Dienes

The polymerization of nonconjugated diene monomers might be expected to afford polymer chains with pendant unsaturation and ultimately, on further reaction of these groups, crosslinked insoluble polymer networks. Thus, the finding by Butler *et al.*,[91] that polymerizations of diallylammonium salts, of general structure (2) [*e.g.* diallyldimethylammonium chloride (3)] gave linear saturated polymers, was initially considered surprising.

The explanation proposed was that propagation involved sequential inter- and intramolecular addition steps. The presence of cyclic structures within the polymer chain was soon confirmed by degradation experiments.[92] However, these experiments did not define the precise nature of the cyclic units unambiguously. Their nature was inferred on the basis of the then prevailing theory, that radical additions proceed so as to give the more stable product (a six-membered ring and a secondary radical). As a consequence, the structure of these cyclopolymers was not firmly established until the 1970s. Spectroscopic studies have shown that 5-membered ring formation is the preferred (kinetic) pathway during cyclopolymerization of simple diallyl compounds (4).[93-96]

Other 1,6-dienes yield varying ratios of 5- and 6-membered ring products depending on the substitution pattern of the starting diene. Substitution of the olefinic methine hydrogen (*e.g.* structure 5, R= CH₃) causes a shift from 5- to 6-membered ring formation. Bulky R substituents prevent efficient cyclization and cross-linked polymers may result.

164   The Chemistry of Free Radical Polymerization

A vast range of symmetrical and unsymmetrical 1,6-diene monomers has now been prepared and polymerized and the generality of the process is well established.[86,95] A summary of symmetrical 1,6-diene structures, known to give cyclopolymerization, is presented in Table 4.4. In most cases, the structure of the repeat units has not been rigorously established. Often the only direct evidence for cyclopolymerization is the solubility of the polymer or the absence of residual unsaturation. In these cases the proposed repeat unit structures are speculative.

The understanding of the mechanism of cyclopolymerization has been one of the driving forces responsible for many studies on the factors controlling the mode of ring closure of 5-hexenyl radicals and other simple model compounds.[97]

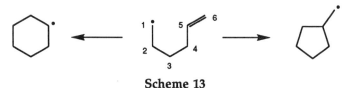

Scheme 13

The preferential 1,5-ring closure of unsubstituted 5-hexenyl radicals has been attributed to various factors; these are discussed in greater detail in chapter 2. The mode and rate of cyclization is strongly influenced by substituents. The results may be summarized as follows (see Scheme 13):

(a) Methyl substitution at C-1 slows the rate of both 1,5 and 1,6-ring closure. Substituents which delocalize the spin into a π-system (CN, CO_2Me) may result in a predominance of six-membered ring products by rendering intramolecular addition readily reversible.
(b) Substitution at C-5 dramatically retards 1,5-ring closure to the extent that 1,6-ring closure may predominate.
(c) Substitution at C-6 retards 1,6-ring closure. If both the 5 and 6 positions are substituted 1,5-ring closure predominates.
(d) Substitution at C-2, C-3, or C-4 facilitates both 1,5- and 1,6- ring closure.
(e) Increased reaction temperatures favor 1,6-ring closure at the expense of 1,5-ring closure.

The presence of heteroatoms and the inclusion of sp^2 centers are also known to affect the rate and mode of cyclization.

Thus, on the basis of model studies, it is possible to reconcile the observation that diallyl monomers that are unsubstituted on the double bond (4, $X=Z=CH_2$, $Y=CR_2$, NR, O, etc.) give predominantly five-membered rings for the intramolecular step. Dimethallyl monomers and other similarly substituted monomers (i.e. 5, R≠H) generally give predominantly 6-membered rings (e.g. 5, $X=Z=CH_2$, $R=CH_3$ or CO_2R - see Table 4.4). Dimethacrylic anhydride (5, Y=O, $X=Z=C=O$, $R=CH_3$) gives 6-membered rings.[98] It is surprising that dimethacrylic imides (**5**, Y=N-alkyl, X=Z=C=O, R=CH$_3$) are reported to give 5-membered rings.[98,99]

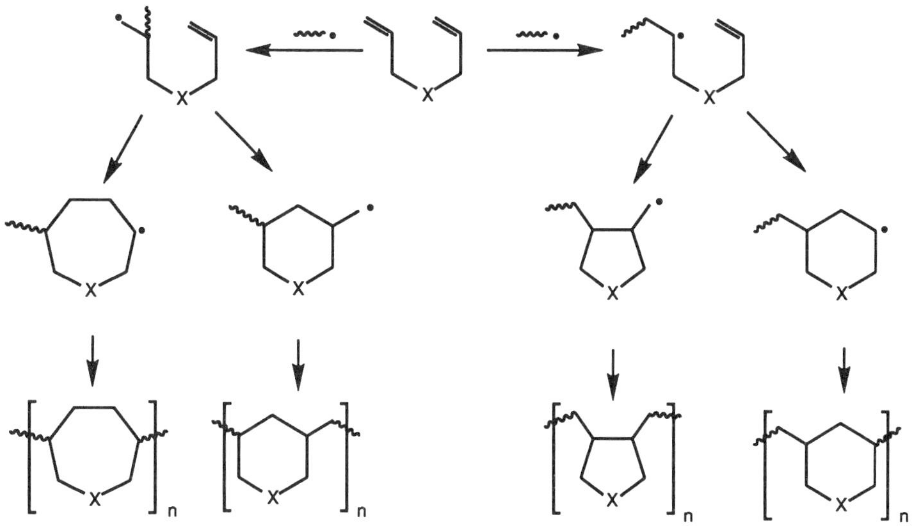

Scheme 14

The observation by Matsumoto *et al.* (see 4.3.1.4) that significant amounts of head addition occur in polymerization of simple allyl monomers brings into question the origin of the small amounts of 6-membered ring products that are formed in cyclopolymerization of simple diallyl monomers (Scheme 14). If the intermolecular addition step were to involve head addition, then the intramolecular step should give predominantly a 6-membered ring product (**7**) (by analogy with chemistry seen for 1,7 dienes - see 4.4.1.4). Note that the repeat units in (**7**) and (**9**) are the same; however, they are oriented differently in the chain. In circumstances where there is significant intermolecular head addition, the formation of seven membered units (**6**) might even be contemplated.[100]

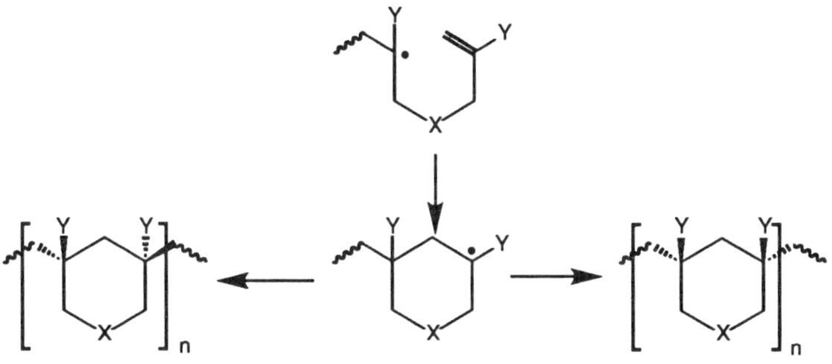

Scheme 15

Table 4.4 Symmetrical 1,6-Diene Monomers for Cyclopolymerization

	Monomer	substituents	Ring size[a]	refs.
(a) all carbon skeleton		X=Ph; Y=Z=H	6	101,102
		X=CO_2H, CO_2R, CN; Y=Z=H	6	103,104
		X=CO_2Me; Y=Z=CN	6	105
		X=CO_2R; Y=Z=CO_2R'	6	106
(b) 4-nitrogen		X=H	5	86,87,93,96
		X=CH_3	6	
		X=H	5	107
		X=CH_3	5	98,99
(c) 4-oxygen		X=H	5	
		X=CH_3	6	70,108,109
		X=CO_2R	6	110-112
(c) 3,5-oxygen		X=Y=H	5+6	113,114
		X=H; Y=C_3H_7		114
(d) other heteroatom substituents		X=H		115
		X=CH_3		116,117
				117
		X=O, NH, NCH_3	5+6+7?	100
			5+6	118
		X=H		119
		X=CH_3		
		X=H		120,121
		X=CH_3		120
		X=H		122

[a] Predominant ring size. If not specified, it has not been unambiguously determined.

The stereospecificity of the cyclization step has been examined both for model systems[97,123] and in a few cyclopolymerizations.[105] In formation of either 5- or 6-membered[105] rings, there appears to be a preference for the polymeric residues to end up *cis*- to each other. Note that for cyclopolymers with 6-membered ring units, the ring stereochemistry is established in the intermolecular addition step (Scheme 15). In the case of 5-membered ring units, ring stereochemistry is established during the intramolecular step (Scheme 16).

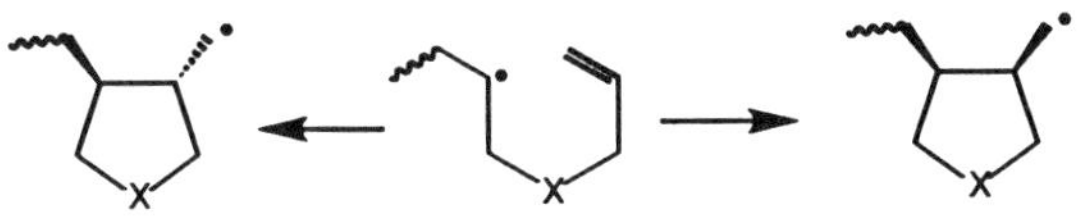

Scheme 16

Unsymmetrical 1,6-dienes known to undergo cyclopolymerization include allyl (meth)acrylate (**10** X=H, CH$_3$; Y=H),[124] (**10** X=CH$_3$; Y=Ph)[125] and (meth)acrylamide derivatives (**11** X=H, CH$_3$)[124,126-128] and o-allyl (**12** X=H)[129] and o-isopropenylstyrene (**12** X=CH$_3$).[130] With these cyclopolymerizations initial addition is to the double bond with the α–phenyl or carbonyl group and residual double bonds are isopropenyl or allyl groups.[127,128] For these examples, the cyclization step is relatively slow and reaction conditions are extremely important in obtaining soluble (uncrosslinked) polymers.

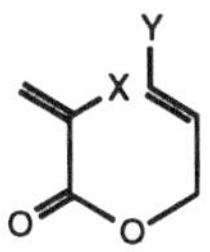

(**10**) 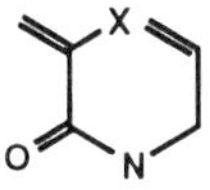   (**11**) 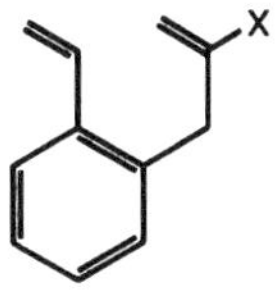   (**12**)

Propagation in cyclopolymerization may be substantially faster than for analogous monoene monomers.[131] The various theories put forward to account for this observation are summarized in Butlers review.[86]

One contributing factor, which seems to have been largely ignored, is that the ring closed radical (in many cases a primary alkyl radical) is likely to be much more reactive towards double bonds than the allyl radical propagating species. This species will also have a different propensity for degradative chain transfer (a particular problem with allylamines and related monomers - see 5.3.3.4) and other processes which complicate polymerizations of the monoenes.

4.4.1.2 Triene monomers

Triallyl monomers [*e.g.* (13) or salts thereof] can potentially undergo two successive intramolecular cyclizations.[132,133] However, in practice these materials give insoluble products.

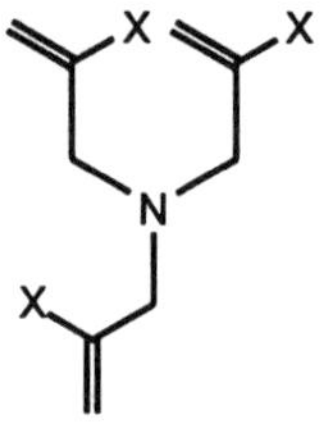

(13)

A model study has demonstrated the pathways shown in Scheme 17. The first cyclization step gave predominantly 5-ring closure, the second a mixture of 6- and 7-ring closure.[134] Relative rate constants for the individual steps were measured. The first cyclization step was found to be some five-fold faster than for the parent 5-hexenyl system. Although originally put forward as evidence for hyperconjugation in 1,6-dienes, further work showed the rate acceleration to be steric in origin.[97]

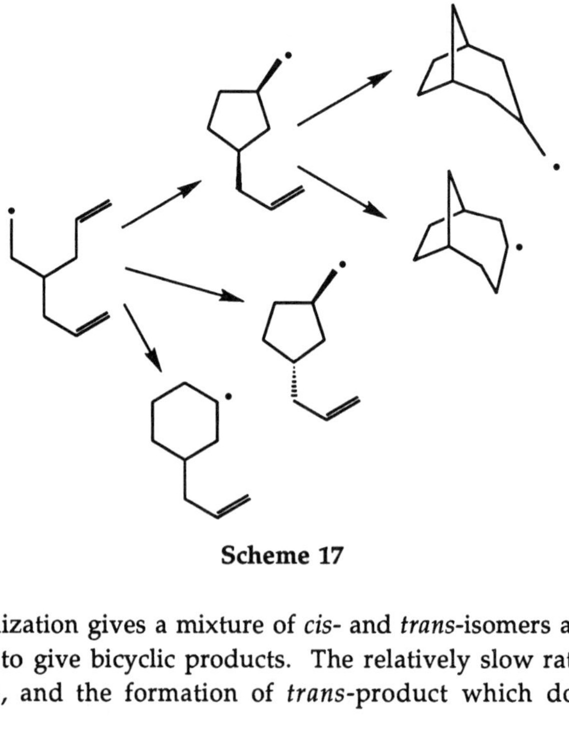

Scheme 17

The first cyclization gives a mixture of *cis*- and *trans*-isomers and only the *cis*-isomer goes on to give bicyclic products. The relatively slow rate of the second cyclization step, and the formation of *trans*-product which does not cyclize, provides an explanation for the observation that radical polymerizations of triallyl monomers often give a crosslinked product.

4.4.1.3 1,4- and 1,5-dienes

Geometric considerations would seem to dictate that 1,4- and 1,5-dienes should not undergo cyclopolymerization readily. However, in the case of 1,4-dienes, a 5-hexenyl system is formed after one propagation step. Cyclization *via* 1,5-backbiting generates a second 5-hexenyl system.

Homopolymerization of divinyl ether is thought to involve such a bicyclization. The polymer contains a mixture of structures including that formed by the pathway shown in Scheme 18.

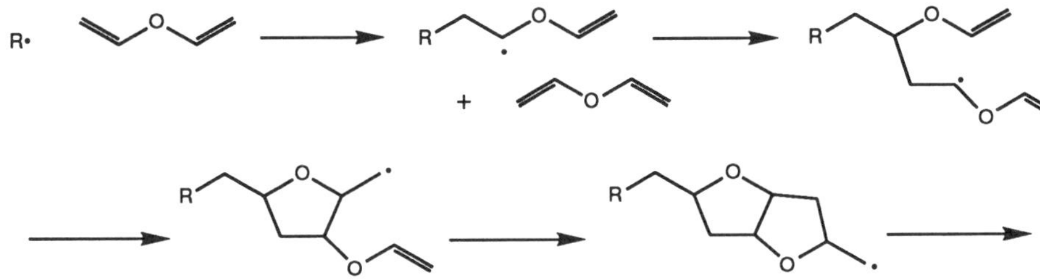

Scheme 18

It has been suggested that certain 1,5-dienes including *o*-divinylbenzene (**14**),[135] vinyl acrylate (**15**, X=H) and vinyl methacrylate (**15**, X=CH$_3$)[124] may also undergo cyclopolymerization with a monomer addition occurring prior to cyclization and formation of a large ring. However, the structures of these cyclopolymers have not been rigorously established.

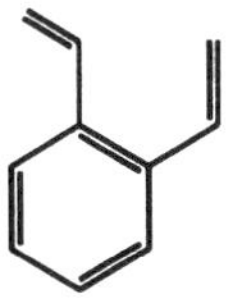

(14) 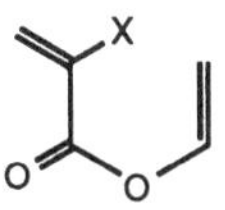   (15)    (16)

Bicyclo[2,2,1]heptadiene derivatives (**16**) are set up to undergo ring closure to form a 3-membered ring and it is proposed that polymers formed from (**16**) contain predominantly nortricyclene units.[136,137]

4.4.1.4 1,7- and higher 1,n-dienes

Several polymerizations of 1,7- and higher diene monomers have been reported. Cyclization to large rings (> 6-membered) has been postulated.[138-143] However, in most examples, cyclization is not quantitative and crosslinked polymers are formed. Evidence for ring formation comes from kinetic data and, in particular, from the delay in the gel point from that expected (based on the assumptions that no cyclization occurs and that all pendant double bonds are available for crosslinking reactions). One common monomer that is thought to show such behavior is methylene-bis-acrylamide (ring structure not proven).[138,139]

1,7-dienes give 6-membered rings in preference to 7-membered rings; examples include ethylene glycol divinyl ether (17) and bis-acryloylhydrazine.[140] This preference is also seen with model 6-heptenyl radicals.[141] One of the first reported examples of a 'cyclopolymerization' was that of the 1,11-diene, diallyl phthalate (18). A significant fraction (30-40%) of repeat units in the low conversion polymer was postulated to have a cyclic structure.[142,143] NMR studies on polymers formed by exhaustive hydrolysis suggest the cyclopolymer contains 11-membered rings.[144]

(17)   (18)   (19)

Various dimethacrylates have been polymerized in an effort to synthesize a poly(methacrylate) with head-to-head linkages.[98,99] Various 1,6- (e.g. dimethacrylamides - see Table 4.4), 1,7- (e.g. dimethacrylhydrazines) and 1,8-dienes (e.g. dimethacryloylureas) are reported to give head-to-head addition (5-, 6- or 7-membered rings respectively) or a mixture of head-to-head and head-to-tail addition. A 1,9-diene, o-dimethacryloylbenzene (19)[108] affords only head-to-tail addition (9-membered ring). Methacrylate derivatives of oligo- and polyhydroxy compounds have been shown to undergo cyclopolymerization to give ladder polymers. These polymerizations are considered further in the section on template polymerization (see 7.3.3.2).

4.4.1.5 Cyclo-copolymerization

Certain 1,4-dienes undergo cyclo-copolymerization with suitable olefins. For example, divinyl ether and MAH are proposed to undergo alternating copolymerization as illustrated in Scheme 19.[145] These cyclo-copolymerizations can be quantitative only for the case of a strictly alternating copolymer. This can be achieved with certain electron donor-electron acceptor pairs, for example divinyl ether-maleic anhydride.

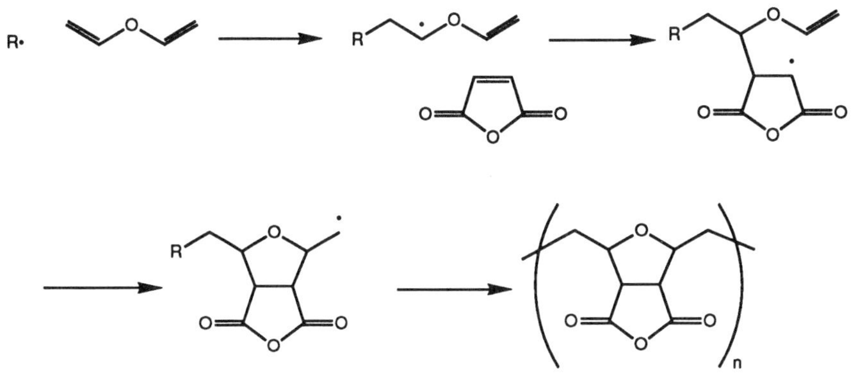

Scheme 19

4.4.2 Ring-Opening Polymerization

Much of the interest in ring-opening polymerizations stems from the fact that the polymers formed may have lower densities than the monomers from which they are derived (*i.e.* volume expansion may accompany polymerization).[146-148] This is in marked contrast with conventional polymerizations which invariably involve a nett volume contraction. Such polymerizations are therefore of particular interest in adhesive, mold filling, and other applications where volume contraction is undesirable.

Ring-opening polymerizations and copolymerizations also offer novel routes to polyesters and polyketones (see 4.4.2.2). These polymers are not otherwise available by radical polymerization. Finally, ring-opening copolymerization can be used to give end functional polymers. For example, copolymerization of ketene acetals with, for example, styrene, and basic hydrolysis of the ester linkages in the resultant copolymer offers a route to α,ω-difunctional polymers (see 6.3.1.3).

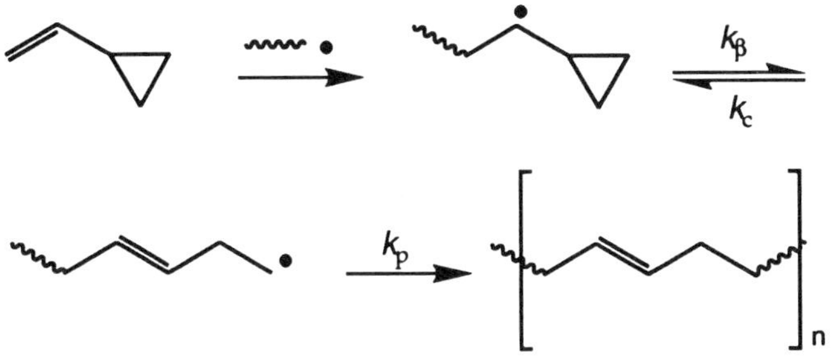

Scheme 20

Recent reviews on radical ring-opening polymerization include those by Bailey,[149] Endo[150] and Stansbury.[148] Monomers used in ring-opening are typically vinyl (*e.g.* vinylcyclopropane - Scheme 20; see 4.4.2.1) or methylene substituted cyclic compounds (*e.g.* ketene acetals - see 4.4.2.2) where addition to the double bond is followed by β-scission.

However, there are also examples of addition across a strained carbon-carbon single bond, as occurs with bicyclobutane[151] and derivatives (Scheme 21).[152]

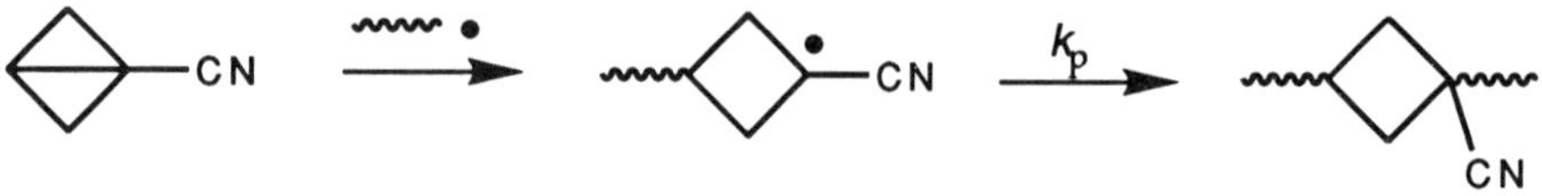

Scheme 21

For ring-opening to compete effectively with propagation, the former must be extremely facile. For example with $k_p \sim 10^2$-10^3 M^{-1} s^{-1} the rate constant for ring opening (k_β) must be at least ~10^5-10^6 s^{-1} to give >99% ring-opening in bulk polymerization. The reaction conditions can be chosen so as to favor ring-opening. Ring-opening will be favored by dilute reaction media and, usually, by higher polymerization temperatures.

The ring-opening reaction usually results in the formation of a new unsaturated linkage. When this is a carbon-carbon double bond, the further reaction of this group during polymerization leads to a crosslinked (and insoluble) structure and can be a serious problem when networks are undesirable.

4.4.2.1 *Vinyl substituted cyclic compounds*

There must be considerable driving force for ring-opening if it is to compete with propagation. In the case of vinylcyclopropane and derivatives (**20**, Scheme 20) this is provided by the relief of strain inherent in the three membered ring. Rates of ring-opening of cyclopropylmethyl radicals are reported to be in the range 10^5-10^8 s^{-1} depending on the substitution pattern.[153-157]

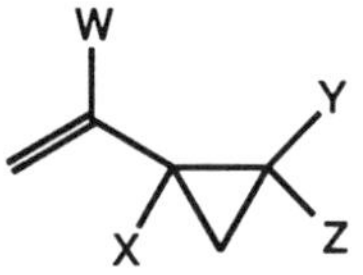

(20) 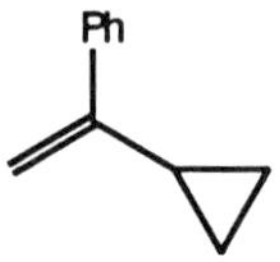   (21)

Many polymerizations of vinylcyclopropane and substituted derivatives (**20**) have now been reported.[158-175] All examples give 100% opening of the cyclopropane ring. However, conversions and polymerization rates are often low, even when the double bond is activated towards addition by a phenyl substituent (**21**).[175-177] For this example, the explanation for low polymerization rates probably lies with the reversibility of ring-opening.[176] The reversibility of cyclopropylmethyl radical ring-opening has been established even for the parent system. The α-phenyl substituent reduces the rate of ring-opening by some two to three orders of magnitude[155,178] and the equilibrium lies in favor of the ring-closed radical.[178] Even though the rate constant for ring-opening is slow in the case of (21), the monomer is unlikely to undergo polymerization without ring-opening. Such a polymerization should have a low ceiling temperature[176] since (21) is structurally analogous to α-methylstyrene (see 4.5.1).

For ring-substituted vinylcyclopropane derivatives (22), two pathways for ring opening are available (Scheme 22). There have been a number of studies on substituent effects on ring-opening of cyclopropylmethyl radicals.[153-157] Steric, polar and stereoelectronic factors are all important in determining the kinetics and preferred mode of ring-opening. Since this is a reversible process, the kinetic and thermodynamic products may be different.[157] In ring-opening polymerizations of unsymmetrically substituted vinylcyclopropanes (22) (Y or Z≠H), units derived from propagation of the more thermodynamically stable radical (25) dominate (see Scheme 22).[179]

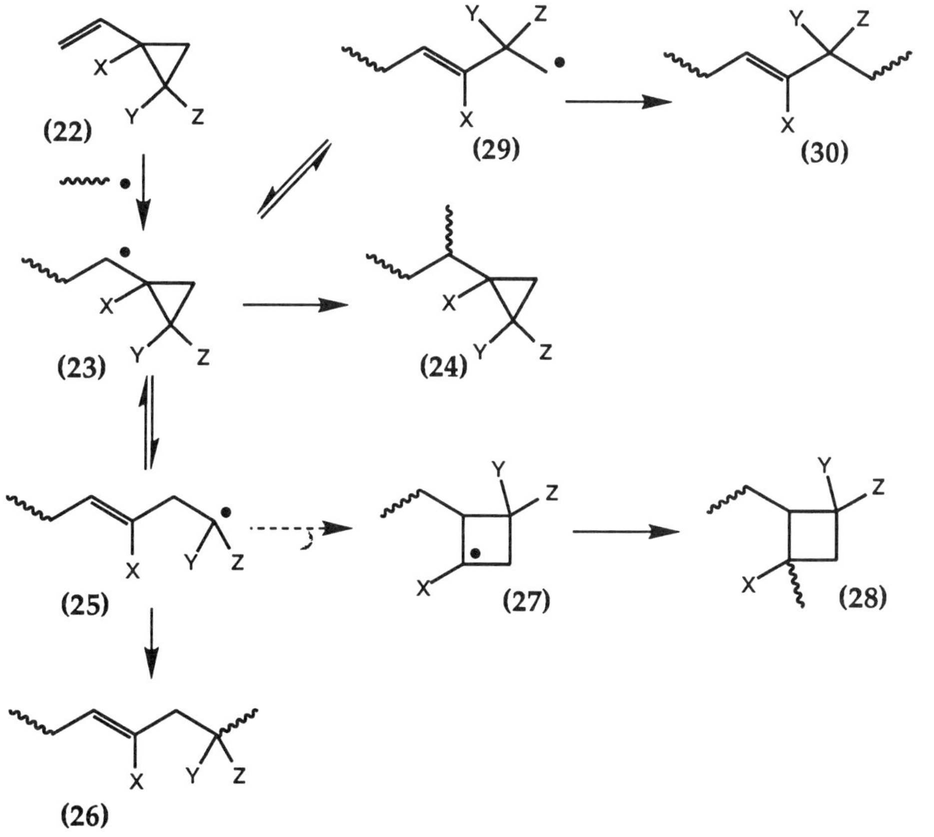

Scheme 22

It has been proposed that ring-opened radicals (25) may undergo ring-closure to a cyclobutane ring (27) in competition with propagation.[172,179] The final polymer then is a copolymer containing units of (26) and (28) (Scheme 22). At this stage the only evidence for this pathway is observation of signals in the NMR spectrum of the polymer that cannot be rationalized in terms of the structures (24), (25) or (30). There is no precedent for 1,4-ring-closure of a butenyl radical in small molecule chemistry and the result is contrary to expectation based on stereoelectronic requirements for intramolecular addition (see 2.2.4). However, an alternate explanation has yet to be proposed.

The vinyloxirane (31, Scheme 23) undergoes ring-opening polymerization to give a polyether structure[180-182] with specific cleavage of the C-C bond. Note that other oxiranylmethyl radicals (without the phenyl substituent) are reported to give specific cleavage of the C-O bond.[183]

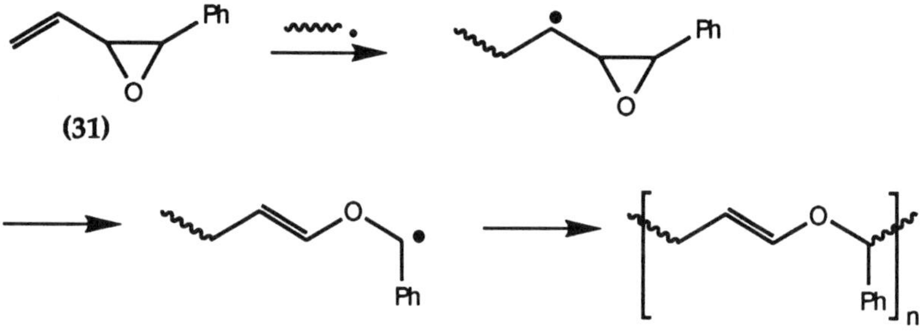

Scheme 23

Rate constants for ring-opening of cyclobutylmethyl radicals[184] are less than those for the corresponding cyclopropylmethyl radicals by a factor of ca. 10^4.[153] This is consistent with the smaller degree of ring strain inherent in the four-membered ring. Model studies have shown that cis-β-substituents on the cyclobutane ring led to a markedly enhanced rate constant for ring-opening.[184] The substituted vinylcyclobutane (32; stereochemistry unspecified, Scheme 24) is reported to give >90% ring-opening on polymerization in bulk at 60°C.[185]

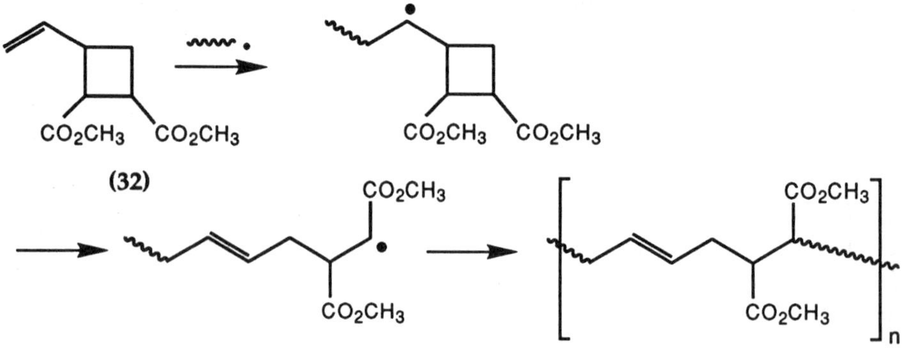

Scheme 24

Propagation

For most vinylcyclopentane (cyclopentylmethyl radical) and vinylcyclohexane (cyclohexylmethyl radical) derivatives, ring-opening is generally not a favorable process (see 4.4.1). However, a number of ring-opening polymerizations involving five- or larger-membered rings have been reported where appropriate substitution is present to provide the driving force for the β-scission step. Examples are the vinylsulfones (33, n=0,1,2),[186-188] which undergo ring-opening polymerization by scission of a relatively weak C-S bond and loss of sulfur dioxide, and the spiro derivatives (34)[189] and (35),[190] where ring-opening is facilitated by the concomitant aromatization of a cyclohexadiene derivative.

(33)   (34)   (35)

Polymerization of compound (35) gives between 43% (85°C, bulk) and 98% (130°C, bulk) ring-opening depending on reaction temperature.[190] Near quantitative ring-opening has been obtained in the case of polymerizations of (36) and (37) where further driving force for ring-opening is provided by formation of a benzylic radical.[191] These monomers (36 and 37) also undergo ring-opening in copolymerization with styrene.

(36)   (37)

4.4.2.2 Methylene substituted cyclic compounds

The ring-opening polymerization of ketene acetals (38, X=O) is a novel route to polyesters and many examples have now been reported (Scheme 25).[192-197] A disadvantage of these systems is the marked acid sensitivity of the monomers which makes them relatively difficult to handle. This area is covered by a series of reviews by Bailey *et al.*[149,198-200]

The main driving force for ring-opening in polymerizations of these compounds is formation of a strong carbon-oxygen double bond. The nitrogen (38, X=N-CH$_3$) and sulfur (38, X=S) analogs undergo ring-opening polymerization (see Table 4.5) with selective cleavage of the C-O bond to give polyamides or polythioesters respectively (Scheme 25). The specificity is most likely a reflection of the greater bond strength of C=O *vs.* the C=S or C=N- double bonds. The corresponding dithianes do not give ring-opening even though ring-opening would involve cleavage of a weaker C-S bond.[201,202]

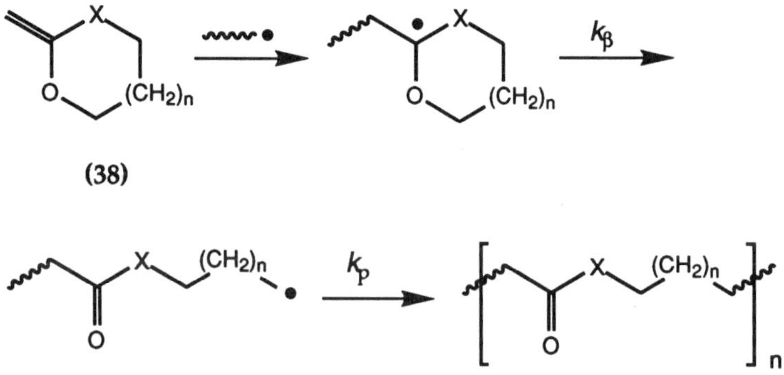

Scheme 25

The competition between ring-opening and propagation is dependent on ring size and substitution pattern. For the 5-membered ring ketene acetal (**38**, n=0) ring-opening is not complete except at very high temperatures. However, with the larger-ring system (**38**, n=2) ring-opening is quantitative. This observation (for the n=2 system) was originally attributed to greater ring strain. However, it may also reflect the greater ease with which the larger ring systems can accommodate the stereoelectronic requirements for β-scission (see 2.2).[97] Substituents (e.g. CH_3, Ph) which lend stabilization to the new radical center, or increase strain in the breaking bond, also favor ring-opening (see Table 4.5).

Table 4.5 Extents of Ring-opening During Polymerizations of 2-Methylene-1,3-dioxolane and Related Species

Monomer	% ring-opening	conditions	ref.[a]
	100	160°C, bulk, tBu_2O_2	
	87	120°C	
	50	60°C	
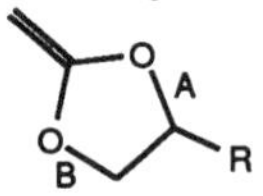	bond A 61[b] bond B 27	110°C, bulk, tBu_2O_2	203
	bond A 100[c]	120°C, bulk, tBu_2O_2	194,204
		30°C hv	205
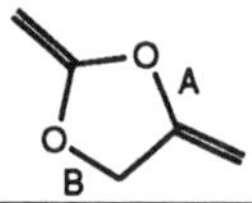	bond B 100	120°C, bulk, tBu_2O_2	206

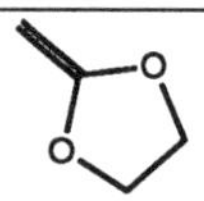

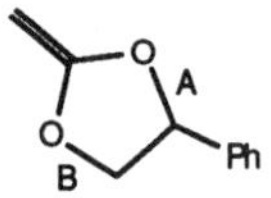

Table 4.5 (continued)

Monomer	% ring-opening	conditions	ref.[a]
(methylene dioxolane with two CH₃)	100	120°C, bulk, tBu_2O_2	
(methylene dioxolane with two Ph)	100	120°C, bulk, tBu_2O_2	195
(cyclohexadiene spiro dioxolane)	100	65-125°C, benzene, various initiators	196
(methylene 1,3-dioxane)	<100	120°C, bulk, tBu_2O_2	
(methyl-substituted methylene dioxane)	100	120°C, bulk, tBu_2O_2	
(methylene dioxane with exocyclic methylene)	<100	120°C, bulk, tBu_2O_2	206
(methylene 1,3-dioxepane)	100	120°C, bulk, tBu_2O_2	207,208
(dimethyl methylene dioxepane)	100	120°C, bulk, tBu_2O_2	209
(benzo-fused methylene dioxepane)	100	120°C, bulk, tBu_2O_2	209
(N-methyl methylene oxazolidine)	100	80°C, bulk, $(PhCO)_2$	210
(methylene oxathiolane)	45	120°C, bulk, tBu_2O_2	211

[a] Where no reference is given, the examples are taken from Bailey's review.[149]
[b] Data for R=n-decyl. Specificity dependent on R, temperature, and monomer concentration.
[c] Racemization accompanies polymerization of optically active monomer.[194]

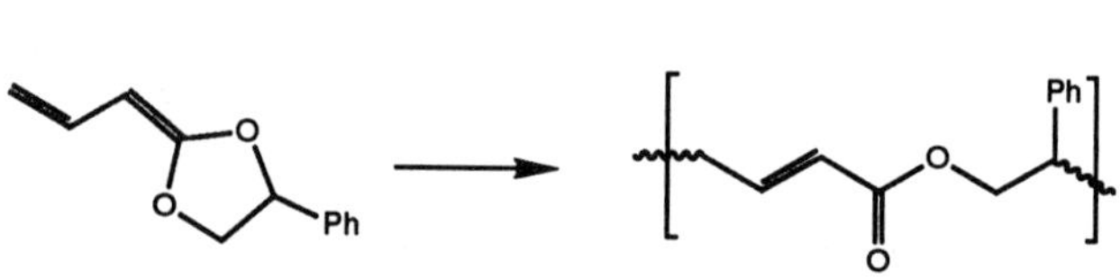

Scheme 26

The diene shown in Scheme 26 is also reported to give 100% ring-opening.[197] However, polymerization had to be carried out in very dilute solution to give a soluble (not crosslinked) product.

Rate constants for ring-opening of dioxolan-2-yl radicals have been measured by Barclay et al.[212] as 10^3-10^4 s^{-1} at 75°C (Scheme 27). There is also evidence that ring-opening is reversible.[212,213] Thus, isomerization of the initially formed product to one more thermodynamically favored is possible if propagation is slow.

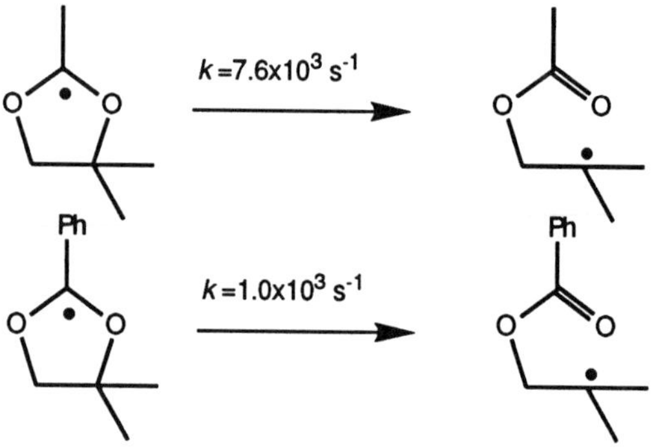

Scheme 27

Bailey et al.[149,214] observed that ring-opening polymerization of the monomers (39) and (40), which can potentially give rise to the same ring-opened radical, give different polymers. That from (39) has pendant vinyl groups, while that from (40) has in-chain double bonds. They proposed that in radical polymerization of ketene acetals, ring-opening might be concerted with addition of the next monomer unit and various experiments were suggested to test the hypothesis.[149] One of these was carried out by Acar et al.,[194] who showed that ring-opening polymerization of optically active 4-phenyl-1,3-dioxolane was accompanied by racemization. This is evidence against concerted ring opening.

It was originally proposed[149] that radical addition to (39) or (40) should occur exclusively at the respective methylene group (Scheme 28).

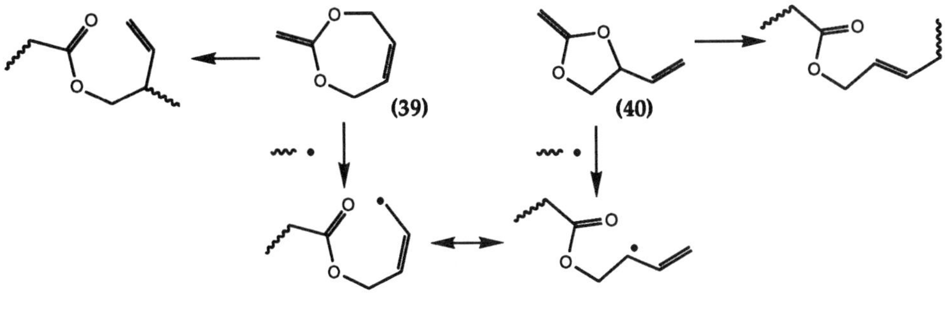

Scheme 28

If, however, radicals add preferentially to the vinyl group of (40), ring opening polymerization would give the polymer with in-chain double bonds specifically (Scheme 29). No other ring-opening polymerizations of vinyl dioxolane derivatives appear to have been reported to date.

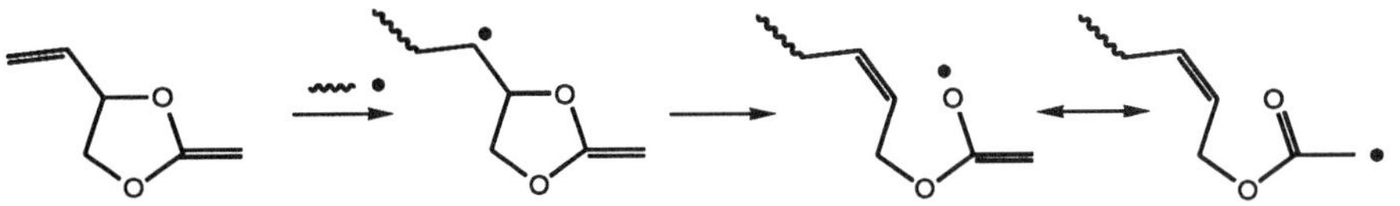

Scheme 29

The 4-methylenedioxolane derivatives also undergo ring-opening. However, the ring-opened radical may undergo a further β-scission (scheme 30). The final product may thus contain ketone units in the backbone.[193,215-221] The importance of the second β-scission step depends on the nature of substituents at the 2-position and the reaction conditions (Table 4.6).

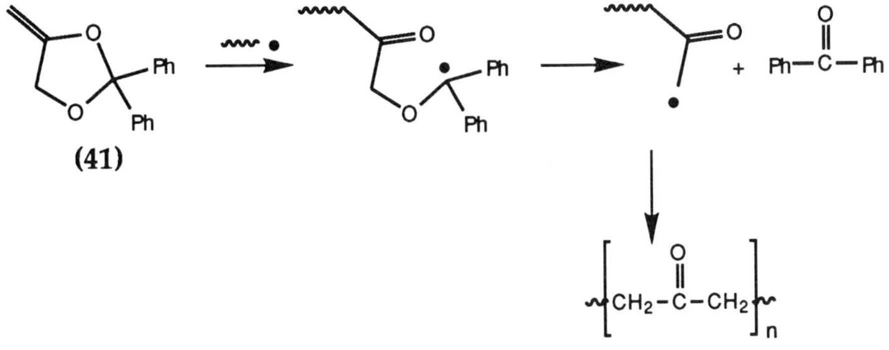

Scheme 30

Table 4.6 Extent of Ring-Opening During Polymerizations of 4-Methylene-1,3-dioxolane and 2-Methylene-1,4-dioxane Derivatives

Monomer	% ring-opening	% elimination	conditions.	ref.[a]
	30	100	130°C, bulk	149
	73	36	120°C, bulk	149,219
	100	0	<30°C, hv	218
	23	100	120°C, bulk	b,193
(41)	18	100	60°C, bulk	217
	100	100	120°C, bulk	215,216
	10	0	140°C, bulk	
	40	0	140°C, bulk	
	20	0	80°C, benzene	
	40	0	80°C, benzene	
	100	0	80°C, benzene	

[a] Where no reference is given, the examples are taken from Bailey's review.[149]
[b] Other 2-phenyl-2-alkyl derivatives reported to give incomplete ring-opening and 100% elimination at 120°C.

Of the 4-methylene-1,3-dioxolanes reported thus far (Table 4.6), only the 2,2-diphenyl derivative (41) is reported to give the polyketone quantitatively (Scheme 29). This requires temperatures in excess of 120°C in bulk polymerization.[215,216] 2-Phenyl-2-alkyl derivatives give incomplete ring-opening, though 100% elimination, at 120°C.[193] The 2-phenyl derivative is reported to afford ring-opening without elimination of benzaldehyde at temperatures less than 30°C (photochemical initiation).[218] At higher temperatures terpolymers are formed that comprise units that are non-ring-opened, ring-opened, and ring-opened with β-scission.

The structurally analogous 5-membered ring α-alkoxyacrylates (Scheme 31) are slow to ring-open and do not undergo β-scission to form an acyl radical propagating species.[149,222] This latter observation is probably a reflection of a higher bond strength for the bond α- to the carbonyl group. More ring-opening is observed for 6-membered ring systems (see Table 4.6).

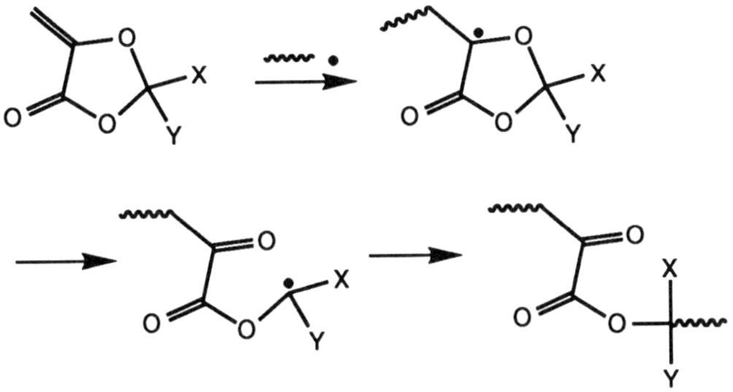

Scheme 31

Table 4.7 Extents of Ring-Opening During Polymerizations of 2-Methylenetetrahydrofuran and Related Compounds

Monomer	% ring-opening	conditions	ref.[a]
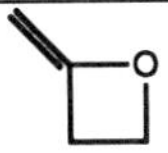	40	120°C, bulk, tBu_2O_2	211
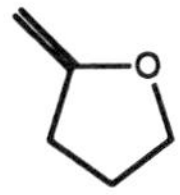	5	120°C, bulk, tBu_2O_2	223
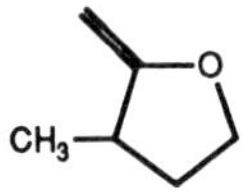	15-20	120°C, bulk, tBu_2O_2	
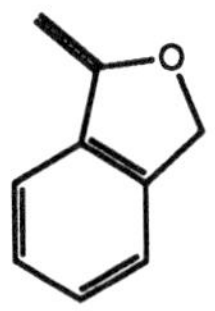	0	120°C, bulk, tBu_2O_2	
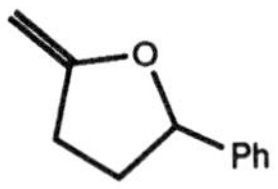	50	120°C, bulk, tBu_2O_2	223
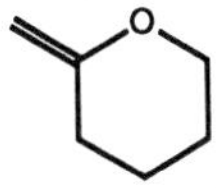	4-8	120°C, bulk, tBu_2O_2	

[a] Where no reference is given, the examples are taken from Bailey's review.[149]

Monomers with only a single ring oxygen-atom give less facile ring-opening. For example, the 2-methylenetetrahydrofuran derivatives (38, X=CH$_2$) give substantially less ring-opening than the corresponding 2- or 4-methylene-1,3-dioxolanes (see Table 4.7).

4.4.2.3 Double ring-opening polymerization

While many factors affect the degree of volume change which accompanies polymerization, any volume increase is directly related to the number of rings opened in the propagation step and is inversely related to the size of the rings being broken. Consideration of these factors leads to the conclusion that appreciable volume expansion on polymerization should only be expected when two or more rings are opened[148] and substantial effort has been put into designing systems where two or more rings are opened on polymerization.

It should also be noted that for many of the applications where volume expansion is required (adhesives, composites, *etc.*) a crosslinked product is desirable and some monomers have been designed with this in mind. This does, however, make the products difficult to characterize. Some monomers with potential for double ring-opening are reported in Table 4.8.

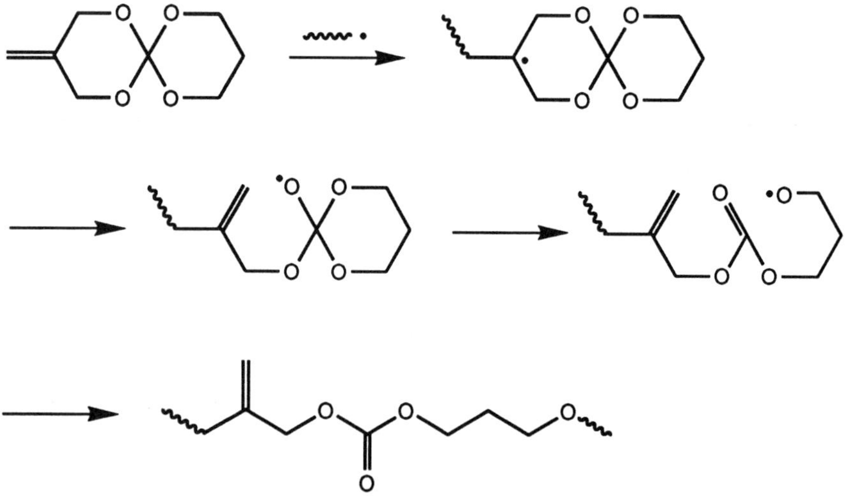

Scheme 32

Various methylene derivatives of spiroorthocarbonates and spiroorthoesters have been reported to give double ring-opening polymerization (*e.g.* Scheme 32). Like the parent monocyclic systems, these monomers can be sluggish to polymerize and reactivity ratios are such that they do not undergo ready copolymerization with acrylic and styrenic monomers. Copolymerizations with vinyl acetate have been reported.[148] These monomers, like other acetals, show marked acid sensitivity.

Table 4.8 Extents of Double Ring-Opening During Polymerization of Polycyclic Monomers

Monomer	% ring-opening	conditions	ref.[a]
spiroorthocarbonates			
	100	130°C, bulk, 30% conv.[b]	
	5-100	130°C, PhCl, <50% conv.[b]	148,177,224
	0	165°C, PhCl	225
spiroorthoesters			
	10[c]	120°C, bulk, tBu_2O_2	226,227
	10[d]	120°C, bulk, tBu_2O_2	228
	100	120°C, bulk, tBu_2O_2	
other systems			
	high	100°C, AIBN	229
	100	130°C, bulk, tBu_2O_2	230
	0[e]	60°C, bulk, AIBN	172,173
	46	60°C, bulk, AIBN	172,173

[a] Where no reference is given, the examples are taken from Bailey's reviews.[149]
[b] Insoluble and presumably crosslinked polymer formed at higher conversions.
[c] 50:50 mixture single and double ring-opened products.
[d] >50% double ring-opened product.
[e] Single ring-opened product only.

4.4.3 Intramolecular Atom Transfer

Intramolecular atom transfer, or backbiting, complicates polymerizations of E (Scheme 33 - see 4.4.3.1), VAc, and VC (see 4.4.3.2). Viswanadhan and Mattice[231] have carried out calculations aimed at rationalizing the relative frequency of backbiting in these and other polymerizations in terms of the ease of adopting the required conformation for intramolecular abstraction (see 2.3.5).

Scheme 33

Recently cases of "addition-abstraction" polymerization have been reported. In these polymerizations, propagation occurs by a mechanism involving sequential addition and intramolecular 1,5-hydrogen atom transfer steps (see 4.4.3.3).

4.4.3.1 Polyethylene

The extent of short-chain branching in PE may be quantitatively determined by a variety of techniques including IR,[232,233] pyrolysis-GC,[234] and γ-radiolysis.[235] The most definitive information comes from ^{13}C NMR studies.[236-240] The typical concentration of branch points in PE formed by radical polymerization chains is 8-25 per 1000 CH_2.[238] These are made up of: ethyl, 1.2-11.3; butyl, 3.9-8.5; pentyl (amyl), 0.6-2.2; hexyl and longer, 0.5-2.8. The range of values for extent and type of short-chain branches arises because the branching process is extremely dependent on the polymerization conditions.[238] High reaction temperatures and low pressures (monomer concentrations) favor the backbiting process.

The backbiting reaction first proposed by Roedel[241] (Scheme 33) is generally accepted as the mechanism for short chain branch formation during polymerization of E (for discussion on alternative mechanisms see[242,243]). The preferential formation of butyl [vs. propyl, pentyl, or longer branches] branches can be rationalized in terms of the stereoelectronic requirements imposed on the transition state (see 2.3.5). The preferred coplanar arrangement of atoms is most readily achieved in a 6-membered chair-like transition state.[244] 1-Undecyl radicals are a simple model of the PE propagating species and give 1,5- and 1,6-H transfer in the ratio 3:1. Other 1,n-H transfers were not detected.[245]

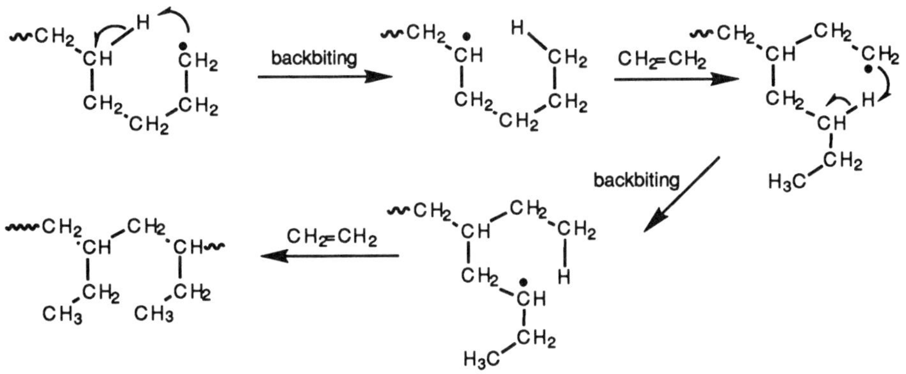

Scheme 34

Direct formation of an ethyl branch would require backbiting *via* a highly strained 4-membered transition state and therefore should have a low probability.[244,246] The relatively large numbers of ethyl branches in PE are accounted for by the occurrence of two successive 1,5-H transfers which leads to either a pair of ethyl branches (Scheme 34) or a 2-ethylhexyl branch depending on the site of abstraction.[247] This mechanism for ethyl branch formation requires that the radical formed by backbiting (secondary alkyl) should be substantially more prone to undertake backbiting than the normal propagating species (primary alkyl). This suggests that the former has a reduced rate of propagation (more sterically hindered radical) and/or an increased rate of intramolecular abstraction (Thorpe-Ingold effect).

4.4.3.2 Vinyl polymers

There is evidence for backbiting during the polymerizations of VC[57] and VAc.[46,248,249] The mechanism is now generally believed to be analogous to that discussed for PE above and should lead to the formation of 2,4-dichlorobutyl or 2,4 diacetoxybutyl branches (Scheme 35) respectively.

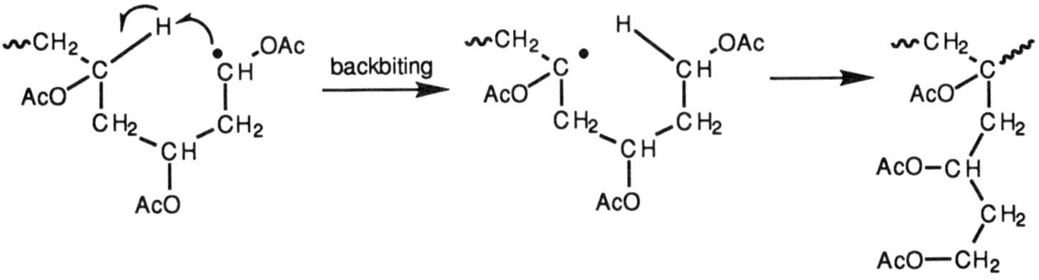

Scheme 35

The process is favored by low monomer concentrations as occurs at high conversions. Theoretical calculations suggest that the incidence of backbiting should be strongly dependent on the tacticity of the penultimate dyad.[250] Double backbiting (as in Scheme 34) in VC or VAc polymerization will lead to 2-chloroethyl or 2-acetoxy ethyl branches respectively.

There are no proven examples of 1,2-hydrogen atom shifts; this can be understood in terms of the stereoelectronic requirements on the process. The same limitations are not imposed on heavier atoms (*e.g.* chlorine). The postulate[251] that ethyl branches in reduced PVC are all derived from chloroethyl branches formed by sequential 1,5-intramolecular hydrogen atom transfers as described for PE (see 4.4.3.1) has been questioned.[47,55] It has been shown that many of these ethyl branches are derived from dichloroethyl groups. The latter are formed by sequential 1,2-chlorine atom shifts which follow a head addition (see 4.3.1.2).

4.4.3.3 *"Addition-abstraction" polymerization*

Recently several examples of "addition-abstraction" polymerization have been reported. In these polymerizations, the monomers are designed to give quantitative rearrangement of the initially formed adduct via 1,5-hydrogen atom transfer (Scheme 36). The monomers (42) are such that the double bond is electron rich (vinyl ether) and the site for 1,5-H transfer is electron deficient. This arrangement favors intramolecular abstraction over addition. Thus compound (42a) is reported[252,253] to undergo quantitative rearrangement during homopolymerization. For (42b) where the site of intramolecular attack is less electron deficient up to 80% of propagation steps involve intramolecular abstraction. As expected, the intramolecular abstraction pathway is favored by higher reaction temperatures and lower monomer concentrations.

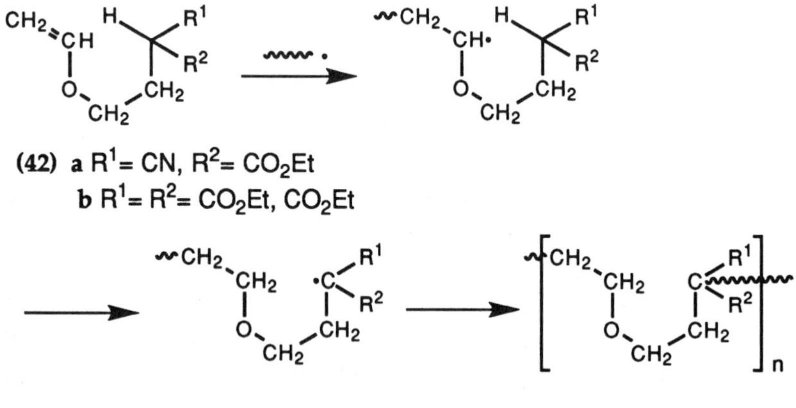

(42) a R^1 = CN, R^2 = CO_2Et
    b R^1 = R^2 = CO_2Et, CO_2Et

Scheme 36

4.5 Propagation Kinetics and Thermodynamics

In this section, we consider the kinetics of propagation and those features of monomer and propagating radical structure which render a monomer polymerizable by radical homopolymerization (see 4.5.1). The reactivities of monomers towards initiator derived species (see chapter 3) and in copolymerization (see chapter 6) are considered elsewhere.

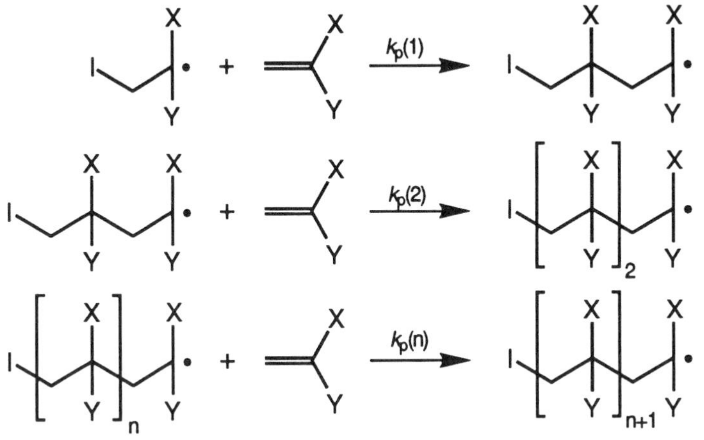

Scheme 37

In the literature on radical polymerization, the rate constant for propagation, k_p, is often taken to have a single value (refer Scheme 37). However, there is now good evidence that the value of k_p is dependent on chain length, at least for the first few propagation steps (4.5.3), and on the reaction conditions (see 7.3).

4.5.1 Polymerization Thermodynamics

Polymerization thermodynamics has been reviewed by Allen and Patrick,[254] Ivin,[255] Ivin and Busfield,[256] Sawada[257] and Busfield.[258] In most radical polymerizations, the propagation steps are facile (k_p typically > 10^2 M^{-1} s^{-1} - see 4.5.2) and highly exothermic. Heats of polymerization (ΔH_p) for addition polymerizations may be measured by analyzing the equilibrium between monomer and polymer or from calorimetric data using standard thermochemical techniques. Data for polymerization of some common monomers are collected in Table 4.9. Entropy of polymerization (ΔS_p) data are more scarce. The scatter in experimental numbers for ΔH_p obtained by different methods appears quite large and direct comparisons are often complicated by effects of the physical state of the monomer and polymers (*i.e.* whether for solid, liquid or solution, degree of crystallinity of the polymer).

Table 4.9 Heats of Polymerization for Selected Monomers ($CH_2=CRX$)

monomer	X	R	ΔH_p (kJ mol^{-1}) a	b	c	ΔS_p c J mol^{-1} K^{-1}		T_c d °C
AA	CO_2H	H	67	-	-	-		-
MAA	CO_2H	CH_3	43	65	-	-		-
MA	CO_2CH_3	H	78	-	-	-		-
MMA	CO_2CH_3	CH_3	56 (58)	55	56[259,260]	118[259,260]		202
EMA	$CO_2C_2H_5$	CH_3	60 (58)	-	60[261]	124[261]		211
BMA	$CO_2C_4H_9$	CH_3	58 (60)	-	-	-		-
MEA[262]	CO_2CH_3	C_2H_5	32e	-	-	-		22
AN	CN	H	75f	-	-	109[258]		415
MAN	CN	CH_3	57	-	64[263]	142g,[263]		177
S	Ph	H	69 (73)	70	73[264]	104[264]		428
AMS	Ph	CH_3	-	35	45[265]	148[265]		31
VAc	O_2CCH_3	H	88 (90)	-	-	-		-
VC	Cl	H	96	112	-	-		-

a   From calorimetry - data are for liquid monomer to amorphous solid polymer or for liquid monomer to polymer in monomer (in parentheses) and are taken from the Polymer Handbook.[258] All data are rounded to the nearest whole number.
b   From heat of combustion monomer and polymer - data are for liquid monomer to amorphous solid polymer and are taken from the Polymer Handbook.[258] All data are rounded to the nearest whole number.
c   From studies of monomer-polymer equilibria - data are for liquid monomer to amorphous solid polymer. All data are rounded to the nearest whole number.
d   Calculated from numbers of ΔH_p (column c except for AN) and DS_p shown and [M] = 1.0.
e   Based on a measured T_c of 82°C in bulk monomer and an assumed value for ΔS_p of -105 J mol^{-1} K^{-1}.[262] A more reasonable value of ΔS_p of 120 kJ mol^{-1} would suggest a ΔH_p of 40 kJ mol^{-1}.
f   Partially crystalline polymer.
g   In benzonitrile solution.

The addition of radicals to unsaturated systems is potentially a reversible process (Scheme 38). Depropagation is entropically favored and its importance therefore increases with increasing temperature (see Figure 2). The temperature at which the rate of propagation and depropagation become equal is known as the ceiling temperature (T_c). Above T_c there will be nett depolymerization.

$$P_n^\bullet + M \underset{}{\overset{K_{eq}}{\rightleftharpoons}} P_{n+1}^\bullet$$

Scheme 38

With most common monomers, the rate of the reverse reaction (depropagation) is negligible at typical polymerization temperatures. However, monomers with alkyl groups in the α-position have lower ceiling temperatures than monosubstituted monomers (see Table 4.9). For MMA at temperatures <100°C, the value of K_{eq} is <0.01 (see Figure 2). AMS has a ceiling temperature of <30°C and is not readily polymerizable by free radical methods. This monomer can, however, be copolymerized successfully (see 6.2.1.4).

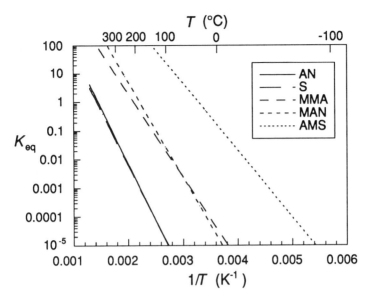

Figure 2 Dependence of K_{eq} on Temperature for Selected Monomers. Based on values of ΔH_p and ΔS_p shown in Table 4.9.

The value of T_c and the propagation/depropagation equilibrium constant (K_{eq}) can be measured directly by studying the equilibrium between monomer and polymer or they can be calculated at various temperatures given values of ΔH_p and ΔS_p

$$K_{eq} = \exp\left(\frac{\Delta H_p}{RT} - \frac{\Delta S_p}{R}\right) = \frac{1}{[M]_{eq}}$$

where $[M]_{eq}$ is the equilibrium monomer concentration.

$$T_c = \frac{\Delta H_p}{\Delta S_p + R \ln [M]}$$

Note that the value of T_c is dependent on the monomer concentration. In the literature, values of T_c may be quoted for [M] = 1.0 M, for [M] = $[M]_{eq}$ or for bulk monomer. Thus care must be taken to note the monomer concentration when comparing values of T_c. One problem with using the above method to calculate K_{eq} or T_c, is the paucity of data on ΔS_p. A further complication is that literature values of ΔH_p show variation of ±2 kJ mol^{-1} which may in part reflect medium effects.[258] This "error" in ΔH_p corresponds to a significant uncertainty in T_c.

Steric factors appear to be dominant in determining ΔH_p and ΔS_p. The resonance energy lost in converting monomer to polymer is of secondary importance for most common monomers. It is thought to account for ΔH_p for VAc and VC being lower than for acrylic and styrenic monomers.

Evidence for the importance of steric factors comes from a consideration of the effect of α-alkyl substituents. It is found that the presence of an α-methyl substituent raises ΔH_p by at least 20 kJ mol^{-1} (see Table 4.9, compare entries for AA and MAA, MA and MMA, AN and MAN, S and AMS). The higher ΔH_p probably reflects the greater difficulty in forming bonds to tertiary centers. This view is supported by the observation that higher alkyl substituents further increase ΔH_p [e.g. ethyl in MEA,[262] see Table 4.9). Increasing the chain length of the α-substituent from methyl to ethyl should not greatly increase the thermodynamic stability of the radical, but steric factors will make the new bond both more difficult to form and easier to break.

Limited data suggest that the entropic term may be as important as the enthalpic term in determining polymerizability. The value of ΔS_p is lowered >20 J mol^{-1} K^{-1} by the presence of an α-methyl substituent (see Table 4.9, compare entries for AN and MAN, S and AMS). This is likely to be a consequence of the polymers from α-methyl vinyl monomers to have a more rigid, more ordered structure than those from the corresponding vinyl monomers.

There have been many studies on the polymerizability of α-substituted acrylic monomers.[262,266-269] It is established that the ceiling temperature for α-alkoxyacrylates decreases with the size of the alkoxy group.[266] However, it is of interest that polymerizations of α-(alkoxymethyl)acrylates[266] and α-(acyloxymethyl)acrylates[268] appear to have much higher ceiling temperatures than the corresponding α-alkylacrylates. For example, methyl α-ethoxymethacrylate[266] readily polymerizes at 110°C whereas methyl ethacylate[262] has a very low ceiling temperature (see Table 4.9). The values of the thermodynamic parameters for these polymerizations have not been reported.

4.5.2 Measurement of Propagation Rate Constants

Methods for measurement of k_p have been reviewed by Stickler[270,271] and critically assessed by an IUPAC working party.[272-274] Traditionally, measurement of k_p has required determination of the rate of polymerization under steady state (to give k_p/k_t^2) and non-steady state conditions (to give k_p/k_t). The most commonly used techniques in this context are the rotating sector,[275-277] spatially intermittent polymerization (SIP)[278] and related methods.

Recently two methods which allow a more direct determination of k_p have become available. EPR methods have been developed which allow the absolute radical concentration to be determined as a function of conversion. With especially sensitive instrumentation this can be done by direct measurement.[279,280] An alternative method, applicable at high conversions involves trapping the propagating species in a frozen matrix[281,282] by rapidly cooling the sample to liquid nitrogen temperatures. The radical concentration, when coupled with information on the rate of polymerization, allows k_p (and k_t) to be calculated. The EPR methods have been applied to various polymerizations including those of MMA,[281,283-287] S[288,289] and VAc.[290] Values for k_p are at this stage not in complete agreement with those obtained by other methods (*e.g.* PLP, SIP) and this may reflect a calibration problem. Problems also arise because of the heterogeneity of the polymerization reaction mixture[285] and insufficient sensitivity for the radical concentrations in low conversion polymerizations.[282]

Pulsed laser photolysis (PLP)[291] has recently emerged as one of the more reliable methods for extracting absolute rate constants for the propagation step of free radical polymerizations.[272,273] The method has been successfully applied to establish rate constants, k_p(overall), for many polymerizations and copolymerizations (see Table 4.10).[291-306] In PLP the sample is subjected to a series of short (<30 ns) laser pulses at intervals τ. Analysis of the molecular weight distribution gives the length of chain formed between successive pulses (ν) and this yields a value for k_p.

$$\nu = k_p [M] \tau$$

A molecular weight distribution for a PS sample obtained from a PLP experiment with S is shown in Figure 3. Olaj *et al.*[291] proposed that ν was best estimated from the point inflection in the molecular weight distribution.

Table 4.10 Kinetic Parameters for Propagation in Selected Free Radical Polymerizations

monomer	A (M^{-1} s^{-1})	E_a(kJ mol^{-1})	method
MMA[307]	4.92×10^5	18.22	SIP
MMA[306]	2.39×10^6	22.3	PLP
EMA[295]	1.5×10^6	20.46	PLP
n-BMA[295]	3.44×10^6	23.3	PLP
t-BMA[302]	2.51×10^7	27.7	PLP
S[a,307]	1.1×10^7	29.5	SIP
S[304]	9.7×10^6	28.95	PLP
B[308]	8.05×10^7	35.7	PLP
VAc[309]	2.7×10^8	27.8	PLP

a   In benzene solution.

Kinetic modeling of PLP has been carried out using Monte Carlo methods[310,311] or by numerical integration.[304,309] These studies confirm that the point of inflection in the molecular weight distribution is a good measure of ν. The values of ν were also shown to be relatively insensitive to the termination rate and mechanism and the occurrence of many side reactions. Some difficulties experienced with monomers with high k_p (acrylates, VAc) appear to have been resolved through the use of low reaction temperatures and dilute media.[309]

Independent determination of the rate of polymerization allows k_p/k_t and hence k_t to be evaluated.³¹²

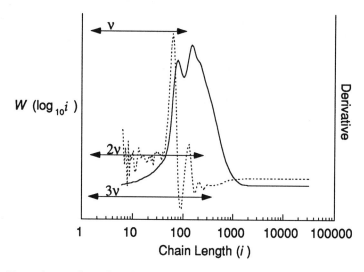

Figure 3 Experimental molecular weight distribution obtained by GPC (———) and its first derivative with respect to chain length (--------) for PS prepared by PLP. The vertical scales are in arbitary units. Polymerization of 4.33 M S at 60°C with benzoin 0.006M and laser conditions: λ=350 nm, 80-100 mJ/pulse, τ=0.05 s.³⁰⁴

There are some reports that values of k_p are conversion dependent and that the value decreases at high conversion due to k_p becoming limited by the rate of diffusion of monomer. While conversion dependence of k_p at extremely high conversions is possible, some early data may need to be reinterpreted, as the conversion dependence of the initiator efficiency was not recognized (see 3.3.1.1.3, 3.3.2.1.3 and 5.2.1.1).

4.5.3 Chain Length Dependence of Propagation Rate Constants

It is usually assumed that propagation rate constants for radical homopolymerization k_p are independent of chain length and, for longer chains (length >20), there is good experimental evidence to support this assumption.²⁷⁸,²⁸⁸ However, there is a growing body of indirect evidence to suggest that the rate constants for the first few propagation steps $k_p(1)$, $k_p(2)$, etc. are substantially greater than k_p(overall) (refer Scheme 37). Evidence comes from a number of sources, for example:

(a) The absolute rate constants for the reaction of small model radicals with monomers are often several orders of magnitude greater than the corresponding values of k_p.³¹³

(b) Chain transfer constants ($k_p/k_{transfer}$) often show a marked chain length dependence for very short chain lengths (see 5.3).[314]

(c) Aspects of the kinetics of emulsion polymerization[315] and molecular weight distributions of polymers formed in pulsed laser photolysis experiments[304] are best explained by invoking chain length dependence of k_p.

There have been attempts at direct measurements of these important kinetic parameters for MA and MAN polymerization.[316,317] These data suggest $k_p(1)$ is as much as 10-fold greater than k_p(overall). The effect can be seen as a special case of a penultimate unit effect (see 6.2.1.2).

References

1. Bovey, F.A., Chain Structure and Conformation of Macromolecules (Wiley: New York 1982).
2. Randall, J.C., Polymer Sequence Determination (Academic Press: New York 1977).
3. Bovey, F.A., Polymer Conformation and Configuration, p. 41 (Academic Press: New York 1969).
4. Koenig, J.L., Chemical Microstructure of Polymer Chains (Wiley: New York 1980).
5. Tonelli, A.E., NMR Spectroscopy and Polymer Microstructure (VCH: New York 1989).
6. Farina, M., *Top. Stereochem.*, 1987, **17**, 1.
7. Bovey, F.A., and Tiers, G.V.D., *J. Polym. Sci.*, 1960, **44**, 173.
8. Hatada, K., Kitayama, T., and Ute, K., in 'Annual Reports in NMR Spectroscopy' (Ed. Webb, G.A.), Vol. 26, p. 100 (Academic Press: London 1993).
9. King, J., Bower, D.I., Maddams, W.F., and Pyszora, H., *Makromol. Chem.*, 1983, **184**, 879.
10. Fox, T.G., and Schnecko, H.W., *Polymer*, 1962, **3**, 575.
11. Khanarian, G., Cais, R.E., Kometani, J.M., and Tonelli, A.E., *Macromolecules*, 1982, **15**, 866.
12. Bovey, F.A., and Mirau, P.A., *Acc. Chem. Res.*, 1988, **21**, 37.
13. Moad, G., in 'Annual Reports in NMR Spectroscopy' (Ed. Webb, G.A.), Vol. 29, p. 287 (Academic Press: London 1994).
14. Besbah, K., *Makromol. Chem.*, 1993, **194**, 3311.
15. Moad, G., Rizzardo, E., Solomon, D.H., Johns, S.R., and Willing, R.I., *Macromolecules*, 1986, **19**, 2494.
16. Berger, P.A., Kotyk, J.J., and Remsen, E.E., *Macromolecules*, 1992, **25**, 7227.
17. Kotyk, J.J., Berger, P.A., and Remsen, E.E., *Macromolecules*, 1990, **23**, 5167.
18. Schilling, F.C., Bovey, F.A., Bruch, M.D., and Kozlowski, S.A., *Macromolecules*, 1985, **18**, 1418.

19. Dong, L., Hill, D.J.T., O'Donnell, J.H., Whittaker, A.K., *Macromolecules*, 1994, **27**, 1830.
20. Hikichi, K., and Yasuda, M., *Polym. J. (Tokyo)*, 1987, **19**, 1003.
21. Gippert, G.P., and Brown, L.R., *Polym. Bull. (Berlin)*, 1984, **11**, 585.
22. Mirau, P.A., and Bovey, F.A., *Macromolecules*, 1986, **19**, 210.
23. Crowther, M.W., Szeverenyi, N.M., and Levy, G.C., *Macromolecules*, 1986, **19**, 1333.
24. Bruch, M.D., Bovey, F.A., and Cais, R.E., *Macromolecules*, 1984, **17**, 2547.
25. Hatada, K., Kitayama, T., Terawaki, Y., and Chujo, R., *Polym. J. (Tokyo)*, 1987, **19**, 1127.
26. Chujo, R., Hatada, K., Kitamaru, R., Kitayama, T., Sato, H., and Tanaka, Y., *Polym. J. (Tokyo)*, 1987, **19**, 413.
27. Chujo, R., Hatada, K., Kitamaru, R., Kitayama, T., Sato, H., Tanaka, Y., Horii, F., and Terawaki, Y., *Polym. J. (Tokyo)*, 1988, **20**, 627.
28. Nakano, T., Mori, M., and Okamoto, Y., *Macromolecules*, 1993, **26**, 867.
29. Tsuruta, T., Makimoto, T., and Kanai, H., *J. Macromol. Chem.*, 1966, **1**, 31.
30. Moad, G., Solomon, D.H., Spurling, T.H., Johns, S.R., and Willing, R.I., *Aust. J. Chem.*, 1986, **39**, 43.
31. Reinmöller, M., and Fox, T.G., *Polym. Prepr. (Am. Chem. Soc., Div. Polym. Chem)*, 1966, **1**, 999.
32. Ferguson, R.C., and Ovenall, D.W., *Polym. Prepr. (Am. Chem. Soc., Div. Polym. Chem.)*, 1985, **26(1)**, 182.
33. Ovenall, D.W., *Macromolecules*, 1984, **17**, 1458.
34. Elias, H.G., in 'Polymer Handbook, 3rd Edition' (Eds. Brandup, J., and Immergut, E.H.), p. II/357 (Wiley: New York 1989).
35. Kamide, K., Yamazaki, H., Okajima, K., and Hikichi, K., *Polym. J. (Tokyo)*, 1985, **17**, 1233.
36. Matsuzaki, K., Kanai, T., Kawamura, T., Matsumoto, S., and Uryu, T., *J. Polym. Sci., Polym. Chem. Ed.*, 1973, **11**, 961.
37. Suzuki, T., Santee, E.R., Jr, Harwood, H.J., Vogl, O., and Tanaka, T., *J. Polym. Sci., Polym. Lett.*, 1974, **12**, 635.
38. Sato, H., Tanaka, Y., and Hatada, K., *J. Polym. Sci., Polym. Phys. Ed.*, 1983, **21**, 1667.
39. Kawamura, T., Uryu, T., and Matsuzaki, K., *Makromol. Chem., Rapid Commun.*, 1982, **3**, 661.
40. Kawamura, T., Toshima, N., and Matsuzaki, K., *Macromol. Rapid Commun.*, 1994, **15**, 479.
41. Bugada, D.C., and Rudin, A., *J. Appl. Polym. Sci.*, 1985, **30**, 4137.
42. Wu, T.K., and Ovenall, D.W., *Macromolecules*, 1974, **7**, 776.
43. Flory, P.J., and Leutner, F.S., *J. Polym. Sci.*, 1950, **5**, 267.
44. Adelman, R.L., and Ferguson, R.C., *J. Polym. Sci., Polym. Chem. Ed.*, 1975, **13**, 891.
45. Vercauteren, F.F., and Donners, W.A.B., *Polymer*, 1986, **27**, 993.

46. Amiya, S., and Uetsuki, M., *Macromolecules*, 1982, **15**, 166.
47. Starnes, W.H., Jr., and Wojciechowski, B.J., *Makromol. Chem., Macromol. Symp.*, 1993, **70/71**, 1.
48. Starnes, W.H., Jr., in 'Developments in Polymer Degradation.' (Ed. Grassie, N.), Vol. 3, p. 135 (Applied Science: London 1981).
49. Caraculacu, A.A., *Pure Appl. Chem.*, 1981, **53**, 385.
50. Hjertberg, T., and Sorvik, E., in 'Degradation and Stabilisation of PVC' (Ed. Owen, E.D.), p. 21 (Elsevier Applied Science: Barking 1984).
51. Starnes, W.H., Jr., Schilling, F.C., Abbas, K.B., Cais, R.E., and Bovey, F.A., *Macromolecules*, 1979, **12**, 556.
52. Darricades-Llauro, M.F., Michel, A., Guyot, A., Waton, H., Petiaud, R., and Pham, Q.T., *J. Macromol. Sci., Chem.*, 1986, **A23**, 221.
53. Rigo, A., Palma, G., and Talamini, G., *Makromol. Chem.*, 1972, **153**, 219.
54. Fossey, J., and Nedelec, J.-Y., *Tetrahedron*, 1981, **37**, 2967.
55. Starnes, W.H., Jr., Wojciechowski, B.J., Velazquez, A., and Benedikt, G.M., *Macromolecules*, 1992, **25**, 3638.
56. Park, G.S., and Saleem, M., *Polym. Bull. (Berlin)*, 1979, **1**, 409.
57. Starnes, W.H., Jr., Schilling, F.C., Plitz, I.M., Cais, R.E., Freed, D.J., Hartless, R.L., and Bovey, F.A., *Macromolecules*, 1983, **16**, 790.
58. Mitani, K., Ogata, T., Awaya, H., and Tomari, Y., *J. Polym. Sci., Polym. Chem. Ed.*, 1975, **13**, 2813.
59. Wilson, C.W., III, and Santee, E.R., Jr., *J. Polym. Sci., Part C*, 1965, **8**, 97.
60. Cais, R.E., and Kometani, J.M., *ACS Symp. Ser.*, 1984, **247**, 153.
61. Ovenall, D.W., and Uschold, R.E., *Macromolecules*, 1991, **24**, 3235.
62. Görlitz, V.M., Minke, R., Trautvetter, W., and Weisgerber, G., *Angew. Macromol. Chem.*, 1973, **29/30**, 137.
63. Cais, R.E., and Kometani, J.M., *Macromolecules*, 1984, **17**, 1887.
64. Cais, R.E., and Sloane, N.J.A., *Polymer*, 1983, **24**, 179.
65. Cais, R.E., and Kometani, J.M., *Macromolecules*, 1984, **17**, 1932.
66. Ferguson, R.C., and Brame, E.G., Jr., *J. Phys. Chem.*, 1979, **83**, 1397.
67. Matsumoto, A., Iwanami, K., and Oiwa, M., *J. Polym. Sci., Polym. Lett. Ed.*, 1980, **18**, 211.
68. Matsumoto, A., Iwanami, K., Kawaguchi, N., and Oiwa, M., *Technol. Rep. Kansai Univ.*, 1983, **24**, 183.
69. Matsumoto, A., Kikuta, M., and Oiwa, M., *J. Polym. Sci., Part C: Polym. Lett.*, 1986, **24**, 7.
70. Matsumoto, A., Terada, T., and Oiwa, M., *J. Polym. Sci., Part A: Polym. Chem.*, 1987, **25**, 775.
71. Matsumoto, A., Iwanami, K., and Oiwa, M., *J. Polym. Sci., Polym. Lett. Ed.*, 1981, **19**, 497.
72. Marvel, C.S., and Cowan, J.C., *J. Am. Chem. Soc.*, 1939, **61**, 3156.
73. Marvel, C.S., Dec, J., Cooke, H.G., Jr., and Cowan, J.C., *J. Am. Chem. Soc.*, 1940, **62**, 3495.

74. McCurdy, K.G., and Laidler, K.J., *Can. J. Chem.*, 1964, **42**, 818.
75. Sawant, S., and Morawetz, H., *J. Polym. Sci., Polym. Lett. Ed.*, 1982, **20**, 385.
76. Minagawa, M., *J. Polym. Sci., Polym. Chem. Ed.*, 1980, **18**, 2307.
77. Kamachi, M., Kajiwara, A., Saegusa, K., and Morishima, Y., *Macromolecules*, 1993, **26**, 7369.
78. Henderson, J.N., in 'Encyclopaedia of Polymer Science and Engineering', 2nd Edn (Eds. Mark, H.F., Bikales, N.M., Overberger, C.G., and Menges, G.), Vol. 2, p. 515 (Wiley: New York 1985).
79. Tate, D.P., and Bethea, T.W., in 'Encyclopaedia of Polymer Science and Engineering', 2nd Edn (Eds. Mark, H.F., Bikales, N.M., Overberger, C.G., and Menges, G.), Vol. 2, p. 537 (Wiley: New York 1985).
80. Senyek, M.L., in 'Encyclopaedia of Polymer Science and Engineering', 2nd Edn (Eds. Mark, H.F., Bikales, N.M., Overberger, C.G., and Menges, G.), Vol. 8, p. 487 (Wiley: New York 1987).
81. Stewart, C.A., Takeshita, T., and Coleman, M.L., in 'Encyclopaedia of Polymer Science and Engineering', 2nd Edn (Eds. Mark, H.F., Bikales, N.M., Overberger, C.G., and Menges, G.), Vol. 3, p. 441 (Wiley: New York 1986).
82. Sato, H., Takebayashi, K., and Tanaka, Y., *Macromolecules*, 1987, **20**, 2418.
83. Coleman, M.M., Tabb, D.L., and Brame, E.G., *Rubber Chem. Technol.*, 1977, **50(1)**, 49.
84. Coleman, M.M., and Brame, E.G., *Rubber Chem. Technol.*, 1978, **51**, 668.
85. Sato, H., Ono, A., and Tanaka, Y., *Polymer*, 1977, **18**, 580.
86. Butler, G.B., in 'Comprehensive Polymer Science' (Eds. Eastmond, G.C., Ledwith, A., Russo, S., and Sigwalt, P.), Vol. 4, p. 423 (Pergamon: London 1989).
87. Butler, G.B., in 'Encylopaedia of Polymer Science and Engineering, 2nd Edition' (Eds. Mark, H.F., Bikales, N.M., Overberger, C.G., and Menges, G.), Vol. 4, p. 543 (Wiley: New York 1986).
88. Butler, G.B., *Acc. Chem. Res.*, 1982, **15**, 370.
89. Butler, G.B., in 'Polymeric Amines and Ammonium Salts' (Ed. Goethals, E.J.), p. 125 (Pergamon: New York 1981).
90. Butler, G.B., Cyclopolymerization and Cyclocopolymerization (Marcel Dekker: New York 1992).
91. Butler, G.B., and Angelo, R.J., *J. Am. Chem. Soc.*, 1957, **79**, 3128.
92. Butler, G.B., Crawshaw, A., and Miller, W.L., *J. Am. Chem. Soc.*, 1958, **80**, 3615.
93. Solomon, D.H., *J. Polym. Sci., Polym. Symp.*, 1975, **49**, 175.
94. Beckwith, A.L.J., Hawthorne, D.G., and Solomon, D.H., *Aust. J. Chem.*, 1976, **29**, 995.
95. Solomon, D.H., *J. Macromol. Sci. Chem.*, 1975, **A9**, 97.
96. Solomon, D.H., and Hawthorne, D.G., *J. Macromol. Sci., Rev. Macromol. Chem.*, 1976, **C15**, 143.
97. Beckwith, A.L.J., *Tetrahedron*, 1981, **37**, 3073.

98. Xi, F., and Vogl, O., *J. Macromol. Sci., Chem.*, 1983, **A20**, 321.
99. Otsu, T., and Ohya, T., *J. Macromol. Sci. Chem.*, 1984, **A21**, 1.
100. Seyferth, D., and Robison, J., *Macromolecules*, 1993, **26**, 407.
101. Field, N.D., *J. Org. Chem.*, 1960, **25**, 1006.
102. Marvel, C.S., and Gall, E.J., *J. Org. Chem.*, 1960, **25**, 1784.
103. Marvel, C.S., and Vest, R.D., *J. Am. Chem. Soc.*, 1959, **81**, 984.
104. Milford, G.N., *J. Polym. Sci.*, 1959, **41**, 295.
105. Tsuda, T., and Mathias, L.J., *Macromolecules*, 1993, **26**, 6359.
106. Thang, S.H., Moad, G., and Rizzardo, E., in 'MakroAkron'94 Abstracts', p. 18 (University of Akron: Akron, Ohio 1994).
107. Miyake, T., *Kogyo Kagaku Zasshi*, 1961, **64**, 359.
108. Ohya, T., and Otsu, T., *J. Polym. Sci., Polym. Chem. Ed.*, 1983, **21**, 3503.
109. Matsumoto, A., Kitamura, T., Oiwa, M., and Butler, G.B., *J. Polym. Sci., Polym. Chem. Ed.*, 1981, **19**, 2531.
110. Stansbury, J.W., *Polym. Prepr. (Am. Chem. Soc., Div. Polym. Chem)*, 1990, **31(1)**, 503.
111. Mathias, L.J., Kusefoglu, S.H., and Ingram, J.E., *Macromolecules*, 1988, **21**, 545.
112. Mathias, L.J., Colletti, R.F., and Bielecki, A., *J. Am. Chem. Soc.*, 1991, **113**, 1550.
113. Aso, C., Kunitake, T., and Ando, S., *J. Macromol. Sci. Chem.*, 1971, **A5**, 167.
114. Raymond, M.A., and Dietrich, H.J., *J. Macromol. Sci. Chem.*, 1972, **A6**, 207.
115. Billingham, N.C., Jenkins, A.D., Kronfli, E.B., and Walton, D.R.M., *J. Polym. Sci., Polym. Chem. Ed.*, 1977, **15**, 675.
116. Butler, G.B., and Stackman, R.W., *J. Org. Chem.*, 1960, **25**, 1643.
117. Butler, G.B., and Stackman, R.W. *J. Macromol. Sci. Chem.*, 1969, **A3**, 821.
118. Kida, S., Nozakura, S., and Murahashi, S., *Polym. J. (Tokyo)*, 1972, **3**, 234.
119. Butler, G.B., Skinner, D.L., Bond, W.C., Jr., and Rogers, C.L., *J. Macromol. Sci., Chem.*, 1970, **A4**, 1437.
120. Berlin, K.D., and Butler, G.B., *J. Org. Chem.*, 1960, **25**, 2006.
121. Benyon, K.I., *J. Polym. Sci., Part A*, 1963, **1**, 3357.
122. Corfield, G.C., and Monks, H.H., *J. Macromol. Sci. Chem.*, 1975, **A9**, 1113.
123. Beckwith, A.L.J., *Chem Soc. Rev.*, 1993, 143.
124. Fukuda, W., Nakao, M., Okumura, K., and Kakiuchi, H., *J. Polym. Sci., Part A-1*, 1972, **10**, 237.
125. Ichihashi, T., and Kawai, W., *Kobunshi Kagaku*, 1971, **28**, 225. (Chem. Abstr. 1971, 75, 46941x)
126. Trossarelli, L., Guaita, M., and Priola, A., *Makromol. Chem.*, 1967, **100**, 147.
127. Kodaira, T., Okumura, M., Urushisaki, M., and Isa, K., *J. Polym. Sci., Part A: Polym. Chem.*, 1993, **31**, 169.
128. Kodaira, T., and Mae, Y., *Polymer*, 1992, **33**, 3500.
129. Yokata, K., and Takada, Y., *Kobunshi Kagaku*, 1969, **26**, 317. (Chem. Abstr. 1969, 71, 22369v)

130. Kaye, H., *Macromolecules*, 1971, **4**, 147.
131. Butler, G.B., and Kimura, S., *J. Macromol. Sci. Chem.*, 1971, **A5**, 181.
132. Matsoyan, S.G., Pogosyan, G.M., and Elliasyan, M.A., *Vysokomol. Soedin*, 1963, **5**, 777. (Chem. Abstr. 1963, 59, 7654f)
133. Hawthorne, D.G., and Solomon, D.H., *J. Macromol. Sci. Chem.*, 1975, **A9**,
134. Beckwith, A.L.J., and Moad, G., *J. Chem. Soc., Perkin Trans. 2*, 1975, **2**, 1726.
135. Costa, L., Chiantore, O., and Guaita, M., *Polymer*, 1978, **19**, 202.
136. Wiley, R.H., Rivera, W.H., Crawford, T.H., and Bray, N.F., *J. Polym. Sci.*, 1962, **61**, 538.
137. Graham, P.J., Buhle, E.L., and Pappas, N., *J. Org. Chem.*, 1961, **26**, 4658.
138. Paulrajan, S., Gopalan, A., Subbaratnam, N.R., and Venkatarao, K., *Polymer*, 1983, **24**, 906.
139. Gopalan, A., Paulrajan, S., Subbaratnam, N.R., and Rao, K.V., *J. Polym. Sci., Polym. Chem. Ed.*, 1985, **23**, 1861.
140. Nishikubo, T., Iizawa, T., Yoshinaga, A., and Nitta, M., *Makromol. Chem.*, 1982, **183**, 789.
141. Beckwith, A.L.J., and Moad, G., *J. Chem. Soc., Chem. Commun.*, 1974, 472.
142. Simpson, W., Holt, T., and Zetie, R.J., *J. Polym. Sci.*, 1953, **10**, 489.
143. Haward, R.N., *J. Polym. Sci.*, 1953, **10**, 535.
144. Matsumoto, A., Iwanami, K., and Oiwa, M., *J. Polym. Sci., Polym. Lett. Ed.*, 1980, **18**, 307.
145. Barton, J.M., Butler, G.B., and Chapin, E.C., *J. Polym. Sci., Part A*, 1965, **3**, 501.
146. Bailey, W.J., and Sun, R.L., *Polym. Prepr. (Am. Chem. Soc., Div. Polym. Chem)*, 1972, **13**, 281.
147. Brady, R.F., Jr., *J. Macromol. Sci., Rev. Macromol. Chem. Phys.*, 1992, **C32**, 135.
148. Stansbury, J.W., in 'Expanding Monomers' (Eds. Sadhir, R.K., and Luck, R.M.), p. 153 (CRC Press: Boca Raton, Florida 1992).
149. Bailey, W.J., in 'Comprehensive Polymer Science' (Eds. Eastmond, G.C., Ledwith, A., Russo, S., and Sigwalt, P.), Vol. 3, p. 283 (Pergamon: London 1989).
150. Endo, T., and Yokozawa, T., in 'New Methods for Polymer Synthesis' (Ed. Mijs, W.J.), p. 155 (Plenum: New York 1992).
151. Hall, H.K., Jr., and Ykman, P.J., *J. Polym. Sci., Macromol. Rev.*, 1976, **11**, 1.
152. Bothe, H., and Schluter, A.-D., *Polym. Prepr. (Am. Chem. Soc., Div. Polym. Chem)*, 1988, **29**, 412.
153. Beckwith, A.L.J., and Moad, G., *J. Chem. Soc., Perkin Trans. 2*, 1980, **2**, 1473.
154. Newcomb, M., *Tetrahedron*, 1993, **49**, 1151.
155. Masnovi, J., Samsel, E.G., and Bullock, R.M., *J. Chem. Soc., Chem. Commun.*, 1989, 1044.
156. Ingold, K.U., Maillard, B., and Walton, J.C., *J. Chem. Soc., Perkin Trans. 2*, 1981, 970.

157. Beckwith, A.L.J., and Bowry, V.W., *J. Org. Chem.*, 1989, **54**, 2681.
158. Takahashi, T., Yamashita, I., and Miyakawa, T., *Bull. Chem. Soc. Japan*, 1964, **37**, 131.
159. Takahashi, T., and Yamashita, I., *J. Polym. Sci., Part B*, 1965, **3**, 251.
160. Cho, I., and Ahn, K.-D., *J. Polym. Sci., Polym. Chem. Ed.*, 1979, **17**, 3169.
161. Lishanskii, I.S., Zak, A.G., Fedorova, E.F., and Khachaturov, A.S., *Vysokomolekul. Soedin*, 1965, **7**, 966 (*Chem. Abstr.* 1965, **63**, 10072h).
162. Endo, T., Watanabe, M., Suga, K., and Yokozawa, T., *J. Polym. Sci., Part A: Polym. Chem.*, 1989, **27**, 1435.
163. Endo, T., and Suga, K., *J. Polym. Sci., Part A: Polym. Chem.*, 1989, **27**, 1831.
164. Endo, T., Watanabe, M., Suga, K., and Yokozawa, T., *Makromol. Chem.*, 1989, **190**, 691.
165. Endo, T., Watanabe, M., Suga, K., and Yokozawa, T., *J. Polym. Sci., Part A: Polym. Chem.*, 1987, **25**, 3039.
166. Cho, I., and Ahn, K.-D., *J. Polym. Sci., Polym. Lett. Ed.*, 1977, **15**, 751.
167. Cho, I., and Lee, J.-Y., *Makromol. Chem., Rapid Commun.*, 1984, **5**, 263.
168. Cho, I., and Song, S.S., *J. Polym. Sci., Part A: Polym. Chem.*, 1989, **27**, 3151.
169. Takahashi, T., *J. Polym. Sci., Part A-1*, 1968, **6**, 403.
170. Kennedy, J.P., Elliot, J.J., and Butler, P.E., *J. Macromol. Sci. Chem.*, 1968, **A2**, 1415.
171. Cho, I., and Song, S.S., *Makromol. Chem., Rapid Commun.*, 1989, **10**, 85.
172. Sanda, F., Takata, T., and Endo, T., *Macromolecules*, 1994, **27**, 1099.
173. Sanda, F., Takata, T., and Endo, T., *J. Polym. Sci., Part A: Polym. Chem.*, 1993, **31**, 2659.
174. Sanda, F., Takata, T., and Endo, T., *Macromolecules*, 1993, **26**, 5748.
175. Sanda, F., Takata, T., and Endo, T., *Macromolecules*, 1992, **25**, 6719.
176. Britten, C., Moad, G., Rizzardo, E., and Thang, S.H., to be submitted.
177. Sanda, F., Takata, T., and Endo, T., *Macromolecules*, 1993, **26**, 729.
178. Bowry, V.W., Lusztyk, J., and Ingold, K.U., *J. Chem. Soc., Chem. Commun*, 1990, 923.
179. Sanda, F., Takata, T., and Endo, T., *Macromolecules*, 1993, **26**, 1818.
180. Cho, I., and Kim, J.-B., *J. Polym. Sci., Polym. Lett. Ed.*, 1983, **21**, 433.
181. Endo, T., and Kanda, N., *Polym. Prepr. Jpn.*, 1987, **36**, 140.
182. Koizumi, T., Nojima, Y., and Endo, T., *J. Polym. Sci., Part A: Polym. Chem.*, 1993, **31**, 3489.
183. Laurie, D., Nonhebel, D.C., Suckling, C.J., and Walton, J.C., *Tetrahderon*, 1993, **49**, 5869.
184. Beckwith, A.L.J., and Moad, G., *J. Chem. Soc., Perkin Trans. 2*, 1980, **2**, 1083.
185. Hiraguri, Y., and Endo, T., *J. Polym. Sci., Part C: Polym. Lett.*, 1989, **27**, 333.
186. Cho, I., and Choi, S.Y., *Makromol. Chem., Rapid Commun.*, 1991, **12**, 399.
187. Cho, I., Kim, S.-K., and Lee, M.-H., *J. Polym. Sci., Polym. Symp.*, 1986, **74**, 219.
188. Cho, I., and Lee, M.-H., *J. Polym. Sci., Part C: Polym. Lett.*, 1987, **25**, 309.

189. Errede, L.A., *J. Polym. Sci.*, 1961, **49**, 253.
190. Bailey, W.J., and Chou, J.L., *Polym. Mater. Sci. Eng.*, 1987, **56**, 30.
191. Bailey, W.J., Amone, M.J., and Chou, J.L., *Polym. Prepr. (Am. Chem. Soc., Div. Polym. Chem)*, 1988, **29(1)**, 178.
192. Klemm, E., and Schulze, T., *Makromol. Chem.*, 1993, **194**, 2087.
193. Hiraguri, Y., and Endo, T., *J. Polym. Sci., Part A: Polym. Chem.*, 1989, **27**, 4403.
194. Acar, M.H., Nambu, Y., Yamamoto, K., and Endo, T., *J. Polym. Sci., Part A: Polym. Chem.*, 1989, **27**, 4441.
195. Bailey, W.J., Gu, J.M., and Zhou, L. -L., *Polym. Prepr. (Am. Chem. Soc., Div. Polym. Chem)*, 1990, **31(1)**, 24.
196. Cho, I., and Song, K.Y., *Makromol. Chem., Rapid Commun.*, 1993, **14**, 377.
197. Cho, I., and Kim, S.-K., *J. Polym. Sci., Part A: Polym. Chem.*, 1990, **28**, 417.
198. Bailey, W.J., Chen, P.Y., Chen, S.-C., Chiao, W.-B., Endo, T., Gapud, B., Kuruganti, Y., Lin, Y.-N., Ni, Z., Pan, C.-Y., Shaffer, S.E., Sidney, L., Wu, S.-R., Yamamoto, N., Yamazaki, N., Yonezawa, K., and Zhou, L.-L., *Makromol. Chem., Macromol. Symp.*, 1986, **6**, 81.
199. Bailey, W.J., *Polym. J. (Tokyo)*, 1985, **17**, 85.
200. Bailey, W.J., Chou, J.L., Feng, P.-Z., Issari, B., Kuruganti, V., and Zhou, L.-L., *J. Macromol. Sci., Chem.*, 1988, **A25**, 781.
201. Kobayashi, S., Kadokawa, J., Shoda, S., and Uyama, H., *Macromol. Reports*, 1991, **A28 (Suppl. 1)**, 1.
202. Kobayashi, S., Kadokawa, J., Matsumura, Y., Yen, I.F., and Uyama, H., *Macromol. Reports*, 1992, **A29 (Suppl. 3)**, 243.
203. Bailey, W.J., Wu, S.-R., and Ni, Z., *J. Macromol. Sci. Chem.*, 1982, **A18**, 973.
204. Bailey, W.J., Wu, S.-R., and Ni, Z., *Makromol. Chem.*, 1982, **183**, 1913.
205. Endo, T., Yako, N., Azuma, K., and Nate, K., *Makromol. Chem.*, 1985, **186**, 1543.
206. Yokozawa, T., Hayashi, R., and Endo, T., *J. Polym. Sci., Part A: Polym. Chem.*, 1990, **28**, 3739.
207. Bailey, W.J., Ni, Z., and Wu, S.-R., *J. Polym. Sci., Polym. Chem. Ed.*, 1982, **20**, 3021.
208. Endo, T., Okawara, M., Bailey, W.J., Azuma, K., Nate, K., and Yokona, H., *J. Polym. Sci., Polym. Lett. Ed.*, 1983, **21**, 373.
209. Bailey, W.J., Ni, Z., and Wu, S.-R., *Macromolecules*, 1982, **15**, 711.
210. Bailey, W.J., Arfaei, P.Y., Chen, P.Y., Chen, S.-C., Endo, T., Pan, C.-Y., Ni, Z., Shaffer, S.E., Sidney, L., Wu, S.-R., and Yamazaki, N., in 'Proc. IUPAC 28th Macromol. Symp.', p. 214 Amherst, MA 1982).
211. Sidney, L.N., Shaffer, S.E., and Bailey, W.J., *Polym. Prepr. (Am. Chem. Soc., Div. Polym. Chem)*, 1981, **22(2)**, 373.
212. Barclay, L.R.C., Griller, D., and Ingold, K.U., *J. Am. Chem. Soc.*, 1982, **104**, 4399.
213. Beckwith, A.L.J., and Thomas, C.B., *J. Chem. Soc., Perkin Trans. 2*, 1973, 861.

214. Bailey, W.J., and Zhou, L.L., *Polym. Prepr. (Am. Chem. Soc., Div. Polym. Chem)*, 1989, **30(1)**, 195.
215. Hiraguri, Y., and Endo, T., *J. Polym. Sci., Part A: Polym. Chem.*, 1992, **30**, 689.
216. Hiraguri, Y., and Endo, T., *J. Am. Chem. Soc.*, 1987, **109**, 3779.
217. Hiraguri, Y., and Endo, T., *J. Polym. Sci., Part A: Polym. Chem.*, 1989, **27**, 2135.
218. Cho, I., Kim, B.-G., Park, Y.-C., Kim, C.-B., and Gong, M.-S., *Makromol. Chem., Rapid Commun.*, 1991, **12**, 141.
219. Pan, C.-Y., Wu, Z., and Bailey, W.J., *J. Polym. Sci., Part C: Polym. Lett.*, 1987, **25**, 243.
220. Sugiyama, J.-I., Yokozawa, T., and Endo, T., *J. Am. Chem. Soc.*, 1993, **115**, 2041.
221. Morariu, S., Buruiana, E.C., and Simionescu, B.C., *Polym. Bull. (Berlin)*, 1993, **30**, 7.
222. Bailey, W.J., and Feng, P.-Z., *Polym. Prepr. (Am. Chem. Soc., Div. Polym. Chem)*, 1987, **28(1)**, 154.
223. Tsang, R., Dickson, J.K., Jr., Pak, H., Walton, R., and Fraser-Reid, B., *J. Am. Chem. Soc.*, 1987, **109**, 3484.
224. Endo, T., and Bailey, W.J., *J. Polym. Sci., Polym. Lett.*, 1975, **13**, 193.
225. Sugiyama, J.-I., Yokozawa, T., and Endo, T., *J. Polym. Sci., Part A: Polym. Chem.*, 1990, **28**, 3529.
226. Bailey, W.J., and Zheng, Z.-F., *J. Polym. Sci., Part A: Polym. Chem.*, 1991, **29**, 437.
227. Endo, T., and Bailey, W.J., *J. Polym. Sci., Polym. Lett. Ed.*, 1980, **18**, 25.
228. Tagoshi, H., and Endo, T., *J. Polym. Sci., Part C: Polym. Lett.*, 1988, **26**, 77.
229. Schulze, T., and Klemm, E., *Polym. Bull. (Berlin)*, 1993, **31**, 409.
230. Issari, B., and Bailey, W.J., *Polym. Prepr. (Am. Chem. Soc., Div. Polym. Chem)*, 1988, **29(1)**, 217.
231. Viswanadhan, V.N., and Mattice, W.L., *Makromol. Chem.*, 1985, **186**, 633.
232. Usami, T., and Takayama, S., *Polym. J. (Tokyo)*, 1984, **16**, 731.
233. Blitz, J.P., and McFaddin, D.C., *J. Appl. Polym. Sci.*, 1994, **51**, 13.
234. Ohtani, H., Tsuge, S., and Usami, T., *Macromolecules*, 1984, **17**, 2557.
235. Bowmer, T.N., and O'Donnell, J.H., *Polymer*, 1977, **18**, 1032.
236. Cutler, D.J., Hendra, P.J., Cudby, M.E.A., and Willis, H.A., *Polymer*, 1977, **18**, 1005.
237. Usami, T., and Takayama, S., *Macromolecules*, 1984, **17**, 1756.
238. Axelson, D.E., Levy, G.C., and Mandelkern, L., *Macromolecules*, 1979, **12**, 41.
239. Bovey, F.A., Schilling, F.C., McCrackin, F.L., and Wagner, H.L., *Macromolecules*, 1976, **9**, 76.
240. Bugada, D.C., and Rudin, A., *Eur. Polym. J.*, 1987, **23**, 809.
241. Roedel, M.J., *J. Am. Chem. Soc.*, 1953, **75**, 6110.
242. Stoiljkovich, D., and Jovanovich, S., *Makromol. Chem.*, 1981, **182**, 2811.

243. Stoiljkovich, D., and Jovanovich, S., *Br. Polym. J.*, 1984, **16**, 291.
244. Huang, X.L., and Dannenberg, J.J., *J. Org. Chem.*, 1991, **56**, 5421.
245. Nedelec, J.Y., and LeFort, D., *Tetrahedron*, 1975, **31**, 411.
246. Beckwith, A.L.J., and Ingold, K.U., in 'Rearrangements in Ground and Excited States' (Ed. de Mayo, P.), Vol. 1, p. 162 (Academic Press: New York 1980).
247. Wilbourn, A.H., *J. Polym. Sci.*, 1959, **34**, 569.
248. Morishima, Y., and Nozakura, S., *J. Polym. Sci., Polym. Chem. Ed.*, 1976, **14**, 1277.
249. Melville, H.W., and Sewell, P.R., *Makromol. Chem.*, 1959, **32**, 139.
250. Mattice, W.L., and Viswanadhan, V.N., *Macromolecules*, 1986, **19**, 568.
251. Hjertberg, T., and Sorvik, E., *ACS Symp. Ser.*, 1985, **280**, 259.
252. Sato, T., Takahashi, H., Tanaka, H., and Ota, T., *J. Polym. Sci., Part A: Polym. Chem.*, 1988, **26**, 2839.
253. Sato, T., Ito, D., Kuki, M., Tanaka, H., and Ota, T., *Macromolecules*, 1991, **24**, 2963.
254. Allen, P.E.M., and Patrick, C.R., Kinetics and Mechanisms of Polymerization Reactions (Ellis Horwood: Chichester 1974).
255. Ivin, K.J., in 'Reactivity, Mechanism and Structure in Polymer Chemistry' (Eds. Jenkins, A.D., and Ledwith, A.), p. 514 (Wiley: London 1974).
256. Ivin, K.J., and Busfield, W.K., in 'Encyclopaedia of Polymer Science and Engineering', 2nd Edn (Eds. Mark, H.F., Bikales, N.M., Overberger, C.G., and Menges, G.), Vol. 12, p. 555 (Wiley: New York 1987).
257. Sawada, H., *J. Macromol. Sci., Rev. Macromol. Chem.*, 1969, **C3**, 313.
258. Busfield, W.K., in 'Polymer Handbook, 3rd Edition' (Eds. Brandup, J., and Immergut, E.H.), p. II/295 (Wiley: New York 1989).
259. Ivin, K.J., *Trans. Faraday Soc.*, 1955, **51**, 1273.
260. Bywater, S., *Trans. Faraday Soc.*, 1955, **51**, 1267.
261. Cook, R.E., and Ivin, K.J., *Trans. Faraday Soc.*, 1957, **53**, 1273.
262. Penelle, J., Collot, J., and Rufflard, G., *J. Polym. Sci., Part A: Polym. Chem.*, 1993, **31**, 2407.
263. Bywater, S., *Can. J. Chem.*, 1957, **34**, 552.
264. Bywater, S., and Worsfold, D.J., *J. Polym. Sci.*, 1962, **58**, 571.
265. Ivin, K.J., and leonard, J., *Eur. Polym. J.*, 1970, **6**, 331.
266. Yamada, B., Satake, M., and Otsu, T., *Makromol. Chem.*, 1991, **192**, 2713.
267. Madruga, E.M., in 'Macromolecules 1992' (Ed. Kahovec, J.), p. 109 (VSP: Utrecht 1992).
268. Avci, D., Kusefoglu, S.H., Thompson, R.D., and Mathias, L.J., *Macromolecules*, 1994, **27**, 1981.
269. Cheng, J., Yamada, B., and Otsu, T., *J. Polym. Sci., Part A: Polym. Chem.*, 1991, **29**, 1837.

270. Stickler, M., in 'Comprehensive Polymer Science' (Eds. Eastmond, G.C., Ledwith, A., Russo, S., and Sigwalt, P.), Vol. 3, p. 85 (Pergamon: London 1989).
271. Stickler, M., in 'Comprehensive Polymer Science' (Eds. Eastmond, G.C., Ledwith, A., Russo, S., and Sigwalt, P.), Vol. 3, p. 59 (Pergamon: London 1989).
272. Buback, M., Garcia-Rubio, L.H., Gilbert, R.G., Napper, D.H., Guillot, J., Hamielec, A.E., Hill, D., O'Driscoll, K.F., Olaj, O.F., Shen, J., Solomon, D., Moad, G., Stickler, M., Tirrell, M., and Winnik, M.A., *J. Polym. Sci., Part C: Polym. Lett.*, 1988, **26**, 293.
273. Buback, M., Gilbert, R.G., Russell, G.T., Hill, D.J.T., Moad, G., O'Driscoll, K.F., Shen, J., and Winnik, M.A., *J. Polym. Sci., Part A: Polym. Chem.*, 1992, **30**, 851.
274. Gilbert, R.G., *Pure Appl. Chem.*, 1992, **64**, 1563.
275. Fukuda, T., Ma, Y.-D., and Inagaki, H., *Macromolecules*, 1985, **18**, 17.
276. Olaj, O.F., Kremminger, P., and Schnöll-Bitai, I., *Makromol. Chem., Rapid Commun.*, 1988, **9**, 771.
277. Olaj, O.F., Schnöll-Bitai, I., and Kremminger, P., *Eur. Polym. J.*, 1989, **25**, 535.
278. O'Driscoll, K.F., and Mahabadi, H.K., *J. Polym. Sci., Polym. Chem. Ed.*, 1976, **14**, 869.
279. Bresler, S.E., Kazbekov, E.N., Fomichev, V.N., and Shadrin, V.N., *Makromol. Chem.*, 1972, **157**, 167.
280. Bresler, S.E., Kazbekov, E.N., and Shadrin, V.N., *Makromol. Chem.*, 1974, **175**, 2875.
281. Carswell, T.G., Hill, D.J.T., Londero, D.I., O'Donnell, J.H., Pomery, P.J., and Winzor, C.L., *Polymer*, 1992, **33**, 137.
282. Otsu, T., and Nayatani, K., *Makromol. Chem.*, 1958, **73**, 225.
283. Kamachi, M., Kohno, M., Kuwae, Y., and Nozakura, S.-I., *Polym. J. (Tokyo)*, 1982, **14**, 749.
284. Shen, J., Tian, Y., Wang, G., and Yang, M., *Makromol. Chem.*, 1991, **192**, 2669.
285. Zhu, S., Tian, Y., Hamielec, A.E., and Eaton, D.R., *Macromolecules*, 1990, **23**, 1144.
286. Carswell, T.G., Hill, D.J.T., Hunter, D.S., Pomery, P.J., O'Donnell, J.H., and Winzor, C.L., *Eur. Polym. J.*, 1990, **26**, 541.
287. Tonge, M.P., Pace, R.J., and Gilbert, R.G., *Macromol. Chem. Phys.*, 1994, **195**, 3159.
288. Yamada, B., Kageoka, M., and Otsu, T., *Polym. Bull. (Berlin)*, 1992, **28**, 75.
289. Yamada, B., Kageoka, M., and Otsu, T., *Macromolecules*, 1991, **24**, 5234.
290. Kamachi, M., Kuwae, Y., Kohno, M., and Nozakura, S.-I., *Polym. J. (Tokyo)*, 1985, **17**, 541.
291. Olaj, O.F., Bitai, I., and Hinkelmann, F., *Makromol. Chem.*, 1987, **188**, 1689.
292. Olaj, O.F., and Bitai, I., *Angew. Makromol. Chem.*, 1987, **155**, 177.

293. Davis, T.P., O'Driscoll, K.F., Piton, M.C., and Winnik, M.A., *Polym. Int.*, 1991, **24**, 65.
294. Davis, T.P., O'Driscoll, K.F., Piton, M.C., and Winnik, M.A., *Macromolecules*, 1989, **22**, 2785.
295. Davis, T.P., O'Driscoll, K.F., Piton, M.C., and Winnik, M.A., *Macromolecules*, 1990, **23**, 2113.
296. Davis, T.P., O'Driscoll, K.F., Piton, M.C., and Winnik, M.A., *J. Polym. Sci., Part C: Polym. Lett.*, 1989, **27**, 181.
297. Schnöll-Bitai, I., and Olaj, O.F., *Makromol. Chem.*, 1990, **191**, 2491.
298. Pascal, P., Napper, D.H., Gilbert, R.G., Piton, M.C., and Winnik, M.A., *Macromolecules*, 1990, **23**, 5161.
299. Olaj, O.F., and Schnöll-Bitai, I., *Makromol. Chem., Rapid Commun.*, 1990, **11**, 459.
300. Olaj, O.F., and Schnöll-Bitai, I., *Eur. Polym. J.*, 1989, **25**, 635.
301. Olaj, O.F., and Schnöll-Bitai, I., *Makromol. Chem., Rapid Commun.*, 1988, **9**, 275.
302. Pascal, P., Winnik, M.A., Napper, D.H., and Gilbert, R.G., *Makromol. Chem., Rapid Commun.*, 1993, **14**, 213.
303. Morrison, B.R., Piton, M.C., Winnik, M.A., Gilbert, R.G., and Napper, D.H., *Macromolecules*, 1993, **26**, 4368.
304. Deady, M., Mau, A.W.H., Moad, G., and Spurling, T.H., *Makromol. Chem.*, 1993, **194**, 1691.
305. Davis, T.P., *J. Photochem. Photobiol., A*, 1994, **77**, 1.
306. Hutchinson, R.A., Aronson, M.T., and Richards, J.R., *Macromolecules*, 1993, **26**, 6410.
307. Mahabadi, H.K., and O'Driscoll, K.F., *J. Macromol. Sci., Chem.*, 1977, **A11**, 967.
308. Deibert, S., Bandermann, F., Schweer, J., and Sarnecki, J., *Makromol. Chem., Rapid Commun.*, 1992, **13**, 351.
309. Hutchinson, R.A., Richards, J.R., and Aronson, M.T., *Macromolecules*, 1994, **27**, 4530.
310. O'Driscoll, K.F., and Kuindersma, M.E., *Macromol. Theory Simul.*, 1994, **3**, 469.
311. Lu, J., Zhang, H., and Yang, Y., *Makromol. Chem., Theory Simul.*, 1993, **2**, 747.
312. Buback, M., Huckestein, B., Kuchta, F.-D., Russell, G.T., and Schmid, E., *Macromol. Chem. Phys.*, 1994, **195**, 2117.
313. Lorand, J.P., in 'Landoldt-Bornstein, New Series, Radical Reaction Rates in Solution' (Ed. Fischer, H.), Vol. II/13a, p. 135 (Springer-Verlag: Berlin 1984).
314. Starks, C.M., Free Radical Telomerization (Academic Press: New York 1974).
315. Morrison, B.R., Maxwell, I.A., Gilbert, R.G., and Napper, D.H., *ACS Symp. Ser.*, 1992, **492**, 28.

316. Krstina, J., Moad, G., Willing, R.I., Danek, S.K., Kelly, D.P., Jones, S.L., and Solomon, D.H., *Eur. Polym. J.*, 1993, **29**, 379.
317. Moad, G., Rizzardo, E., Solomon, D.H., and Beckwith, A.L.J., *Polym. Bull. (Berlin)*, 1992, **29**, 647.

5
Termination

5.1 Introduction

In this chapter we consider all processes which lead to the cessation of growth of one or more polymer chains. Four main processes should be distinguished:

(a) The self-reaction of propagating radicals by combination and/or disproportionation (*e.g.* Scheme 1) (see 5.2).

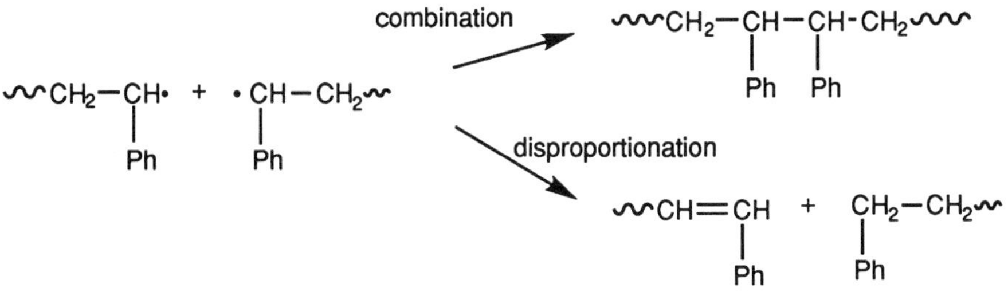

$$\sim\sim CH_2-CH\bullet \;+\; \bullet CH-CH_2\sim\sim$$
$$||$$
$$PhPh$$

combination → $\sim\sim CH_2-CH-CH-CH_2\sim\sim$ with Ph, Ph substituents

disproportionation → $\sim\sim CH=CH-Ph \;+\; CH_2-CH_2\sim\sim$ with Ph substituent

Scheme 1

(b) Primary radical termination (see 3.2.8, 3.4, and 5.2.3); the reaction of a propagating radical with a radical derived from the initiator or transfer agent (*e.g.* I•, see Scheme 2). This process is highly dependent on the structure of the initiator-derived radical.

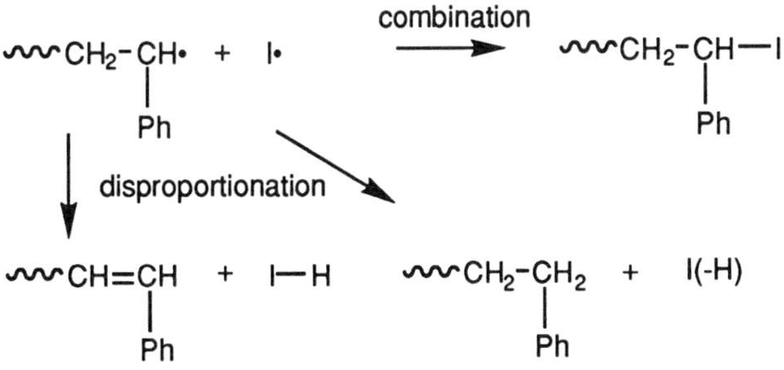

$$\sim\sim CH_2-CH\bullet \;+\; I\bullet$$

combination → $\sim\sim CH_2-CH-I$ with Ph substituent

disproportionation → $\sim\sim CH=CH-Ph \;+\; I-H$ and $\sim\sim CH_2-CH_2-Ph \;+\; I(-H)$

Scheme 2

(c) Chain transfer (see 5.3); the reaction of a propagating radical with a non-radical substrate to produce a dead polymer chain and a radical capable of initiating a new polymer chain (e.g. X-Y, see Scheme 3). The transfer agent may be a deliberate additive (e.g. a thiol) or it may be the initiator, monomer, polymer, solvent or an adventitious impurity.

$$\sim\!\!\sim\!\!CH_2\text{-}\underset{Ph}{CH\bullet} + X\text{—}Y \xrightarrow{\text{transfer}} \sim\!\!\sim\!\!CH_2\text{-}\underset{Ph}{CH}\text{—}X + Y\bullet$$

Scheme 3

(d) Inhibition (see 5.4); the reaction of a propagating radical with another species to give a dead polymer chain (e.g. S•, see Scheme 4). The inhibitor may be a "stable" radical (e.g. a nitroxide, oxygen), a non-radical species which reacts to give a "stable" radical (e.g. a phenol, a quinone) or a transition metal salt.

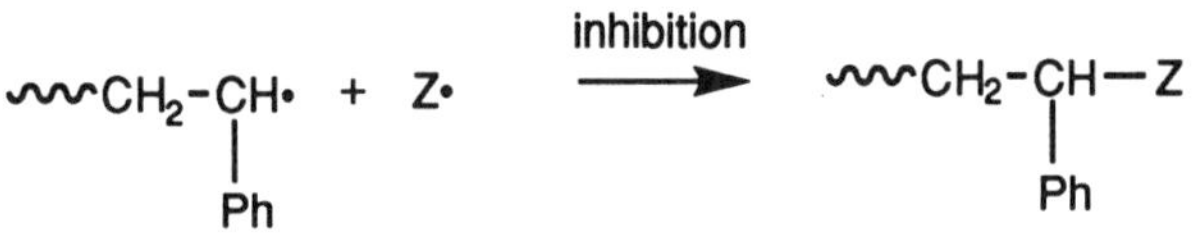

$$\sim\!\!\sim\!\!CH_2\text{-}\underset{Ph}{CH\bullet} + Z\bullet \xrightarrow{\text{inhibition}} \sim\!\!\sim\!\!CH_2\text{-}\underset{Ph}{CH}\text{—}Z$$

Scheme 4

There are also situations intermediate between (c) and (d) where the reaction produces a dead polymer chain and a radical which is less reactive than the propagating radical but still capable of reinitiating polymerization. The process is then termed retardation or degradative chain transfer.

5.2 Radical-Radical Termination

The most important mechanism for the destruction of propagating species in radical polymerization is radical-radical reaction by combination or disproportionation (e.g. Scheme 1). This process is sometimes simply referred to as bimolecular termination. However, this term is misleading since all of the chain termination processes referred to in this chapter are bimolecular reactions.

Before any chemistry can take place the radical centers of propagating species must come into appropriate proximity and it is now generally accepted that the self-reaction of propagating radicals is a diffusion-controlled process. For this reason there is no single rate constant for termination in radical polymerization. The average rate constant usually quoted is a composite term that depends on the nature of the medium and the chain lengths of the two propagating species. Diffusion mechanisms and other factors which affect the absolute rate constants for termination are discussed in section 5.2.1.

Even though the absolute rate constant for reactions between propagating species may be determined largely by diffusion, this does not mean that there is no specificity in the termination process or that the activation energies for combination and disproportionation are zero or the same. It simply means that this chemistry is not involved in the rate determining step of the overall termination process.

The significance of the termination mechanism on the course of polymerization and on the properties of polymers is discussed briefly in section 5.2.2 and 7.2.2. The relative importance of combination and disproportionation in relevant model systems and in polymerizations of some common monomers is considered in sections 5.2.3 and 5.2.4 respectively.

5.2.1 Termination Kinetics

Recent reviews of the kinetics of radical-radical termination of propagating species include those by North[1] and O'Driscoll.[2] Values of and methods of treating termination rate constants are currently being critically assessed by an IUPAC working party.[3-5] A detailed treatment of termination kinetics is beyond the scope of this book. However some knowledge of this process is important in understanding the chemistry described in subsequent sections.

The overall rate constant for radical-radical termination can be defined in terms of the rate of consumption of propagating radicals:[6]

$$\frac{-d[P\bullet]}{dt} = 2k_t[P\bullet]^2$$

where $[P\bullet]$ is the total concentration of propagating radicals and $k_t = k_{tc} + k_{td}$. However, in many works on free radical polymerization the factor 2 is by convention incorporated into the rate constant.[7,8] Thus:

$$\frac{-d[P\bullet]}{dt} = k_t[P\bullet]^2$$

It should also be noted that the termination rate constant is often then expressed as $k_t = k_{tc}/2 + k_{td}$ to reflect the fact that only one polymer chain is formed when two propagating radicals combine. In reading the literature and comparing values of k_t, care must be taken to establish which definition has been used.

Measurement of the rate of polymerization in conventional polymerization processes yields a value of the ratio k_p^2/k_t. To obtain an absolute value of k_t requires evaluation of the rate under non-steady state conditions (yields k_p/k_t), information on the value of k_p (e.g. from PLP, see 4.5.2) or the radical concentrations (e.g. from EPR). Experimental methods for extracting k_p are discussed in section 4.5.2.

In extracting k_t from rate of polymerization data, attention must be paid to the reaction conditions. For low conversions, values of the rate constants k_t typically lie in the range 10^6-10^8 M^{-1} s^{-1}. This scatter in literature values of k_t is not wholly a consequence of experimental error. It is, in part, a reflection of the dependence of k_t on experimental conditions and differences in the chain length population of the propagating radicals (see 5.2.1.1).

5.2.1.1 Homopolymerization

Termination (Scheme 5) is a diffusion controlled process. The rate of reaction is dependent on how fast the radical centers of the propagating chains ($P_i\bullet$ and $P_j\bullet$) come together.

Termination:

$$P_i\bullet \ + \ P_j\bullet \ \xrightarrow{k_{t(i,j)}} \ \text{dead polymer}$$

Scheme 5

Simple center of mass or translational diffusion is known to be the rate determining step for small radicals and may also be important for larger species at very low conversions. However, other diffusion mechanisms are also operative and are often the major term in the apparent rate constant for the case of macromolecular species. These include:

(a) Segmental motion. The internal reorganization of the chain that is required to bring the reactive ends together.
(b) Reptation. The snaking of the chain through a viscous medium.
(c) Reaction diffusion (also called residual termination). Chain end motion by addition of monomer to the chain end.

The relative importance of these mechanisms, and the value of the overall k_t, depends on the molecular weight and polydispersity of the propagating species, the medium and the degree of conversion. The value of k_t is not constant!

In dealing with radical-radical termination in bulk polymerization it is common practice to divide the polymerization timeline into three or more conversion regimes.[2,9] Within each regime, expressions for the termination rate constant are defined according to the dominant mechanism for chain end diffusion. The usual division is as follows:

(a) Low conversion - prior to the onset of the gel or Trommsdorff effect and characterized by highly mobile propagating species. Center of mass and segmental diffusion are the rate determining mechanisms for chain end movement.
(b) Medium to high conversion - after the onset of the gel effect. The diffusion mechanism is complex. Large chains become immobile, however the chain ends may move by reptation or reaction diffusion. Monomeric species and short chains may still diffuse rapidly.
(c) Very high conversion - the polymerization medium is a glassy matrix. Most chains are immobile and reaction diffusion is the rate determining diffusion mechanism. New chains are rapidly immobilized. Initiator efficiencies are very low.

The precise conversion ranges are determined by a variety of factors including the nature of the monomer, the molecular weight of polymeric species and the solvent (if any). For bulk polymerization (a) is typically <10%, (b) is 15-90% and (c) is >90%. In solution polymerization, or for polymerizations carried out in the presence of chain transfer agents, the duration of the low conversion regime is extended and the very high conversion regime may not occur.

Logic dictates that simple center of mass diffusion and overall chain movement by reptation or other mechanisms will be chain length dependent. It has been suggested that the overall rate constant for termination should be defined as an average of the individual rate constants weighted by the concentrations of the propagating species.[10]

$$k_t = \sum_{i=1}^{\infty} \sum_{j=1}^{\infty} \frac{k_{t(i,j)}[P_i\bullet][P_j\bullet]}{[P\bullet]^2}$$

where $k_{t(i,j)}$ is the rate constant for reaction between species of chain lengths i and j, and $[P\bullet]$ is the total radical concentration.

Various expressions have been proposed for estimating how the overall rate constant k_t and the individual rate constants $k_{t(i,j)}$ vary with the chain lengths of the reacting species.[2,10-17] For the situation where the chain length of one or both of the species is "small" (not entangled with itself or other chains) and conversions of monomer to polymer are low, the termination kinetics should be dominated by the rate of diffusion of the shorter chain. The time required for the chain reorganization to bring the reacting centers together will be insignificant and center of mass diffusion can be the rate determining step. Simple relationships of the following forms have been applied:[13,17-19]

geometric mean: $\qquad k_{t(i,j)} = k_{to} (i \cdot j)^{-\alpha/2}$

harmonic mean: $\qquad k_{t(i,j)} = k_{to} \left(\frac{2 \cdot i \cdot j}{i+j}\right)^{-\alpha}$

where k_{to} is a constant and at least one of i and j is "small". Values of $k_{t(1,1)}$ have been determined for a variety of monomeric radicals to be ca. 10^9 M^{-1} s^{-1}.[20] Taking k_{to} as $k_{t(1,1)}$ and α as 1.0 in the geometric expression yields values of $k_{t(i,j)}$ as shown in Figure 1.[21]

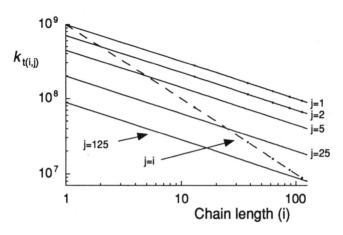

Figure 1 Chain length dependence of $k_{t(i,j)}$ predicted by the geometric mean approximation with $\alpha=1.0$; i and j are the lengths of the reacting chains.

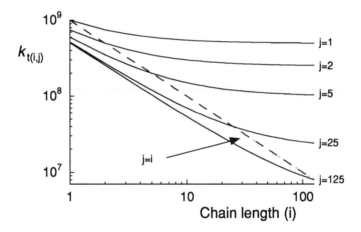

Figure 2 Chain length dependence of $k_{t(i,j)}$ predicted by the harmonic mean approximation[17] with $\alpha=1.0$; i and j are the lengths of the reacting chains.

Use of the harmonic mean approximation predicts a shallower dependence of $k_{t(i,j)}$ on the chain length (Figure 2). There is as yet no appropriate experimental data to test how closely these approximations model the chain length dependence of $k_{t(i,j)}$.

For larger species, even though the chains themselves may be in contact, chain end diffusion by segmental motion, reptation, or reactive diffusion will be required to bring the radical centers together. These terms are likely be more important than center of mass diffusion. North[1] has argued that diffusion of the reactive chain end of longer chains by a segmental diffusion should be independent of chain length and has presented some experimental evidence for this hypothesis.

Mahabadi and O'Driscoll[10] considered that segmental motion and center of mass diffusion should be the dominant mechanisms at low conversion. They proposed expressions which indicate that $k_{t(i,j)}$ should be dependent on chain length such that the overall rate constant obeys the expression:

$$k_t \propto \bar{X}_n^\alpha$$

where $\bar{X}_n$ is the number average degree of polymerization and $\alpha = -0.5$ for short $\bar{X}_n$ reducing to -0.1 for large $\bar{X}_n$. Bamford[16,22-26] has proposed a general treatment for solving polymerization kinetics with chain length dependent k_t and considered in some detail the ramifications with respect to molecular weight distributions and the kinetics of chain transfer, retardation, *etc*.

Since shorter, more mobile, chains diffuse more rapidly (by center of mass diffusion or other mechanisms), they are more likely to be involved in termination. For this reason, most termination involves reaction of a long species with a short species. The lower mobility of long chains ensures that they are unlikely to react with each other. Cardenas and O'Driscoll[27] proposed that propagating species be considered as two populations; those with chain length below the entanglement limit and those above. This basic concept has now been adopted by other authors.[28-31] Russell[28] has provided a detailed critique of these concepts. Direct experimental evidence for the importance of polydispersity of the propagating radicals on termination kinetics has recently been reported by Faldi *et al*.[29]

Empirical relationships defining k_t or the rate of diffusion of long chains in terms of either the viscosity[1,32] or the free volume[9,15,33-36] have been proposed. Tulig and Tirrell,[37] Ito[38] and de Gennes[39] have proposed expressions for k_t based on a reptation mechanism. In most of these works there has been no allowance for chain length dependence of k_t or any variation in f with conversion.

Long chains are essentially immobile. The concept of reaction diffusion (also called residual termination) has been incorporated into a number of treatments.[40,41] If polymer concentrations and chain lengths are sufficient that all propagating species become entangled, then reaction diffusion is likely to be the most important diffusion mechanism for bringing the reactive chain ends together. This situation requires that the rate of growth of new chains and the concentration of propagating radicals are such that the new chains reach the entanglement length before encountering another reactive chain end.

If reaction diffusion is dominant, the termination rate constant is determined by the value of k_p and the monomer concentration. In these circumstances the rate constant for termination $k_{t(i,j)}$ may be independent of the chain lengths i and j and should obey an expression of the form:[40]

$$k_{t(i,j)} = k_{t1} k_p [M]$$

where k_{tl} is a constant and i and j are "large". It follows that the overall termination rate constant may also be independent of chain length in some circumstances.

Changes in the population of propagating species and the increase in the polymer concentration mean that the effective rate constant for termination will usually decrease with conversion. Most early treatments of polymerization kinetics assume the initiator efficiency f and rate constant for initiator decomposition k_d to be constant with conversion. In fact, depending on the initiator and the polymerization conditions, the values of f and k_d can both decrease with conversion due to an increase in the importance of cage reactions.* It has been proposed that the propagation rate constant k_p may also decrease due to the rate of monomer diffusion becoming the rate limiting step. However recent findings suggest that this is unlikely except at extremely high conversions.[42] The number of conversion dependent variables complicate any exact treatment of polymerization kinetics.

With the use of supercomputers for kinetic simulations, it is now possible to deal with chains as discrete species for chain lengths ≤500 in a modest amount of computer time.[21,43,44] Chains above this limit are likely to be entangled, thus handling species with chain length > 100 as a composite species should not lead to serious error. Therefore, a more exact treatment of polymerization kinetics is feasible and advances in computer technology are likely to place this methodology on the desktop within a very short time. Further developments in theory and more experimental measurements of chain length dependent rate constants are required to take full advantage of this method.

5.2.1.2 Copolymerization

Several theories have been developed to predict the rate of copolymerization based on the terminal model for chain propagation (see 6.2.1.1). In early work, it was assumed that the rate constant for termination was determined by the nature of the monomer unit at the reacting chain ends (see 5.2.1.2.1). This "chemical control" model fell from favor with the recognition that chain diffusion is the rate determining step.

North et al.[45,46] proposed a "diffusion control" model for copolymerization (see 5.2.1.2.2). This treatment, while it may be more correct, offers no mechanistic insight into the process. More complex models for diffusion controlled termination in copolymerization have also appeared.[2]

Recently, measurements of the absolute value of the average rate constants for propagation and termination have shown that the propagation kinetics of copolymerization cannot be explained in terms of the simple terminal

---

* Cage return (reformation of initiator in the solvent cage) is particularly important with peroxide initiators and leads to a reduction in k_d. Formation of by-products leads to a reduction in f (see 3.3.1.1.3 and 3.2.2.1.3).

model.[47,48] The kinetics of copolymerization can be explained by diffusion controlled termination and by assuming that propagation is subject to remote unit effects (see 6.2.1.2).

Finally it is important to realize that, while the rate of termination may be determined largely by the rates of chain diffusion, the chain end composition and the ratio of combination and disproportionation are not. Prediction of the overall rate of termination offers little insight into the detailed chemistry of the process. The rate constant for termination in copolymerization is dependent on the same factors as that for homopolymerization (see 5.2.1.1) with reaction diffusion as the principal mechanism for chain-end diffusion at high conversion.

5.2.1.2.1 Chemical control model

According to the terminal model, the kinetics of copolymerization are dictated by the rate of initiation and the rates of the four propagation reactions and three termination reactions (Scheme 6).[49]

propagation:

$$P_A\bullet + M_A \xrightarrow{k_{pAA}} P_A\bullet$$
$$P_A\bullet + M_B \xrightarrow{k_{pAB}} P_B\bullet$$
$$P_B\bullet + M_A \xrightarrow{k_{pBA}} P_A\bullet$$
$$P_B\bullet + M_B \xrightarrow{k_{pBB}} P_B\bullet$$

termination:

$$P_A\bullet + P_A\bullet \xrightarrow{k_{tAA}}$$
$$P_A\bullet + P_B\bullet \xrightarrow{k_{tAB}} \text{dead polymer}$$
$$P_B\bullet + P_B\bullet \xrightarrow{k_{tBB}}$$

Scheme 6

The instantaneous rate of monomer consumption in binary copolymerization is then given by the following expression:

$$\frac{-d[M_A+M_B]}{dt} = k_{pAA}[P_A\bullet][M_A] + k_{pBA}[P_B\bullet][M_A] + k_{pAB}[P_A\bullet][M_B] + k_{pBB}[P_B\bullet][M_B]$$

Use of the steady state approximation:

$$R_i = R_t = 2k_{tAA}[P_A\bullet]^2 + 2k_{tAB}[P_A\bullet][P_B\bullet] + 2k_{tBB}[P_B\bullet]^2$$

(where R_i is the rate of initiation) allows the concentrations of the active species to be eliminated. Thus:

$$\frac{-d[M_A+M_B]}{dt} = \frac{k_{pAA}k_{pBA}[M_A]^2 + 2k_{pAB}k_{pBA}[M_A][M_B] + k_{pBB}k_{pAB}[M_B]^2}{2k_{tAA}k_{pBA}^2[M_A]^2 + 2k_{tAB}k_{pAB}k_{pBA}[M_A][M_B] + 2k_{tBB}k_{pAB}^2[M_B]^2}$$

which can be rewritten as:[49]

$$\frac{-d[M_A+M_B]}{dt} = \frac{(r_A[M_A]^2 + 2[M_A][M_B] + r_B[M_B]^B) R_i^{1/B}}{\delta_A^2 r_A^2[M_A]^2 + 2\phi\delta_A\delta_B r_A r_B[M_A][M_B] + \delta_B^2 r_B^2[M_B]^2}$$

where:

$$\phi = \frac{k_{tAB}}{2(k_{tAA} \cdot k_{tBB})^{0.5}} \qquad \delta_A = \frac{2k_{tAA}^{0.5}}{k_{pAA}} \qquad \delta_B = \frac{2k_{tBB}^{0.5}}{k_{pBB}}$$

$$r_A = \frac{k_{pAA}}{k_{pAB}} \qquad r_B = \frac{k_{pBB}}{k_{pBA}}$$

In evaluating the kinetics of copolymerization according to the chemical control model, it is assumed that the termination rate constants k_{tAA} and k_{tBB} are known from studies on homopolymerization. The only unknown in the above expression is the rate constant for cross termination (k_{tAB}). The rate constant for this reaction in relation to k_{tAA} and k_{tBB} is given by the parameter ϕ.

In applying these relationships to explain the observed dependence of the rate of copolymerization on the monomer feed composition, it is found that:

(a) The overall rate constant for termination in copolymerization is substantially faster than that in homopolymerization of any of the component monomers. Values of ϕ required to fit the rate of copolymerization are typically in the range 5-50. For S-MMA copolymerization ϕ is calculated to be in the range 5-14 depending on the monomer feed ratio.
(b) The rate constants are monomer feed dependent (see 6.2.1.2). This concept cannot be encompassed within the provisions of the terminal model.

Experiments with model compounds indicate that there is specificity in termination (see 5.2.4.8). For example, in S-MMA copolymerization cross termination is 2-3 times faster than either homotermination process. However, this should not be regarded as a proof of the chemical control model.

5.2.1.2.2 Classical diffusion control model

According to the diffusion control model, the rate of the termination step is limited only by the rates of diffusion of the polymer chains (at low conversion or in dilute solution). This may be dependent on the overall polymer chain composition. It should not depend solely on the nature of the chain end.[1,45]

North et al.[45,46] proposed the following expression in which it is assumed that all chains terminate with the same rate constant which is determined by the rate of diffusion (diffusion control model):

$$\frac{-d[M_A+M_B]}{dt} = \frac{(r_A[M_A]^2 + 2[M_A][M_B] + r_B[M_B]^2) R_i^{1/2}}{\varepsilon_1^2 r_A^2 [M_A]^2 + 2\varepsilon_1\varepsilon_2 r_A r_B [M_A][M_B] + \varepsilon_2^2 r_B^2 [M_B]^2}$$

where:

$$\varepsilon_1 = \frac{k_{t(AB)}^{0.5}}{k_{pAA}} \qquad \varepsilon_2 = \frac{k_{t(AB)}^{0.5}}{k_{pBB}}$$

and $k_{t(AB)}$ is the copolymer-composition dependent rate constant for termination. In this scheme, the value of $k_{t(AB)}$ is obtained by fitting the experimental data. It is not a constant.

5.2.2 Disproportionation vs. Combination - General Considerations

Combination involves the coupling of two propagating radicals (see Scheme 1). The resulting polymer chain has a molecular weight equal to the sum of the molecular weights of the reactant species. If it is assumed that all chains are formed by initiator-derived radicals adding to monomer, then the combination product will have two initiator-derived ends (assuming chain transfer is unimportant). Disproportionation involves the transfer of a β-hydrogen from one propagating radical to the other. This results in the formation of two polymer molecules. Both chains will have one initiator-derived end and while one chain has an unsaturated end, the other has a saturated end (see Scheme 1).

Since the mode of termination clearly plays an important part in determining the polymer end groups and the molecular weight distribution, a knowledge of the disproportionation:combination ratio (k_{td}/k_{tc}) is vital to the understanding of structure-property relationships. Unsaturated linkages at the ends of polymer chains, as may be formed by disproportionation, have long been thought to contribute to polymer instability and it has been demonstrated that both head-to-head linkages and unsaturated ends are weak links during the thermal degradation of PMMA.[50-53] Polymer chains with unsaturated ends may also be reactive during polymerization. Copolymerization of macromonomers formed by disproportionation is a possible mechanism for the formation of long chain branches.[54-56] Such macromonomers may also function as transfer agents.[56]

A knowledge of k_{td}/k_{tc} is also important in designing polymer syntheses. For example, in the preparation of block copolymers using polymeric or multifunctional initiators (see 6.3), ABA or AB blocks may be formed depending on whether termination involves combination or disproportionation respectively. The relative importance of combination and disproportionation is also important in the analysis of polymerization kinetics and, in particular, in the derivation of absolute rate parameters.

5.2.3 Disproportionation vs. Combination - Model Studies

The determination of k_{td}/k_{tc} by direct analysis of a polymerization or the resultant polymer often requires data on aspects of the polymerization mechanism that are not readily available. For this reason, it is appropriate to consider the self-reactions of low molecular weight radicals which are structurally analogous to the propagating species. These model studies provide valuable insights by demonstrating the types of reaction that are likely to occur during polymerization and the factors influencing k_{td}/k_{tc}. These have been discussed in general terms in section 2.4.

In these model studies, evaluation of k_{td}/k_{tc} is simplified because reactions which compete with disproportionation or combination are more readily detected and allowed for. However, by their very nature, model studies cannot exactly simulate all aspects of the polymerization process. Consequently, a number of factors must be borne in mind when using model studies to investigate the termination process. These stem from differences inherent in polymerization vs. simple organic reactions and include:

(a) There may be additional pathways open to the poly- or oligomeric radicals which are not available to the simple model species.[57]
(b) In polymerization particular propagating species have only transient existence since they are scavenged by the addition of monomer or other reactions. Model studies are usually designed such that the self-reaction is the only process. This can lead to a very different and sometimes misleading product distribution. A knowledge of the reaction kinetics is extremely important in analyzing the results.
(c) Reaction conditions (solvent, viscosity, etc.) chosen for the model experiment and the polymerization experiment are often very different.

Model carbon-centered radicals are conveniently generated from azo-compounds. These have the advantage that radicals are generated in pairs and that transfer to initiator is generally not a serious problem. All of the major products from thermal or photochemical decomposition in an inert solvent are the products from radical-radical reaction. One frequently observed complication is polymerization of the unsaturated by-products of disproportionation. This problem may be circumvented by conducting experiments in the presence of an inhibitor, the concentration of which can be chosen such that all radicals which escape the solvent cage are trapped and reactions of the initiator derived radicals with other species are eliminated.[55] The value of k_{td}/k_{tc} is determined by analyzing the products of cage recombination. Most data indicate no difference in specificity between the cage and encounter (*i.e.* non-cage) processes.[55]

Radical-radical termination in binary copolymerization involves three reactions. There are two homotermination processes and a cross termination reaction (Scheme 6). If penultimate unit effects are unimportant then k_{td}/k_{tc} for the homotermination reactions can be assumed to be similar to those in the corresponding homopolymerizations.

There are three pieces of information to be gained from model studies on cross termination:

(a) The value of k_{td}/k_{tc} for the cross reaction.
(b) The specificity for hydrogen transfer in disproportionation (*i.e.* from monomer A to monomer B or vice versa).
(c) The relative rates of the homo- and cross-termination processes.

Three types of model study have been reported. The first approach to the problem has been to decompose a mixture of two initiators (*i.e.* one to generate radical A, the other to generate radical B). With this method experimental difficulties arise because the two radicals may not be generated at the same rate and analysis is complicated by homotermination products from cage recombination.

A second approach has been to use an unsymmetrical initiator which allows the two radicals of interest to be generated simultaneously.[58] In this case, analysis of the cage recombination products provides information on cross termination uncomplicated by homotermination. Analysis of products of the encounter reaction can also give information on the relative importance of cross and homotermination. However, copolymerization of unsaturated products causes severe analytical problems.

A third technique is to examine the products of primary radical termination in polymerizations carried out with high concentrations of initiator.[59,60]

5.2.3.1 Polystyrene and derivatives

The self reaction of phenylethyl radicals (1) has been widely investigated.[61-65] The findings of these studies are summarized in Table 5.1. Unless R^2 is very bulky (*e.g.* *t*-butyl, see below), combination is by far the dominant process with the value k_{td}/k_{tc} typically in the range 0.05-0.16. However, a small amount of disproportionation is always observed.

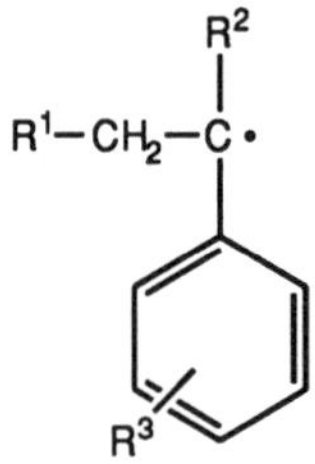

R¹−CH₂−C(R²)(•)−C₆H₄−R³

(1)

The value of k_{td}/k_{tc} shows no significant dependence on chain length for oligostyryl radicals (**4a, b**).[64,65] On the basis of these findings, k_{td}/k_{tc} for PS• should also be small and non-zero.

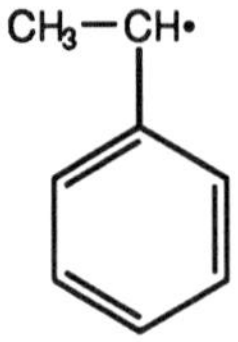

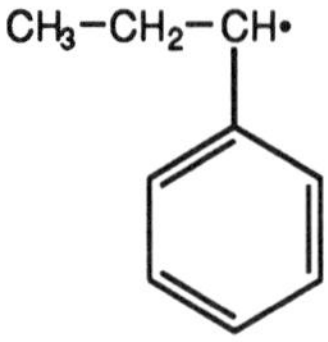

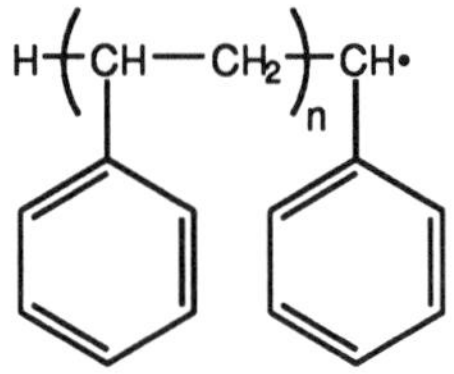

(2) CH₃−CH•−Ph   (3) CH₃−CH₂−CH•−Ph   (4a) n=1, (4b) n=2 H−(CH(Ph)−CH₂)ₙ−CH•−Ph

For radicals (1), k_{td}/k_{tc} shows a marked dependence on the bulk of the substituent (R²). While phenylethyl radicals (**2-4**) and cumyl radicals (**5**) afford predominantly combination, the radicals (**6**), which have an α-t-butyl group, give predominantly disproportionation. On this basis, it would not be surprising if, in polymerization of AMS there was more disproportionation than is observed with (**5**).

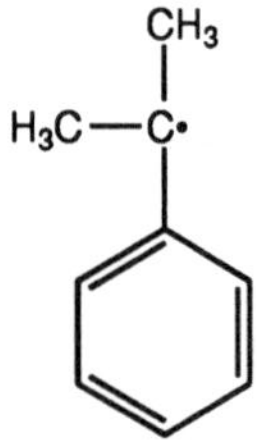

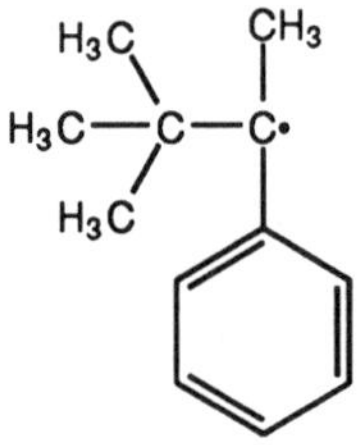

(5) H₃C−C•(CH₃)−Ph   (6) (H₃C)₃C−C•(CH₃)−Ph

The value of k_{td}/k_{tc} for oligostyryl radicals (**4**) is reported to decrease with increasing temperature. With 1,3,5-triphenylpropyl radicals (**4b**) k_{td}/k_{tc} halves on increasing the temperature from 80°C to 160°C (see Table 5.1).[65]

Table 5.1 Values of k_{td}/k_{tc} for Styryl Radical Model Systems

System	Structure	Temperature	k_{td}/k_{tc}	ref.
S	(2)	20	0.073	64
S	(2)	80	0.081	64
S	(2)	118	0.097	62
S	(3)	20	0.141	64
S	(3)	80	0.146	64
S	(3)	118	0.107	62
S	(4a)	80	0.156	64
S	(4a)	90	0.146	65
S	(4a)	90	0.141	65
S	(4a)	100	0.130	65
S	(4a)	120	0.109	65
S	(4a)	141	0.090	65
S	(4a)	161	0.078	65
S	(4b)	80	0.159	65
S	(4b)	90	0.150	65
S	(4b)	100	0.134	65
S	(4b)	120	0.119	65
S	(4b)	141	0.097	65
S	(4b)	161	0.082	65
AMS	(5)	20-60	0.05	66
AMS	(5)	55	0.1	67
AMS	(6)	55	∞	68

The result indicates that the activation energy for combination is higher than that for disproportionation by ca. 10 kJ mol^{-1}. A similar inverse temperature dependence is seen for other small radicals (see 2.4). However, markedly different behavior is reported for polymeric radicals (see below).

Benzyl radicals and α- and β- substituted derivatives also undergo unsymmetrical coupling through the aromatic ring (see 2.4). The formation of the α–o and α–p coupling products is reversible. Consequently, these materials are usually only observed as transient intermediates.

Direct aromatization of the quinonoid intermediates is a photochemically allowed but thermally forbidden rearrangement (Scheme 7). When phenylethyl radicals are generated photochemically at 20°C there is evidence[64] of α–o coupling by way of the aromatized product (7). The products derived from these pathways can be trapped in thermal reactions by radical[67] or acid[69] catalyzed aromatization. With benzyl radicals the ratio of α–o:α–p and [α–o + α–p]:α–α has been shown to increase with increasing temperature.[69] A transient species, presumed to be a quinonoid intermediate, has also been observed when oligomeric radicals (4) are generated thermally.[65]

The formation of the quinonoid species is favored by substitution at the radical center (see 2.4). Cumyl radicals (5)[66,67,70] are reported to give α–α, α–o and α–p coupling products in the ratio 77:8:15. Several studies have examined the reactions of p-substituted phenylethyl radicals. Electron withdrawing substituents favor disproportionation over combination. However, the effect is small.

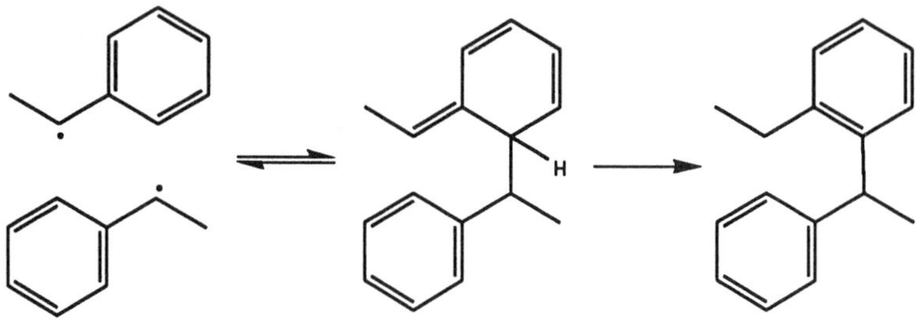

Scheme 7

5.2.3.2 Poly(alkyl methacrylates)

The self-reaction of 2-carboalkoxy-2-propyl radicals (8-10) has been examined.[55,71,72] The results of these studies are reported in Table 5.2. Combination is slightly favored over disproportionation. The value of k_{td}/k_{tc} for (8) was found to be essentially independent of temperature.

Disproportionation increases in the series where the ester is methyl<ethyl<butyl suggesting that this process is favored by increasing the bulk of the ester alkyl group. This trend is also seen for polymeric radicals.

Bizilj et al.[55] reported that disproportionation may be slightly more important for oligomeric radicals. While combination products were unequivocally identified, analytical difficulties prevented a precise determination of the disproportionation products. Accordingly, they were only able to state a maximum value of k_{td}/k_{tc}. Their data show that $k_{td}/k_{tc} \leq 1.85$ for the self reaction of (11) and ≤1.50 for reaction between (8) and (11).

The value of k_{td}/k_{tc}(80°C) in the cross-reaction between radicals (8) and (10) has also been examined.[58] This system is a model for cross-termination in MMA-BMA copolymerization. The value of k_{td}/k_{tc} (1.22) is similar to that found for the self-reaction of (10) (1.17) and much larger than that for the self-reaction of (8) (0.78). There is a small preference (ca. 1.4 fold) for the transfer of hydrogen from the butyl ester (10) to the methyl ester (8).

An early report[73] indicated that the self reaction of 2-carbomethoxy-2-propyl radicals (8), like cyanoisopropyl radicals (12) (see 5.2.3.3), affords an unstable coupling product (analogous to a ketenimine). Precedent for a reversible unsymmetrical C-O coupling mode for radicals with a α-carbonyl group has recently been established for the case where normal C-C coupling is sterically very hindered.[74] However, the more recent studies on reactions of 2-carbomethoxy-2-propyl radicals (8) and related species provide no evidence for this pathway.[55,71]

Table 5.2 Values of k_{td}/k_{tc} for Methacryate Ester Model Systems

System	Structure	Temperature	k_{td}/k_{tc}	ref.
MMA	(8)	70-90	0.78	55
MMA	(8)	90	0.62	71
MMA	(8)	115	0.61	55,71
MMA	(8)	140	0.60	55,71
MMA	(8)	165	0.59	71
MMA	(11)	80	≤1.85	55
EMA	(9)	80	0..72	55
BMA	(10)	80	1.17	55
MMA-co-BMA	(8, 10)	80	1.22	58

$$\begin{array}{ccc}
\text{CH}_3 & \text{CH}_3 & \text{CH}_3 \\
| & | & | \\
\text{H}_3\text{C}-\text{C}\bullet & \text{H}_3\text{C}-\text{C}\bullet & \text{H}_3\text{C}-\text{C}\bullet \\
| & | & | \\
\text{CO}_2\text{CH}_3 & \text{CO}_2\text{C}_2\text{H}_5 & \text{CO}_2\text{C}_4\text{H}_9 \\
(8) & (9) & (10)
\end{array}$$

$$\begin{array}{c}
\text{CH}_3 \quad\quad \text{CH}_3 \\
| \quad\quad\quad\quad | \\
\text{CH}_3-\text{C}-\text{CH}_2-\text{C}\bullet \\
| \quad\quad\quad\quad | \\
\text{CO}_2\text{CH}_3 \quad \text{CO}_2\text{CH}_3 \\
(11)
\end{array}$$

Bizilj et al.[55] demonstrated that during disproportionation of oligomeric radicals, the abstraction of a methyl hydrogen (to generate a terminal methylene group - see Scheme 8) is preferred ≥ 10-fold over abstraction of a methylene hydrogen (to afford an internal double bond). The simplest explanation for this behavior is that the methyl hydrogens are more sterically accessible than the methylene hydrogens.

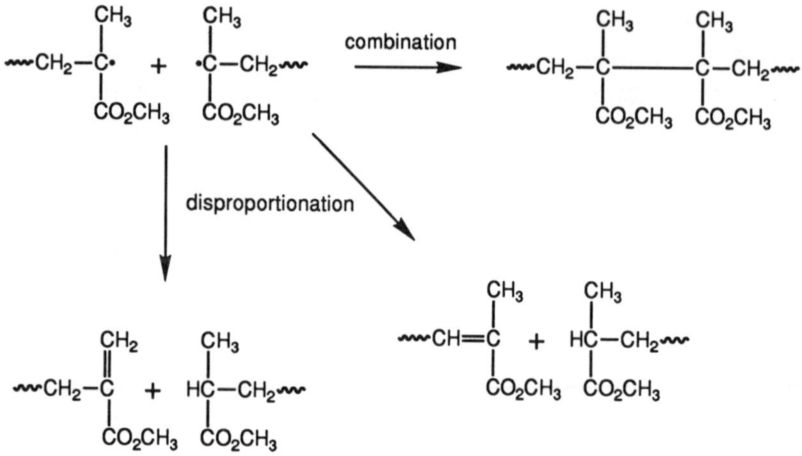

[Scheme 8 – combination and disproportionation reactions of PMMA-type radicals]

Scheme 8

5.2.3.3 Poly(methacrylonitrile)

The simplest model for the propagating species in MAN polymerization is the cyanoisopropyl radical (12). The reactions of these radicals (from AIBN; see Scheme 9) have been extensively studied. In contrast with analogous esters (8-10) (see 5.2.3.2), combination is by far the dominant process (see Table 5.3).

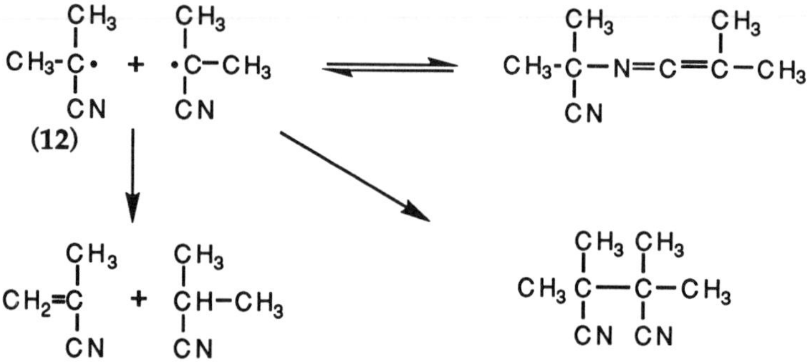

[Scheme 9 – reactions of cyanoisopropyl radical (12): combination, disproportionation, and ketenimine formation]

Scheme 9

Serelis and Solomon[75] found that primary radical termination of oligo(MAN) radicals (13) with (12) also gives predominantly combination. The ratio k_{td}/k_{tc} was found to have little, if any, dependence on the oligomer chain length ($n \leq 4$). As with PMMA•, disproportionation involves preferential abstraction of a methyl hydrogen and chains terminated in this way will, therefore, possess a potentially reactive terminal methylene (14).

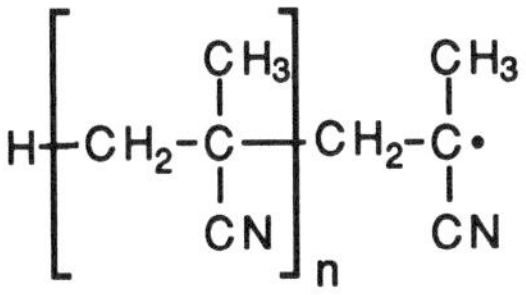

Termination

(13) H⫟CH₂-C(CH₃)(CN)⫠ₙ CH₂-C•(CH₃)(CN)

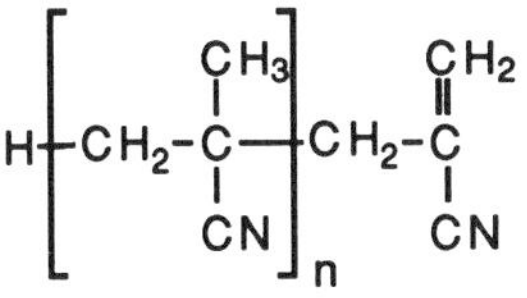

(14) H⫟CH₂-C(CH₃)(CN)⫠ₙ CH₂-C(=CH₂)(CN)

Table 5.3 Values of k_{td}/k_{tc} for Reactions of Cyanoisopropyl Radicals (12)

System	Structure	Temperature (°C)	k_{td}/k_{tc}	ref.
MAN	(12)	80	0.05-0.1	75-77
MAN	(13)	80	0.1	75
MAN-co-S	(4a)	90	0.61	78
MAN-co-S	PS•	98	a	79
MAN-co-BMA	PBMA•	25	b	80
MAN-co-E	PE•	80	b	60

a Predominantly combination.
b Predominantly disproportionation.

Cyanoisopropyl radicals undergo unsymmetrical C-N coupling in preference to C-C coupling.[81] The preferential formation of the ketenimine is a reflection of the importance of polar and steric influences.[82] However, the ketenimine is itself thermally unstable and a source of cyanoisopropyl radicals, thus the predominant isolated product is often from C-C coupling.

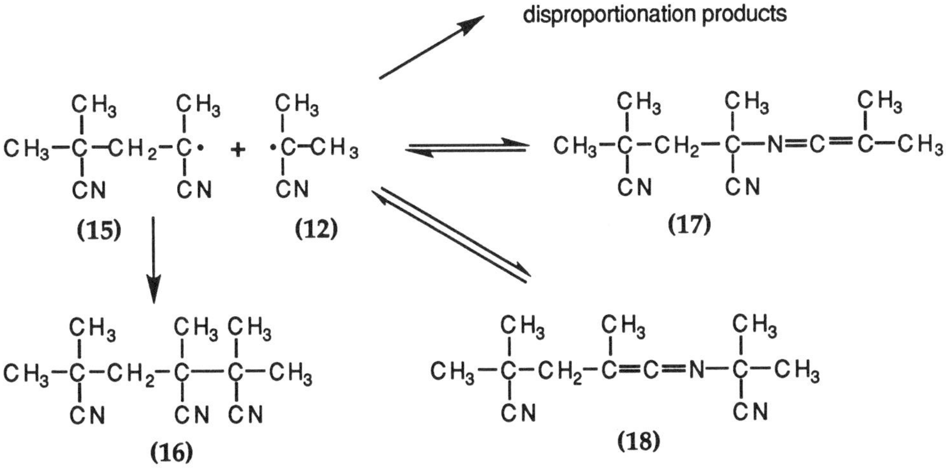

Scheme 10

Preferential C-N coupling is also observed for oligomeric radicals (see Scheme 10).[83] A ketenimine (17) is the major product from the reaction of the "dimeric" MAN radical (15) with cyanoisopropyl radicals (12). Only one of the two possible ketenimines was observed; a result which is attributed to the thermal lability of (18). If this explanation is correct then, although C-N coupling may occur during MAN polymerization, ketenimine structures are unlikely to be found in PMAN.

5.2.3.4 Polyethylene

The self reaction of primary alkyl radicals gives mainly combination.[84] For primary alkyl radicals [$CH_3(CH_2)_nCH_2\bullet$], k_{td}/k_{tc} is reported to lie in the range 0.12-0.14, apparently independent of chain length (n=0-3).[84,85]

5.2.3.5 Poly(styrene-co-methyl methacrylate)

The reaction between the PMMA (8) and PS (4a) model radicals (generated from the unsymmetrical azo-compound) has been studied as a model for cross-termination in S-MMA copolymerization (Scheme 11).[78,86]

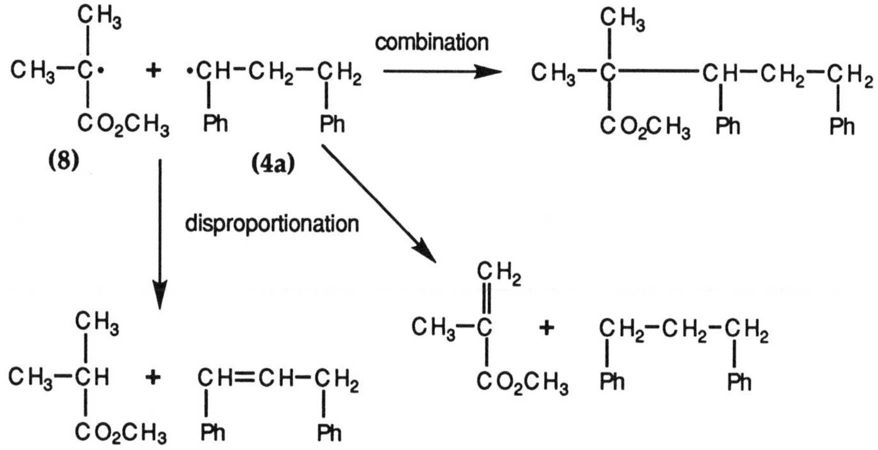

Scheme 11

The value for k_{td}/k_{tc}(90°C) for the cross reaction was 0.56. In disproportionation, transfer of hydrogen from the PS• model (4a) to the PMMA• radical (8) was ca. 5.1 times more prevalent than transfer in the reverse direction (i.e. from 8 to 4a). The value of k_{td}/k_{tc}(90°C) is between those of k_{td}/k_{tc}(90°C) for the self-reaction of these radicals under similar conditions (0.13 and 0.78 for (4a) and (8) respectively). Analysis of the encounter products indicated a small preference for cross termination over either homotermination process.[86]

On the other hand, Ito found that primary radical termination in S polymerization initiated by AIBMe[59,85] [i.e. PS• + (8)] and in MMA polymerization initiated by 1,1'-azobis-1-phenylethane[59] [i.e. PMMA• + (2)] have been reported to give predominantly combination. Ito[59] concluded that cross termination was not particularly favored over homotermination in S-MMA copolymerization.

5.2.3.6 Poly(styrene-co-methacrylonitrile)

Analysis of the products from the thermal decomposition of the mixed azo compound showed that in the cross-reaction of radicals (4a) and (12) $k_{td}/k_{tc}(90°C)$ is 0.61.[78] This study also found that in disproportionation, hydrogen transfer from (4a) to (12) is *ca.* 2.2 times more frequent than transfer from (12) to (4a).

Both self-reactions involve predominantly combination (Scheme 12). The values of $k_{td}/k_{tc}(80°C)$ are 0.18 and 0.05 for radicals (4a) (see 5.2.3.1) and (12) (see 5.2.3.3) respectively. Clearly, values of k_{td}/k_{tc} for homotermination cannot be used as a guide to the value for k_{td}/k_{tc} in cross-termination.

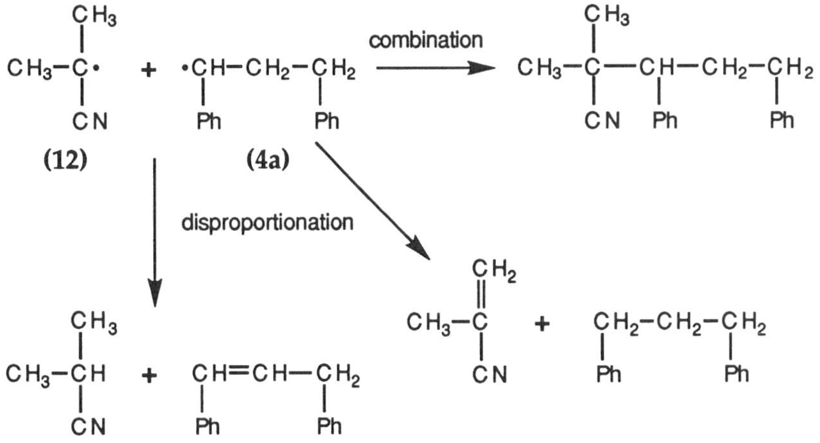

Scheme 12

The reaction of oligostyrene radicals with cyanoisopropyl radicals (12) has been studied by several groups and reported to give exclusively combination (98°C, toluene),[79,85] mainly combination (60°C, ethyl acetate;[87] 98°C, toluene[88]). Moad *et al.*[88] examined styrene oligomerization in toluene at 98°C using high concentrations of AIBN as initiator. While the major products arose from combination, they also isolated and identified small amounts of disproportionation products thus demonstrating that some disproportionation does occur.

5.2.3.7 Other copolymers

Disproportionation:combination ratios in primary radical termination have been determined for other polymerizations carried out with AIBN initiation (see Table 5.3). Guth and Heitz[60] have reported that primary radical termination between PE• radicals and cyanoisopropyl radicals (12) involves substantial disproportionation. Both homotermination processes involve largely combination (see 5.2.3.3 and 5.2.3.4). Barton *et al.*[80] have indicated that primary radical termination between PBMA• and cyanoisopropyl radicals (12) involves largely disproportionation.

5.2.4 Disproportionation vs. Combination - Polymerization

A substantial number of studies give information on k_{td}/k_{tc} for polymerizations of S (5.2.4.1) and MMA (5.2.4.2). There has been less work on other systems. The main problem in assessing k_{td}/k_{tc} is in assessing the importance of other termination mechanisms (*i.e.* transfer to initiator, solvent, *etc.*, primary radical termination).

Three major techniques have been applied in assessing the relative importance of disproportionation and combination:

(a) The Gelation technique. This method was developed by Bamford et al.[89] In graft copolymerization, termination by combination will give rise to a crosslink while disproportionation (and most other termination reactions) will lead to graft formation. The initiation system based on a polymeric halo-compound [poly(vinyl trichloroacetate)/$Mn_2(CO)_{10}$/hv] was used to initiate polymerization and the time for gelation was used to calculate k_{td}/k_{tc}. In the original work, the results were calibrated with reference to data for S polymerization for which a k_{td}/k_{tc} of 0.0 was assumed. Recent studies suggest that, in S polymerization, disproportionation may account for 10-20% of chains (see 5.2.4.1). Thus the data may require minor adjustment. Systems studied with this technique include AN, MAN, MA, MMA, and S.

(b) Molecular weight measurement. The mode of termination can be calculated by comparing the kinetic chain length (the ratio of the rate of propagation to the rate of initiation or termination) with the measured number average molecular weight.[90-92]

(c) Polydispersity evaluation. This method relies on a precise evaluation of the molecular weight distribution.[93-95] The mode of termination has a significant influence on the shape of the molecular weight distribution with the instantaneous polydispersity ($\bar{M}_w/\bar{M}_n$) being ~1.5 if termination occurs exclusively by disproportionation of propagating radicals and ~2.0 if termination involves only combination.[96] The value of $\bar{M}_w/\bar{M}_n$ is also conversion dependent.

(d) End group determination. Polymer chains terminated by combination possess two initiator-derived chain ends. Disproportionation affords chains with only one such end. The value of k_{td}/k_{tc} can therefore be determined by evaluating the initiator-derived polymer end groups/molecule and number average molecular weight by applying the relationship: $k_{td}/k_{tc} = (2-x)/2(x-1)$; where x is the number of initiator fragments per molecule. The errors inherent in this technique are large since the polymer end groups typically comprise only a very small fraction of a polymer sample. The initiator-derived ends may be labeled for ease of detection. These techniques are described in section 3.6. It is necessary to allow for side reactions. If there is transfer to monomer, solvent, *etc.*, the value of k_{td}/k_{tc} will be overestimated. The occurrence of transfer to initiator, primary radical termination, or copolymerization of initiator by-products will lead to k_{td}/k_{tc} being underestimated.

5.2.4.1 Polystyrene

Hensley et al.[97] have reported the only direct experimental observation of head-to-head linkages in PS by 2D INADEQUATE NMR on ^{13}C-enriched PS. The method did not enable these groups to be quantified with sufficient precision for evaluation of k_{td}/k_{tc}.

A wide range of less direct methods has been applied to determine k_{td}/k_{tc} in S polymerization. Most indicate predominant combination.[88,91,94,98-112] However, distinction between a k_{td}/k_{tc} of 0.0 and one which is non-zero but ≤0.2, as predicted by model studies (5.2.3.1), is extremely difficult even with the precision achievable with the most modern instrumentation. Therefore, it is not surprising that many have taken the step of interpreting the experimental finding of predominantly combination as exclusively combination.

Olaj et al.[93] have proposed that termination of S polymerization involves substantial disproportionation. They analyzed the molecular weight distribution of PS samples prepared with either BPO or AIBN as initiator at temperatures in the range 20-90°C and estimated k_{td}/k_{tc} to be ca. 0.2. In a more recent study, Olaj et al.[113] have determined the molecular weight distribution of PS samples prepared with photoinitiation at 60 and 85°C and estimated values of k_{td}/k_{tc} of 0.5 and 0.67 respectively. Dawkins and Yeadon[94] have discussed the problems associated with estimating k_{td}/k_{tc} on the basis of polydispersity measurements and determined that k_{td}/k_{tc} should be "substantially smaller" than suggested by Olaj et al.[113]

Berger and Meyerhoff[114] have also reported that termination involves substantial disproportionation. They determined the initiator fragments per molecule in PS prepared with radiolabeled AIBN and conducted a detailed kinetic analysis of the system. They also found a marked temperature dependence for k_{td}/k_{tc}. Values of k_{td}/k_{tc} ranged from 0.168 at 30°C to 0.663 at 80°C.

Other determinations of k_{td}/k_{tc} based on end group determination are at variance with these findings. End group analyses by NMR,[88,111] radiotracer techniques,[107-109] or chemical analysis[110] on PS formed with appropriately labeled initiators all indicate predominantly combination. Moad et al.[88,111] used ^{13}C NMR to define and quantify the end groups in samples of PS prepared at 60°C with either ^{13}C-labeled BPO or AIBN as initiator. This method has the advantage that the end groups from primary radical termination, transfer to initiator, residual initiator and any copolymerized initiator by-products can be distinguished from the end groups formed by initiation (see 3.5.3.2). They showed that, under the conditions employed (60°C, bulk), there are 1.7±0.2 initiator-derived end groups per molecule.

The influence of substituents (p-Cl, p-OMe) on k_{td}/k_{tc} has been investigated by Ayrey et al.[112] They found disproportionation was favored by the p-OMe substituent and that the extent of disproportionation increased with increasing temperature. This result is contrary to the model studies (see 5.2.3.1) which show that k_{td}/k_{tc} has little dependence on substituents and, indeed, suggest the opposite trend.

5.2.4.2 Poly(alkyl methacrylates)

The nature of the termination reaction in MMA polymerization has been investigated by a number of groups using a wide range of techniques (Table 5.4). There is general agreement that there is substantial disproportionation. However, there is considerable discrepancy in the precise values of k_{td}/k_{tc} (most are in the range 1-3 at 60°C). In some cases the difference has been attributed to variations in the way molecular weight data are interpreted or to the failure to allow for other modes of termination under the polymerization conditions (chain transfer, primary radical termination).[115] In other cases the reasons for the discrepancies are less clear.

Four studies suggest that k_{td}/k_{tc} has a significant temperature dependence (see Table 5.4). Although not agreeing on the precise value of k_{td}/k_{tc}, all four studies indicate that the proportion of disproportionation increases with increasing temperature. These results are at variance with model studies which suggest that k_{td}/k_{tc} should be independent of temperature. It has also been proposed that the preferred termination mechanism is solvent dependent and that disproportionation is favored in more polar media.[116]

Table 5.4 Determinations of k_{td}/k_{tc} for MMA Polymerization

Temperature	Method[a]								
°C	E[98,107]	E[99]	E[117]	G[118]	E[119]	P[120]	E[121]	M[90]	E[51]
-25	-	-	-	-	-	0.14	-	-	-
0	1.50	-	-	-	-	0.50	-	-	-
15	-	-	-	-	-	0.76	-	-	-
25	2.13	-	-	2.0	-	-	-	-	-
30	-	-	-	-	-	1.18	-	-	-
40	-	-	0.45	-	-	-	-	-	-
45	-	-	-	-	-	∞	-	-	-
60	5.67	1.35	0.75	2.7	2.62	-	2.57	0.44	1.28
80	-	-	1.32	4.0	-	-	-	-	-

[a] Methods used (see 5.2.4): G-gelation technique, M-molecular weight measurement, P-polydispersity evaluation, E-end group determination.

Hatada et al.[122] have shown that the disproportionation-derived unsaturated ends in PMMA can be determined directly by ^{1}H NMR. Unfortunately, the technique does not allow saturated ends or head-to-head linkages to be determined simultaneously. They have also demonstrated the preference for transfer of a methyl vs. a methylene hydrogen in disproportionation. This is in line with the studies on model radicals (see 5.2.3.2).

Values of k_{td}/k_{tc} for polymerizations of EMA and BMA and higher methacrylate esters have been determined.[80,89,119,121] The extent of disproportionation increases with the size of the ester alkyl group.

5.2.4.3 Poly(methacrylonitrile)

Bamford et al.[89] have examined MAN polymerization (25°C, DMSO) using the gelation technique (see 5.2.4) and have estimated that termination occurs predominantly by disproportionation (k_{td}/k_{tc} = 1.86). This result is at variance with the model studies (see 5.2.3.3).

5.2.4.4 Poly(alkyl acrylates)

The termination mechanism in MA polymerization has been variously determined to be predominantly disproportionation[102,119] or predominantly combination.[89,121,123]

Ayrey et al.[123] have suggested that transfer reactions may have led to erroneous conclusions being drawn in some of the earlier studies. They concluded that termination is almost exclusively by combination (25°C, benzene). Bamford et al.[89] have confirmed this finding with the gelation technique (25°C, bulk) and determined that the polymerizations of higher acrylate esters also terminate predominantly by combination.

5.2.4.5 Poly(acrylonitrile)

There appears to be general agreement that termination in AN polymerization under a variety of conditions (10-90°C, DMSO, DMF, H$_2$O) involves mainly combination.[89,92,124,125] It has been suggested that this may involve either C-N or C-C coupling.[126]

5.2.4.6 Poly(vinyl acetate)

Early reports[102,119,127] have suggested that termination during VAc polymerization involved predominantly disproportionation. It is thought that these investigations did not adequately allow for the occurrence of transfer to monomer and/or polymer, which are extremely important during VAc polymerization (see 5.3.2.2, 5.3.3.4). These problems were addressed by Bamford et al.[89] who used the gelation technique (see 5.2.4.1) to show that the predominant radical-radical termination mechanism is combination (25°C).

5.2.4.7 Poly(vinyl chloride)

Studies on VC polymerization are also complicated by the fact that only a small proportion of termination events may involve radical-radical reactions (see 4.3.1.2). Early studies on the termination mechanism which do not allow for this probably overestimate the importance of disproportionation.[128,129]

Park and Smith [130] attempted to allow for chain transfer in their examination of the termination mechanism during VC polymerization at 30 and 40°C in chlorobenzene. They determined the initiator-derived ends in PVC prepared with radiolabeled AIBN and concluded that $k_{td}/k_{tc} = 3.0$. However, questions have been raised regarding the reliability of these measurements.[131,132] Atkinson et al.[132] applied the gelation technique (see 5.2.4.1) to VC polymerization and proposed that termination involves predominantly combination.

5.2.4.8 Poly(styrene-co-methyl methacrylate)

In termination, the rate determining step is the rate of chains diffusing together. Since propagation is rapid with respect to termination, the relative radical concentrations are generally more important than the termination rate constants in determining the products of termination.[86] The relative radical concentrations are in turn determined by the values of the reactivity ratios and the propagation rate constants. These considerations ensure that during S-MMA copolymerization the instantaneous concentration of chains ending in S is significantly greater than those with a terminal MMA unit.[86] Therefore, most termination events involve these chains. Homotermination of chains ending in S and cross termination are the most important processes. There should be very little homotermination between chains ending in MMA (see Table 5.5).

Table 5.5 Identity of Chain End Units Involved in Radical-Radical Termination in MMA-S Copolymerization[a]

Reaction	'Chemical Control'			'Diffusion Control'
	$\phi=13$	$\phi=3$	$\phi=1$	
-S• + -S•	0.18	0.47	0.72	0.57
-S• + -MMA•	0.81	0.51	0.26	0.37
-MMA• + MMA•	0.01	0.02	0.02	0.06

[a] Calculated by kinetic modeling[86] for discussion of models and the ϕ parameter (see 5..2.1.2).

Several experimental studies on S-MMA copolymerization have appeared. Bevington et al.[98] have examined S-MMA copolymerization (60°C, benzene) using the radiotracer method and suggested that the cross termination reaction involves predominantly combination (k_{td}/k_{tc} for the homotermination processes were taken to be 0 and 5.67 for chains ending in S and MMA respectively).

Chen et al.[133] conducted an analysis of polymerization kinetics and came to a similar conclusion. However, these data are based on the assumption of a "chemical control model" for termination (see 5.2.1.2).

5.2.5 Disproportionation vs. Combination - Summary

Unequivocal numbers for k_{td}/k_{tc} are not yet available for most polymerizations and there is only qualitative agreement between values obtained in model studies and real polymerizations.

It is tempting to attribute problems in reconciling data from model studies and actual polymerizations to difficulties associated with the interpretation of the results of the polymerizations. These are often complicated by other termination pathways, in particular transfer reactions, which, while not fully characterized, must be allowed for when assessing the results. It is notable in this context that the discrepancies are most evident for reactions carried out at higher temperatures (see 5.2.3.1 and 5.2.3.2).

However, some of the differences may be explicable in terms of an effect of molecular size. For many of the model systems at least one of the reaction partners is monomeric [i.e. (2), (5), (8-10), (12)]. Since combination is known to be more sensitive to steric factors than disproportionation (see 2.4.3.2), k_{td}/k_{tc} may be anticipated to be higher for the corresponding propagating species. The values of k_{td}/k_{tc} reported for (4) or (3) are significantly greater than those for (2). Similarly, (6) gives much more disproportionation than (5). Thus values of k_{td}/k_{tc} seen for systems involving a monomeric model radical [(2), (5), (8-10) or (12)] should be considered only as a lower limit for the polymeric system.

Despite these problems in assessing k_{td}/k_{tc}, it is possible to make some generalizations:

(a) Termination of polymerizations involving vinyl monomers involves predominantly combination.
(b) Termination of polymerizations involving α−methylvinyl monomers always involves a measurable proportion of disproportionation.
(c) During disproportionation of radicals bearing an α−methyl substituent (for example, those derived from MMA), there is a strong preference for transfer of a hydrogen from the α−methyl group rather than the methylene group.
(d) Within a series of vinyl or α−methylvinyl monomers, k_{td}/k_{tc} appears to decrease according to the ability of the substituent to stabilize a radical center in the series Ph > CN >> CO_2R.

5.3 Chain Transfer

The general mechanism of chain transfer as first proposed by Flory,[134,135] may be written schematically as follows (Scheme 13). The overall process involves a propagating chain ($P_i\bullet$) reacting with a transfer agent (T) to terminate one polymer chain and produce a radical (T$\bullet$) which initiates a new chain ($P_1\bullet$).

Initiation:

$$I_2 \xrightarrow{k_d} I\bullet$$

$$I\bullet + M \xrightarrow{k_i} P_1\bullet$$

Propagation:

$$P_i\bullet + M \xrightarrow{k_p} P_{i+1}\bullet$$

Termination:

$$P_i\bullet + P_j\bullet \xrightarrow{k_{tc}} P_{i+j}$$

$$P_i\bullet + P_j\bullet \xrightarrow{k_{td}} P_i + P_j$$

Chain Transfer:

$$P_i\bullet + T \xrightarrow{k_{tr}} P_i + T\bullet$$

$$T\bullet + M \xrightarrow{k_s} P_1\bullet$$

Scheme 13

Transfer agents find widespread use in both industrial and laboratory polymer syntheses. They are used to control (a) the molecular weight of polymers, (b) the polymerization rate (by controlling the gel or Trommsdorff effect) and/or (c) the nature of the polymer end groups. General aspects of chain transfer have been reviewed by Barson,[136] Farina,[137] Eastmond[138] and Palit et al.[139] and their use in producing telechelic and other functional polymers has been reviewed by Boutevin,[140] Heitz,[85] Corner[141] and Starks[142] (see 6.3.1.2).

Even in the absence of added transfer agents, all polymerizations are likely to be complicated by transfer to initiator (see 3.2.9, 3.3), solvent (5.3.2.5), monomer (5.3.3), polymer (5.3.4), etc. The significance of transfer reactions is dependent upon the exact nature of the species involved and the polymerization conditions.

For efficient chain transfer, the rate constant for reinitiation (k_s; refer Scheme 13) must be greater than that for propagation (k_p). In these circumstances, the presence of the transfer agent reduces the molecular weight of the polymer without directly influencing the rate of polymerization. If, however, $k_s<k_p$ then polymerization will be retarded and the likelihood that the transfer agent derived radical (T$\bullet$) will undergo side reactions is increased. Thus, retardation is much more likely in polymerizations of high k_p monomers (e.g. MA, VAc) than it is with low k_p monomers (e.g. S, MMA).

To understand the effect of the transfer process on the molecular weight, note that the number average degree of polymerization ($\bar{X}_n$) of polymer formed at any given instant during the polymerization can be expressed simply as the rate of monomer usage in propagation divided by the rate of formation of polymer molecules (the overall rate of termination). Thus, if termination is only by radical-radical reaction of chain transfer:

$$\bar{X}_n = -\frac{\frac{d[M]}{dt}}{\frac{d[P]}{dt}} = \frac{R_p}{R_t + R_{tr}} \tag{1}$$

$$= \frac{k_p[M][P\bullet]}{2k_t[P\bullet]^2 + k_{tr}[T][P\bullet]} \tag{2}$$

which can be rewritten as:

$$\frac{1}{\bar{X}_n} = \frac{2k_t[P\bullet]}{k_p[M]} + \frac{k_{tr}[T]}{k_p[M]} \tag{3}$$

Thus:

$$\frac{1}{\bar{X}_n} = \frac{1}{\bar{X}_{n_0}} + \frac{k_{tr}[T]}{k_p[M]} \tag{4}$$

where $\bar{X}_{n_0}$ is the number average degree of polymerization of polymer formed in the absence of a transfer agent. The ratio k_{tr}/k_p is called the transfer constant (C_S). Thus,

$$\frac{1}{\bar{X}_n} = \frac{1}{\bar{X}_{n_0}} + C_S\frac{[T]}{[M]} \tag{5}$$

This equation is commonly known as the Mayo equation.[143]

If transfer to monomer, initiator and polymer also occur, then equation (3) becomes:

$$\frac{1}{\bar{X}_n} = \frac{2k_t[P\bullet]}{k_p[M]} + C_M + C_I\frac{[I]}{[M]} + C_P\frac{[P]}{[M]} + C_S\frac{[T]}{[M]} \tag{6}$$

where C_M, C_I, C_P are the transfer constants for transfer to monomer, initiator, and polymer respectively.

The magnitude of a transfer constant depends on structural features of the attacking radical and the transfer agent. A C_S of unity has been called ideal. In these circumstances, the transfer agent:monomer ratio will remain constant throughout the polymerization.[141] This means that $\bar{X}_n$ should remain constant with conversion and the polydispersity of the product is minimized (close to 2.0). If C_S is high (>>1), the transfer agent will be consumed rapidly during the early stages of polymerization and the polymerization will be unregulated at higher conversion. If, on the other hand, C_S is low (<<1), the relative concentration of the transfer agent ([S]:[M]) will increase as the polymerization progresses and there will be a corresponding decrease in $\bar{X}_n$. In both circumstances a broad molecular weight distribution will result. It may be possible to overcome these problems by establishing an incremental protocol for monomer addition such that the ratio [S]:[M] is maintained at a constant value throughout the polymerization.

The rate constants for chain transfer and propagation may well have a different temperature dependence (*i.e.* the two reactions may have different activation parameters) and, as a consequence, transfer constants are temperature dependent. The temperature dependence of C_S has not been determined for most transfer agents. Appropriate care must therefore be taken when using literature values of C_S if the reaction conditions are different from those employed for the measurement of C_S. For cases where the transfer constant is close to 1.0, it may be possible to choose a reaction temperature such that the transfer constant is 1.0 and thus obtain ideal behavior.[144]

The value of C_S in homopolymerization can show significant chain length dependence for chain lengths ≤ 5 (see Table 5.6).[142] The magnitude of the remote unit effect is dependent on the particular transfer agent. The variation in C_S with chain length could reflect variations in either k_p, k_{tr} or (most likely) both. However, the experimental procedures typically do not allow chain length dependence of the individual rate constants to be determined.

The magnitude of the effect varies according to the particular transfer agent. This indicates that k_{tr} is sensitive to chain length (the magnitude of any remote unit effect on k_p should be independent of the transfer agent, given otherwise similar reaction conditions). It is also evidence that k_p is dependent on chain length for at least the first few propagation steps (see 4.5.3).

Table 5.6 Chain Length Dependence of Transfer Constants (C_S)

Transfer Agent	Monomer	T (°C)[a]	C_1	C_2	C_3	C_4	$C_{5-\infty}$	Refs.
C_2H_5SH[b]	MA	50	0.94±.07	1.65±.12	1.57±.09	1.52±.06	1.57±.18	145
i-C_3H_7SH[b]	MA	50	0.54±.08	0.67±.07	0.70±.08	0.66±.08	-	146
C_2H_5SH[b]	Styrene	50	7.1±.3	30±10	-	-	17±1	147
CCl_3Br	Styrene	30	0.52±.14	9.4±4.6	37±3	96±12	460±61	148
CCl_4	Styrene	76	0.0006	0.0025	0.0069	0.0115	-	149
CCl_4	VAc	60	-	0.13	0.47	0.67	0.80	150
CCl_4	VC	60	0.00284	0.0184	0.0280	-	-	151
$CHCl_3$	VC	60	0.006	0.0141	0.0292	-	-	152

[a] Bulk polymerization, medium comprises monomer + transfer agent.
[b] The variation between C_2, C_3, and $C_{4-\infty}$ is within experimental error.

Bamford[153] has provided evidence that, in copolymerization, penultimate unit effects can be important in determining the reactivity of propagating radicals toward transfer agents. The magnitude of this effect also depends on the particular monomers and transfer agent involved. The finding that the most pronounced remote unit effects are observed for the most bulky transfer agents (see 5.3.2.4), suggests that the magnitude of the remote unit effect is determined at least in part by steric factors.

5.3.1 Mechanisms

There are at least two basic mechanisms for chain transfer which should be considered in any discussion of chain transfer. They are: (a) atom or group transfer and (b) addition-elimination.

5.3.1.1 Atom or group transfer

Chain transfer most commonly involves transfer of an atom or group from the transfer agent to the propagating radical by a homolytic substitution (S_H2) mechanism. The general factors influencing the rate and specificity of these reactions have been dealt with in Chapter 2 (see 2.3). Rate constants are determined by a combination of bond strength, steric and polar factors. The 'Patterns of Reactivity' scheme was introduced as a method for predicting reactivities in chain transfer and reactivity ratios in copolymerization (see 6.3).

The moiety transferred will most often be a hydrogen atom, for example, when the transfer agent is a thiol (Scheme 14) (see 5.3.2.1), a hydroperoxide (see 3.3.2.5), the solvent (5.3.2.5), etc.

$$\sim\sim\cdot\ +\ H-S(CH_2)_3CH_3 \longrightarrow \sim\sim H\ +\ \cdot S(CH_2)_3CH_3$$
Scheme 14

It is also possible to transfer a heteroatom (e.g. a halogen atom from bromotrichloromethane - Scheme 15) (see 5.3.2.4).

$$\sim\sim\cdot\ +\ Br-CCl_3 \longrightarrow \sim\sim Br\ +\ \cdot CCl_3$$
Scheme 15

or a group of atoms (e.g. from diphenyl disulfide - Scheme 16) (see 5.3.2.2).

$$\sim\sim\cdot\ +\ PhS-SPh \longrightarrow \sim\sim-SPh\ +\ \cdot SPh$$
Scheme 16

Group transfer processes are of particular importance in the production of telechelic polymers.

5.3.1.2 Addition-elimination

Some transfer agents react with radicals by an addition-elimination mechanism. This involves the formation of a short-lived intermediate $(P_iT)\bullet$. The reaction scheme can be summarized schematically as follows (Scheme 17).

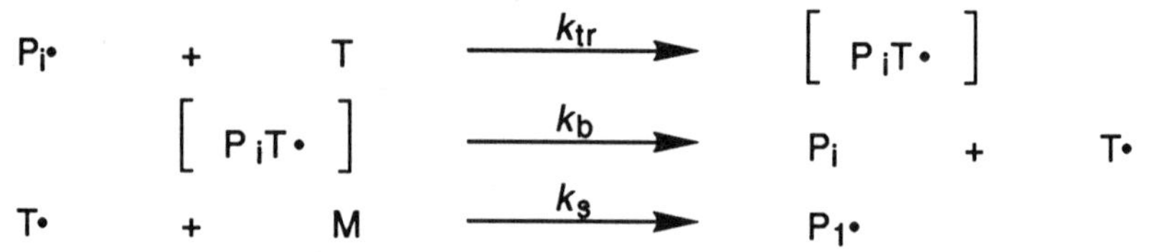

$$P_i\bullet + T \xrightarrow{k_{tr}} [P_iT\bullet]$$

$$[P_iT\bullet] \xrightarrow{k_b} P_i + T\bullet$$

$$T\bullet + M \xrightarrow{k_s} P_1\bullet$$

Scheme 17

The reactivity of the transfer agent (T) towards the propagating species and the properties of the adduct $(P_iT\bullet)$ are both important in determining the effectiveness of the transfer agent: if the lifetime of the intermediate $(P_iT\bullet)$ is significant (k_b slow), it may react by other pathways than β–scission; if it $(P_iT\bullet)$ undergoes coupling or disproportionation with another radical species it will inhibit or retard polymerization; if it adds to monomer (T copolymerizes) it will be an inefficient transfer agent.

5.3.1.3 Measurement of Transfer Constants

The most used method of evaluating transfer constants is by application of the Mayo equation (equation 5). For low conversion polymerizations carried out in the presence of added transfer agent T, it follows from equation 5 (see 5.3.1.1) that a plot of $1/\bar{X}_n$ vs. [T]/[M] should yield a straight line with slope C_S.[143] Thus a typical experiment procedure involves evaluation of the degree of polymerization for low conversion polymerizations carried out in the presence of various (at least three) concentrations of added transfer agent. The usual way of obtaining $\bar{X}_n$ values is by GPC analysis of the entire molecular weight distribution.

It has recently been shown that equivalent information can be obtained by analysis of the limiting slope of log(number molecular weight distribution) plots.[154-156] It was pointed out that:

$$\lim_{i\to\infty, [I]\to 0} [P_i] = \exp\left\{-\frac{k_{tr\,T}[T]}{k_p[M]} \cdot i\right\} \qquad (7)$$

It follows that for sufficiently large i the slope of a plot of ln [Pi] vs. i will be:

$$-\frac{k_{tr\,T}[T]}{k_p[M]} = -C_T\frac{[T]}{[M]}$$

This method offers better quality data as it is less sensitive to experimental noise and has application to measuring transfer constant to polymeric species where the distributions of the transfer agent and the polymer being formed overlap.[156]

Problems can arise during the measurement of transfer constants for very active transfer agents. In response, Bamford[157-159] proposed the technique of moderated copolymerization. In these experiments the monomer of interest is copolymerized with an excess of a moderating monomer which has a much lower (preferably negligible) transfer constant. This method has been applied to evaluate penultimate unit effects on the transfer constant.

Another method follows from equation 8 and involves evaluation of the usage of transfer agent (or better the incorporation of transfer agent fragments into the polymer) and monomer.[160] This method does not rely on molecular weight measurements.

$$\frac{d[T]}{d[M]} = \frac{k_{tr}[T]}{k_p[M]} = C_S \frac{[T]}{[M]} \tag{8}$$

The above treatments can only be applied to systems where the conversion is small and the ratio [T]/[M] is close to the initial value. A number of authors have provided integrated forms of the Mayo equation which have application when the conversion of monomer to polymer is non-zero. Cardenas and O'Driscoll[161] and Stickler[90] have proposed the expression:

$$\frac{1}{\bar{X}_n} = C_M - 2(fk_d)^{1/2} \frac{k_t^{1/2} [I]_0^{1/2}}{k_p [M]_0} \frac{\ln(1-x)}{x} - C_S \frac{[T]_0}{[M]_0} \frac{\ln(1-x)}{x} \tag{9}$$

where x is the fractional conversion of monomer into polymer. Provided that the term due to chain formation by radical-radical termination is small, a plot of

$$\frac{1}{\bar{X}_n} \text{ vs } \frac{-[T]_0}{[M]_0} \frac{\ln(1-x)}{x}$$

should yield a straight line with slope C_S. This expression is only applicable when the transfer agent is in vast excess or the transfer constant is small since consumption of transfer agent is neglected.

Integration of equation 8 provides the following relationship:

$$\frac{[T]}{[T]_0} = \left(\frac{[M]}{[M]_0}\right)^{C_S} \tag{10}$$

which enables substitution for [T] in the integrated form of the Mayo equation.[162]

$$\ln\left(1 - \frac{[M]_0 x}{[T]_0}\left(\frac{1}{\bar{X}_n} - \frac{1}{\bar{X}_{n0}}\right)\right) = C_S \ln(1-x) \tag{11}$$

Thus a plot of

$$\ln\left(1 - \frac{[M]_0 x}{[T]_0}\left(\frac{1}{\bar{X}_n} - \frac{1}{\bar{X}_{n0}}\right)\right) \quad vs. \quad C_S \ln(1-x)$$

should yield a straight line passing through the origin with slope C_S. Bamford has reported the derivation of the similar equation for copolymerization.[159] This method is highly dependent on the precision of the conversion measurements since errors in conversions are magnified in C_S.

5.3.2 Transfer Agents

The following sections detail the chemistry undergone by specific transfer agents. Transfer constant data have not been critically assessed and are provided only to show the order of magnitude of these values and to provide a guide to the relative reactivity of the various reagents.

5.3.2.1 Thiols

Thiols are perhaps the most commonly used transfer agents in radical polymerization (see Scheme 14). Some typical transfer constants are presented in Table 5.7. The values of the transfer constants depend markedly on the particular polymerization system and can depend on reaction conditions.[163]

Table 5.7 Transfer Constants (60°C, bulk) for Thiols (RSH) With Various Monomers[a]

Transfer agent R	C_S				
	MMA	MA	AN	S	VAc
H	-	-	0.30	5[b]	-
n-C$_4$H$_9$-	0.67[160]	1.7[c,160]	-	22[160]	48[d,160]
n-C$_{12}$H$_{25}$-	-	-	0.73	19[160]	-
HO-CH$_2$CH$_2$-	0.45[e]	-	-	-	-
HO$_2$CCH$_2$-	0.38[e]	-	-	-	-
H$_3$N$^+$-CH$_2$CH$_2$-	0.11[f,163]	-	-	11[163]	-
Ph-	2.7	-	-	-	-

[a] Numbers are taken from the Polymer Handbook[164] unless otherwise stated and have been rounded to two significant figures. Where a choice of numbers is available the average value has been quoted.
[b] At 70°C.
[c] In ethyl acetate solvent.
[d] Substantial retardation observed.[165]
[e] Extrapolated to 60°C from the data given. The activation energies quoted by Roy et al.[166] appear to be calculated incorrectly.
[f] The free amine has a very low transfer constant with MMA.[163] It may be consumed in a Michael reaction with the monomer.

Thiols react more rapidly with nucleophilic radicals than with electrophilic radicals. They have very large C_S with S and VAc, but near ideal transfer constants with acrylic monomers (see Table 5.7). Aromatic thiols have higher C_S than aliphatic thiols but also give more retardation. The substitution pattern of alkanethiols appears to have only a small (<2-fold) effect on the transfer constant. The thio radicals produced have electrophilic character and in copolymerization they react preferentially with electron-rich monomers (see 3.4.3.1).

Bamford et al.[167] have investigated the importance of penultimate unit effects on the reactivity of n-butanethiol in a number of copolymerizations (S-MMA, S-MA) using the technique of "moderated copolymerization". Their data indicate that penultimate unit effects are unimportant in these systems. This contrasts with findings for transfer to carbon tetrabromide (see 5.3.2.4). It has also been found, again in contrast with halocarbons, that C_S for various primary and secondary thiols is essentially independent of chain length for chain lengths ≥ 2 (see Table 5.6). The practical implication of this is that transfer constants measured in homopolymerization can be used to predict molecular weights in copolymerization.

A range of functional thiols [e.g. mercaptoethanol (19), thioglycolic acid (20)] has been used to produce monofunctional polymers (see 6.3.1.2).[141,168-170]

HO–CH$_2$–CH$_2$–SH   HO–C(=O)–CH$_2$–SH

(19)   (20)

5.3.2.2 Disulfides

A wide range of dialkyl[171] and diaryl disulfides,[172,173] diaroyl disulfides,[174] and xanthogens[175] has been used as transfer agents (see Scheme 16). Their use ideally leads to the incorporation of functionality at both ends of the polymer chain, thus they find application in the synthesis of telechelics (see 7.5.1).

The C-S bond of the sulfide end groups is relatively weak and susceptible to thermal and photo- or radical-induced homolysis. This means that certain disulfides [including (21-25)] may act as iniferters in living radical polymerization and they can be used as precursors to block copolymers (see 7.4.1).

H$_3$C–(CH$_2$)$_n$–S–S–(CH$_2$)$_n$–CH$_3$ X–C$_6$H$_4$–S–S–C$_6$H$_4$–X

(21)   (22)

X–C$_6$H$_4$–C(=O)–S–S–C(=O)–C$_6$H$_4$–X   (H$_3$C)$_2$N–C(=O)–S–S–C(=O)–N(CH$_3$)$_2$

(23)   (24)

Aliphatic disulfides (**21**) are not particularly reactive in chain transfer towards MMA and S (see Table 5.8). However, they appear to be ideal transfer agents ($C_S \sim 1.0$) for VAc polymerizations.

The reactivity of diphenyl (**22**) and dibenzoyl (**23**) disulfide derivatives is higher than aliphatic derivatives. The value of C_S depends markedly on the substituents, X, and on the pattern of substitution. Electron withdrawing substituents (e.g. X = p-CN or p-NO$_2$) may increase C_S by an order of magnitude.[173,174] However, these compounds also give marked retardation.

Table 5.8 Transfer Constants for Disulfides (R-S-S-R) With Various Monomers[a]

Transfer agent	C_S		
R	MMA	S	VAc
C$_2$H$_5$-	0.00013	-	-
n-C$_4$H$_9$-	-	0.0024[171]	1.0
PHCH$_2$-	0.0063	0.01	-
EtO$_2$CCH$_2$-	0.00065	0.015	1.5
Ph-	0.0085[173]	0.15	-
PhC(=O)-[b]	0.0010	0.0036	-
p-CNC$_6$H$_4$C(=O)-[b]	0.029	0.32	-
(CH$_3$)$_2$NC(=S)-[c]	0.014[176]	0.012[176]	-

[a] 60°C, bulk unless indicated otherwise. Numbers are taken from the Polymer Handbook[164] unless otherwise stated, and have been rounded to two significant figures. Where a choice of numbers is available the average value has usually been quoted.
[b] Tsuda and Otsu give the data shown.[174] These numbers are reported incorrectly in the Polymer Handbook and many other compilations.
[c] 70°C.

The dithiauram disulfide (**25**) has a much higher transfer constant than most other disulfides [an order of magnitude higher than that for the oxygen analogue (**24**)].[177] This may be due to another mechanism for induced decomposition being available (Scheme 18) that involves addition to the C=S double bond. This mechanism has been demonstrated for related compounds (see 5.3.2.6).

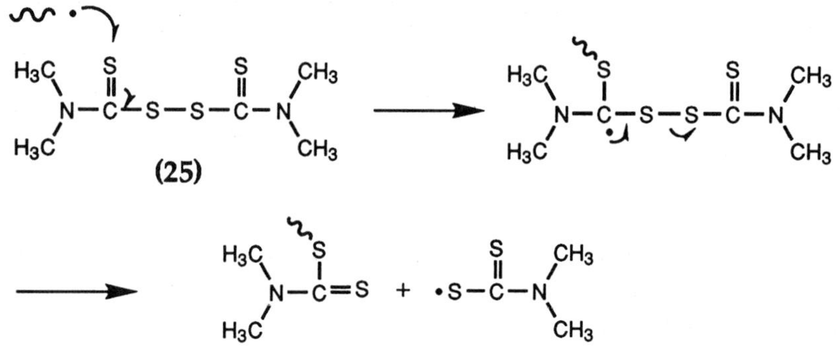

Scheme 18

5.3.2.3 Monosulfides

Most monosulfides have low transfer constants when considered in relation to disulfides. This is probably a reflection both of steric factors and the relative strength of the C-S bond; it is significantly stronger than the S-S bond. Exceptions to this rule are the allyl sulfides (see 5.3.2.6).

t-Butanesulfide (**26**) has a substantially higher transfer constant than other monosulfides. Pryor and Pickering[171] proposed that this compound may react by hydrogen atom transfer (Scheme 19).

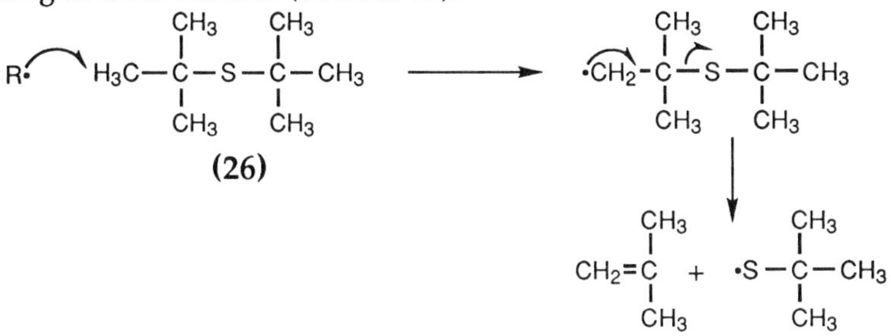

Scheme 19

5.3.2.4 Halocarbons

Halocarbons including carbon tetrachloride, chloroform, bromotrichloromethane[178] (see Scheme 15) and carbon tetrabromide have been widely used for the production of telomers and transfer to these compounds has been the subject of a large number of investigations.[142] Representative data are shown in Table 5.9. Telomerization involving halocarbons has also been developed as a means of studying the kinetics and mechanism of radical additions.[179]

Table 5.9 Transfer Constants (80°C, bulk) for Halocarbons With Various Monomers[a]

Transfer agent	C_S				
	MMA	MA	AN	S	VAc
CBr_4	0.27	0.41	0.19	2.2	>39
CCl_4	2.4	1.3	0.85	130	0.96[b,165]
$CHCl_3$	1.4	2.3	5.6	0.5	0.015[b,165]

[a] Numbers are taken from the Polymer Handbook[164] unless otherwise stated, and have been rounded to two significant figures.
[b] 60°C.

The perhalocarbons react with carbon-centered radicals by halogen-atom transfer to form a perhaloalkyl radical. Halogen atom abstractability decreases in the series iodine>bromine>chlorine.

Halohydrocarbons may in principle react by either hydrogen-atom, halogen-atom transfer or both. The preferred pathway can often be predicted by considering the relative C-X bond strengths (see chapter 2). For $CHCl_3$, transfer of a hydrogen atom is favored.

The halocarbons react more rapidly with nucleophilic radicals than with electrophilic radicals. Thus values of C_S with S and VAc are substantially higher than those with acrylic monomers (see Table 5.9). The haloalkyl radicals formed have electrophilic character (see 2.2.2).

Bamford et al.[153] have shown that C_S for transfer to carbon tetrabromide is subject to penultimate unit effects. They found $C_{S.S}$ = 368, $C_{MA.S}$ = 302, $C_{MMA.S}$ = 60 (compare behavior observed with thiols - see 5.3.2.1). The finding ($C_{MA.S} \sim C_{S.S} >> C_{MMA.S}$) suggests that steric factors are more important than either polar or electronic factors in determining the magnitude of the remote unit effect on C_S. Bamford et al.[153] proposed that k_{tr} is more sensitive to remote unit effects than k_p. That C_S in MA and S polymerizations is chain length dependent for chain lengths ≤3 units (see Table 5.6) is further evidence for remote unit effects. A variation in C_S with chain length for ethylene polymerization has been attributed to polar effects (i.e. electron donating ability of the alkyl chain increases in the series: ethyl>butyl>hexyl).[142]

Transfer to the halocarbons may be effectively catalyzed by certain transition metal salts and complexes.[180] In these cases, initiation typically involves a redox reaction between the metal and the halocarbon. An inhibitor reacts with the propagating radical by group transfer to regenerate the metal in its original oxidation state. Transition metal species which are effective in this context include copper salts and $RuCl_2(PPh_3)$. Effective transfer constants are substantially higher than when the transfer agent is used alone. However, little quantitative work on the kinetics of these polymerizations has been reported.

5.3.2.5 Solvents and other reagents

Many solvents and additives have measurable transfer constants (see Table 5.10). The accuracy of much of the transfer constant data in the literature is questionable with values for a given system often spanning an order of magnitude. In some cases the discrepancies may be real and reflect differences in experimental conditions. In other cases they are less clear and may be due to difficulties in molecular weight measurements or other problems.

Nonetheless, it is clear that the reactivity of solvents in transfer reactions depends on the nature of the propagating species and some general conclusions can be drawn. The propagating species derived from MMA has relatively little tendency to undertake transfer. That derived from VAc appears extremely reactive towards solvents and other transfer agents (note, however that many reagents give marked retardation with VAc[165]). The factors influencing reactivity in hydrogen atom abstraction reactions are discussed in general terms in chapter 2 (see 2.3).

Table 5.10 Transfer Constants (60°C, bulk) for Selected Solvents and Additives With Various Monomers[a]

Solvent	$C_S \times 10^4$ for monomer				
	MMA	MA	AN	Styrene	VAc
benzene	0.04	0.3[b]	2.5	0.023	3.0
toluene	0.20	2.7	5.8	0.12	21
acetone	0.20	0.23	1.1	0.32	12
butan-2-one	0.45	3.2[b]	6.4	5.0	74
ethyl acetate	0.15	-	2.5	5.7	3.0
triethylamine	8.3	400	790	7.1	370

[a] Numbers have been selected from the Polymer Handbook[164] or referernces given therein have been rounded to two significant figures.
[b] 80°C.

Mechanisms for chain transfer depend on the solvent. Many solvents have abstractable hydrogens (*e.g.* acetone, butanone, toluene) and may react by loss of those hydrogens (Scheme 20).

Scheme 20

Benzene may react by addition as shown in Scheme 21 (this pathway is also open to other aromatic solvents). The cyclohexadienyl radical may then terminate a second chain by hydrogen atom transfer.

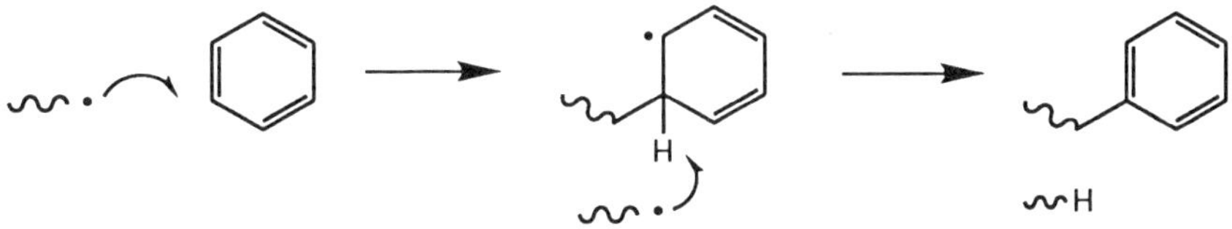

Scheme 21

In the case of S, it has been proposed that initiation occurs by hydrogen-atom transfer to monomer (Scheme 22).[143,181]

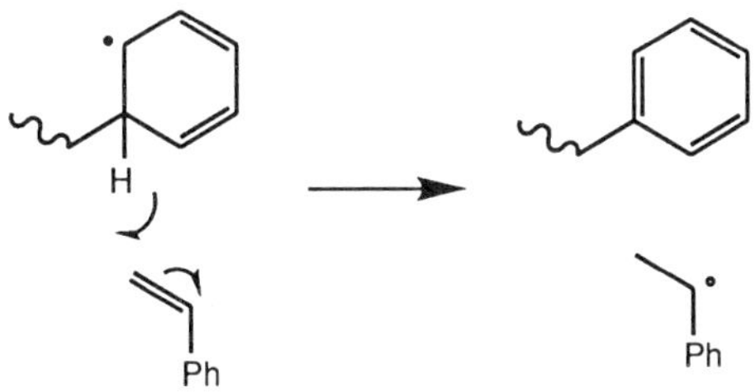

Scheme 22

5.3.2.6 Unsaturated compounds

Certain unsaturated compounds may act as transfer agents by an addition-elimination mechanism. These compounds have the general structure (27) where R is a group chosen to give the transfer agent an appropriate reactivity with respect to the monomer(s) and Y or Z is a radical leaving group.

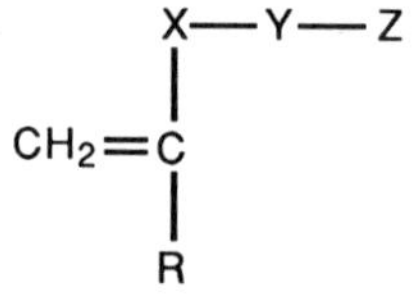

$$CH_2=C(R)-X-Y-Z$$

(27)

Examples of this class of transfer agent include vinyl ethers (Scheme 23).[182-184] The driving force for fragmentation is provided by formation of a strong carbonyl double bond.

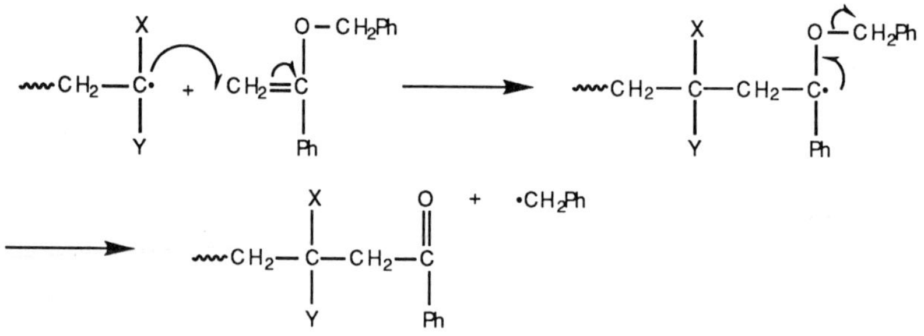

Scheme 23

In other cases, including allyl halides,[185-191] sulfides (e.g. Scheme 24),[192-194] sulfones,[185] silanes[185] and similar compounds,[185] the main driving force is the weak X-Y bond [see (27)].

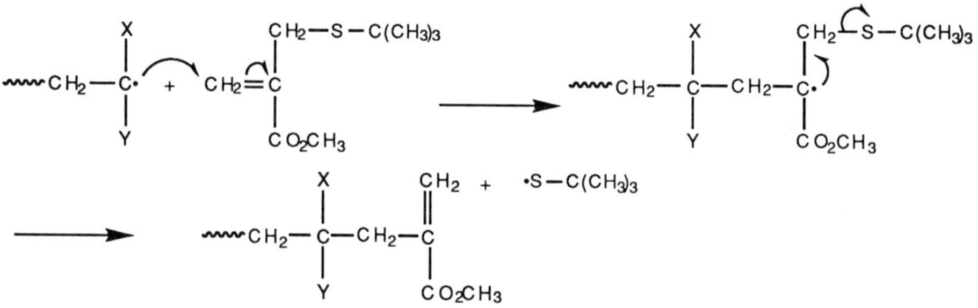

Scheme 24

In the case of allyl peroxides (e.g. Scheme 25)[195] intramolecular homolytic substitution on the O-O bond gives an epoxide end group.

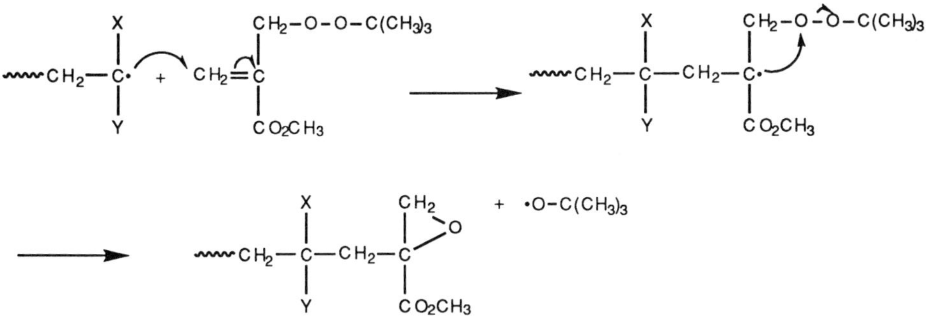

Scheme 25

Other examples of this chemistry are provided by MMA macromonomer (see 5.3.4.2)[56,196,197] and VC (see 5.3.3.3).

The double bonds of the transfer agents shown in Schemes 23-25 have a reactivity towards propagating radicals which is comparable with that of the common monomers they resemble (*i.e.* transfer constants with acrylic and styrenic monomers are close to unity - see Table 5.11). However, the radicals formed by addition have low reactivity towards further propagation and other intermolecular reactions because of steric crowding about the radical center.

The radicals undergo facile β-scission to form a new radical which may reinitiate polymerization. The driving force for fragmentation of the intermediate radical is provided by cleavage of a weak Y-Z bond and/or formation of a strong C=X bond. Note that, for examples with X=carbon (*e.g.* Scheme 24), the product of the reaction is itself a potential transfer agent or macromonomer.[56,196] If β-scission is slow then retardation may result. This is seen for some high k_p monomers (*e.g.* VAc, MA).

Some typical transfer constants for these reagents are given in Table 5.10. The values of C_S for these reagents are less dependent on the particular monomer than those for halocarbons or thiol transfer agents (compare tables 5.7 and 5.9 above).

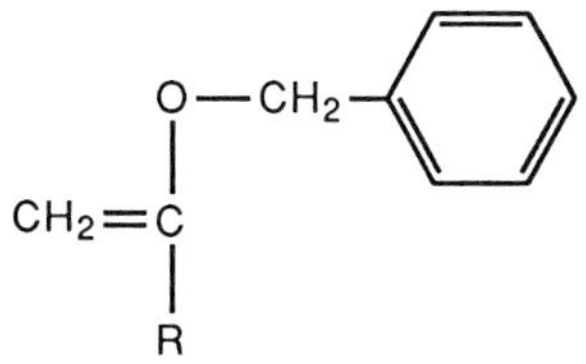

(28) 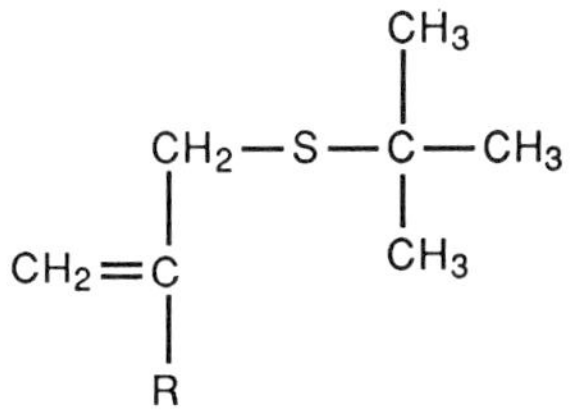 (29)

Table 5.11 Transfer Constants for Some Benzyloxy Ether and Allyl Sulfide Transfer Agents[a]

Transfer agent	Temperature (°C)	C_S for monomer				References
		St	MMA	MA	VAc	
28, R=Ph	60°C	0.26	0.76	5.7[b]	9.7[b]	182
28, R=CN	60°C	0.036	0.081	0.3[b]	12[b]	183
28, R=CO$_2$Me	60°C	0.046	0.16	0.54[b]	20[b]	183
28, R=CO$_2$NH$_2$	60°C	0.14	0.47	1.1[b]	-	183
29, R=Ph	60°C	0.80	1.24	3.95[b]	20[b]	194
29, R=CN	60°C	1.75	1.36	1.64[b]	60[b]	194
29, R=CO$_2$Et	60°C	0.95	0.74	2.23[b]	-	194
30, R=CH$_2$CO$_2$H	60°C	0.95	1.13	-	-	192
30, R=CH$_2$CH$_2$NH$_2$	60°C	0.79	0.91	-	-	192
30, R=CH$_2$CH$_2$OH	60°C	0.77	1.04	-	-	192
31	60°C	1.72	0.64	-	-	192

[a] Bulk, medium comprises only monomer and transfer agent.
[b] Significant retardation observed.

Since functionality can be introduced in either or both the reinitiation and termination steps, suitably substituted derivatives of benzyloxystyrenes, allyl sulfides, *etc.* offer a route to a variety of end-functional polymers including telechelics. Examples include the following compounds (30-31).

Other transfer agents which react with propagating species by an addition-elimination mechanism include various thione derivatives (32-34).[198,199]

5.3.2.7 Cobalt complexes

Enikolopyan et al.[200] established that Co^{II} porphyrin complexes (e.g. 35) can function as catalytic chain transfer agents.

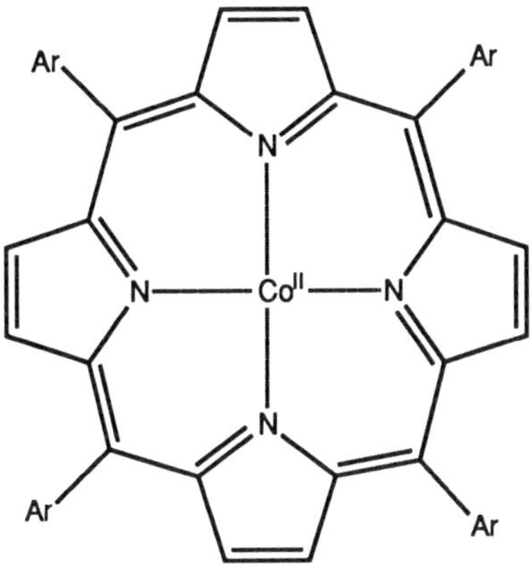

(35)

Other square planar cobalt complexes [e.g. (36-38)] are also effective transfer agents.[201,202]

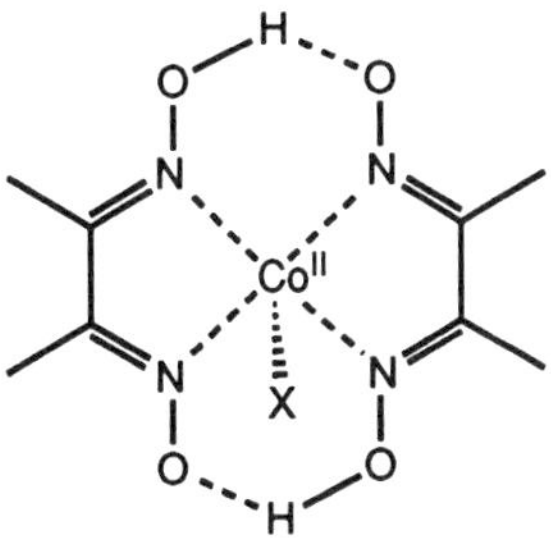

(36) 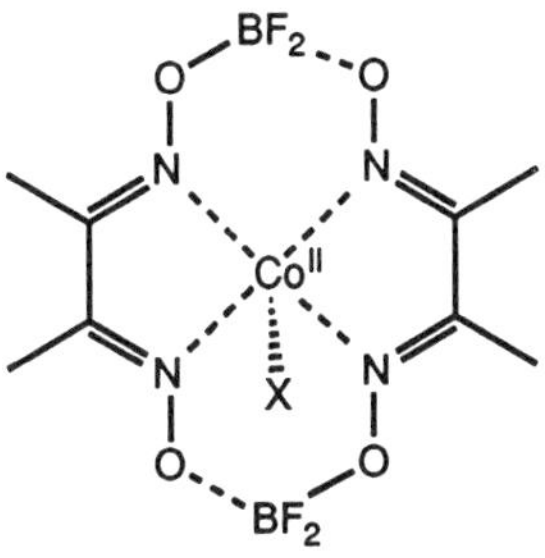   (37) 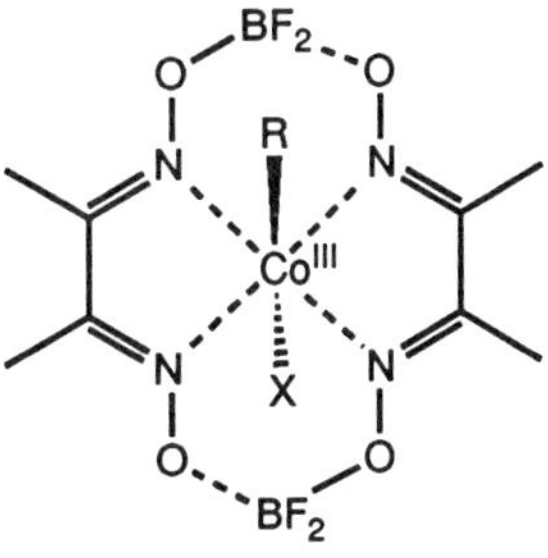   (38)

The scope and utility of the process has been reviewed by Karmilova,[203] Davis,[204] and Parshall and Ittel.[205] The major applications of these compounds are in molecular weight control and in synthesis of macromonomers based on methacrylate esters.[56]

The mechanism proposed for catalytic chain transfer[201] is shown in Scheme 26. The square planar cobalt(II) complex (39) rapidly and reversibly combines with carbon-centered radicals. The product, the alkyl cobalt(III) complex (40), may eliminate the macromonomer (41) to form the cobalt hydride (42). Alternatively the cobalt(II) complex (39) may undergo disproportionation with the carbon-centered radical to give (41) and (42).

The cobalt hydride (42) reinitiates polymerization by donating a hydrogen atom to monomer and in doing so regenerates the cobalt complex (39). All chains (41) formed in the presence of these reagents will have one unsaturated end group.

Chain transfer:

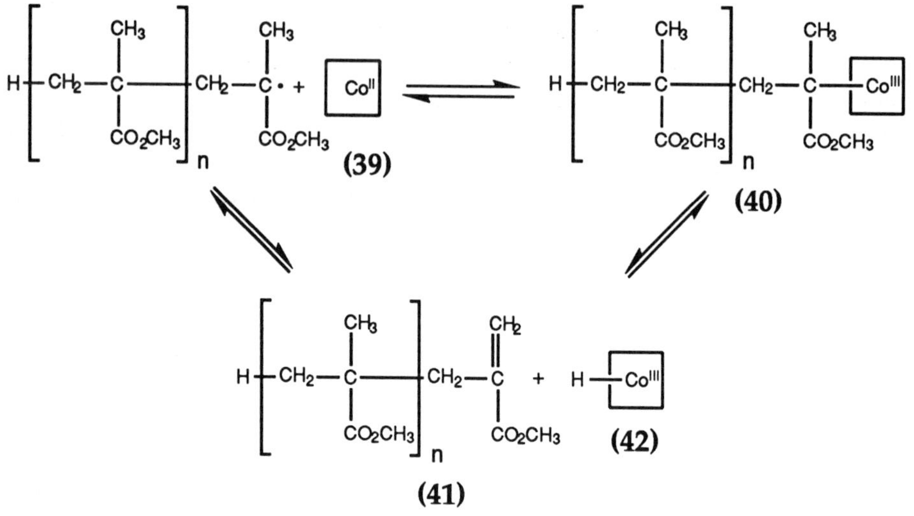

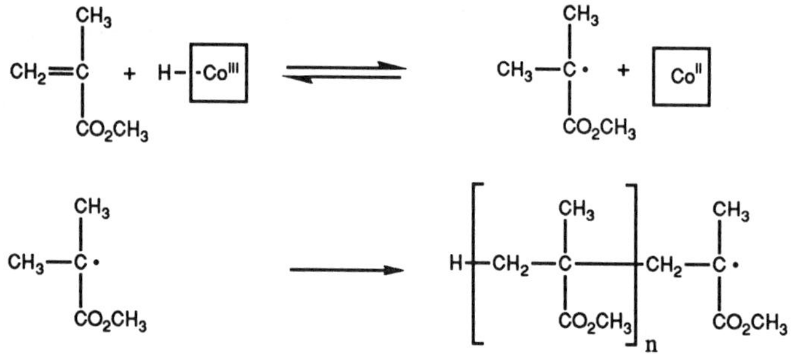

Scheme 26

A major advantage of catalytic transfer agents over conventional agents is that they are not used up during polymerization. Furthermore, they have very high transfer constants. The value of C_S is typically in the range 10^3-10^4, thus only very small amounts are required to bring about a large reduction in molecular weight. Exact values for C_S are dependent on the reaction conditions[200,201,206,207] and, for chain lengths ≤6, on the molecular weight of the propagating species.[207,208]

The application of these transfer agents is unfortunately not universal.[200] They are effective in polymerizations of methacrylate esters, MAAm, MAN, S (slow poisoning of catalyst occurs), and copolymerizations of these monomers. However, homopolymerizations involving acrylate esters and VAc are inhibited probably due to the higher stability of the alkyl cobalt intermediates and/or combination rather than disproportionation with propagating species. The latter chemistry has led to cobalt complexes being used in living polymerization of acrylate esters (see 7.5.4).

Although the complexes can be applied under a wide range of reaction conditions, selection of solvent and reaction conditions is important. Catalytic inhibition is reported for MMA and MAAm polymerizations with DMF solvent.[209] Other potential problems associated with the use of these reagents are sensitivity to oxygen and hydrolytic stability. The BF_2-bridged derivatives (37 and 38) show greater stability to hydrolysis over (7). Alkyl-Co^{III} complexes (38) show enhanced air stability.[210] The latter complexes (38) generate the active Co^{II} complexes (37) on heating or on photolysis.[202,210]

5.3.3 Transfer to Monomer

Non-zero transfer constants (C_M) can be found in the literature for most monomers. Values of C_M for some common monomers are given in Table 5.12. Transfer to monomer is usually described as a process involving hydrogen atom transfer. While this mechanism is reasonable for those monomers possessing aliphatic hydrogens (*e.g.* MMA, VAc, allyl monomers), it is less acceptable for monomers possessing only vinylic or aromatic hydrogens (*e.g.* VC, S). The details of the mechanisms by which transfer occurs are, in most cases, not proven. Mechanisms for transfer to monomer that involve loss of vinylic hydrogens seem unlikely given the high strength of the bonds involved.

Irrespective of the mechanism by which transfer to monomer occurs, the process will usually produce an unsaturated radical as a by-product. This species may initiate polymerization to afford a macromonomer which may be reactive under typical polymerization conditions.

Table 5.12 Transfer Constants to Monomer[a]

Monomer	Temperature (°C)	$C_M \times 10^4$	Ref.
S	60	0.6	103
MMA	60	0.1	211
MA	60	0.4	212
AN	60	0.3	213
VAc	60	1.8	214
VC	100	50	215,216
allyl acetate	80	1600	217
allyl chloride	80	700	217

[a] Numbers are taken from the Polymer Handbook[164] and have been rounded to two significant figures.

5.3.3.1 Styrene

The value of C_M has been determined by a number of groups as 6×10^{-5} (see Table 5.12). However, the mechanism of transfer has not been firmly established.

A mechanism involving direct hydrogen abstraction seems unlikely. The observed value of C_M is slightly higher than that for ethylbenzene.

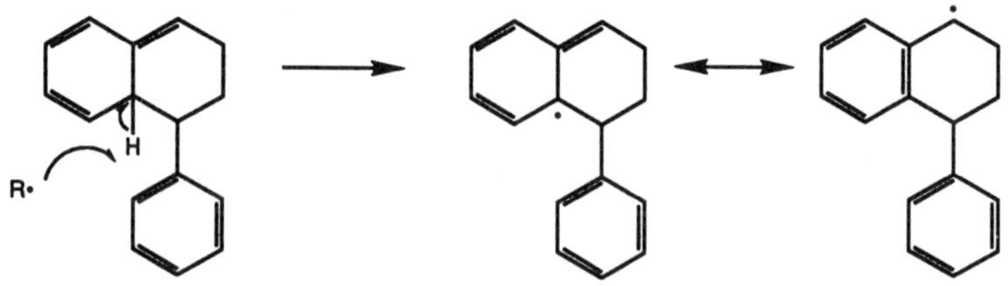

Scheme 27

Transfer to monomer may not involve the monomer directly but rather the intermediate (43) formed by Diels-Alder dimerization (Scheme 27).[218] Since (43) is formed during the course of polymerization, its involvement could be confirmed by analysis of the polymerization kinetics.

5.3.3.2 Vinyl acetate

There is a considerable body of evidence (kinetic studies, chemical and NMR analysis) indicating that transfer to VAc monomer involves largely, if not exclusively, the acetate methyl hydrogen to give radical (44) (see Scheme 28).[219,220] This radical (44) initiates polymerization to yield a reactive macromonomer (44).

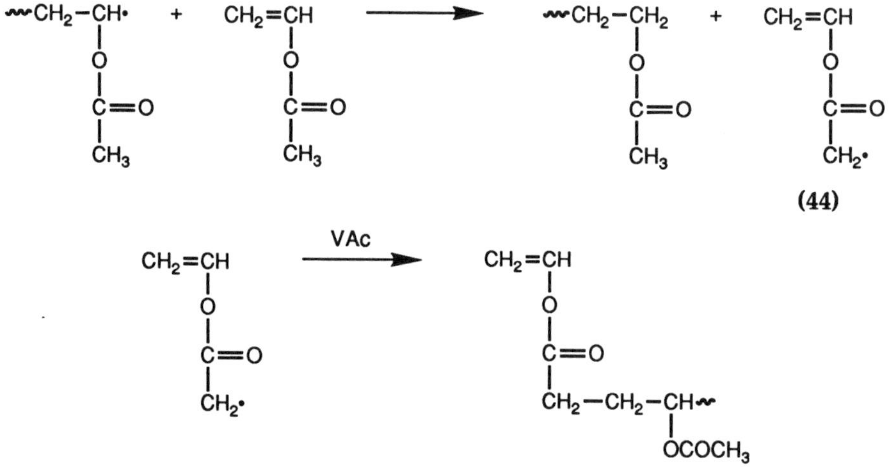

Scheme 28

Starnes et al.[221] have provided support for the above mechanism (Scheme 28) by determining the unsaturated chain ends (45) in low conversion PVAc by ^{13}C NMR. They were able to distinguish (45) from chain ends which might have been formed if transfer involved abstraction of a vinylic hydrogen. The number of unsaturated chain ends (45) was found to equate with the number of -CH$_2$OAc ends suggesting that most chains are formed by transfer to monomer. Starnes *et al.*[221] also found an isotope effect k_H/k_D of 2.0 for the abstraction reaction with CH$_2$=CHO$_2$CCD$_3$ as monomer. This result is consistent with the mechanism shown in Scheme 28 but is contrary to an earlier finding.[222]

Stein[214] has indicated that the reactivity of the terminal double bond of the macromonomer (45) is 80% that of VAc monomer. The kinetics of incorporation of (45) have also been considered by Wolf and Burchard[223] who concluded that (45) played an important role in determining the time of gelation in VAc homopolymerization in bulk.

5.3.3.3 Vinyl chloride

It has been proposed that chain transfer to monomer determines the length of the polymer molecules formed during VC polymerization.[224]

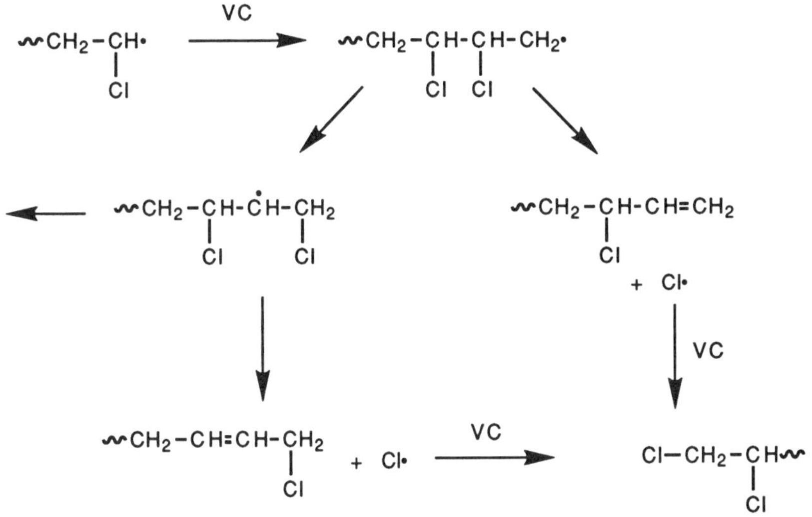

Scheme 29

The mechanism for transfer, involving an addition-elimination sequence consequent on head addition to monomer (Scheme 3 - see also 4.3.1.2), was first proposed by Rigo *et al.*[225] Direct evidence for this pathway has been provided by Starnes *et al.*[226] and Park and Saleem.[227] This pathway (Scheme 29) accounts for C_M for VC being much greater than C_M for other commercially important monomers (see Table 5.12) where the analogous pathway is not available. Starnes *et al.*[228] have reported kinetic data which suggest that the chlorine atom does not have a discrete existence but is transferred directly from the β-chloroalkyl radical to VC.

5.3.3.4 Allyl monomers

Transfer to monomer is of particular importance during the polymerization of allyl esters (**46**, $X=O_2CR$), ethers (**46**, $X=OR$), amines (**46**, $X=NR_2$) and related monomers.[217,229,230] The allylic hydrogens of these monomers are activated towards abstraction by both the double bond and the heteroatom substituent (Scheme 30). These groups lend stability to the radical formed (**47**) and are responsible for this radical adding monomer only slowly. This, in turn, increases the likelihood of side reactions (*i.e.* degradative chain transfer) and cause the allyl monomers to retard polymerization.

For allyl acetate a significant deuterium isotope effect supports the hydrogen abstraction mechanism (Scheme 30).[231] Allyl compounds with weaker CH_2-X bonds (**46** $X=SR, SO_2R, Br, etc.$) may also give chain transfer by an addition-elimination mechanism (see 5.3.1.2).

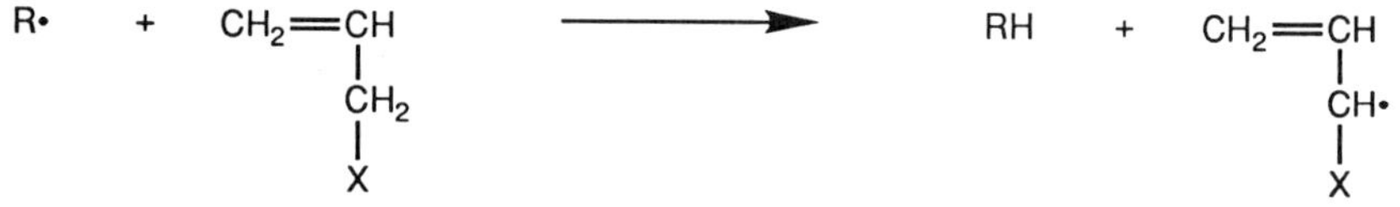

$$R\cdot \ + \ \underset{(46)}{\underset{X}{\underset{|}{\underset{CH_2}{\underset{|}{CH_2=CH}}}}} \ \longrightarrow \ RH \ + \ \underset{(47)}{\underset{X}{\underset{|}{\underset{CH\cdot}{\underset{|}{CH_2=CH}}}}}$$

Scheme 30

Diallyl monomers find significant use in cyclopolymerization (see 4.4.2). Transfer to monomer is of greater importance in polymerizations of allyl than it is in diallyl monomers.[232] This may reflect differences in the nature of the propagating species [*e.g.* a secondary alkyl (**48**) *vs.* a primary alkyl radical (**49**)].

$$\underset{(48)}{\underset{X}{\underset{|}{\underset{CH_2}{\underset{|}{\sim\sim CH_2-CH\cdot}}}}} \qquad \underset{(49)}{\overset{\sim\sim CH_2 \diagdown \diagup CH_2\cdot}{\underset{X}{\bigcirc}}}$$

The polymerizability of these monomers is thus thought to be directly related to the abstractability of α–hydrogens.[233]

5.3.4 Transfer to Polymer

Two forms of transfer to polymer should be distinguished:

(a) Intramolecular reaction or backbiting, which gives rise to short chain branches (length usually ≤5 carbons).
(b) Intermolecular reaction, which generally results in the formation of long chain branches.

The intramolecular process does not give rise to a new polymer chain and is considered in section 4.4.3. It will not be considered further in this section.

Available evidence suggests that transfer to vinyl polymers (*e.g.* PMA, PVAc, PVC, PVF) usually involves abstraction of a methine hydrogen (Scheme 31) (see 5.4.3, 5.4.5, 5.4.6, 5.4.7). However, definitive evidence for the mechanism is currently only available for a few polymers (*e.g.* PVAc, PVF).

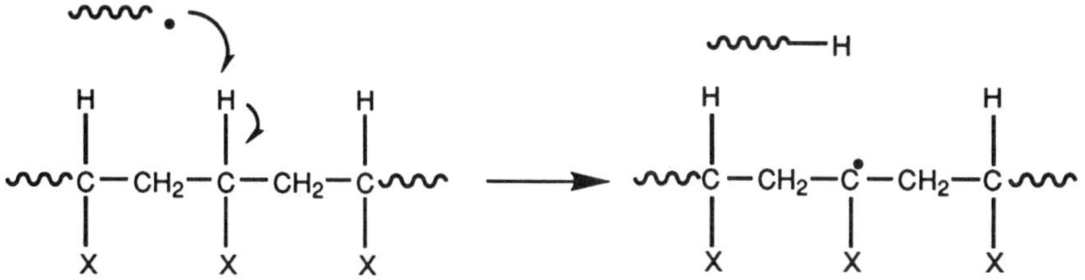

Scheme 31

Table 5.13 Transfer Constants to Polymer[a]

Monomer	Temperature (°C)	$C_P \times 10^4$
S	60	1.9-15.8
MMA	60	0.1-360
MA	60	0.5-1.0
AN	60	3.5
VAc	60	1.4-47
VC	50	5

[a] Numbers are taken from the Polymer Handbook[164] and have been rounded to two significant figures.

Transfer constants to polymer (C_P) are not as readily determined as other transfer constants because the process need not lead to an overall lowering of molecular weight. If transfer occurs by hydrogen-atom abstraction from the polymer backbone then, for every polymer chain terminated by transfer, another branched chain is formed. In these circumstances the overall molecular weight remains constant. The extent of chain transfer can then be estimated by measuring the number of long chain branches or by analyzing the molecular weight distribution. As NMR measurement of long chain branching relies on determining the branch points, a major analytical problem is distinguishing the long chain branches from the short chain branches formed by backbiting.

The values of C_P to added polymer are measurable in circumstances where the added material is readily distinguishable from that being formed *in situ*, for example, if it is of significantly different molecular weight or if it is uniquely labeled.[98] Studies with model compounds suggest that oligomers of chain length ≥3 can be used to provide a good estimate of the transfer constant.[234,235]

For some polymers, the value of C_P depends on the polymer molecular weight (*e.g.* see 5.3.4.2). This may account for the wide range of values for C_P in the literature (see Table 5.13).

5.3.4.1 Polyethylene

The presence of long chain branches in low density polyethylene (LDPE) accounts for the difference in properties (*e.g.* better melt strength, greater toughness for the same average molecular weight) between LDPE and linear low density polyethylene (LLDPE, made by coordination polymerization).

Long chain branching (>8 carbons) in polyethylene can be detected by ^{13}C NMR analysis.[236-239] However, the length and distribution of the branches are more difficult to determine. Measurements of long chain branching have been made by GPC-light scattering[240-242] or GPC-viscometry.[242-244] The extent of long chain branching is known to be strongly dependent on the reactor design and the reaction conditions employed. These studies indicate that, for a given sample, the branch frequency appears to decrease with increasing molecular weight of PE.[242] An explanation was not given.

5.3.4.2 Poly(alkyl methacrylates)

ω-Unsaturated PMMA is produced during radical polymerization of MMA through termination by disproportionation (see 5.2.3.2 & 5.2.4.3). Schulz *et al.*[245] were the first to suggest that reactions of these species may complicate MMA homopolymerization. Cacioli *et al.*[56] have shown that ω-unsaturated oligo(MMA) may act as a chain transfer agent in MMA polymerization. Tanaka *et al.*[197] have suggested that these species may also act as retarders or inhibitors of radical polymerization.

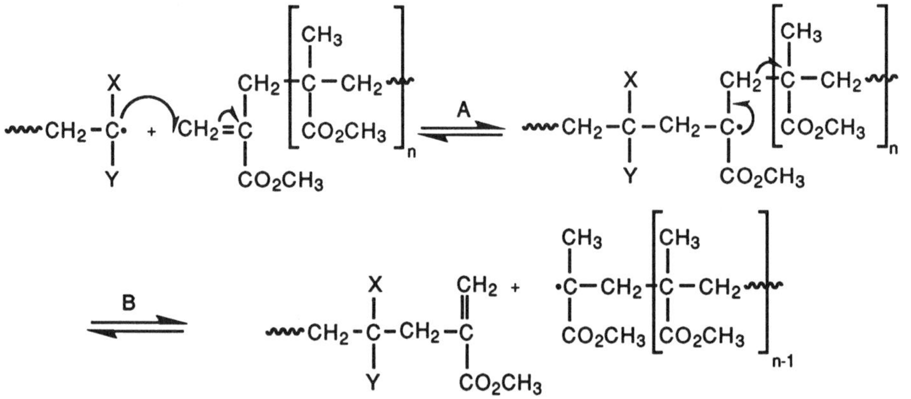

Scheme 32

The mechanism proposed for chain transfer is shown in Scheme 32. In MMA polymerization, adducts formed by addition to the macromonomer radicals are relatively unreactive towards adding further monomer and most undergo β-scission. There are two possible pathways for β-scission: pathway A leads back to starting materials; pathway B gives a new MMA propagating radical and a macromonomer. Transfer is catalytic in macromonomer.

If this mechanism operates, values of C_P measured in the presence of added PMMA will depend on how the PMMA was prepared and its molecular weight (*i.e.* on the concentration of unsaturated ends). PMMA formed by radical polymerization in the presence of a H-donor transfer agent (or by anionic polymerization) would have saturated chain ends. These should have a different transfer constant to that formed by normal radical polymerization where termination occurs by a mixture of combination and disproportionation. This could account for some of the variation in the values of C_P for this polymer.

5.3.4.3 *Poly(alkyl acrylates)*

Chain transfer to polymer is a major complication and is thought to be unavoidable in the polymerization of alkyl acrylates.[246,247] The mechanism is believed to involve abstraction of a tertiary backbone hydrogen (Scheme 31). It has been proposed that this process and the consequent formation of branches may contribute to the early onset of the gel or Trommsdorff effect in the polymerization of these monomers.

5.3.4.4 *Poly(vinyl acetate)*

The degree of branching in PVAc is strongly dependent on the polymerization conditions. Differences in the degree of branching are thought to be one of the main factors responsible for substantial differences in properties between various commercial samples of PVAc or PVA.[248-250]

PVAc is know to contain a significant number of long chain branches. Branches to the acetate methyl may arise by copolymerization of the VAc macromonomer produced as a consequence of transfer to monomer (see 5.3.3.2). Transfer to polymer may involve either the acetate methyl hydrogens (Scheme 33) or the methine (Scheme 34) or methylene hydrogens of the polymer backbone.

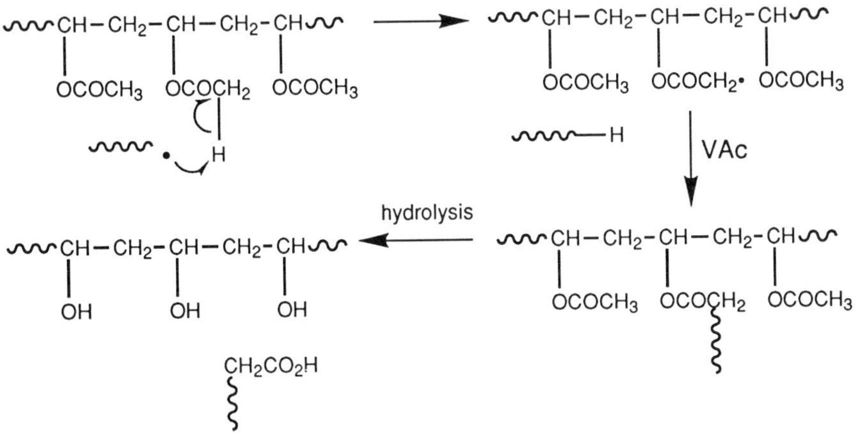

Scheme 33

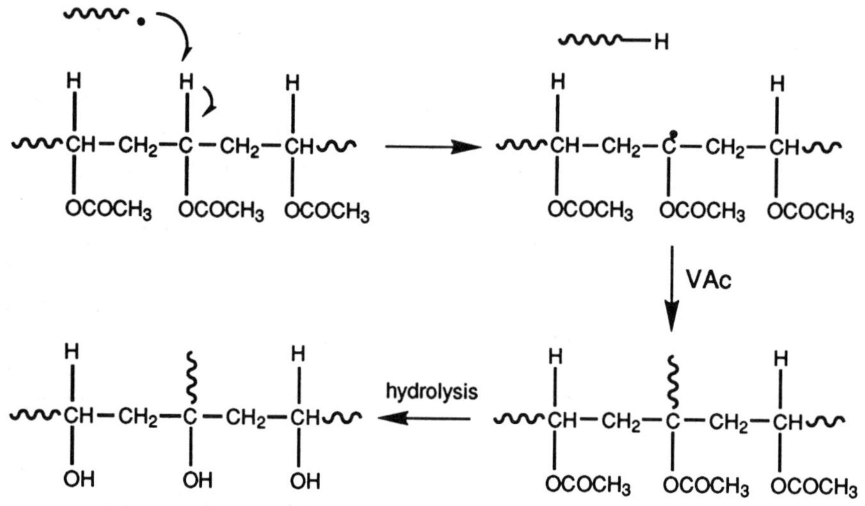

Scheme 34

The presence of hydrolyzable long chain branches in PVAc was established by McDowell and Kenyon[251] in 1940. They observed a reduction in molecular weight obtained on successively hydrolyzing and reacetylating samples of PVAc. Only branches to the acetate methyl will be lost on hydrolysis of the polymer; i.e. on conversion of PVAc to PVA.

The proposal that PVAc also has non-hydrolyzable long chain branches stems from the finding that PVA also possesses long chain branches. Nozakura et al.[219,252] suggested on the basis of kinetic measurements coupled with chemical analysis, that chain transfer to PVAc involves preferential abstraction of backbone (methine) hydrogens (ca. 5:1 vs. the acetate methyl hydrogens at 60°C).

^{1}H and ^{13}C NMR studies on PVAc or PVA also provide information on the nature of branches.[248,249,253,254] Dunn and Naravane[248] and Bugada and Rudin[249] proposed that the difference in intensity of the methylene and methine regions of the ^{13}C NMR spectrum could be used as a quantitative measure of the non-hydrolyzable branches (short chain + long chain) in PVA. However, this approach has been questioned by Vercauteren and Donners[249] because of the relatively large errors inherent in the method.

In order to prove that non-hydrolyzable long chain branches are present in a pre-existing sample of PVA, it is required that long chain branches can be distinguished from short chain branches. This distinction cannot be made solely on the basis of the ^{13}C NMR data. Extents of long chain branching can be obtained from GPC coupled with viscometry, ultracentrifugation or low angle laser light scattering on PVAc or reacetylated PVA.[250,255]

The extent of branching, of whatever type, is dependent on the polymerization conditions and, in particular, on the solvent and temperature employed and the degree of conversion. Nozakura et al.[219] found that, during bulk polymerization of VAc, the extent of transfer to polymer increased and the selectivity (for abstraction of a backbone *vs.* an acetoxy hydrogen) decreases with increasing temperature.

Adelman and Ferguson[253] have suggested, on the basis of ^{1}H NMR data (detection of $CH_3CH(OH)CH(OH)CH_2-$ ends) and chemical analyses (formation of acetaldehyde on periodate cleavage of 1,2-glycol units) on PVA, that the radical formed by head addition to VAc may be responsible for a high proportion of transfer events. Their PVAc was prepared in methanol at 60-75°C and much of the transfer involves the solvent. ^{13}C NMR[254,256] studies on several commercial PVA samples showed that those materials had equal numbers of head-to-head and tail-to-tail linkages (see 4.3.1.1) and indicated the presence of $-CH_2OH$ ends (*i.e.* most transfer involves the normal propagating species). These polymers are likely to have been prepared by emulsion polymerization, thus most transfer will involve monomer or polymer.

Hatada *et al.*[257] have indicated that PVAc prepared in aromatic solvents (benzene, chlorobenzene) at 60°C has fewer branch points than the polymer prepared in ethyl acetate under similar conditions. They attributed this observation to complexation of the propagating radical in the aromatic solvents and the different reactivity of this complexed radical. They have also reported that VAc polymerization is substantially slowed in aromatic solvents and this was also attributed to complexation of the propagating radical[258] (see 7.3.1.1).

5.3.4.5 Poly(vinyl chloride)

The microstructure of PVC has been the subject of numerous studies (see 4.3.1.2 and 5.3.3.3). Starnes *et al.*[216] determined the long chain branch points by NMR studies on polyethylene formed by Bu_3SnH reduction of PVC. They concluded that the probable mechanism for the formation of these branches involved transfer to polymer which occurred by hydrogen abstraction of a backbone methine by the propagating radical (Scheme 31).

5.3.4.6 Poly(vinyl fluoride)

Ovenall and Uschold[259] have recently measured the concentration of branch points (tertiary F, see Scheme 31) in PVF by ^{19}F NMR. These were found to account for between 0.5 to 1.5% of monomer units depending on reaction conditions. Branching was found to be favored by lower reactor pressures or higher reactor temperatures. More branching was observed for polymers produced in batch as opposed to continuous reactors. This effect was attributed to longer residence time of the polymer in the reactor.

5.4 Inhibition and Retardation

An inhibitor is a species which is able to rapidly and efficiently scavenge propagating and/or initiator derived radicals and thus prevent polymer chain formation. The species formed either do not or are slow to initiate polymerization (Scheme 35). Inhibitors which stop all polymerization until such time as they are completely consumed (*i.e.* during the induction period) and then allow polymerization to proceed at the normal rate, are called ideal inhibitors.

$$P_i\bullet \; + \; Z \; \xrightarrow{k_z} \; P_i$$

Scheme 35

The kinetics and mechanism of inhibition has been reviewed by Bamford,[260] Eastmond,[138] Goldfinger et al.[261] and Bovey and Kolthoff.[262] Common inhibitors include stable radicals (5.4.1), captodative olefins (5.4.3), phenols (5.4.4), quinones (5.4.5), oxygen (5.4.2), and certain transition metal salts (5.4.7). Some rate constants for the reactions of these species with simple carbon-centered radicals are summarized in Table 5.14.

Whether a given species functions as an inhibitor, retarder, or transfer agent in polymerization is dependent on the monomer(s) and the reaction conditions. For example, oxygen acts as an inhibitor or a retarder in many polymerizations yet it readily copolymerizes with S. Reactivity ratios for VAc-S copolymerization are such that small amounts of S are an effective inhibitor of VAc polymerization (r_S =0.02, r_{VAc} = 22.3). The propagating chain with a terminal VAc is very active towards S and adds even when S is present in small amounts. The propagating radical with S adds to VAc only slowly.

Table 5.14 Absolute Rate Constants for the Reaction of Carbon-Centered Radicals with Some Common Inhibitors

Inhibitor	Radical	T (°C)	k_z (M^{-1}s^{-1})	Refs.
TEMPO (51)	prim. alkyl	60	~1 x 10^9	263-265
oxygen	benzyl	27	2.9 x 10^9	266
p-benzoquinone (62)	prim. alkyl	69	2.0 x 10^7	267
CuCl$_2$	prim. alkyl	25	6.5 x 10^5	267

5.4.1 'Stable' Radicals

The kinetics and mechanism of inhibition by stable radicals has been reviewed by Rozantsev et al.[268] Ideally, for radicals to be useful inhibitors in radical polymerization they should have the following characteristics:

(a) They should not add to, abstract from, or otherwise react with the monomer, solvent, *etc*.

(b) They should not undergo self reaction or unimolecular decomposition.
(c) They must react rapidly with the propagating and/or the initiator derived radicals to terminate polymer chains.

Examples of radicals which are reported to meet these criteria are diphenylpicrylhydrazyl [DPPH, (50)], Koelsch radical, nitroxides [*e.g.* TEMPO (51)], triphenylmethyl (52), galvinoxyl (53), and verdazyl radicals [*e.g.* triphenylverdazyl (54)]. These reagents have seen practical application in a number of contexts. They have been widely utilized in the determination of initiator efficiency (see 3.3.1.1.3) and in mechanistic investigations (see 3.5.1).

Stable radicals can show selectivity for particular radicals. For example, nitroxides do not trap oxygen-centered radicals yet react with carbon-centered radicals at or near diffusion controlled rates.[264,269] This capability was utilized by Rizzardo and Solomon[270] to develop a technique for characterizing free radical reactions and has been extensively used in the examination of initiation of radical polymerization (see 3.5.1.4).

The efficiency of these inhibitors may depend on reaction conditions. For example: the reaction of radicals with stable radicals (*e.g.* nitroxides) may be reversible at elevated temperatures (see 7.5.3); triphenylmethyl may initiate polymerizations (see 7.5.2). A further complication is that the products may be capable of undergoing further radical chemistry. In the case of DPPH (50) this is attributed to the fact that the product is an aromatic nitro-compound (see 5.4.5).

Certain adducts may undergo induced decomposition to form a stable radical which can then scavenge further.

5.4.2 Oxygen

The role of oxygen in radical and other polymerizations has been reviewed by Bhanu and Kishore.[271] Rate constants for the reaction of carbon-centered radicals with oxygen are extremely fast, generally $\geq 10^9$ M^{-1} s^{-1}.[266,272] The initially formed species are peroxy radicals (**55**). These may abstract hydrogen or add monomer (Scheme 36).

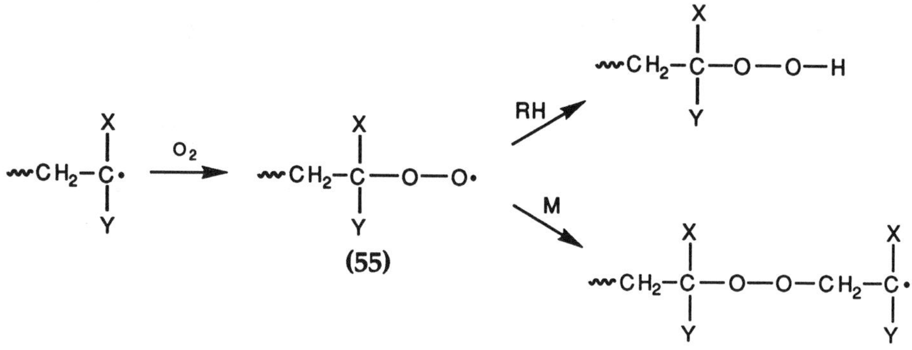

Scheme 36

Thus oxygen is an efficient scavenger of both initiating and propagating species in radical polymerization and usually steps must be taken to exclude oxygen from polymerizations or to minimize its effects. Typically, this involves conducting the experiment under an inert atmosphere (*e.g.* nitrogen) or a refluxing solvent. Oxygen may act as an inhibitor or retarder of polymerization, copolymerize (*e.g.* S polymerization), and/or facilitate chain transfer (*e.g.* VAc polymerization).

The effect observed is dependent on the reactivity of the monomer and other agents present in the polymerization medium towards hydroperoxy radicals (**61**). If addition of (**55**) to monomer is slow, in relation to normal propagation, then retardation or inhibition will be observed. It should also be noted that one of the products of reaction with oxygen, polymeric peroxides, are themselves potential sources of free radicals. These may complicate polymerization and can impair the properties of the final polymer.

5.4.3 Captodative olefins

Certain captodative olefins (**56**) rapidly scavenge radicals to give new radicals (**57**) which are unable or slow to reinitiate polymerization (Scheme 37).[273] Termination is believed to occur exclusively by combination, thus telechelic polymers are available by appropriate choice of the initiator. The head to head coupling product (**58**) is stable at normal polymerization temperatures. However, at higher temperatures (58) undergoes reversible homolysis and radicals (57) may initiate polymerization.[274]

It is reported that oxygen analogs of (56) do not inhibit polymerization but enter into copolymerization.[274-276]

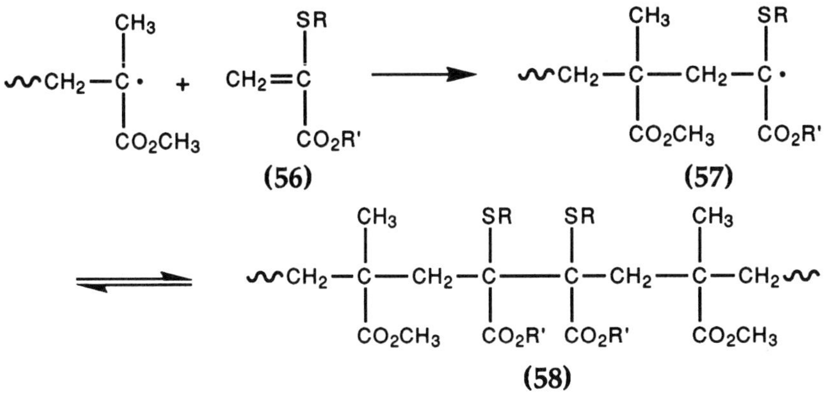

Scheme 37

5.4.4 Phenols

Phenolic inhibitors are commonly added to many commercial monomers to prevent polymerization during transport and storage. These include hydroquinone (59), monomethylhydroquinone (60) and 3,5-di-t-butylcatechol (61).

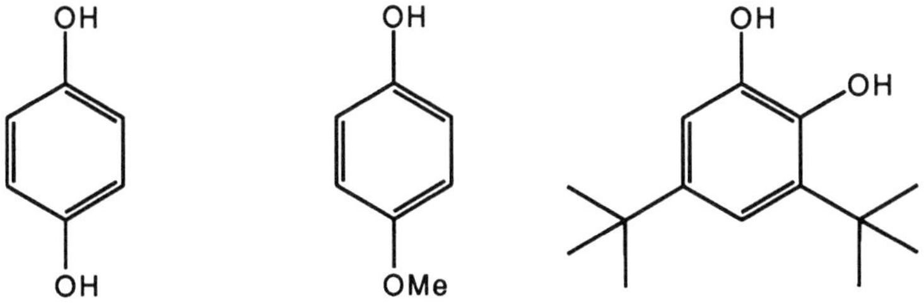

Studies with simple radicals show that carbon-centered radicals react with phenols by abstracting a phenolic hydrogen (Scheme 38). The phenoxy radicals may then scavenge a further radical by C-C or C-O coupling or (in the case of hydroquinones) by loss of a hydrogen atom to give a quinone. The quinone may then react further (see 5.4.4). Thus two or more propagating chains are terminated for every mole of phenol.[277]

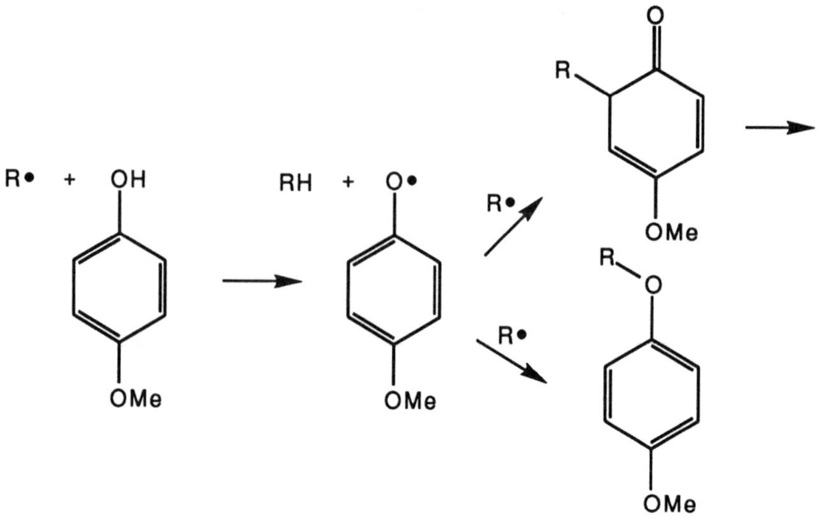

Scheme 38

5.4.5 Quinones

Quinones may react with carbon-centered radicals by addition at oxygen or at carbon (Scheme 39) or by electron transfer.[267,277-279] The preferred reaction pathway depends both on the attacking radical and the particular quinone. The radical formed may then scavenge another radical. There is evidence that certain quinones [*e.g.* chloranil, benzoquinone (62)] may copolymerize under some conditions.[280]

The absolute rate constants for attack of carbon-centered radicals on *p*-benzoquinone (62) and other quinones have been determined to be in the range 10^7-10^8 M^{-1} s^{-1}.[267,281] This rate shows a strong dependence on the electrophilicity of the attacking radical and there is some correlation between the efficiency of various quinones as inhibitors of polymerization and the redox potential of the quinone.

The complexity of the mechanism means that the stoichiometry of inhibition by these compounds is often not straight forward. Measurements of moles of inhibitor consumed for each chain terminated for common inhibitors of this class give values in the range 0.05-2.0.[261]

It has been reported that inhibition by certain phenols (*e.g.* hydroquinone) is more effective in the presence of oxygen.[282] This has been attributed to the phenols being better at scavenging hydroperoxy radicals (formed by the reaction of oxygen with propagating species) than with propagating radicals themselves. Alternatively, oxygen may oxidize the intermediate phenoxy radical to a quinone which is a more effective inhibitor.

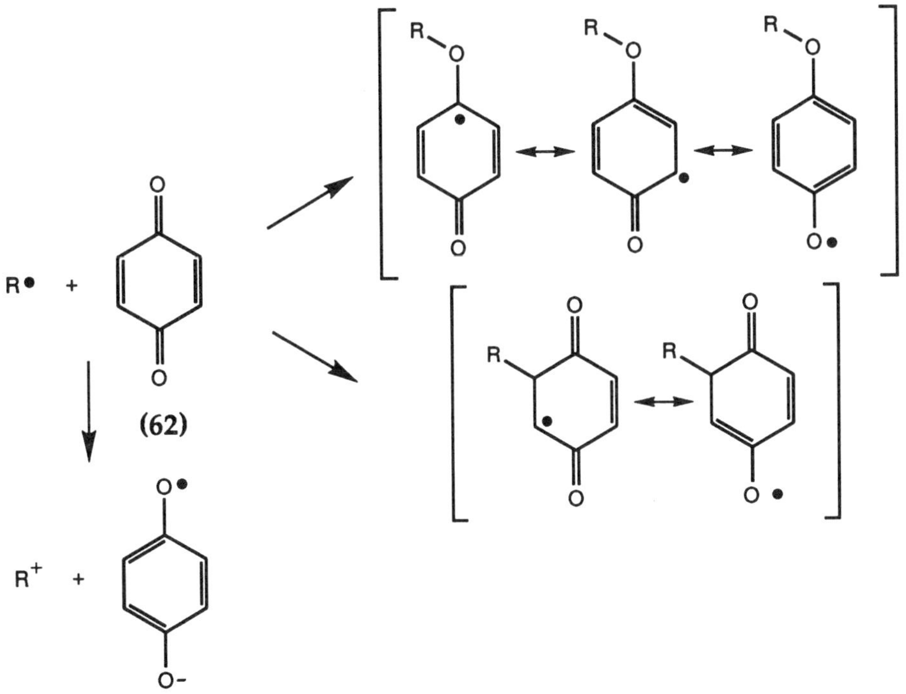

Scheme 39

5.4.6 Nitrones, nitro- and nitroso-compounds

Many nitrones and nitroso-compounds have been exploited as spin traps in elucidating radical reaction mechanisms by EPR spectroscopy (see 3.5.1.1). The initial adducts are nitroxides which can trap further radicals (see Scheme 40).

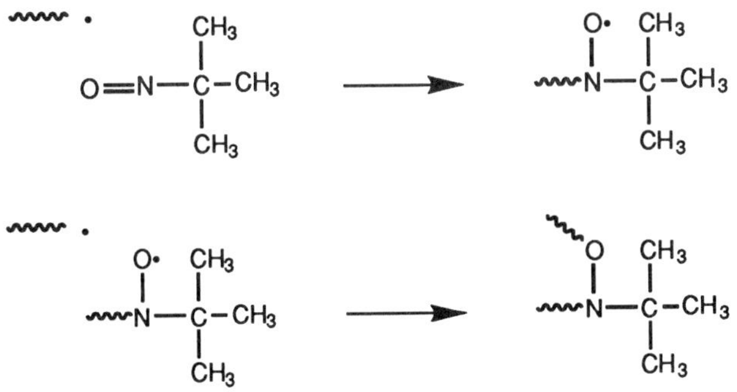

Scheme 40

Aromatic nitro-compounds have also seen use as inhibitors in polymerization and as additives in free radical reactions. The reactions of these compounds with radicals are very complex and may involve nitroso-compounds and nitroxide intermediates.[283,284] Up to four moles of radicals may be consumed per mole of nitro-compound.

5.4.7 Transition Metal Salts

Transition metal salts trap radicals by electron transfer or by ligand transfer. These reagents often show high specificity for reaction with specific radicals and the rates of trapping may be correlated with the nucleophilicity of the radical. For example, PS• radicals are much more reactive towards ferric chloride than acrylic propagating species.[285] Various transition metal salts have been applied in quantitative determination of initiation reactions and the chemistry of these inhibitors is described in section 3.5.1.2.

References

1. North, A.M., in 'Reactivity, Mechanism and Structure in Polymer Chemistry' (Eds. Jenkins, A.D., and Ledwith, A.), p. 142 (Wiley: London 1974).
2. O'Driscoll, K.F., in 'Comprehensive Polymer Science' (Eds. Eastmond, G.C., Ledwith, A., Russo, S., and Sigwalt, P.), Vol. 3, p. 161 (Pergamon: London 1989).
3. Buback, M., Garcia-Rubio, L.H., Gilbert, R.G., Napper, D.H., Guillot, J., Hamielec, A.E., Hill, D., O'Driscoll, K.F., Olaj, O.F., Shen, J., Solomon, D., Moad, G., Stickler, M., Tirrell, M., and Winnik, M.A., *J. Polym. Sci., Part C: Polym. Lett.*, 1988, **26**, 293.
4. Buback, M., Gilbert, R.G., Russell, G.T., Hill, D.J.T., Moad, G., O'Driscoll, K.F., Shen, J., and Winnik, M.A., *J. Polym. Sci., Part A: Polym. Chem.*, 1992, **30**, 851.
5. Gilbert, R.G., *Pure Appl. Chem.*, 1992, **64**, 1563.
6. Odian, G., Principles of Polymerization; 3rd Edn, p. 207 (Wiley-Interscience: New York 1991).
7. Eastmond, G.C., in 'Comprehensive Chemical Kinetics' (Eds. Bamford, C.H., and Tipper, C.F.H.), Vol. 14A, p. 1 (Elsevier: Amsterdam 1976).
8. Bamford, C.H., in 'Encyclopaedia of Polymer Science and Engineering', 2nd Edn (Eds. Mark, H.F., Bikales, N.M., Overberger, C.G., and Menges, G.), Vol. 13, p. 708 (Wiley: New York 1988).
9. Soh, S.K., and Sundberg, D.C., *J. Polym. Sci., Polym. Chem. Ed.*, 1982, **20**, 1345.
10. Mahabadi, H.K., and O'Driscoll, K.F., *J. Polym. Sci., Polym. Chem. Ed.*, 1977, **15**, 283.
11. Olaj, O.F., and Zifferer, G., *Macromolecules*, 1987, **20**, 850.
12. Zhu, S., and Hamielec, A.E., *Macromolecules*, 1989, **22**, 3093.
13. Yasukawa, T., and Murakami, K., *Macromolecules*, 1981, **14**, 227.
14. Yasukawa, T., and Murakami, K., *Polymer*, 1980, **21**, 1423.

15. Marten, F.L., and Hamielec, A.E., *J. Appl. Polym. Sci.*, 1982, **27**, 489.
16. Bamford, C.H., *Polymer*, 1990, **31**, 1720.
17. Olaj, O.F., Zifferer, G., and Gleixner, G., *Makromol. Chem.*, 1986, **187**, 977.
18. Benson, S.W., and North, A.M., *J. Am. Chem. Soc.*, 1962, **84**, 935.
19. Olaj, O.F., and Zifferer, G., *Makromol. Chem., Rapid Commun.*, 1982, **3**, 549.
20. Griller, D., in 'Landoldt-Bornstein, New Series, Radical Reaction Rates in Solution' (Ed. Fischer, H.), Vol. II/13a, p. 5 (Springer-Verlag: Berlin 1984).
21. Deady, M., Mau, A.W.H., Moad, G., and Spurling, T.H., *Makromol. Chem.*, 1993, **194**, 1691.
22. Bamford, C.H., *Eur. Polym. J.*, 1989, **25**, 683.
23. Bamford, C.H., *Eur. Polym. J.*, 1993, **29**, 313.
24. Bamford, C.H., *Eur. Polym. J.*, 1990, **26**, 719.
25. Bamford, C.H., *Eur. Polym. J.*, 1991, **27**, 1289.
26. Bamford, C.H., *Eur. Polym. J.*, 1990, **26**, 1245.
27. Cardenas, J.N., and O'Driscoll, K.F., *J. Polym. Sci., Polym. Chem. Ed.*, 1976, **14**, 883.
28. Russell, G.T., *Macromol. Theory Simul.*, 1994, **3**, 439.
29. Faldi, A., Tirrell, M., and Lodge, T.P., *Macromolecules*, 1994, **27**, 4176.
30. O'Shaughnessy, B., and Yu, J., *Macromolecules*, 1994, **27**, 5079.
31. O'Shaughnessy, B., and Yu, J., *Macromolecules*, 1994, **27**, 5067.
32. Verravalli, M.S., and Rosen, S.L., *J. Polym. Sci., Part B: Polym. Phys.*, 1990, **28**, 775.
33. Balke, S.T., and Hamielec, A.E., *J. Appl. Polym. Sci.*, 1973, **17**, 905.
34. Soh, S.K., and Sundberg, D.C., *J. Polym. Sci., Polym. Chem. Ed.*, 1982, **20**, 1299.
35. Soh, S.K., and Sundberg, D.C., *J. Polym. Sci., Polym. Chem. Ed.*, 1982, **20**, 1315.
36. Soh, S.K., and Sundberg, D.C., *J. Polym. Sci., Polym. Chem. Ed.*, 1982, **20**, 1331.
37. Tulig, T.J., and Tirrell, M., *Macromolecules*, 1981, **14**, 1501.
38. Ito, K., *Polym. J. (Tokyo)*, 1980, **12**, 499.
39. de Gennes, P.G., *J. Chem. Phys.*, 1982, **76**, 3316.
40. Buback, M., Huckestein, B., and Russell, G.T., *Macromol. Chem. Phys.*, 1994, **195**, 539.
41. Chiu, W.Y., Carrat, G.M., and Soong, D.S., *Macromolecules*, 1983, **16**, 348.
42. Faldi, A., Tirrell, M., Lodge, T.P., and von Meewwall, E., *Macromolecules*, 1994, **27**, 4184.
43. Johnson, C.H.J., Moad, G., Solomon, D.H., Spurling, T.H., and Vearing, D.J., *Aust. J. Chem.*, 1990, **43**, 1215.
44. Spurling, T.H., Deady, M., Krstina, J., and Moad, G., *Makromol. Chem., Macromol. Symp.*, 1991, **51**, 127.
45. Atherton, J.N., and North, A.M., *Trans. Faraday. Soc.*, 1962, **58**, 2049.

46. North, A.M., and Postlethwaite, D., *Polymer*, 1964, **5**, 237.
47. Fukuda, T., Ma, Y.-D., and Inagaki, H., *Macromolecules*, 1985, **18**, 17.
48. Fukuda, T., Kubo, K., Ma, Y.D., and Inagaki, H., *Polym. J. (Tokyo)*, 1987, **19**, 523.
49. Mayo, F.R., and Walling, C., *Chem. Rev.*, 1950, **46**, 191.
50. Cacioli, P., Moad, G., Rizzardo, E., Serelis, A.K., and Solomon, D.H., *Polym. Bull. (Berlin)*, 1984, **11**, 325.
51. Kashiwagi, T., Inaba, A., Brown, J.E., Hatada, K., Kitayama, T., and Masuda, E., *Macromolecules*, 1986, **19**, 2160.
52. Meisters, A., Moad, G., Rizzardo, E., and Solomon, D.H., *Polym. Bull. (Berlin)*, 1988, **20**, 499.
53. Manring, L.E., *Macromolecules*, 1989, **22**, 2673.
54. Bamford, C.H., and White, E.F.T., *Trans. Faraday Soc.*, 1958, **54**, 268.
55. Bizilj, S., Kelly, D.P., Serelis, A.K., Solomon, D.H., and White, K.E., *Aust. J. Chem.*, 1985, **38**, 1657.
56. Cacioli, P., Hawthorne, D.G., Laslett, R.L., Rizzardo, E., and Solomon, D.H., *J. Macromol. Sci., Chem.*, 1986, **A23**, 839.
57. Morawetz, H., *J. Polym. Sci., Polym. Symp.*, 1978, **62**, 271.
58. Kelly, D.P., Serelis, A.K., Solomon, D.H., and Thompson, P.E., *Aust. J. Chem.*, 1987, **40**, 1631.
59. Ito, K., *Polymer*, 1985, **26**, 1253.
60. Guth, W., and Heitz, W., *Makromol. Chem.*, 1976, **177**, 1835.
61. Overberger, C.G., and Finestone, A.B., *J. Am. Chem. Soc.*, 1956, **78**, 1638.
62. Gibian, M.J., and Corley, R.C., *J. Am. Chem. Soc.*, 1972, **94**, 4178.
63. Shelton, J.R., and Liang, C.K., *J. Org. Chem.*, 1973, **38**, 2301.
64. Gleixner, G., Olaj, O.F., and Breitenbach, J.W., *Makromol. Chem.*, 1979, **180**, 2581.
65. Schreck, V.A., Serelis, A.K., and Solomon, D.H., *Aust. J. Chem.*, 1989, **42**, 375.
66. Nelsen, S.F., and Bartlett, P.D., *J. Am. Chem. Soc.*, 1966, **88**, 137.
67. Neuman, R.C., Jr., and Amrich, M.J., Jr., *J. Org. Chem.*, 1980, **45**, 4629.
68. Fraenkel, G., and Geckle, M.J., *J. Chem. Soc., Chem. Commun.*, 1980, 55.
69. Langhals, H., and Fischer, H., *Chem. Ber.*, 1978, **111**, 543.
70. Skinner, K.J., Hochster, H.S., and McBride, J.M., *J. Am. Chem. Soc.*, 1974, **96**, 4301.
71. Trecker, D.J., and Foote, R.S., *J. Org. Chem.*, 1968, **33**, 3527.
72. Kodaira, K., Ito, K., and Iyoda, S., *Polym. Commun.*, 1987, **28**, 86.
73. Mackie, J.S., and Bywater, S., *Can. J. Chem.*, 1957, **35**, 570.
74. Neumann, W.P., and Stapel, R., *Chem. Ber.*, 1986, **119**, 3422.
75. Serelis, A.K., and Solomon, D.H., *Polym. Bull. (Berlin)*, 1982, **7**, 39.
76. Bickel, A.F., and Waters, W.A., *Rec. Trav. Chim.*, 1950, **69**, 1490.
77. Barbe, W., and Rüchardt, C., *Makromol. Chem.*, 1983, **184**, 1235.
78. Serelis, A.K., *Personal Communication*,

79. Konter, W., Bömer, B., Köhler, K.H., and Heitz, W., *Makromol. Chem.*, 1981, **182**, 2619.
80. Barton, J., Capek, I., Juranicova, V., and Riedel, S., *Makromol. Chem., Rapid Commun.*, 1986, **7**, 521.
81. Jaffe, A.B., Skinner, K.J., and McBride, J.M., *J. Am. Chem. Soc.*, 1972, **94**, 8510.
82. Minato, T., Yamabe, S., Fujimoto, H., and Fukui, K., *Bull. Chem. Soc. Japan*, 1978, **51**, 1.
83. Krstina, J., Moad, G., Willing, R.I., Danek, S.K., Kelly, D.P., Jones, S.L., and Solomon, D.H., *Eur. Polym. J.*, 1993, **29**, 379.
84. Gibian, M.J., and Corley, R.C., *Chem. Rev.*, 1973, **73**, 441.
85. Heitz, W., in 'Telechelic Polymers: Synthesis and Applications' (Ed. Goethals, E.J.), p. 61 (CRC Press: Boca Raton, Florida 1989).
86. Moad, G., Serelis, A.K., Solomon, D.H., and Spurling, T.H., *Polym. Commun.*, 1984, **25**, 240.
87. Kodaira, T., Ito, K., and Iyoda, S., *Polym. Commun.*, 1988, **29**, 83.
88. Moad, G., Solomon, D.H., Johns, S.R., and Willing, R.I., *Macromolecules*, 1984, **17**, 1094.
89. Bamford, C.H., Dyson, R.W., and Eastmond, G.C., *Polymer*, 1969, **10**, 885.
90. Stickler, M., *Makromol. Chem.*, 1979, **180**, 2615.
91. Burnett, G.M., and North, A.M., *Makromol. Chem.*, 1964, **73**, 77.
92. Bamford, C.H., Jenkins, A.D., and Johnston, R., *Trans. Faraday Soc.*, 1959, **55**, 179.
93. Olaj, O.F., Breitenbach, J.W., and Wolf, B., *Monatsh. Chem.*, 1964, **95**, 1646.
94. Dawkins, J.V., and Yeadon, G., *Polymer*, 1979, **20**, 981.
95. Baker, C.A., and Williams, R.J.P., *J. Chem. Soc.*, 1956, 2352.
96. Rudin, A., in 'Comprehensive Polymer Science' (Eds. Eastmond, G.C., Ledwith, A., Russo, S., and Sigwalt, P.), Vol. 3, p. 239 (Pergamon: London 1989).
97. Hensley, D.R., Goodrich, S.D., Harwood, H.J., and Rinaldi, P.L., *Macromolecules*, 1994, **27**, 2351.
98. Bevington, J.C., Melville, H.W., and Taylor, R.P., *J. Polym. Sci.*, 1954, **14**, 463.
99. Ayrey, G., and Moore, C.G., *J. Polym. Sci.*, 1959, **36**, 41.
100. Hakozaki, J., and Yamada, N., *J. Chem. Soc., Japan*, 1967, **70**, 1560 (*Chem. Abstr.* 96263r (1968)).
101. O'Driscoll, K.F., and Bevington, J.C., *Eur. Polym. J.*, 1985, **21**, 1039.
102. Bamford, C.H., and Jenkins, A.D., *Nature*, 1955, **176**, 78.
103. Mayo, F.R., Gregg, R.A., and Matheson, M.S., *J. Am. Chem. Soc.*, 1951, **73**, 1691.
104. Johnson, D.H., and Tobolsky, A.V., *J. Am. Chem. Soc.*, 1952, **74**, 938.
105. Braks, J.G., and Huang, R.Y.M., *J. Appl. Polym. Sci.*, 1978, **22**, 3111.
106. Henrici-Olive, G., and Olive, S., *J. Polym. Sci.*, 1960, **48**, 329.

107. Bevington, J.C., Melville, H.W., and Taylor, R.P., *J. Polym. Sci.*, 1954, **12**, 449.
108. Kolthoff, I.M., O'Connor, P.R., and Hansen, J.L., *J. Polym. Sci.*, 1955, **15**, 459.
109. Arnett, L.M., and Peterson, J.H., *J. Am. Chem. Soc.*, 1952, **74**, 2031.
110. Bessiere, J.-M., Boutevin, B., and Loubet, O., *Polym. Bull. (Berlin).*, 1993, **31**, 673.
111. Moad, G., Solomon, D.H., Johns, S.R., and Willing, R.I., *Macromolecules*, 1982, **15**, 1188.
112. Ayrey, G., Levitt, F.G., and Mazza, R.J., *Polymer*, 1965, **6**, 157.
113. Olaj, O.F., Kaufmann, H.F., Breitenbach, J.W., and Bieringer, H., *J. Polym. Sci., Polym. Lett. Ed.*, 1977, **15**, 229.
114. Berger, K.C., and Meyerhoff, G., *Makromol. Chem.*, 1975, **176**, 1983.
115. Allen, P.W., Ayrey, G., Merrett, F.M., and Moore, C.G., *J. Polym. Sci.*, 1956, **22**, 549.
116. Boudevska, H., Brutchkov, C., and Platchkova, S., *Makromol. Chem.*, 1981, **182**, 3257.
117. Schulz, G.V., Henrice-Olive, G., and Olive, S., *Makromol. Chem.*, 1959, **31**, 88.
118. Bamford, C.H., Eastmond, G.C., and Whittle, D., *Polymer*, 1969, **10**, 771.
119. Chaudhuri, A.K., and Palit, S.R., *J. Polym. Sci., Part A-1*, 1968, **6**, 2187.
120. Braks, J.G., Mayer, G., and Huang, R.Y.M., *J. Appl. Polym. Sci.*, 1980, **25**, 449.
121. Ayrey, G., and Haynes, A.C., *Eur. Polym. J.*, 1973, **9**, 1029.
122. Hatada, K., Kitayama, T., and Masuda, E., *Polym. J. (Tokyo)*, 1986, **18**, 395.
123. Ayrey, G., Humphrey, M.J., and Poller, R.C., *Polymer*, 1977, **18**, 840.
124. Bailey, B.E., and Jenkins, A.D., *Trans Faraday Soc.*, 1960, 903.
125. Bevington, J.C., and Eaves, D.E., *Trans Faraday Soc.*, 1959, **55**, 1777.
126. Patron, L., and Bastianelli, U., *Appl. Polym. Symp.*, 1974, **25**, 105.
127. Funt, B.L., and Paskia, W., *Can. J. Chem.*, 1960, **38**, 1865.
128. Danusso, F., Pajaro, G., and Sianesi, D., *Chim. Ind (Milan)*, 1959, **41**, 1170.
129. Talamini, G., G., *Chim. Ind (Milan).*, 1964, **46**, 16. (Chem. Abstr. 1964, 60, 10804a)
130. Park, G.S., and Smith, D.G., *Makromol. Chem.*, 1970, **131**, 1.
131. Starnes, W.H., Jr., Plitz, I.M., Schilling, F.C., Villacorta, G.M., Park, G.S., and Saremi, A.H., *Macromolecules*, 1984, **17**, 2507.
132. Atkinson, W.H., Bamford, C.H., and Eastmond, G.C., *Trans. Faraday Soc.*, 1970, **66**, 1446.
133. Chen, C.Y., Wu, Z.Z., and Kuo, J.F., *Polym. Eng. Sci.*, 1987, **27**, 553.
134. Flory, P.J., *J. Am. Chem. Soc.*, 1937, **59**, 241.
135. Flory, P.J., Principles of Polymer Chemistry, p. 106 (Cornell University Press: Ithaca, N.Y. 1953).
136. Barson, C.A., in 'Comprehensive Polymer Science' (Eds. Eastmond, G.C., Ledwith, A., Russo, S., and Sigwalt, P.), Vol. 3, p. 171 (Pergamon: London 1989).

137. Farina, M., *Makromol. Chem., Macromol. Symp.*, 1987, **10/11**, 255.
138. Eastmond, G.C., in 'Comprehensive Chemical Kinetics' (Eds. Bamford, C.H., and Tipper, C.F.H.), Vol. 14A, p. 153 (Elsevier: Amsterdam 1976).
139. Palit, S.R., Chatterjee, S.R., and Mukherjee, A.R., in 'Encyclopaedia of Polymer Science and Technology' (Eds. Mark, H., F., Gaylord, N.G., and Bikales, N.M.), Vol. 3, p. 575 (Wiley: New York 1966).
140. Boutevin, B., *Adv. Polym. Sci.*, 1990, **94**, 69.
141. Corner, T., *Adv. Polym. Sci.*, 1984, **62**, 95.
142. Starks, C.M., Free Radical Telomerization (Academic Press: New York 1974).
143. Mayo, F.R., *J. Am. Chem. Soc.*, 1943, **65**, 2324.
144. Clouet, G., and Knipper, M., *Makromol. Chem.*, 1987, **188**, 2597.
145. Scott, G.P., and Foster, F.J., *Macromolecules*, 1969, **2**, 428.
146. Scott, G.P., and Elghoul, A.M.R., *J. Polym. Sci., Part A-1*, 1970, **8**, 2255.
147. Scott, G.P., and Wang, J.C., *J. Org. Chem.*, 1963, **28**, 1314.
148. Barson, C.A., Mather, R.R., and Robb, J.C., *Trans. Faraday Soc.*, 1970, **66**, 2585.
149. Mayo, F.R., *J. Am. Chem. Soc.*, 1948, **70**, 3689.
150. Asahara, T., and Makishima, T., *Kogyo Kagaku Zasshi*, 1966, **69**, 2173.
151. Englin, B.A., and Onishchenko, T.A., *Izv. Akad. Nauk. SSSR, Ser. Khim.*, 1960, 1906.
152. Englin, B.A., Onishchenko, T.A., and Freidlina, R.K., *Izv. Akad. Nauk. SSSR, Ser. Khim.*, 1968, **11**, 2489.
153. Bamford, C.H., *Polym. Commun.*, 1989, **30**, 36.
154. Whang, B.Y.C., Ballard, M.J., Napper, D.H., and Gilbert, R.G., *Aust. J. Chem.*, 1991, **44**, 1133.
155. Clay, P.A., and Gilbert, R.G., *Macromolecules*, 1995, **28**, 552.
156. Moad, G., and Winzor, C.L., *Makromol. Chem., Theory Simul.*, 1995, in preparation.
157. Bamford, C.H., *J. Chem. Soc., Faraday Trans. 1*, 1976, **72**, 2805.
158. Bamford, C.H., and Basahel, S.N., *J. Chem. Soc., Faraday Trans. 1*, 1978, **74**, 1020.
159. Bamford, C.H., and Basahel, S.N., *Polymer*, 1978, **19**, 943.
160. Walling, C., *J. Am. Chem. Soc.*, 1948, **70**, 2561.
161. Cardenas, J.N., and O'Driscoll, K.F., *J. Polym. Sci., Polym. Chem. Ed.*, 1977, **15**, 2097.
162. Harwood, H.J., Medsker, R.E., and Rapo, A., in 'MakroAkron 94 Abstracts', p. 16 (IUPAC: 1994).
163. Nair, C.P.R., Richou, M.C., Chaumont, P., and Clouet, G., *Eur. Polym. J.*, 1990, **26**, 811.
164. Berger, K.C., and Brandup, G., in 'Polymer Handbook, 3rd Edition' (Eds. Brandup, J., and Immergut, E.H.), p. II/81 (Wiley: New York 1989).

165. Clarke, J.T., Howard, R.O., and Stockmayer, W.H., *Makromol. Chem.*, 1961, **44**, 427.
166. Roy, K.K., Pramanick, D., and Palit, S.R., *Makromol. Chem.*, 1972, **153**, 71.
167. Bamford, C.H., and Basahel, S.N., *J. Chem. Soc., Faraday Trans. 1*, 1980, **76**, 112.
168. Boutevin, B., El Idrissi, A., and Parisi, J.P., *Makromol. Chem.*, 1990, **191**, 445.
169. Boutevin, B., Lusinchi, J.-M., Pietrasanta, Y., and Robin, J.-J., *Eur. Polym. J.*, 1994, **30**, 615.
170. Boutevin, B., and Pietrasanta, Y., *Makromol. Chem.*, 1985, **186**, 817.
171. Pryor, W.A., and Pickering, T.L., *J. Am. Chem. Soc.*, 1962, **84**, 2705.
172. Costanza, A.J., Coleman, R.J., Pierson, R.M., Marvel, C.S., and King, C., *J. Polym. Sci.*, 1955, **17**, 319.
173. Otsu, T., Kinoshita, Y., and Imoto, M., *Makromol. Chem.*, 1964, **73**, 225.
174. Tsuda, K., and Otsu, T., *Bull. Chem. Soc. Japan*, 1966, **39**, 2206.
175. Otsu, T., and Nayatani, K., *Makromol. Chem.*, 1958, **73**, 225.
176. Ferington, T.E., and Tobolsky, A.V., *J. Am. Chem. Soc.*, 1955, **77**, 4510.
177. Popielarz, R., and Clouet, G., *Makromol. Chem.*, 1993, **194**, 2897.
178. Kimura, T., Kodaira, T., and Hamashima, M., *Polym. J. (Tokyo)*, 1983, **15**, 293.
179. Tedder, J.M., *Angew. Chem., Int. Ed. Engl.*, 1982, **21**, 401.
180. Ameduri, B., and Boutevin, B., *Macromolecules*, 1990, **23**, 2433.
181. Mayo, F.R., *J. Am. Chem. Soc.*, 1953, **75**, 6133.
182. Meijs, G.F., and Rizzardo, E., *Makromol. Chem., Rapid Commun.*, 1988, **9**, 547.
183. Meijs, G.F., and Rizzardo, E., *Makromol. Chem.*, 1990, **191**, 1545.
184. Dais, V.A., Priddy, D.B., Bell, B., Sikkema, K.D., and Smith, P., *J. Polym. Sci., Part A: Polym. Chem.*, 1993, **31**, 901.
185. Meijs, G.F., Rizzardo, E., and Thang, S.H., *Polym. Bull. (Berlin)*, 1990, **24**, 501.
186. Yamada, B., Kobatake, S., and Aoki, S., *Macromol. Chem. Phys.*, 1994, **195**, 581.
187. Yamada, B., and Otsu, T., *Makromol. Chem.*, 1991, **192**, 333.
188. Yamada, B., and Otsu, T., *Makromol. Chem., Rapid Commun.*, 1990, **11**, 513.
189. Yamada, B., Satake, M., and Otsu, T., *Polym. J. (Tokyo)*, 1992, **24**, 563.
190. Yamada, B., Kato, E., Kobatake, S., and Otsu, T., *Polym. Bull. (Berlin)*, 1991, **25**, 423.
191. Yamada, B., Kobatake, S., and Aoki, S., *Polym. Bull. (Berlin)*, 1993, **31**, 263.
192. Meijs, G.F., Morton, T.C., Rizzardo, E., and Thang, S.H., *Macromolecules*, 1991, **24**, 3689.
193. Mathias, L.J., Thompson, R.D., and Lightsey, A.K., *Polym. Bull. (Berlin)*, 1992, **27**, 395.
194. Meijs, G.F., Rizzardo, E., and Thang, S.H., *Macromolecules*, 1988, **21**, 3122.

195. Meijs, G.F., Rizzardo, E., and Thang, S.H., *Polym. Prepr. (Am. Chem. Soc., Div. Polym. Chem)*, 1992, **33(1)**, 893.
196. Rizzardo, E., Laslett, R.L., Harrison, D., Meijs, G.F., Morton, T.C., and Thang, S.H., *Prog. Pacific Polym. Sci.*, 1991, **2**, 77.
197. Tanaka, H., Kawa, H., Sato, T., and Ota, T., *J. Polym. Sci., Part A: Polym. Chem.*, 1989, **27**, 1741.
198. Meijs, G.F., Morton, T.C., and Le, T.P.T., *Polym. Int.*, 1991, **26**, 239.
199. Meijs, G.F., and Rizzardo, E., *Polym. Bull. (Berlin)*, 1991, **26**, 291.
200. Enikolopyan, N.S., Smirnov, B.R., Ponomarev, G.V., and Belgovskii, I.M., *J. Polym. Sci., Polym. Chem. Ed.*, 1981, **19**, 879.
201. Burczyk, A.F., O'Driscoll, K.F., and Rempel, G.L., *J. Polym. Sci., Polym. Chem. Ed.*, 1984, **22**, 3255.
202. Gridnev, A.A., *Polym. Sci. USSR (Engl. Transl.)*, 1989, **31**, 2369.
203. Karmilova, L.V., Ponomarev, G.V., Smirnov, B.R., and Bel'govskii, I.M., *Russ. Chem. Rev. (Engl. Transl.)*, 1984, **53**, 132.
204. Davis, T.P., Haddleton, D.M., and Richards, S.N., *J. Macromol. Sci., Rev. Macromol. Chem. Phys.*, 1994, **C34**, 243.
205. Parshall, G.W., and Ittel, S.D., Homogeneous Catalysis (Wiley: New york 1992).
206. Sanayei, R.A., and O'Driscoll, K.F., *J. Macromol. Sci. Chem.*, 1989, **A26**, 1137.
207. Smirnov, B.R., Marchenko, A.P., Plotnikov, V.D., Kuzayev, A.I., and Yenikolopyan, N.S., *Polym. Sci. USSR (Engl. Transl.)*, 1981, **23**, 1169.
208. Sanayei, R.A., and O'Driscoll, K.F., *J. Macromol. Sci. Chem.*, 1989, **A26**, 1137.
209. Suddaby, K.G., O'Driscoll, K.F., and Rudin, A., *J. Polym. Sci., Part A: Polym. Chem.*, 1992, **30**, 643.
210. Hawthorne, D.G., European Patent EP 0 249 614 B 1
211. Baysal, B., and Tobolsky, A.V., *J. Polym. Sci.*, 1952, **8**, 529.
212. Mahadevan, V., and Santhappa, M., *Makromol. Chem.*, 1955, **16**,
213. Das, S.K., Chatterjee, S.R., and Palit, S.R., *Proc. R. Soc., London, Ser. A*, 1955, **227**, 252.
214. Stein, D.J., *Makromol. Chem.*, 1964, **76**, 170.
215. Kuchanov, S.I., and Olenin, A.V., *Polym. Sci. USSR (Engl. Transl.)*, 1973, **15**, 2712.
216. Starnes, W.H., Jr., Schilling, F.C., Plitz, I.M., Cais, R.E., Freed, D.J., Hartless, R.L., and Bovey, F.A., *Macromolecules*, 1983, **16**, 790.
217. Bartlett, P.D., and Altschul, R., *J. Am. Chem. Soc.*, 1945, **67**, 816.
218. Pryor, W.A., and Coco, J.H., *Macromolecules*, 1970, **3**, 500.
219. Nozakura, S.-I., Morishima, Y., and Murahashi, S., *J. Polym. Sci., Part A-1*, 1972, **10**, 2853.
220. Melville, H.W., and Sewell, P.R., *Makromol. Chem.*, 1959, **32**, 139.
221. Starnes, W.H., Jr., Chung, H., and Benedikt, G.M., *Polym. Prepr. (Am. Chem. Soc., Div. Polym. Chem.)*, 1993, **34(1)**, 604.
222. Litt, M., and Chang, K.H.S., *ACS Symp. Ser.*, 1981, **165**, 455.

223. Wolf, C., and Burchard, W., *Makromol. Chem.*, 1976, **177**, 2519.
224. Vidotto, G., Crosato-Arnaldi, A., and Talamini, G., *Makromol. Chem.*, 1968, **114**, 217.
225. Rigo, A., Palma, G., and Talamini, G., *Makromol. Chem.*, 1972, **153**, 219.
226. Starnes, W.H., Jr., Schilling, F.C., Abbas, K.B., Cais, R.E., and Bovey, F.A., *Macromolecules*, 1979, **12**, 556.
227. Park, G.S., and Saleem, M., *Polym. Bull. (Berlin)*, 1979, **1**, 409.
228. Starnes, W.H., Jr., and Wojciechowski, B.J., *Makromol. Chem., Macromol. Symp.*, 1993, **70/71**, 1.
229. Zubov, V.P., Kumar, M.V., Masterova, M.N., and Kabanov, V.A., *J. Macromol. Sci., Chem.*, 1979, **A13**, 111.
230. Butler, G.B., in 'Comprehensive Polymer Science' (Eds. Eastmond, G.C., Ledwith, A., Russo, S., and Sigwalt, P.), Vol. 4, p. 423 (Pergamon: London 1989).
231. Bartlett, P.D., and Tate, F.A., *J. Am. Chem. Soc.*, 1953, **75**, 91.
232. Butler, G.B., Cyclopolymerization and Cyclocopolymerization (Marcel Dekker: New York 1992).
233. Vaidya, R.A., and Mathias, L.J., *J. Polym. Sci., Polym. Symp.*, 1986, **74**, 243.
234. Schulz, G.V., and Stein, D.J., *Makromol. Chem.*, 1962, **52**, 1.
235. Lim, D., and Wichterle, O., *J. Polym. Sci.*, 1958, **29**, 579.
236. Usami, T., and Takayama, S., *Macromolecules*, 1984, **17**, 1756.
237. Axelson, D.E., Levy, G.C., and Mandelkern, L., *Macromolecules*, 1979, **12**, 41.
238. Bovey, F.A., Schilling, F.C., McCrackin, F.L., and Wagner, H.L., *Macromolecules*, 1976, **9**, 76.
239. Bugada, D.C., and Rudin, A., *Eur. Polym. J..*, 1987, **23**, 809.
240. Rudin, A., Grinshpun, V., and O'Driscoll, K., *J. Liq. Chromatog.*, 1984, **7**, 1809.
241. Grinshpun, V., Rudin, A., Russell, K.E., and Scammell, M.V., *J. Polym. Sci., Part B: Polym. Phys.*, 1986, **24**, 1171.
242. Pang, S., and Rudin, A., *Polym. Mat. Sci. Eng.*, 1991, **65**, 95.
243. Bugada, D.C., and Rudin, A., *Eur. Polym. J.*, 1987, **23**, 847.
244. Martin, J., *J. Appl. Polym. Sci.*, 1990, **40**, 1801.
245. Schulz, G.V., Henrici, G., and Olive, S., *J. Polym. Sci.*, 1955, **17**, 45.
246. Fox, T.G., and Gratch, S., *Ann. New York Acad. Sci.*, 1953, **57**, 367.
247. Lovell, P.A., Shah, T.H., and Heatley, F., *Polym. Commun.*, 1991, **32**, 98.
248. Dunn, A.S., and Naravane, S.R., *Br. Polym. J.*, 1980, 75.
249. Bugada, D.C., and Rudin, A., *Polymer*, 1984, **25**, 1759.
250. Bugada, D.C., and Rudin, A., *J. Appl. Polym. Sci.*, 1985, **30**, 4137.
251. McDowell, W.H., and Kenyon, W.O., *J. Am. Chem. Soc.*, 1940, **62**, 415.
252. Nozakura, S.-I., Morishima, Y., and Murahashi, S., *J. Polym. Sci., Part A-1*, 1972, **10**, 2781.
253. Adelman, R.L., and Ferguson, R.C., *J. Polym. Sci., Polym. Chem. Ed.*, 1975, **13**, 891.

254. Ovenall, D.W., *Macromolecules*, 1984, **17**, 1458.
255. Agarwal, S.H., Jenkins, R.F., and Porter, R.S., *J. Appl. Polym. Sci.*, 1982, **27**, 113.
256. Vercauteren, F.F., and Donners, W.A.B., *Polymer*, 1986, **27**, 993.
257. Hatada, K., Terawaki, Y., Kitayama, T., Kamachi, M., and Tamaki, M., *Polym. Bull. (Berlin)*, 1981, **4**, 451.
258. Kamachi, M., Liaw, D.J., and Nozakura, S.-I., *Polym. J. (Tokyo)*, 1979, **11**, 921.
259. Ovenall, D.W., and Uschold, R.E., *Macromolecules*, 1991, **24**, 3235.
260. Bamford, C.H., in 'Comprehensive Polymer Science' (Eds. Agarwal, S.L., and Russo, S.), Vol. Suppl. 1, p. 1 (Pergamon: London 1992).
261. Goldfinger, G., Yee, W., and Gilbert, R.D., in 'Encyclopaedia of Polymer Science and Technology' (Eds. Mark, H.F., Gaylord, N.M., and Bikales, N.M.), Vol. 7, p. 644 (Wiley: New York 1967).
262. Bovey, F.A., and Kolthoff, I.M., *Chem. Rev.*, 1948, **42**, 491.
263. Bowry, V.W., and Ingold, K.U., *J. Am. Chem. Soc.*, 1992, **114**, 4992.
264. Beckwith, A.L.J., Bowry, V.W., and Moad, G., *J. Org. Chem.*, 1988, **53**, 1632.
265. Beckwith, A.L.J., Bowry, V.W., and Ingold, K.U., *J. Am. Chem. Soc.*, 1992, **114**, 4983.
266. Maillard, B., Ingold, K.U., and Scaiano, J.C., *J. Am. Chem. Soc.*, 1983, **105**, 5095.
267. Citterio, A., Arnoldi, A., and Minisci, F., *J. Org. Chem.*, 1979, **44**, 2674.
268. Rozantsev, E.G., Gol'dfein, M.D., and Trubnikov, A.V., *Russ. Chem. Rev. (Engl. Transl.)*, 1986, **55**, 1070.
269. Chateauneuf, J., Lusztyk, J., and Ingold, K.U., *J. Org. Chem.*, 1988, **53**, 1629.
270. Rizzardo, E., and Solomon, D.H., *Polym. Bull. (Berlin)*, 1979, **1**, 529.
271. Bhanu, V.A., and Kishore, K., *Chem. Rev.*, 1991, **91**, 99.
272. Neta, P., Huie, R.E., and Ross, A.B., *J. Chem. Phys. Ref. Data*, 1990, **19**, 413.
273. Mignani, S., Janousek, Z., Merenyi, R., Viehe, H.G., Riga, J., and Verbist, J., *Tetrahedron Lett.*, 1984, **25**, 1571.
274. Tanaka, H., Teraoka, Y., Sato, T., and Ota, T., *Makromol. Chem.*, 1993, **194**, 2719.
275. Hageman, H.J., Oosterhoff, P., Overeem, T., Polman, R.J., and van der Werf, S., *Makromol. Chem.*, 1985, **186**, 2483.
276. Tanaka, H., Kameshima, T., Sasai, K., Sato, T., and Ota, T., *Makromol. Chem.*, 1991, **192**, 427.
277. Kharasch, M.S., Kawahara, F., and Nudenberg, W., *J. Org. Chem.*, 1954, **19**, 1977.
278. Bevington, J.C., Ghanem, N.A., and Melville, H.W., *Trans. Faraday Soc.*, 1955, **51**, 946.
279. Price, C.C., and Read, D.H., *J. Polym. Res.*, 1946, **1**, 44.
280. Yassin, A.A., and Rizk, N.A., *Eur. Polym. J.*, 1977, **13**, 441.
281. Golubev, V.B., Mun, G.A., and Zubov, V.P., *Russ. J. Phys. Chem. (Engl. Transl.)*, 1986, **60**, 347.

282. Chen, S., and Tsai, L., *Makromol. Chem.*, 1986, **187**, 653.
283. Chalfont, G.R., Hey, D.H., Liang, K.S.Y., and Perkins, M.J., *J. Chem. Soc. (B)*, 1971, 233.
284. Chalfont, G.R., Hey, D.H., Liang, K.S.Y., and Perkins, M.J., *J. Chem. Soc., Chem. Commun.*, 1967, 367.
285. Bamford, C.H., Jenkins, A.D., and Johnston, R., *Proc. R. Soc., London, Ser. A*, 1957, **239**, 214.

6
Copolymerization

6.1 Introduction

Copolymerizations are processes which lead to the formation of polymers containing two or more discrete types of monomer units. Two basic types of copolymers will be considered in this chapter:

(a) Statistical copolymers. These are formed when a mixture of two or more monomers is polymerized in a single process and where the arrangement of the monomers within the chains is dictated purely by kinetic factors (see 6.2).

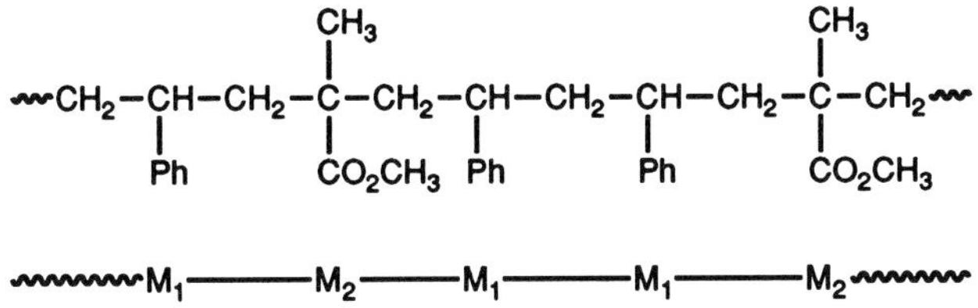

$$\text{\textasciitilde}CH_2-CH-CH_2-\underset{\underset{CO_2CH_3}{|}}{\overset{\overset{CH_3}{|}}{C}}-CH_2-CH-CH_2-CH-CH_2-\underset{\underset{CO_2CH_3}{|}}{\overset{\overset{CH_3}{|}}{C}}-CH_2\text{\textasciitilde}$$
$$\text{Ph} \qquad\qquad \text{Ph} \qquad \text{Ph}$$

$$\text{\textasciitilde}M_1\text{———}M_2\text{———}M_1\text{———}M_1\text{———}M_2\text{\textasciitilde}$$

statistical copolymer

(b) Block, segmented, or graft copolymers. These are typically prepared by multi-step processes aimed at achieving a required architecture (see 6.3).

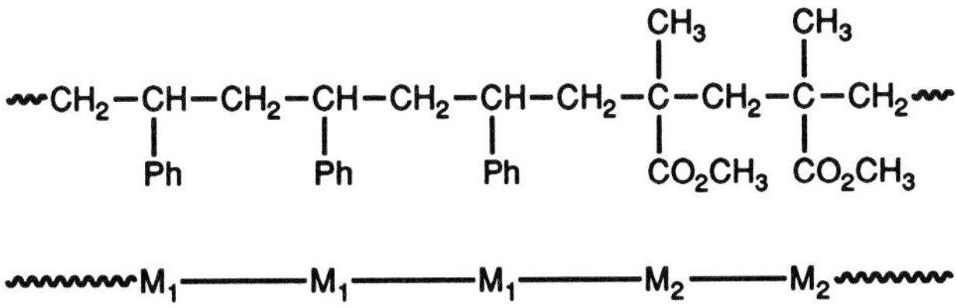

$$\text{\textasciitilde}CH_2-\underset{\underset{Ph}{|}}{CH}-CH_2-\underset{\underset{Ph}{|}}{CH}-CH_2-\underset{\underset{Ph}{|}}{CH}-CH_2-\underset{\underset{CO_2CH_3}{|}}{\overset{\overset{CH_3}{|}}{C}}-CH_2-\underset{\underset{CO_2CH_3}{|}}{\overset{\overset{CH_3}{|}}{C}}-CH_2\text{\textasciitilde}$$

$$\text{\textasciitilde}M_1\text{———}M_1\text{———}M_1\text{———}M_2\text{———}M_2\text{\textasciitilde}$$

block copolymer

$$\text{\textasciitilde}M_2-M_1-M_1-M_1-M_1-M_1-M_2-M_2-M_2-M_2-M_2-M_1-M_1-M_1\text{\textasciitilde}$$

segmented or multiblock copolymer

277

```
          M₂
          |
          M₂                    M₂—M₂    M₂
          |                         \   /
          M₂                         M₂
          |                          |
 ~M₁—M₁—M₁—M₁—M₁—M₁—M₁—M₁—M₁—M₁—M₁—M₁—M₁—M₁~
                  |
                  M₂
                  |
                  M₂
                  |
                  M₂
```

graft copolymer

6.2 Statistical Copolymerization

Statistical copolymers are formed when mixtures of two or more monomers are polymerized by a free radical process. Many reviews of the kinetics and mechanism of statistical copolymerization have appeared.[1-5] The term 'random copolymer', often used to describe these materials, is not appropriate since the incorporation of monomer units is seldom a random process.

The arrangement of monomer units in the chains is dictated by the inherent reactivities of the monomers and radicals involved which may, in turn, be influenced by the reaction conditions (solvent, temperature, *etc.*). These factors mean that it is only in special circumstances (see 6.2.1.1), when monomer reactivities are equal, that the copolymer composition will be a direct reflection of the ratio of monomers in the feed. In most copolymerizations, the monomers usually are consumed at different rates. Consequently both the monomer feed and copolymer composition will drift with conversion. Batch copolymers will generally not be homogeneous in composition.

The detailed microstructure and compositional heterogeneity of copolymers can have a determining influence on copolymer properties. This has been recognized for many years,[6] though the full implications are often not fully appreciated. When copolymers of specific functionality are required, it is generally not sufficient to know only the average number of functional groups/per polymer molecule.[7-9] It is often important to have the functionality distributed in a particular manner along the individual chains (monomer sequence distribution) and amongst the chains (chemical heterogeneity). The degree of heterogeneity can be controlled by skewing the monomer feed and/or, in some cases, by making an appropriate choice of the functional monomers.

Unfortunately, the heterogeneity of copolymers is not always readily measurable; most techniques only give the average composition. However, with knowledge of the polymerization mechanism and values of the kinetic parameters (reactivity ratios, rate constants, *etc.*), the rate and course of copolymerization and the composition and heterogeneity of the product may be predicted (see 7.4).[7,10]

An understanding of the kinetics of copolymerization and the structure of copolymers requires a knowledge of the dependence of the initiation, propagation and termination reactions on the chain composition, the nature of the monomers and radicals, and the polymerization medium. Aspects of the chemistry relating to the specificity of these reactions have already been discussed in the chapters on initiation, propagation and termination. The following sections discuss factors which affect the kinetics of copolymerization, and the way these influence the composition and monomer sequence distribution of the copolymer.

6.2.1 Copolymerization Mechanisms

The number of parameters required to describe a given copolymerization accurately is potentially very large. Studies on radical copolymerization and related model systems have demonstrated that many factors influence the rate and course of copolymerization. These include:

(a) The structure of the propagating species and the likelihood of significant remote unit effects.
(b) The possibility of complex formation between monomers, between monomer and solvent, *etc.*
(c) The kinetics and thermodynamics of copolymerization and the possibility that depropagation is competitive with propagation.
(d) The nature of the medium and the manner in which it changes during the course of the copolymerization.

The main features of the most popular models employed in studies of copolymerization and their implications as regards copolymer structure are discussed in the following sections. These sections are principally concerned with propagation and effects of monomer reactivity on composition and monomer sequence distribution. Termination in copolymerization and the influence of the termination mechanism on the kinetics of the process is discussed in section 5.2.1.2. The effects of initiation and termination processes on compositional heterogeneity are considered in section 7.4.

The various models of copolymerization (terminal, penultimate, complex dissociation, complex participation) should not be considered as alternative descriptions. They are approximations made through necessity to reduce complexity and should at best be considered as a subset of some overall scheme for copolymerization. Any unified theory (if such is possible) would have to take into account all of the factors mentioned above. The models used to describe copolymerization reaction mechanisms are chosen as the simplest possible model capable of explaining a set of experimental data. They do not necessarily provide a complete description of the mechanism.

6.2.1.1 Terminal model

The simplest model for describing binary copolymerization of two monomers, M_A and M_B, is the terminal model. The model has been applied to a vast number of systems and, in most cases, appears to give an adequate description of the overall copolymer composition; at least for low conversions. The limitations of the terminal model generally only become obvious when attempting to describe the monomer sequence distribution or the polymerization kinetics.

Even though the terminal model does not always provide an accurate description of the copolymerization process, it is very useful as a starting point for parameter estimation and is simple to apply. The terminal model (and the derivation of the Mayo-Lewis equation[11] for describing copolymer composition) involves a number of approximations:

(a) It is assumed that the copolymer composition is dictated by the relative rates of four propagation reactions (Scheme 1). It is implicit in the model that the nature of the last added monomer unit determines reactivity of the propagating radicals. Note that $P_A\bullet$ and $P_B\bullet$ are propagating species where the terminal (last added) monomer units are M_A and M_B respectively.

propagation:

$$P_A\bullet + M_A \xrightarrow{k_{pAA}} P_A\bullet$$
$$P_A\bullet + M_B \xrightarrow{k_{pAB}} P_B\bullet$$
$$P_B\bullet + M_A \xrightarrow{k_{pBA}} P_A\bullet$$
$$P_B\bullet + M_B \xrightarrow{k_{pBB}} P_B\bullet$$

Scheme 1

(b) It is assumed that chains are long and therefore the influence of the initiation and termination steps on the rate of monomer consumption can be neglected. The rates of monomer disappearance can then be written as:

$$\frac{d[M_A]}{dt} = k_{pAA}[P_A\bullet][M_A] + k_{pBA}[P_B\bullet][M_A]$$

$$\frac{d[M_B]}{dt} = k_{pAB}[P_A\bullet][M_B] + k_{pBB}[P_B\bullet][M_B]$$

The ratio of these equations is an expression for the instantaneous copolymer composition:

$$\frac{d[M_A]}{d[M_B]} = \frac{k_{pAA}[P_A\bullet][M_A] + k_{pBA}[P_B\bullet][M_A]}{k_{pAB}[P_A\bullet][M_B] + k_{pBB}[P_B\bullet][M_B]}$$

(c) A third assumption is that the concentrations of the two propagating species, $P_A\bullet$ and $P_B\bullet$, achieve a steady state such that:

$$k_{pAB}[P_A\bullet][M_B] = k_{pBA}[P_B\bullet][M_A]$$

This allows elimination of the radical concentrations from the above equation and the Mayo-Lewis equation can now be derived:

$$\frac{d[M_A]}{d[M_B]} = \frac{[M_A]}{[M_B]}\cdot\frac{r_A[M_A]+[M_B]}{[M_A]+r_B[M_B]}$$

where r_A and r_B are the reactivity ratios:

$$r_A = \frac{k_{pAA}}{k_{pAB}} \quad r_B = \frac{k_{pBB}}{k_{pBA}}$$

Other convenient forms of the Mayo-Lewis equation are:

$$y = \frac{F_A}{F_B} = \frac{1+r_A x}{1+r_B/x}$$

and

$$F_A = \frac{1+r_A x}{(1+r_B/x)(2+r_A x)}$$

where $x=[M_A]/[M_B]=f_A/f_B$ and F_A and f_A are the instantaneous mole fractions of monomer A in the polymer and monomer feeds respectively.

(d) It is also implicit in this treatment that medium effects are negligible and that there is no participation by monomer-monomer or monomer-solvent complexes.

Thus the terminal model allows the copolymer composition for a given monomer feed to be predicted on the basis of just two parameters; the reactivity ratios r_A and r_B. Some values of terminal model reactivity ratios for some common monomer pairs are given in Table 6.1. No critical assessment has been made of this data and inclusion in this table is not meant to imply that the system is adequately described by the terminal model. Values for other monomers can be found in data compilations.[12,13]

Table 6.1 Reactivity Ratios for Some Common Monomer Pairs[a]

Monomer B	Monomer A						
	S	MMA	MA	AN	VC	MAH	VAc
S	\	0.51	0.77	0.40	17.3	0.002	22.3
MMA	0.49	\	-	2.03	-	5.195	27.3
MA	0.12	-	\	1.02	-	2.798	9.0
AN	0.05	0.25	0.80	\	3.33	6.00	5.01
VC	0.04	-	-	0.057	\	0.296	1.36
MAH	0.021	0.018	0.011	0.00	0.008	\	0.003
VAc	0.02	0.03		0.02	0.73	0.055	\

[a] r_A tabulated vertically, r_B horizontally. Values taken from Laurier et al.[14] or from Greenley's compilations.[12,13]

It is informative to consider some of the implications of the terminal model and, in particular, how the relative magnitudes of the reactivity ratios affect the copolymer composition.

(a) For the special case where $r_A = r_B = 1.0$, the monomers have equal reactivity in propagation. Therefore, they will be utilized according to their respective proportions in the monomer feed. The product is a random copolymer. The value of F_A always equals f_A irrespective of the starting f_A. Copolymerizations of structurally similar monomers come closest to achieving this ideal. Examples are, copolymerizations of isotopically labeled monomers or mixtures of acrylic esters (with non-bulky ester groups) e.g. MMA and BMA).

(b) Where $r_A > 1$ and $r_B < 1$ (or $r_A < 1$ and $r_B > 1$), the copolymer will always be richer in one monomer than it is in the other. The special case where $r_A.r_B = 1$ has been called 'ideal'. These copolymerizations have no azeotropic composition (see Figure 1).

(c) For many copolymerizations (e.g. S-MMA, S-AN) $r_A < 1$ and $r_B < 1$. In these cases, because cross-propagation is favored over homopropagation, there is a tendency towards alternation. In the extreme, where the values of both r_A and r_B approach zero (e.g. S-MAH), cross propagation occurs to the virtual exclusion of homopropagation and the product is an alternating copolymer.

(d) The converse situation, where both r_A and r_B are greater than one, is very rarely encountered. In this case, homopropagation is always favored over cross-propagation and, as a consequence, there will be a degree of blockiness in the copolymer.

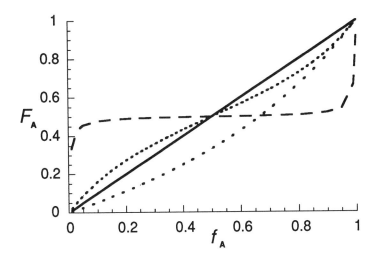

Figure 1 Plot of instantaneous copolymer composition (F_A) vs. monomer feed composition (f_A) for the situation where (a) $r_A=r_B=1.0$ (——), (b) $r_A=r_B=0.5$ (········), (c) $r_A=r_B=0.01$ (- - - - -), (d) $r_A=0.5, r_B=2.0$ (· · · · ·).

In cases where $r_A>1$ and $r_B>1$ or $r_A<1$ and $r_B<1$ there will always be exactly one 'azeotropic composition' or 'critical point' where the copolymer composition will exactly reflect the monomer feed composition (see Figure 1).

$$i.e. \quad \frac{d[M_A]}{d[M_B]} = \frac{[M_A]}{[M_B]} = x \quad \text{or} \quad F_A = f_A$$

Substitution into Mayo-Lewis equation shows that this condition is satisfied when:

$$x = \frac{1-r_A}{1-r_B}$$

The existence of an azeotropic composition has practical significance. By conducting a polymerization with the monomer feed ratio equal to the azeotropic composition, a high conversion batch copolymer can be prepared that has no compositional heterogeneity caused by drift in copolymer composition with conversion. Thus, the complex incremental addition protocols that are otherwise required to achieve this end, are unnecessary.

Conditions for azeotropic compositions in ternary and quaternary copolymerizations have also been defined.[15,16]

6.2.1.2 Penultimate model

It is well-established that β- and more remote substituents can have a significant influence on radical conformation and reactivity. On this basis, it is expected that the nature of the penultimate unit of the propagating chain should modify its reactivity towards monomer and other species. The magnitude of the effect will be dependent on the exact nature of the β-substituent.

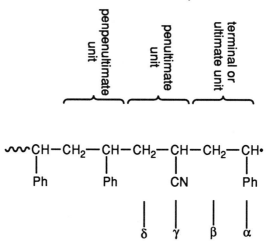

Recent studies on model propagating radicals support this view.[17-23] Of particular interest is the data of Tirrell et al.[19,20] They investigated the relative reactivity of S and AN towards various γ-substituted propyl radicals (see Scheme 2 and Table 6.2). They found that:

(a) There is minimal effect on radical reactivity when the γ-substituent is a styryl unit (at a PSAN chain end), a phenyl, or an alkyl group.
(b) A γ-cyano substituent has a marked effect on radical reactivity.
(c) The relative reactivities of simple model radicals correlate well with the reactivities of propagating species estimated from copolymerization data assuming a penultimate model.

Scheme 2

Table 6.2 Relative Rates for Addition of Substituted Propyl Radicals to AN and S (refer Scheme 2)

R¹	R²	R³	$\dfrac{k_{AN}}{k_S}$
H	H	CH$_3$	24.5±1.1[19]
H	H	C$_3$H$_7$	26.2±2.4[19]
H	H	Ph	22.6±2.0[19]
Ph	H	Ph	4.8±0.3[20]
Ph	PSAN[a]	Ph	4.2[22]
H	H	CN	6.8±0.6[19]
Ph	H	CN	1.9±0.1[20]
Ph	PSAN	CN	1.7[22]

a   Poly(styrene-co-acrylonitrile) chain. Data from reactivity ratios of AN and S.

There is also clear evidence that penultimate group effects are important in determining the stereochemistry of addition in many homopolymerizations. This is made apparent by the fact that most homopolymers have tacticity (i.e. $P(m) \neq 0.5$, see 4.2). Indeed, for some homopolymerizations there is evidence that the configuration of the penpenultimate unit may also influence the stereochemistry of addition. If penultimate units can influence the stereospecificity of addition, it is also reasonable to expect that they could affect the rate and chemospecificity of addition.

Thus, it would seem unusual if reactivity of the propagating species in copolymerization were not sensitive to the nature of the last added monomer units. However, while there is ample experimental data to suggest that copolymerizations should be subject to penultimate unit effects which might affect the rate and/or composition of copolymers, the origin of the effect is not always clear. Some theoretical studies have been performed. Fukuda et al.[24] have suggested that penultimate units affect principally the stability of the propagating radical. Work by Heuts et al.,[25] while not disputing an effect on the activation energy, indicates that penultimate unit effects of the magnitude seen in the S-AN and other systems (i.e. 2-3 fold) could also be explained by variations in the entropy of activation for the process.

The influence of penultimate units on the kinetics of copolymerization was first considered in a formal way by Merz et al.[26] and Ham.[27] In their treatments, eight propagation reactions are considered (Scheme 3). The copolymer composition is then determined by the values of four reactivity ratios.

$$r_{AA} = \frac{k_{pAAA}}{k_{pAAB}}, \quad r_{BA} = \frac{k_{pBAA}}{k_{pBAB}}, \quad r_{AB} = \frac{k_{pABB}}{k_{pABA}}, \quad r_{BB} = \frac{k_{pBBB}}{k_{pBBA}}.$$

A further two reactivity ratios are required to define the polymerization kinetics:[28]

$$s_A = \frac{k_{pAAA}}{k_{pBAA}} \text{ and } s_B = \frac{k_{pBBB}}{k_{pABB}}$$

$$P_{AA}\bullet + M_A \xrightarrow{k_{pAAA}} P_{AA}\bullet$$
$$P_{AA}\bullet + M_B \xrightarrow{k_{pAAB}} P_{AB}\bullet$$
$$P_{AB}\bullet + M_A \xrightarrow{k_{pABA}} P_{BA}\bullet$$
$$P_{AB}\bullet + M_B \xrightarrow{k_{pABB}} P_{BB}\bullet$$
$$P_{BA}\bullet + M_A \xrightarrow{k_{pBAA}} P_{AA}\bullet$$
$$P_{BA}\bullet + M_B \xrightarrow{k_{pBAB}} P_{AB}\bullet$$
$$P_{BB}\bullet + M_A \xrightarrow{k_{pBBA}} P_{BA}\bullet$$
$$P_{BB}\bullet + M_B \xrightarrow{k_{pBBB}} P_{BB}\bullet$$

Scheme 3

It is implicit in traditional treatments of copolymerization kinetics that the values of the ratios s_A and s_B are equal to unity (see 5.2.1.2). Since they contain no terms from cross propagation these parameters have no direct influence on either the overall copolymer composition or the monomer sequence distribution; they only influence the rate of polymerization.

The expression for the instantaneous copolymer composition in systems subject to penultimate unit effects is:

$$\frac{F_A}{F_B} = \frac{1+\dfrac{1+r_{AA}x}{1+r_{BA}/x}}{1+\dfrac{1+r_{BB}/x}{1+x/r_{AB}}}$$

With the penultimate model there may be between zero and three azeotropic compositions, depending on the values of the reactivity ratios.[29]

Cases have been reported where the application of the penultimate model provides a significantly better fit to experimental composition data. These include many copolymerizations of AN[22,30,31] B,[32] MAH[33,34] and VC.[35] In these cases there is no doubt that the penultimate model (or some scheme other than the terminal model) applies. However, the vast majority of copolymerizations have been analyzed only in terms of the terminal model; the possible significance of remote unit effects on copolymer composition and polymerization kinetics has not been considered.

For many copolymerizations, statistical tests may show that the penultimate model does not provide a significantly better fit to experimental data than the terminal model. This, however, should not be construed as evidence that penultimate unit effects are unimportant.[36] For this, it is necessary to test for model discrimination, not simply for fit to a given model.

Recently the MMA-S system has been reinvestigated by several groups.[28,36] Fukuda et al.[28] established that the overall composition of MMA-S copolymers can be satisfactorily accounted for by the terminal model ($r_{AA}=r_{BA}=0.52$, $r_{BB}=r_{AB}=0.46$). They also determined absolute values of the overall propagation and termination rate constants. The data showed that the dependence of the rate of copolymerization on monomer feed composition was due to a dependence of the overall propagation rate constant on the monomer feed ratio. This had been previously rationalized in terms of an effect of chain composition on the kinetics of termination. Fukuda et al.[28] proposed that this could be accommodated by invoking penultimate group effects and values of the parameters s_A (=0.30) and s_B (=0.53) were derived.

If this treatment is accepted at face value, it means that the nature of the penultimate monomer has a strong influence on the relative values of the rate constants for homopropagation but no effect whatsoever on those for cross propagation. It was further proposed that this may be a general phenomenon, notwithstanding the work for the S-AN and other systems (vide infra). The suggestion that the parameters s_A and s_B are not equal to unit may have generality. It is also reasonable that the effect of the penultimate unit on absolute reactivity of a radical might in many cases be larger than any effect on specificity. However, it seems unlikely as a rule that there should be a large affect on reactivity but no effect on specificity.

In this context, it is important to remember that composition data are of very low power when it comes to model discrimination. For MMA-S copolymerization, even though experimental precision is high, the penultimate model confidence intervals are quite large; $0.4 \leq r_{AA}/r_{BA} \leq 2.7$, $0.3 \leq r_{BB}/r_{AB} \leq 2.2$.[36] Thus, the terminal model ($r_{AA}=r_{BA}$, $r_{BB}=r_{AB}$) is only one of a number of possible solutions and the experimental composition data do not rule out the possibility of quite substantial penultimate unit effects.

Triad information is more powerful but typically is subject to more experimental error and signal assignments are often ambiguous (see 6.2.2.1). Triad data for the MMA-S system are consistent with the terminal model and support the view that penultimate unit effects on specificity are small.[37-39]

Penpenultimate and higher order remote unit effect models may also affect the outcome of copolymerizations. However, in most cases, experimental data which typically are not sufficiently powerful to test the penultimate model, offer little hope of testing higher order models. The importance of remote unit effects on copolymerization will only be fully resolved when more powerful experimental data become available.

6.2.1.3 Models involving monomer complexes

Mechanisms for copolymerization involving complexes between the monomers were first proposed to explain the high degree of alternation observed in some copolymerizations. They have also been put forward, as alternatives to the penultimate model, to explain anomalous (not consistent with the terminal model) composition data in certain copolymerizations.[5,40-47]

While there is clear evidence for complex formation between certain electron donor and electron acceptor monomers, the evidence for participation of such complexes in copolymerization is often less compelling. One of the most studied systems is S-MAH copolymerization.[33,48] However, the models have been applied to many copolymerizations of donor-acceptor pairs. A partial list of donor and acceptor monomers follows[5]:

Donors:
    dienes (*e.g.* B, isoprene)
    heterocyclic dienes (*e.g.* furan, indole, thiophene)
    vinylbenzene derivatives (*e.g.* S, AMS)
    vinyl esters (*e.g.* VAc)
    vinyl ethers (*e.g.* ethyl vinyl ether)
    vinyl sulfides

Acceptors:
    acrylate esters (*e.g.* MA)
    cinnamate esters
    cyanoethylenes (*e.g.* AN, 1,1-dicyanoethylene)
    fumarate esters
    MAH
    maleimides (*e.g.* N-phenylmaleimide)

Common features of polymerizations involving such monomer pairs are:

(a) A high degree of monomer alternation in the chain is observed.
(b) The copolymer composition cannot be rationalized on the basis of the terminal model (see 6.2.1.1).
(c) The rate of copolymerization is usually very much faster than that of either homopolymerization.
(d) Many of the monomers do not readily undergo homopolymerization or copolymerization with monomers of like polarity.
(e) For most systems there is spectroscopic evidence for some form of donor-acceptor interaction.

However, these observations are not proof of the role of a donor-acceptor complex in the copolymerization mechanism. Even with the availability of sequence information it is often not possible to discriminate between this and the penultimate model (see 6.2.1.2) or other higher order models.[33] A further problem in analyzing the kinetics of these copolymerizations is that many donor-acceptor systems give spontaneous initiation (see 3.3.6).

Equilibrium constants for complex formation (K) have been measured for many donor-acceptor pairs. Donor-acceptor interaction can lead to formation of highly colored charge-transfer complexes and the appearance of new absorption bands in the UV-visible spectrum are observed. More often spectroscopic evidence for complex formation takes the form of small chemical shift differences in NMR spectra or shifts in the positions of the UV absorption maxima. In analyzing these systems it is important to take into account that some solvents might also interact with donor or acceptor monomers.

Since intermediates usually can not be observed directly, the exact nature of the donor-acceptor complex and the mechanisms for their interaction with radicals are speculative. At least three ways may be envisaged whereby complex formation may affect the course of polymerization:

(a) The complex participation model.[48-50] A binary complex is formed that is much more reactive than either of the non-complexed monomers. The monomers are incorporated into the chain in pairs (Scheme 4). If reaction with the complexed monomer competes with addition to uncomplexed monomer, the mechanism may be described in terms of six reactivity ratios and one equilibrium constant.

$$M_A + M_B \underset{}{\overset{K}{\rightleftharpoons}} M_A M_B$$

$$P_A\cdot + M_A \xrightarrow{k_{pAA}} P_A\cdot$$
$$P_A\cdot + M_B \xrightarrow{k_{pAB}} P_B\cdot$$
$$P_B\cdot + M_A \xrightarrow{k_{pBA}} P_A\cdot$$
$$P_B\cdot + M_B \xrightarrow{k_{pBB}} P_B\cdot$$

$$P_A\cdot + M_B M_A \xrightarrow{k_{pABA}} P_A\cdot$$
$$P_A\cdot + M_A M_B \xrightarrow{k_{pAAB}} P_B\cdot$$
$$P_B\cdot + M_B M_A \xrightarrow{k_{pBBA}} P_A\cdot$$
$$P_B\cdot + M_A M_B \xrightarrow{k_{pBAB}} P_B\cdot$$

Scheme 4

(b) The complex dissociation model.[51-53] A binary complex is formed that is much more reactive than either of the non-complexed monomers. The complex dissociates after addition and only a single monomer unit is incorporated on reaction with the complex (Scheme 5).

$$M_A + M_B \xrightleftharpoons{K} M_A M_B$$

$$P_A\cdot + M_A \xrightarrow{k_{pAA}} P_A\cdot$$
$$P_A\cdot + M_B \xrightarrow{k_{pAB}} P_B\cdot$$
$$P_B\cdot + M_A \xrightarrow{k_{pBA}} P_A\cdot$$
$$P_B\cdot + M_B \xrightarrow{k_{pBB}} P_B\cdot$$

$$P_A\cdot + M_B M_A \xrightarrow{k_{pABA}} P_B\cdot + M_A$$
$$P_A\cdot + M_A M_B \xrightarrow{k_{pAAB}} P_A\cdot + M_B$$
$$P_B\cdot + M_B M_A \xrightarrow{k_{pBBA}} P_B\cdot + M_A$$
$$P_B\cdot + M_A M_B \xrightarrow{k_{pBAB}} P_A\cdot + M_B$$

Scheme 5

(c) Formation of a less reactive complex. This could have the effect of reducing the overall monomer concentration and perhaps altering the ratio of reactive monomers in the feed. However, the fraction of monomer complexed is typically small.

Several studies on the reactivities of small radicals with donor-acceptor monomer pairs have been carried out to provide insight into the mechanism of copolymerizations of donor-acceptor pairs. Tirrell et al.[54-56] reported on the reaction of n-butyl radicals with mixtures of N-phenylmaleimide and various donor monomers (e.g. S, 2-chloroethyl vinyl ether). Jenkins et al.[57,58] have examined the reaction of t-butoxy radicals with mixtures of AN and VAc. Both groups have examined the S-AN system (see also 6.2.1.2). In each of these donor-acceptor systems only simple (one monomer) adducts are observed. Incorporation of monomers as pairs is not an important pathway (i.e. the complex participation model is not applicable). Furthermore, the product mixtures can be predicted on the basis of what is observed in single monomer experiments. The reactivity of the individual monomers (towards initiating radicals) is unaffected by the presence of the other monomer (i.e. the complex dissociation model is not applicable). Unless propagating species are shown to behave differently, these results suggest that neither the complex participation nor complex dissociation models apply to these systems.

6.2.1.4 Copolymerization with depropagation

Propagation reactions in free radical polymerization and copolymerization are generally highly exothermic and can be assumed to be irreversible. Exceptions to this general rule are those involving monomers with low ceiling temperatures (see 4.5.1). The thermodynamics of copolymerization has been reviewed by Sawada.[59]

Some of the most important systems known to involve reversible propagation steps are:

(a) Copolymerizations of AMS. Studies on copolymerizations of AMS with AN,[60,61] MMA[61,62] and S[60,63] have been reported.
(b) Copolymerizations with sulfur dioxide[59] and carbon monoxide.

Copolymerizations of other monomers may also be subject to similar effects given sufficiently high reaction temperatures (at or near their ceiling temperatures). For example, O'Driscoll and Gasparro[60] have reported on the copolymerization of MMA with S at 250°C.

These systems cannot be analyzed in terms of the terminal model. The analysis requires, in addition to reactivity ratios, equilibrium constants for any reversible propagation steps. Theory covering copolymerization with depropagation involving only one monomer was developed by Lowry in 1960.[64] Howell *et al.*[65] have carried out a more general treatment, allowing for all propagation steps being reversible, and provided expressions predicting sequence distribution for these systems.

6.2.1.5 *Chain statistics*

The arrangement of monomer units in copolymer chains is determined by the monomer reactivity ratios and perhaps by the reaction medium. The average sequence distribution to the triad level can often be measured by NMR (see 6.2.2.2) and in special cases by other techniques.[66,67] Longer sequences are difficult to determine experimentally, however, they can be predicted.[2,68] Where sequence distributions can be accurately determined they provide, in principle, a powerful method for determining monomer reactivity ratios (see 6.2.2.2).

If chains are long and initiation and termination reactions have a negligible effect on the overall sequence distribution, then according to the terminal model, P_{AA}, the probability that a chain ending in monomer unit M_A adds another unit M_A, is given by the expression:

$$P_{AA} = \frac{k_{pAA}[P_A\bullet][M_A]}{k_{pAA}[P_A\bullet][M_A]+k_{pAB}[P_A\bullet][M_B]} = \frac{r_A}{r_A + [M]_B/[M_A]}$$

while P_{AB}, the probability that a chain ending in monomer unit M_A adds another unit M_B, is then simply $1-P_{AA}$. P_{BB} and P_{BA} can be similarly defined.

The probability of a given sequence is the product of the probabilities of the individual steps that give rise to that sequence. Thus, the fraction of isolated sequences of monomer M_A which are of length n is:

$$N_n^A = P_{AA}^{n-1} P_{AB}$$

Expressions for the dyad, triad and higher order n-ad fractions can also be derived in terms of these probabilities. Thus the dyad fractions are:

$$(AA) = \frac{P_{BA}(1-P_{AB})}{(P_{AB}+P_{BA})} \qquad (AB) = \frac{2 P_{BA} P_{AB}}{(P_{AB}+P_{BA})} \qquad (BB) = \frac{P_{AB}(1-P_{BA})}{(P_{AB}+P_{BA})}$$

and the six triad fractions are:

$$(AAA) = \frac{P_{BA}(1-P_{AB})^2}{(P_{AB}+P_{BA})} \qquad (AAB) = \frac{2 P_{BA} P_{AB}(1-P_{AB})}{(P_{AB}+P_{BA})} \qquad (BAB) = \frac{P_{AB}^2 P_{BA}}{(P_{AB}+P_{BA})}$$

$$(BBB) = \frac{P_{AB}(1-P_{BA})^2}{(P_{BA}+P_{AB})} \qquad (BBA) = \frac{2 P_{AB} P_{BA}(1-P_{BA})}{(P_{BA}+P_{AB})} \qquad (ABA) = \frac{P_{BA}^2 P_{AB}}{(P_{BA}+P_{AB})}$$

With the penultimate model, the probability that a chain with a terminal M_{BA} dyad will add a M_A unit is:

$$P_{BAA} = \frac{k_{pBAA}[P_{BA}\bullet][M_A]}{k_{pBAA}[P_{BA}\bullet][M_A] + k_{pBAB}[P_{BA}\bullet][M_B]} = \frac{[M_A]}{[M_A] + [M]_B/r_{BA}}$$

and the probability that a chain with a terminal M_{AA} dyad will add a M_A unit is:

$$P_{AAA} = \frac{k_{pAAA}[P_{AA}\bullet][M_A]}{k_{pAAA}[P_{AA}\bullet][M_A] + k_{pAAB}[P_{AA}\bullet][M_B]} = \frac{[M_A]}{[M_A] + [M]_B/r_{AA}}$$

Expressions for predicting monomer sequence distribution for monomer complex and other models have also been proposed. There are at least two additional complications that need to be considered when attempting to predict sequence distribution or measure reactivity on the basis of sequence data:

(a) The effects of chain tacticity. Chain ends of differing tacticity may have different reactivity towards monomers.[67] When tacticity is imposed on top of monomer sequence distribution there are then 6 different dyads and twenty different triads to consider; analytical problems are thus severe. The tacticity of copolymers is usually described in terms of the coisotacticity parameters σ_{AB} and σ_{BA};[69] σ_{AB} is the probability of generating a meso dyad when a chain ending in A adds monomer B. Coisotacticity parameters have to date been reported for only a few copolymers including MMA-S,[70] MMA-MA,[71] and MMA-MAA.[72,73] These data are subject to change due to the complexities associated with data analysis and NMR signal assignment (see also 6.2.2.2). Copolymers involving only monosubstituted monomers are often assumed to have random tacticity (i.e. $\sigma_{AB} = \sigma_{BA} = 0.5$).
(b) The effects of the reaction medium. Harwood[74,75] observed that copolymers of the same composition have the same monomer sequence distribution irrespective of the solvent used for the copolymerization and apparently

different reactivity ratios. He termed this the 'bootstrap effect'. This applies even though the monomer reactivity ratios estimate on the basis of composition data may be significantly different. Solvent effects on copolymerization and the 'bootstrap effect' are considered in more detail in section 7.3.1.2.

The full picture of the factors affecting copolymer sequence distribution and their relative importance still needs to be filled in.

6.2.2 Estimation of Reactivity Ratios

Methods for evaluation of reactivity ratios comprise a significant proportion of the literature on copolymerization. There are two basic types of information that can be analyzed to yield reactivity ratios. These are (a) copolymer composition/conversion data (see 6.2.2.1) and (b) the monomer sequence distribution (see 6.2.2.2). The methods used to analyze these data are summarized in the following sections.

6.2.2.1 Composition data

The traditional method for determining reactivity ratios involves determinations of the overall copolymer composition for a range of monomer feeds at 'zero conversion'. Various methods have been applied to analyze this data.

Early methods (*e.g.* Intersection,[11] Fineman-Ross[76]) do not give equal weighting to the experimental points and do not allow for a non-linear dependence of the error on the composition. Consequently, they can give erroneous results. These problems were addressed by Tidwell and Mortimer[77,78] who advocated numerical analysis by non-linear least squares and Kelen and Tüdos[79,80] who proposed an improved graphical method for data analysis.

The Kelen-Tüdos equation is as follows:

$$\eta = \left(r_A + \frac{r_B}{\alpha}\right)\xi - \frac{r_A}{\alpha}$$

where $\eta = \dfrac{G}{\alpha+F}$, $\xi = \dfrac{F}{\alpha+F}$, $G = \dfrac{x}{y-1}$, $F = \dfrac{x^2}{y}$ and α is a constant.

A plot of η vs. ξ should yield a straight line with intercepts of $-r_B/\alpha$ and r_A at $\xi=0$ and $\xi=1$ respectively. A value of α of $(F_{max}.F_{min})^{1/2}$ results in a symmetrical distribution of experimental data on the plot. Greenley[12] has re-evaluated pre-1980 data using the Kelen-Tüdos method and has provided a compilation of his results.

It is also possible to derive reactivity ratios by analyzing the monomer (or polymer) feed composition vs. conversion and solving the integrated form of the Mayo Lewis equation. The following expression has been proposed:[6,81]

$$\text{conversion} = 1 - \left(\frac{f_A}{f_{A0}}\right)^\alpha \left(\frac{f_B}{f_{B0}}\right)^\beta \left(\frac{f_{A0}-\delta}{f_A-\delta}\right)^\gamma$$

where $\alpha = \dfrac{r_B}{1-r_B}$ $\quad \beta = \dfrac{r_A}{1-r_A} \quad \delta = \dfrac{1-r_A r_B}{(1-r_A)(1-r_B)} \quad \gamma = \dfrac{1-r_B}{2-r_A-r_B}$

Francis et al.[82] have described a purely numerical approach for estimating reactivity ratios by solution of the integrated rate equation.

Potential difficulties associated with the application of these methods based on the integrated form of the Mayo-Lewis equation have been discussed.[82,83] One is that the expressions become undefined under certain conditions, for example, when r_A or r_B is close to unity or when the composition is close to the azeotropic composition. The problem of dependence of experimental error on composition also applies to this method. O'Driscoll et al.[84,85] and van der Meer et al.[86] have recommended the use of "error in variable" methods which take into account the error structure of the experimental data. A further complication is that medium effects may lead to reactivity ratios being conversion dependent.

Clearly great care must be taken in the estimation of reactivity ratios from composition/conversion data. Many papers have been written on the merits of various schemes and comparisons of the various methods for reactivity ratio calculation have appeared.[87-90] Given appropriate design of the experiment graphical methods for the estimation of reactivity ratios can give reasonable values. In addition, they have the virtue of simplicity and do not require the aid of a computer. However, as a general rule, the use of such methods is not recommended except as an initial guide. It is more appropriate to use some form of non-linear least squares regression analysis to derive the reactivity ratios.

It is also possible to process copolymer composition data to obtain reactivity ratios for higher order models (e.g. penultimate model or complex participation, etc.). However, composition data have low power in model discrimination (see 6.2.1.2 and 6.2.1.3). There has been much published on the subject of the design of experiments for reactivity ratio determination and model discrimination.[36,78,91,92] Attention must be paid to the information that is required, the optimal design for obtaining terminal model reactivity ratios may not be ideal for model discrimination.[36]

One final point should be made. The observation of significant solvent effects on k_p in homopolymerization and on reactivity ratios in copolymerization (see 7.3.1) calls into question the methods for reactivity ratio measurement which rely on evaluation of the polymer composition for various monomer feed ratios (see 6.2.2). If solvent effects are significant, it would seem to follow that reactivity ratios in bulk copolymerization should be a function of the feed composition.[93] Also, since the reaction medium alters with conversion, the reactivity ratios are expected to vary with conversion. This is an area in need of further research.

6.2.2.2 Monomer sequence distribution

NMR spectroscopy has made possible the characterization of copolymers in terms of their monomer sequence distribution (see 6.2.1.5). The area has been reviewed by Randall,[66] Bovey,[94] Tonelli[67] and others. Information on monomer sequence distribution is substantially more powerful than simple composition data with respect to model discrimination.[22,36] Although many authors have used the distribution of triad fractions to confirm the adequacy or otherwise of various models, only a few[22,39,95] have used triad fractions to calculate reactivity ratios directly.

While sequence distributions are usually subject to more experimental noise than composition data, this is often outweighed by a greater information content. In principle, the reactivity ratios can be calculated given a single triad concentration and feed composition. A more serious problem is that unambiguous assignments of NMR signals to monomer sequences are, as yet, only available for a few systems. Assignments are complicated by the fact that the sensitivity of chemical shifts to tacticity may be equal or greater than their sensitivity to monomer sequence.[96]

The usual experiment is to prepare a series of copolymers each containing a different ratio of the monomers. A correlation of expected and measured peak intensities may then enable peak assignment.[22,30] However, this method is not foolproof and papers on signal reassignment are not uncommon.[37,70,96] 2D NMR methods,[96] decoupling experiments,[37] special pulse sequences[33] and analyses of isotopically labeled[23,97] or regioregular[37] polymers have greatly facilitated analysis of complex systems. In principle, these methods allow a "mechanism-free" assignment.

6.2.3 Prediction of Reactivity Ratios

Various schemes for the prediction of reactivity ratios have been proposed.[98] These schemes are largely empirical although some have offered a theoretical basis for their function. They typically do not allow for the possibility of variation in reactivity ratios with solvent and reaction conditions. They also presuppose a terminal model. The most popular methods are the Q-e (see 6.2.3.1) and 'patterns of reactivity' schemes (6.2.3.2).

6.2.3.1 Q-e scheme

The scheme in most widespread usage for the prediction of reactivity ratios is the Q-e scheme.[14,99] This scheme was devised in 1947 by Alfrey and Price[100] who proposed that the rate constant for reaction of radical (R) with monomer (M) should be dependent on polarity and resonance terms according to the following expression

$$k_{RM} = P_R Q_M e^{-e_R e_M}$$

where P_R and Q_M are the 'general reactivity' of the radical and monomer respectively. It has been proposed that these take into account resonance factors. The e values are related to the polarity of the radical or monomer (e_R and e_M are assumed to be the same). An expression for the reactivity ratios r_A and r_B can then be derived in terms of Q and e.

$$r_A = (Q_A/Q_B)e^{-e_A(e_A-e_B)}$$

$$r_B = (Q_B/Q_A)e^{-e_B(e_B-e_A)}$$

S is taken as the reference monomer with $Q=1.0$ and $e = -0.8$. Values for other monomers are derived by regression analysis based on literature or measured reactivity ratios. The Q-e values for some common monomers are presented in Table 6.3.

The accuracy of Q-e parameters[101-104] is limited by the quality of the reactivity ratio data and can also suffer from inappropriate statistical treatment employed in their derivation.[14] A further problem is that the data analysis makes no allowance for the dependence of reactivity ratios on reaction conditions. Reactivity ratios can be dependent on solvent (see 7.3.1.2), reaction temperature, pH, *etc.* It follows that values of e and perhaps Q for a given monomer should depend on the medium, the monomer ratio and the particular comonomer. This is especially true for monomers which contain ionizable groups (*e.g.* MAA, AA, vinyl pyridine) or are capable of forming hydrogen bonds (*e.g.* HEMA, HEA).

Table 6.3 Q-e values for Some Common Monomers[14]

Monomer	Q	e
S	1.0	-0.8
MAN	0.85	0.69
AN	0.68	1.33
MMA	0.76	0.38
MA	0.38	0.75
BA	0.41	1.06
VC	0.033	-0.1
VAc	0.024	-0.02

The Q-e scheme is wholly empirical. There have, however, been attempts to correlate Q-e values and hence reactivity ratios to, for example, ^{13}C NMR chemical shifts[105] or the results of MO calculations[106,107] and to provide some theoretical basis for the parameters. The NMR method is attractive since spectra can be measured under the particular reaction conditions (solvent, temperature, pH). Thus it may be possible to predict the dependence of the Q-e values and reactivity ratios on the reaction medium.[105]

6.2.3.2 Patterns of reactivity scheme

Bamford et al.,[108-110] realizing that many of the limitations of the Q-e scheme stem from its empirical nature, proposed a new scheme containing a radical reactivity term, based on experimentally measured values of the rate constant for abstraction of benzylic hydrogen from toluene ($k_{3,T}$), a polar term (the Hammett σ value) and two constants α and β which are specific for a given substrate.[98]

According to this scheme the rate constant for radical-substrate reaction is:

$$\log k = \log k_{3,T} + \alpha\sigma + \beta$$

and the reactivity ratios are then defined by the following expression:

$$\log r_A = \sigma_A(\alpha_A - \alpha_B) + \beta_A - \beta_B$$

This scheme has been used to predict reactivity in transfer and copolymerization. A revised Patterns scheme for copolymerization where rates of addition to styrene are used as the reference reaction has also been described.[111]

Davis and Rogers[106] have applied *ab initio* methods to estimate the electronegativities of common monomers and radicals. They showed that there is a correlation between the Patterns α parameter and electronegativity of the propagating species. On the other hand, there was a poor correlation between the energy of formation of the radical and the Patterns β parameter. This is not surprising in view of the complex interplay of factors known to influence radical addition (see 2.2).

6.3 Block & Graft Copolymerization

Many block and graft copolymer syntheses involve radical polymerization at some stage of the overall preparation. This section has been divided into two main sections. The first deals with syntheses (by radical polymerization) of end-functional polymers useful as building blocks in block and graft copolymer syntheses (see 6.3.1). The second deals with direct syntheses of block and graft copolymers by free radical techniques (see 6.3.2). There is some overlap between these sections.

Graft copolymers may also be called branched polymers or comb polymers. In the standard nomenclature [poly(M_A-*graft*-M_B)] the first named monomer(s) form the backbone while those named second are the grafts. Thus, poly(MMA-*graft*-S) indicates a backbone of MMA and grafts of S.

Graft copolymerizations can be categorized according to three main types:[112]

(a) Grafting from, where active sites are created on the polymer chain from which new polymerization is initiated. Transfer to polymer causes branching in a grafting from process.
(b) Grafting onto, where reactive functionality on one polymer chain reacts with functionality on a second chain. Condensation of polymer bound functionality with end-functional polymers is a grafting onto process.
(c) Grafting through, where a propagating species reacts with pendant unsaturation on another polymer chain. The copolymerization of macromonomers is a grafting through process.

Block copolymerizations might be similarly classified.

6.3.1 End-Functional Polymers

End-functional polymers, including telechelic polymers, can be produced by free radical polymerization with the aid of functional initiators, chain transfer agents, and/or inhibitors. Recent advances in our understanding of free radical polymerization offer greater control of these reactions and hence of the polymer functionality. Recent reviews on the synthesis of end-functional polymers include those by Ebdon,[113] Boutevin,[114] Heitz,[115] Nguyen and Maréchal,[116] and Brosse et al.[117]

Methods for the synthesis of these materials involve conducting radical polymerizations with functional initiators (see 6.3.1.1), transfer agents (6.3.1.2), monomers (6.3.1.3) or inhibitors (6.3.1.4).

6.3.1.1 Functional initiators

End-functional polymers may be prepared by conducting polymerizations with high concentrations of a functional initiator. These conditions have been described by Tobolsky[118] as dead-end polymerization.[119] When a polymer is prepared by radical polymerization under these conditions, the functionality will depend on the relative significance and specificity of the various chain end forming reactions.

Thus, for the formation of telechelic polymers:

(a) The reaction of the initiating species with monomer must involve double bond addition (*i.e.* no primary radical transfer).
(b) Secondary radical formation should either be negligible (*e.g.* β-scission of acyloxy radicals), or not involve loss of the desired functionality.

(c) Chain end formation by chain transfer to monomer, polymer, solvent, etc. must be minimal. Chain transfer to initiator may be tolerated if the initiator functionality is transferred.

(d) All radical-radical termination (reaction with primary or propagating radicals) should involve combination.

These conditions severely limit the range of initiators and monomers that can be used and mean that attention to reaction conditions is of paramount importance. The relatively low incidence of side reactions with azo-compounds (see 3.3.1) has led to these initiators being favored for this application. The use of functional azo compounds in telechelic syntheses has been reported [e.g. (1), (2)[120,121] and (3)].[122,123] The acylazide end groups formed with initiator (3) may be thermally transformed to isocyanate ends.

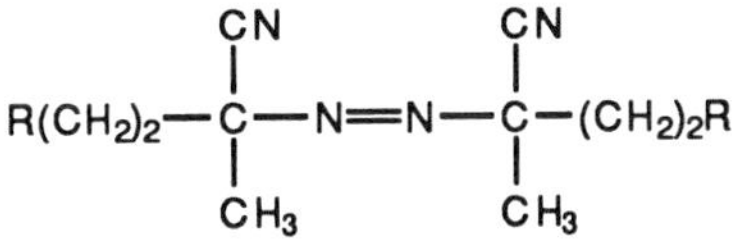

$$R(CH_2)_2-\underset{\underset{CH_3}{|}}{\overset{\overset{CN}{|}}{C}}-N=N-\underset{\underset{CH_3}{|}}{\overset{\overset{CN}{|}}{C}}-(CH_2)_2R$$

(1), R=CO$_2$H
(2), R=CH$_2$OH
(3), R=CON$_3$

Simple azo-compounds (AIBN, AIBMe) have also been used to produce telechelic polymers.[124,125] Nitrile and ester functions can be elaborated to reactive groups and used for polyester or polyurethane formation (see Scheme 6). Functionalities of 1.7 for PE and 2.0 for PS have been reported.

The latter number seems high given that PS• is known to give some disproportionation both in reaction with cyanoisopropyl radicals (k_{td}/k_{tc}(90°C) = 0.61, see 5.2.3.6) and in self reaction (see 5.2.1.1). A possible explanation is that the unsaturated by-product from cage-disproportionation (e.g. MAN from AIBN, see 3.3.1.1.3)[126] may copolymerize. This may result in an apparent functionality of ≥2.

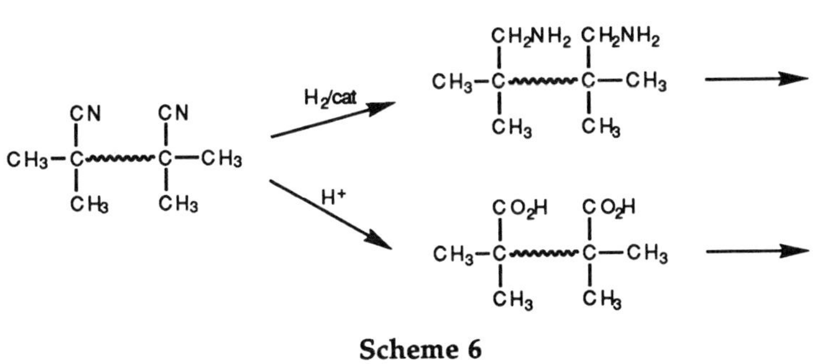

Scheme 6

There have been several studies on the applications of peroxide initiators to the synthesis of α,ω-dihydroxy oligomers. This use of peroxides is complicated by the tendency of acyloxy radicals and alkoxy radicals to undergo β-scission and by the various pathways that may compete with double bond addition (see 3.4.2). However, alkoxycarbonyloxy radicals undergo β-scission relatively slowly (see 3.4.2.2.2) and peroxydicarbonates have been used to form polymers with carbonate end groups.[127] Guth and Heitz[124] have reported that ethylene polymerized with peroxydicarbonate initiator has a functionality of only *ca.* 1.1. As explanation, they proposed that primary radical termination involving the alkoxycarbonyloxy radical involves disproportionation rather than coupling. The carbonate ends can be readily hydrolyzed to hydroxy groups.[127]

The use of ring substituted diacyl peroxides has also been reported.[128] Both the aryl and aroyloxy ends possess the desired functionality. Other initiators used in this context include peroxides (*e.g.* hydrogen peroxide), disulfide derivatives (see 7.5.1) and hexasubstituted ethanes[129] (see 7.5.2).

6.3.1.2 *Functional transfer agents*

Suitably functionalized transfer agents offer a route to end-functional polymers.[114,115,130-132] In these radical polymerizations, it must be remembered that the initiation and termination steps will always be responsible for a fraction of the chain ends. Therefore, to achieve the highest degree of functionality, an initiator should be chosen which gives the same type of end group as the transfer agent.

Chains with undesired functionality from termination by combination or disproportionation cannot be totally avoided. In attempts to prepare a monofunctional polymer, any termination by combination will give rise to a difunctional impurity. Similarly, when a difunctional polymer is required, termination by disproportionation will afford a monofunctional impurity. The amount of termination by radical-radical reactions can be minimized by using the lowest practical rate of initiation (and of polymerization). Computer modeling has been used as a means of predicting the sources of chain ends during polymerization and examining their dependence on reaction conditions.[10,133] The main limitations on accuracy are the precision of rate constants which characterize the polymerization.

Depending on the choice of transfer agent mono- or di-end-functional polymers may be produced. Meijs *et al.*[134] have reported on the use of functional allyl sulfides (Scheme 7) and benzyl ethers in this context (see 5.3.2.6). Boutevin *et al.*[135,136] have described the telomerization of unsaturated alcohols with mercaptoethanol or dithiols to produce telechelic diols in high yield. Dithiols and dienes react spontaneously to afford telechelic dithiols or dienes depending on the monomer:dithiol ratio.[137]

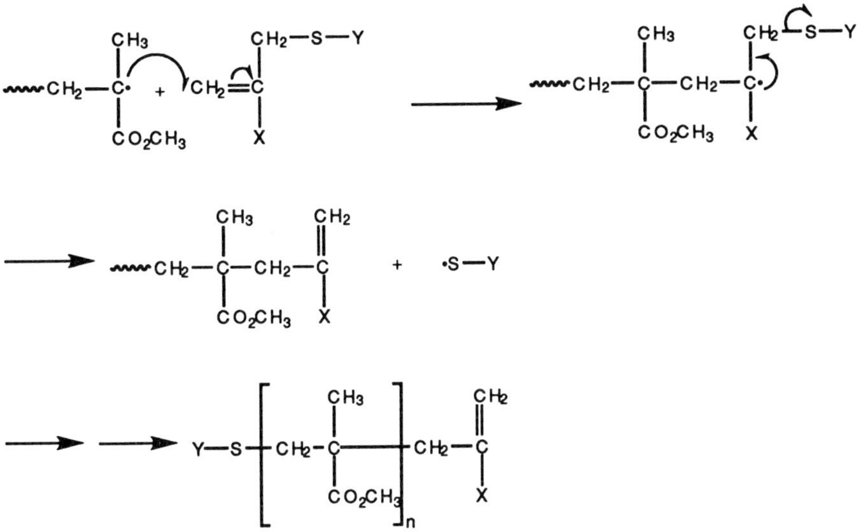

Scheme 7

The degree of functionality of the polymer will depend on the polymerization conditions. Factors which may limit this include: (a) the proportion of end groups generated by initiation or radical-radical termination reactions; (b) the occurrence of transfer to other species (monomer, polymer, initiator).

A novel method for telechelic production, based on the radical initiated reaction of difunctional transfer agents with dienes (*e.g.* divinyl benzene, dimethacrylate esters), has recently been described by Nuyken and Völkel.[138,139] A polymer chain is built up by a sequence of thiyl radical addition and chain transfer steps.

6.3.1.3 *Functional monomers*

Ketene acetals and related monomers undergo ring opening polymerization to produce polyesters (see 4.4.3). Copolymerization of such monomers with, for example, S (see Scheme 8), and basic hydrolysis of the ester linkages in the resultant copolymer offers a route to α,ω-difunctional polymers.[140]

A limitation on this approach is that these particular monomers are relatively unreactive towards propagating radicals (*e.g.* PS•) and rates of copolymerization are slow.

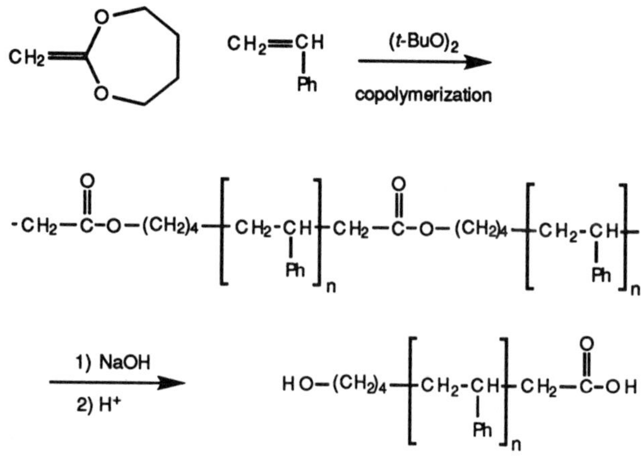

Scheme 8

[Scheme 8 depicts a copolymerization of a cyclic ketene acetal (CH₂= with 7-membered O-C-O ring) with CH₂=CH(Ph) using (t-BuO)₂ to give a copolymer:

−CH₂−C(=O)−O−(CH₂)₄−[CH₂−CH(Ph)]ₙ−CH₂−C(=O)−O−(CH₂)₄−[CH₂−CH(Ph)]ₙ−

followed by 1) NaOH, 2) H⁺ to give:

HO−(CH₂)₄−[CH₂−CH(Ph)]ₙ−CH₂−C(=O)−OH]

6.3.1.4 Functional inhibitors

Inhibitors (see 5.4), including transition metal complexes and nitroxides (see 7.5.3), may be used to prepare mono-end-functional polymers. If an appropriate initiator is employed, di-end-functional polymers are also possible.

Only one polymer molecule is produced per mole of inhibitor. The inhibitor must be at least equimolar with the number of chains formed. Concentrations must be chosen (usually very low) to give the desired molecular weight.

6.3.2 Block and Graft Copolymer Synthesis

In this section methods for block and graft copolymer synthesis based on radical polymerization are described.

6.3.2.1 Polymeric and multifunctional initiators

Multifunctional initiators contain two or more radical generating functions within the one molecule. The chemistry of these initiators has been the subject of several reviews.[141-143] As long as the radical generating functions are sufficiently remote their decompositions are independent events. These initiators can then be used to form polymers with end groups that contain initiator moieties. These end groups can be subsequently utilized to afford higher molecular weight polymers, to achieve higher degrees of conversion, and in the production of block and graft copolymers. The multifunctional initiators may be di- and tri-, azo- or peroxy- compounds of defined structure or they may be polymeric azo- or peroxy- compounds where the radical generating functions may be present as side chains[144] or as part of the polymer backbone (Scheme 9).[145-148]

The success of the multifunctional initiators in the preparation of block copolymers depends critically on the kinetics and mechanism of radical production. In particular, the initiator efficiency, the susceptibility to and mechanism of transfer to initiator, and the relative stability of the various radical generating functions all have a substantial influence on the nature and homogeneity of the polymer formed.

Features of the kinetics of polymerizations initiated by these compounds have been modeled by O'Driscoll and Bevington[149] and Choi and Lei.[150] Block copolymers may be synthesized from a α,ω-diol and AIBN.[151] Another example of this strategy can be found in section 6.3.2.2.

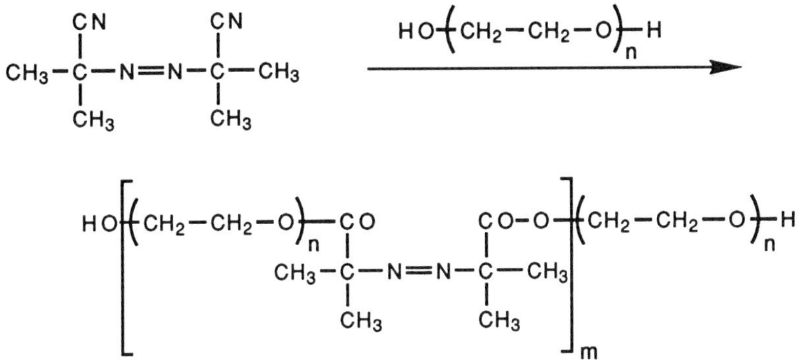

Scheme 9

6.3.2.2 Transformation reactions

Block and graft copolymer syntheses by what have come to be known as 'transformation reactions' involve the preparation of polymeric species by some mechanism which leaves a terminal functionality that allows polymerization to be continued by another mechanism (Scheme 10). Examples of transformation of anionic, cationic, Ziegler-Natta, and group transfer polymerization to radical polymerization have been reported. Examples of transformation of radical to ionic polymerization are also known.

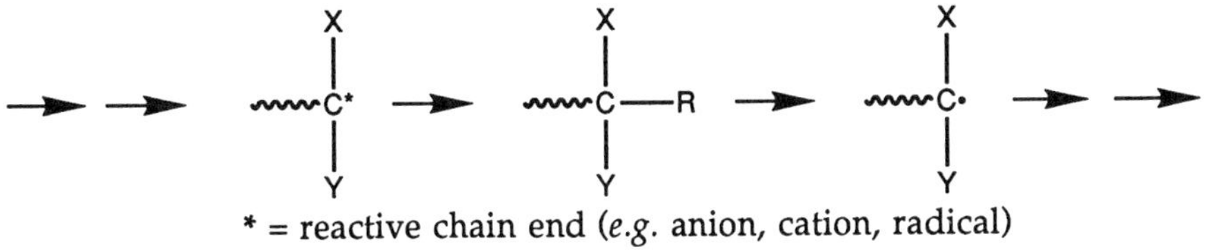

* = reactive chain end (*e.g.* anion, cation, radical)

Scheme 10

The success depends on the efficiency of the transformation process and the avoidance of processes which might lead to concurrent homopolymerization.

The general area of block polymer synthesis through 'transformation reactions' has been reviewed by Stewart[152], Schue,[153] Eastmond[154] and Abadie and Ourahmoune.[155]

The mechanism of termination also plays an important role in determining the type of block copolymers that may be formed. If standard polymerization conditions are employed, an ABA or AB block may be produced depending on whether termination occurs by combination or disproportionation.

One of the earliest examples of this methodology involves the reaction of a polymeric anion (formed by living anionic polymerization) with molecular oxygen to form a polymeric hydroperoxide which can be decomposed either thermally or, preferably, in a redox reaction to initiate block polymer formation with a second monomer (Scheme 11). However, the usual complications associated with initiation by hydroperoxides apply (see 3.3.2.5).

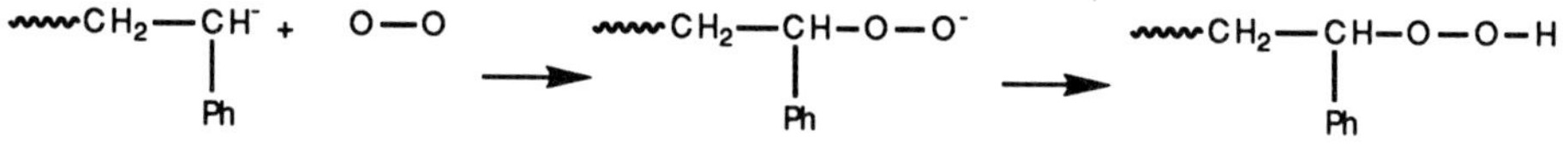

$$\sim\sim CH_2-CH^- + O-O \longrightarrow \sim\sim CH_2-CH-O-O^- \longrightarrow \sim\sim CH_2-CH-O-O-H$$
$$Ph Ph Ph$$

Scheme 11

The reactions of polymeric anions with appropriate azo-compounds or peroxides to form polymeric initiators provide other examples of anion-radical transformation (*e.g.* Scheme 12).[155-158] However, the polymeric azo and peroxy compounds have limited utility in block copolymer synthesis because of the poor efficiency of radical generation from the polymeric initiators (see 6.3.2.1).

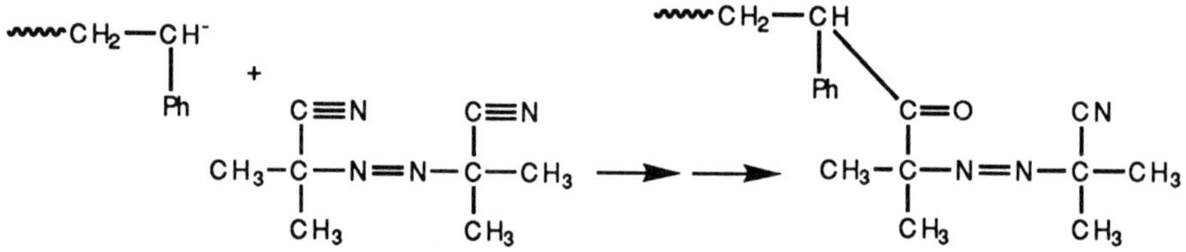

Scheme 12

Tung et al.[159] have reported on the use of a polymeric thiol transfer agent for use in block copolymer production. Various methods have been used for the anion→thiol conversion. Near quantitative yields of thiol are reported to have been obtained by terminating anionic polymerization with ethylene sulfide and derivatives (Scheme 13). Transfer constants for the polymeric thiols are reported to be similar to those of analogous low molecular weight compounds.[159]

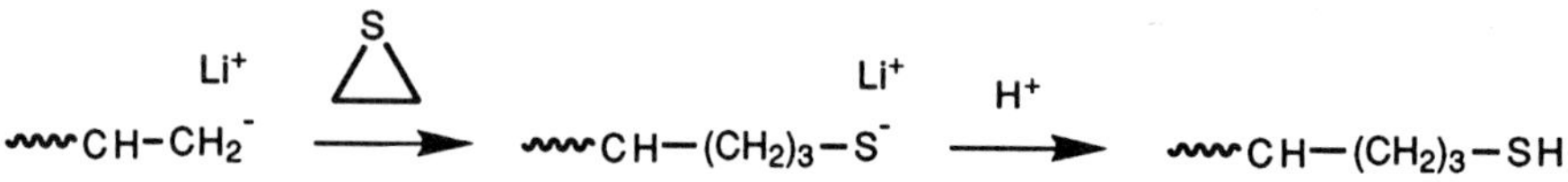

$$\sim\sim CH-CH_2^- \xrightarrow{\overset{S}{\triangle}\ Li^+} \sim\sim CH-(CH_2)_3-S^-\ Li^+ \xrightarrow{H^+} \sim\sim CH-(CH_2)_3-SH$$

Scheme 13

Copolymerization   305

The preparation of ABA triblock polymers requires use of a telechelic bisthiol prepared by termination of anionic polymerization initiated by a difunctional initiator. The relative yields of homopolymer, di- and triblock obtained in these experiments depend critically on conversion.[159]

Richards et al. have carried out extensive studies on the use of mercury,[160,161] lead[162,163] and silver compounds to terminate anionic polymerization and form polymeric organometallic species which can be used to initiate polymerization.

Bamford, Eastmond et al.[164-169] have employed metal complex-polymeric halide redox systems to initiate block and graft copolymerization. The polymeric halides can be synthesized by a variety of techniques, including radical[165] and anionic polymerization (Scheme 14),[164]

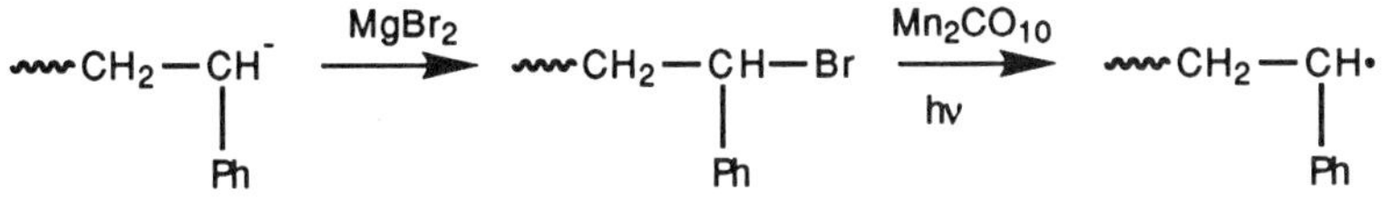

$$\text{\textasciitilde CH}_2-\text{CH}^-\text{(Ph)} \xrightarrow{MgBr_2} \text{\textasciitilde CH}_2-\text{CH-Br (Ph)} \xrightarrow[h\nu]{Mn_2CO_{10}} \text{\textasciitilde CH}_2-\text{CH}\cdot \text{(Ph)}$$

Scheme 14

group transfer polymerization (Scheme 15),[168]

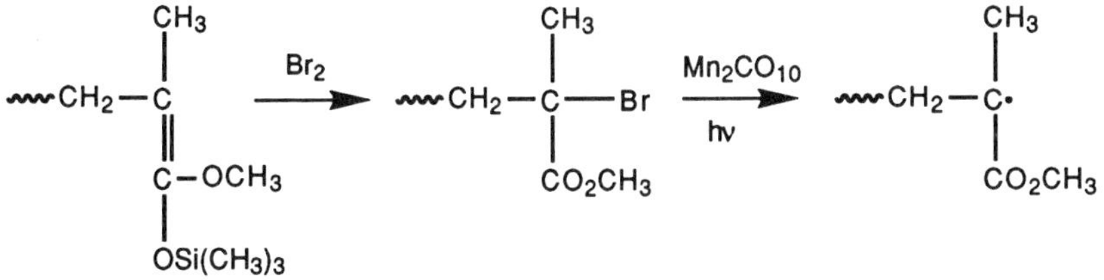

$$\text{\textasciitilde CH}_2-\underset{\underset{OSi(CH_3)_3}{|}}{\overset{\overset{CH_3}{|}}{\underset{\|}{C}}}-OCH_3 \xrightarrow{Br_2} \text{\textasciitilde CH}_2-\underset{\underset{CO_2CH_3}{|}}{\overset{\overset{CH_3}{|}}{C}}-Br \xrightarrow[h\nu]{Mn_2CO_{10}} \text{\textasciitilde CH}_2-\underset{\underset{CO_2CH_3}{|}}{\overset{\overset{CH_3}{|}}{C}}\cdot$$

Scheme 15

cationic polymerization (Scheme 16),[167]

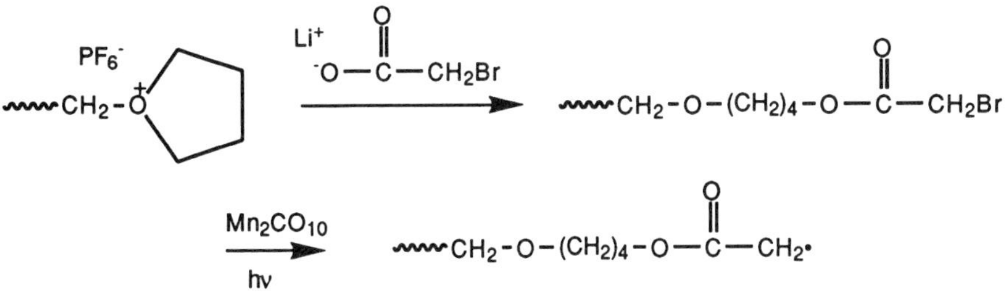

$$\text{\textasciitilde CH}_2-\overset{+}{O}\text{(ring)} \; PF_6^- \xrightarrow{Li^+ \; ^-O-\overset{O}{\underset{\|}{C}}-CH_2Br} \text{\textasciitilde CH}_2-O-(CH_2)_4-O-\overset{O}{\underset{\|}{C}}-CH_2Br$$

$$\xrightarrow[h\nu]{Mn_2CO_{10}} \text{\textasciitilde CH}_2-O-(CH_2)_4-O-\overset{O}{\underset{\|}{C}}-CH_2\cdot$$

Scheme 16

and functionalization of a polymer with carboxylic acid, hydroxy, amino, or ether-urethane groups with a haloisocyanate (Scheme 17).[170]

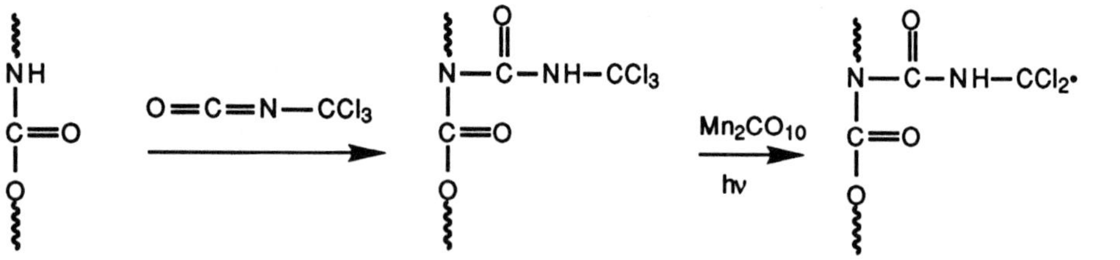

Scheme 17

The efficiency of the halide→radical transformation is reported to be near quantitative. The yield of block or graft is then limited by the efficiency of the halide synthesis. Whether AB or ABA blocks are formed depends on the termination mechanism.

6.3.2.3 Macromonomers

A macromonomer is defined as an oligomer or polymer chain terminated with a double bond or other group such that the material is able to act as a comonomer in radical copolymerization. Copolymerization of macromonomers with low molecular weight monomers gives a graft copolymer. Thus, an important use of these compounds is in the formation of graft copolymers with defined graft lengths. The chain length of the graft is determined by that of the macromonomer.

Various macromonomers have been described in the literature; most are derivatives of S or acrylate esters [e.g. (4-6)]. The relative merits of these have been assessed in recent reviews.[171-173]

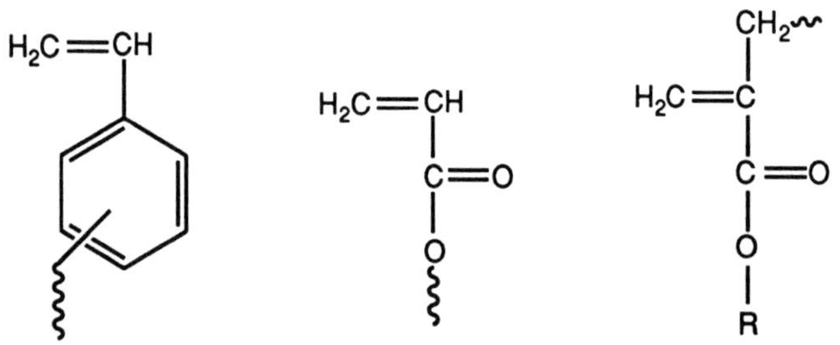

Most macromonomers do not readily undergo homopolymerization. Some that do are derivatives of styrene (4) and methyl acrylate (5). Macromonomers (6) do not undergo homopolymerization or copolymerization with methacrylate esters because of addition-elimination chain transfer (Scheme 18, see also 5.3.1.2). Graft copolymers can be formed with acrylates and S.[174]

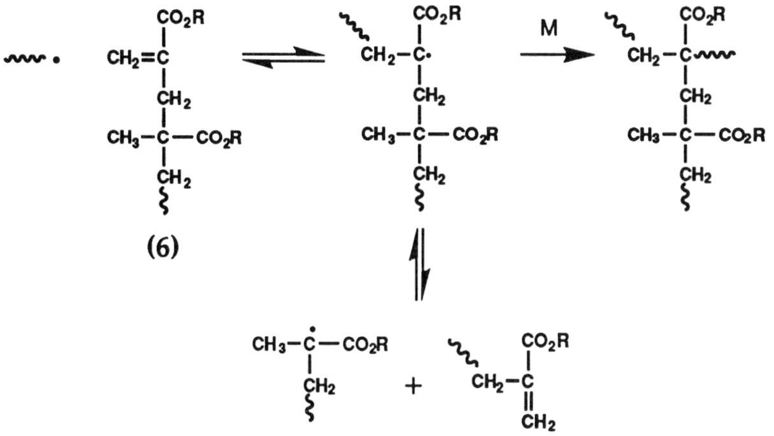

Scheme 18

The reactivity of macromonomers in copolymerization is strongly dependent on the particular comonomer-macromonomer pair. Solvent effects are also important (see 7.3.1.2). The reactivity of the double bonds is in many cases similar to that of the low molecular weight monomers they resemble. However, the propagating species generated often have low reactivity due to adverse steric factors.

Other factors of importance in polymer syntheses involving macromonomers relate to the viscosity of the polymerization medium. Propagation could well be diffusion controlled and the propagation rate constant will then depend on the molecular weight of the macromonomer and the viscosity or, more accurately, the free volume of the system.

Primary radical transfer may complicate the initiation process. Due to the low concentration of reactive double bonds, it is even more important than usual to select initiators with a low propensity for hydrogen atom abstraction. The greater viscosity of reaction media containing high concentrations of macromonomer will also cause reduced initiator efficiencies as compared to conventional polymerizations. Low rates of diffusion of propagating species may lead to reduced rates of termination. The nature of the solvent is also an important consideration. A good solvent for macromonomer and polymer will facilitate interpenetration of the polymer chains. The particular balance between these factors can lead to overall rates of polymerization being higher or lower than conventional polymerizations.

References

1. Kuchanov, S.I., *Adv. Polym. Sci.*, 1992, **103**, 1.
2. Tirrell, D.A., in 'Encyclopaedia of Polymer Science and Engineering', 2nd Edn (Eds. Mark, H.F., Bikales, N.M., Overberger, C.G., and Menges, G.), Vol. 4, p. 192 (Wiley: New York 1985).
3. Tirrell, D.A., in 'Comprehensive Polymer Science' (Eds. Eastmond, G.C., Ledwith, A., Russo, S., and Sigwalt, P.), Vol. 3, p. 195 (Pergamon: London 1989).
4. Braun, D., and Czerwinski, W.K., in 'Comprehensive Polymer Science' (Eds. Eastmond, G.C., Ledwith, A., Russo, S., and Sigwalt, P.), Vol. 3, p. 207 (Pergamon: London 1989).
5. Cowie, J.M.G., in 'Comprehensive Polymer Science' (Eds. Eastmond, G.C., Ledwith, A., Russo, S., and Sigwalt, P.), Vol. 4, p. 377 (Pergamon: London 1989).
6. Skeist, I., *J. Am. Chem. Soc.*, 1946, **68**, 1781.
7. Galbraith, M.N., Moad, G., Solomon, D.H., and Spurling, T.H., *Macromolecules*, 1987, **20**, 675.
8. O'Driscoll, K.F., *J. Coat. Technol.*, 1983, **55**, 57.
9. Hill, L.W., and Wicks, Z.W., *Prog. Org. Coat.*, 1982, **10**, 55.
10. Deady, M., Mau, A.W.H., Moad, G., and Spurling, T.H., *Makromol. Chem.*, 1993, **194**, 1691.
11. Mayo, F.R., and Lewis, F.M., *J. Am. Chem. Soc.*, 1944, **66**, 1594.
12. Greenley, R.Z., *J. Macromol. Sci., Chem*, 1980, **A14**, 445.
13. Greenley, R.Z., in 'Polymer Handbook, 3rd Edition' (Eds. Brandup, J., and Immergut, E.H.), p. II/153 (Wiley: New York 1989).
14. Laurier, G.C., O'Driscoll, K.F., and Reilly, P.M., *J. Polym. Sci., Polym. Symp.*, 1985, **72**, 17.
15. Moad, G., Solomon, D.H., Spurling, T.H., and Vearing, D.J., *Aust. J. Chem.*, 1986, **39**, 1877.
16. Ham, G.E., *J, Macromol. Sci., Chem,*, 1991, **A28**, 733.
17. Giese, B., and Engelbrecht, R., *Polym. Bull. (Berlin)*, 1984, **12**, 55.
18. Tanaka, H., Sasai, K., Sato, T., and Ota, T., *Macromolecules*, 1988, **21**, 3534.
19. Jones, S.A., Prementine, G.S., and Tirrell, D.A., *J. Am. Chem. Soc.*, 1985, **107**, 5275.
20. Cywar, D.A., and Tirrell, D.A., *J. Am. Chem. Soc.*, 1989, **111**, 7544.
21. Tanaka, H., Sakai, I., Sasai, K., Sato, T., and Ota, T., *J. Polym. Sci., Part C: Polym. Lett.*, 1988, **26**, 11.
22. Hill, D.J.T., Lang, A.P., O'Donnell, J.H., and O'Sullivan, P.W., *Eur. Polym. J.*, 1989, **25**, 911.
23. Moad, G., in 'Annual Reports in NMR Spectroscopy' (Ed. Webb, G.A.), Vol. 29, p. 287 (Academic Press: London 1994).

24. Fukuda, T., Ma, Y.D., Kubo, K., and Inagaki, H., *Macromolecules*, 1991, **24**, 370.
25. Heuts, J.P.A., Clay, P.A., Christie, D.I., Piton, M.C., Hutovic, J., Kable, S.H., and Gilbert, R.G., *Macromol. Symp.*, 1994, in press.
26. Merz, E., Alfrey, T., and Goldfinger, G., *J. Polym. Sci.*, 1946, **1**, 75.
27. Ham, G.E., *J. Polym. Sci.*, 1954, **14**, 87.
28. Fukuda, T., Ma, Y.-D., and Inagaki, H., *Macromolecules*, 1985, **18**, 17.
29. Moad, G., Solomon, D.H., Spurling, T.H., and Vearing, D.J., *Aust. J. Chem.*, 1985, **38**, 1287.
30. Hill, D.J.T., O'Donnell, J.H., and O'Sullivan, P.W., *Macromolecules*, 1982, **15**, 960.
31. Lin, J., Petit, A., and Neel, J., *Makromol. Chem.*, 1987, **188**, 1163.
32. Van Der Meer, R., Alberti, J.M., German, A.L., and Linssen, H.N., *J. Polym. Sci., Polym. Chem. Ed.*, 1979, **17**, 3349.
33. Hill, D.J.T., O'Donnell, J.H., and O'Sullivan, P.W., *Macromolecules*, 1985, **18**, 9.
34. Brown, A.S., Fujimora, K., and Craven, I., *Makromol. Chem.*, 1988, **189**, 1893.
35. Guillot, J., Vialle, J., and Guyot, A., *J. Macromol. Sci. Chem.*, 1971, **A5**, 735.
36. Moad, G., Solomon, D.H., Spurling, T.H., and Stone, R.A., *Macromolecules*, 1989, **22**, 1145.
37. Aerdts, A.M., de Haan, J.W., and German, A.L., *Macromolecules*, 1993, **26**, 1965.
38. Maxwell, I.A., Aerdts, A.M., and German, A.L., *Macromolecules*, 1993, **26**, 1956.
39. Uebel, J.J., and Dinan, F.J., *J. Polym. Sci., Polym. Chem. Ed.*, 1983, **21**, 917.
40. Cowie, J.M.G., in 'Alternating Copolymers' (Ed. Cowie, J.M.G.), p. 19 (Plenum: New York 1985).
41. Cowie, J.M.G., in 'Alternating Copolymers' (Ed. Cowie, J.M.G.), p. 1 (Plenum: New York 1985).
42. Hill, D.J.T., O'Donnell, J.H., and O'Sullivan, P.W., *Prog. Polym. Sci.*, 1982, **8**, 215.
43. Furakawa, J., in 'Encyclopaedia of Polymer Science', 2nd Edn (Eds. Mark, H.F., Bikales, N.M., Overberger, C.G., and Menges, G.), Vol. 4, p. 233 (Wiley: New York 1985).
44. Shirota, Y., in 'Encyclopaedia of Polymer Science and Engineering', 2nd Edn (Eds. Mark, H.F., Bikales, N.M., Overberger, C.G., and Menges, G.), Vol. 3, p. 327 (New York: Wiley 1985).
45. Shirota, Y., and Mikawa, H., *J. Macromol. Sci., Rev. Macromol. Chem.*, 1977, **C16**, 129.
46. Ebdon, J.R., Towns, C.R., and Dodgson, K., *J. Macromol. Sci., Rev. Macromol. Chem. Phys.*, 1986, **26**, 523.

47. Rätzsch, M., and Vogl, O., *Prog. Polym. Sci.*, 1991, **16**, 279.
48. Cais, R.E., Farmer, R.G., Hill, D.J.T., and O'Donnell, J.H., *Macromolecules*, 1979, **12**, 835.
49. Seiner, J.A., and Litt, M., *Macromolecules*, 1971, **4**, 308.
50. Pittman, C.U., and Rounsefell, T.D., *Macromolecules*, 1975, **8**, 46.
51. Hill, D.J.T., O'Donnell, J.H., and O'Sullivan, P.W., *Macromol.*, 1983, **16**, 1295.
52. Tsuchida, E., and Tomono, T., *Makromol. Chem.*, 1971, **141**, 265.
53. Karad, P., and Schneider, C., *J. Polym. Sci., Polym. Chem. Ed.*, 1978, **16**, 1137.
54. Prementine, G.S., Jones, S.A., and Tirrell, D.A., *Macromolecules*, 1989, **22**, 52.
55. Saito, J., and Tirrell, D.A., *Eur. Polym. J.*, 1993, **29**, 343.
56. Jones, S.A., and Tirrell, D.A., *J. Polym. Sci., Part A: Polym. Chem.*, 1987, **25**, 3177.
57. Bottle, S.E., *Macromol. Symp.*, 1994, in press.
58. Druliner, J.D., Krusic, P.D., Lehr, G.F., and Tolman, C.A., *J. Org. Chem.*, 1985, **50**, 5838.
59. Sawada, H., *J. Macromol. Sci., Rev. Macromol. Chem.*, 1974, **C11**, 257.
60. O'Driscoll, K.F., and Gasparro, F.P., *J. Macromol. Sci., Chem.*, 1967, **A1**, 643.
61. Wittmer, P., *Makromol. Chem.*, 1967, **103**, 188.
62. Izu, M., O'Driscoll, K.F., Hill, R.J., Quinn, M.J., and Harwood, H.J., *Macromolecules*, 1972, **5**, 90.
63. Fischer, J.P., *Makromol. Chem.*, 1972, **155**, 211.
64. Lowry, G.G., *J. Polym. Sci.*, 1960, **17**, 463.
65. Howell, J.A., Izu, M., and O'Driscoll, K.F., *J. Polym. Sci., Part A-1*, 1970, **8**, 699.
66. Randall, J.C., Polymer Sequence Determination (Academic Press: New York 1977).
67. Tonelli, A.E., NMR Spectroscopy and Polymer Microstructure (VCH: New York 1989).
68. Koenig, J.L., Chemical Microstructure of Polymer Chains (Wiley: New York 1980).
69. Bovey, F.A., *J. Polym. Sci.*, 1962, **62**, 197.
70. Kale, L.T., O'Driscoll, K.F., Dinan, F.J., and Uebel, J.J., *J. Polym . Sci. Part A*, 1986, **24**, 3145.
71. Lopez-Gonzalez, M.M.C., Fernandez-Garcia, M., Barreles-Rienda, J.M., Madruga, E.M., and Arias, C., *Polymer*, 1993, **34**, 3123.
72. Klesper, E., Johnsen, A., Gronski, W., and Wehrli, F.W., *Makromol. Chem.*, 1975, **176**, 1071.
73. Johnsen, A., Klesper, E., and Wirthlin, T., *Makromol. Chem.*, 1976, **177**, 2397.
74. Park, K.Y., Santee, E.R., and Harwood, H.J., *Eur. Polym. J.*, 1989, **25**, 651.

75. Harwood, H.J., *Makromol. Chem., Macromol. Symp.*, 1987, **10/11**, 331.
76. Fineman, M., and Ross, S.D., *J. Polym. Sci.*, 1950, **5**, 259.
77. Tidwell, P.W., and Mortimer, G.A., *J. Macromol. Sci., Rev. Macromol. Chem.*, 1970, **C4**, 261.
78. Tidwell, P.W., and Mortimer, G.A., *J. Polym. Sci., Part A*, 1965, **3**, 369.
79. Kelen, T., Tüdos, F., and Turcsányi, B., *Polym. Bull. (Berlin)*, 1981, **2**, 71.
80. Kelen, T., and Tüdos, F., *J. Macromol. Sci., Chem.*, 1975, **A9**, 1.
81. Meyer, V.E., and Lowry, G.G., *J. Polym. Sci., Part A*, 1965, **3**, 369.
82. Francis, A.P., Solomon, D.H., and Spurling, T.H., *J. Macromol. Sci. Chem.*, 1974, **A8**, 469.
83. Plaumann, H.P., and Branston, R.E., *J. Polym. Sci., Part A: Polym. Chem.*, 1989, **27**, 2819.
84. O'Driscoll, K.F., Kale, L.T., Garcia Rubio, L.H., and Reilly, P.M., *J. Polym. Sci., Polym. Chem. Ed.*, 1984, **22**, 2777.
85. Patino-Leal, H., Reilly, P.M., and O'Driscoll, K.F., *J. Polym. Sci., Polym. Lett. Ed.*, 1980, **18**, 219.
86. Van Der Meer, R., Linssen, H.N., and German, A.L., *J. Polym. Sci., Polym. Chem. Ed.*, 1978, **16**, 2915.
87. Hautus, F.L.M., Linssen, H.N., and German, A.L., *J. Polym. Sci., Polym. Chem. Ed.*, 1984, **22**, 3661.
88. Hautus, F.L.M., Linssen, H.N., and German, A.L., *J. Polym. Sci., Polym. Chem. Ed.*, 1984, **22**, 3487.
89. Leicht, R., and Fuhrman, J., *J. Polym. Sci., Polym. Chem. Ed.*, 1983, **21**, 2215.
90. McFarlane, R.C., Reilly, P.M., and O'Driscoll, K.F., *J. Polym. Sci., Polym. Chem. Ed.*, 1980, **18**, 251.
91. Burke, A.L., Duever, T.A., and Penlidis, A., *J. Polym. Sci., Part A: Polym. Chem.*, 1993, **31**, 3065.
92. Kelen, T., and Tüdos, F., *Makromol. Chem.*, 1990, **191**, 1863.
93. Hill, D.J.T., Lang, A.P., Munro, P.D., and O'Donnell, J.H., *Eur. Polym. J.*, 1992, **28**, 391.
94. Bovey, F.A., Chain Structure and Conformation of Macromolecules (Wiley: New York 1982).
95. Rudin, A., O'Driscoll, K.F., and Rumack, M.S., *Polymer*, 1981, **22**, 740.
96. Moad, G., and Willing, R.I., *Polym. J. (Tokyo).*, 1991, **23**, 1401.
97. Moad, G., *Chem. Aust.*, 1991, **58**, 122.
98. Jenkins, A.D., in 'Reactivity, Mechanism and Structure in Polymer Chemistry' (Eds. Jenkins, A.D., and Ledwith, A.), p. 117 (Wiley: London 1974).
99. Semchikov, Y.D., *Polym. Sci. USSR (Engl. Transl.)*, 1990, **32**, 177.
100. Alfrey, T., and Price, C.C., *J. Polym. Sci.*, 1947, **2**, 101.
101. Greenley, R.Z., *J. Macromol. Sci., Chem*, 1980, **A14**, 427.
102. Young, L.J., *J. Polym. Sci.*, 1961, **54**, 411.

103. Greenley, R.Z., *J. Macromol. Sci., Chem.*, 1975, **A9**, 505.
104. Greenley, R.Z., in 'Polymer Handbook, 3rd Edition' (Eds. Brandup, J., and Immergut, E.H.), p. II/267 (Wiley: New York 1989).
105. Borchardt, J.K., *J. Macromol. Sci. Chem.*, 1985, **A22**, 1711.
106. Davis, T.P., and Rogers, S.C., *Eur. Polym. J.*, 1993, **29**, 1311.
107. Rogers, S.C., Mackrodt, W.C., and Davis, T.P., *Polymer*, 1994, **35**, 1258.
108. Bamford, C.H., Jenkins, A.D., and Johnston, R., *Trans. Faraday Soc.*, 1959, **55**, 418.
109. Bamford, C.H., and Jenkins, A.D., *J. Polm. Sci.*, 1961, **59**, 530.
110. Bamford, C.H., and Jenkins, A.D., *Trans. Faraday Soc.*, 1963, **59**, 530.
111. Jenkins, A.D., *Eur. Polym. J.*, 1989, **25**, 721.
112. Remmp, P.F., and Lutz, P.J., in 'Comprehensive Polymer Science' (Eds. Eastmond, G.C., Ledwith, A., Russo, S., and Sigwalt, P.), Vol. 4, p. 403 (Pergamon: London 1989).
113. Ebdon, J.R., in 'New methods of Polymer Synthesis' (Ed. Ebdon, J.R.), p. 162 (Blackie: Glasgow 1991).
114. Boutevin, B., *Adv. Polym. Sci.*, 1990, **94**, 69.
115. Heitz, W., in 'Telechelic Polymers: Synthesis and Applications' (Ed. Goethals, E.J.), p. 61 (CRC Press: Boca Raton, Florida 1989).
116. Nguyen, H.A., and Marechal, E., *J. Macromol. Sci. Rev. Macomol. Chem. Phys.*, 1988, **C28**, 187.
117. Brosse, J.C., Derouet, D., Epaillard, F., Soutif, J.-C., Legeay, G., and Dusek, K., *Adv. Polym. Sci.*, 1987, **81**, 167.
118. Tobolsky, A.V., *J. Am. Chem. Soc.*, 1958, **80**, 5927.
119. Heitz, W., *Angew. Makromol. Chem.*, 1986, **145/146**, 37.
120. Reed, S.F., *J. Polym. Sci., Polym. Chem. Ed.*, 1973, **11**, 55.
121. Bamford, C.H., Jenkins, A.D., and Johnston, R., *Trans. Faraday Soc.*, 1959, **55**, 179.
122. Idage, B.B., Vernekar, S.P., and Ghatge, N.D., *J. Polym. Sci., Polym. Chem. Ed.*, 1983, **21**, 385.
123. Ghatge, N.D., Vernekar, S.P., and Wadgaonkar, P.P., *Makromol. Chem., Rapid Commun.*, 1983, **4**, 307.
124. Guth, W., and Heitz, W., *Makromol. Chem.*, 1976, **177**, 1835.
125. Konter, W., Bömer, B., Köhler, K.H., and Heitz, W., *Makromol. Chem.*, 1981, **182**, 2619.
126. Moad, G., Solomon, D.H., Johns, S.R., and Willing, R.I., *Macromolecules*, 1984, **17**, 1094.
127. Friedlander, H.N., *J. Polym. Sci.*, 1962, **58**, 455.
128. Bresler, L.S., Barantsevich, E.N., and Polyansky, V.I., *Makromol. Chem.*, 1982, **183**, 2479.
129. Edelmann, D., and Ritter, H., *Makromol. Chem.*, 1993, **194**, 2775.
130. Corner, T., *Adv. Polym. Sci.*, 1984, **62**, 95.

131. Starks, C.M., Free Radical Telomerization (Academic Press: New York 1974).
132. Boutevin, B., Lusinchi, J.-M., Pietrasanta, Y., and Robin, J.-J., *Eur. Polym. J.*, 1994, **30**, 615.
133. Pryor, W.A., and Coco, J.H., *Macromolecules*, 1970, **3**, 500.
134. Meijs, G.F., Morton, T.C., Rizzardo, E., and Thang, S.H., *Macromolecules*, 1991, **24**, 3689.
135. Boutevin, B., El Idrissi, A., and Parisi, J.P., *Makromol. Chem.*, 1990, **191**, 445.
136. Boutevin, B., and Pietrasanta, Y., *Makromol. Chem.*, 1985, **186**, 817.
137. Klemm, E., and Sensfuss, S., *J. Macromol. Sci., Chem.*, 1991, **A28**, 875.
138. Nuyken, O., and Völkel, T., *Makromol. Chem.*, 1990, **191**, 2465.
139. Nuyken, O., and Völkel, T., *Makromol. Chem., Rapid Commun.*, 1990, **11**, 365.
140. Bailey, W.J., Endo, T., Gapud, B., Lin, Y.-N., Ni, Z., Pan, C.-Y., Shaffer, S.E., Wu, S.-R., Yamazaki, N., and Yonezawa, K., *J. Macromol. Sci., Chem.*, 1984, **A21**, 979.
141. Simionescu, C., Comanita, E., Pastravanu, M., and Dumitriu, S., *Prog. Polym. Sci.*, 1986, **12**, 1.
142. Nuyken, O., and Weidner, R., *Adv. Polym. Sci.*, 1986, **73**, 145.
143. Kuchanov, S.I., in 'Comprehensive Polymer Science' (Eds. Agarwal, S.L., and Russo, S.), Vol. Suppl. 1, p. 23 (Pergamon: London 1992).
144. Nukyen, O., and Weidner, R., *Makromol. Chem.*, 1988, **189**, 1331.
145. Qiu, X.-Y., Ruland, W., and Heitz, W., *Angew. Makromol. Chem.*, 1984, **125**, 69.
146. Yagci, Y., Tunca, U., and Biçak, N., *J. Polym. Sci., Polym. Lett. Ed.*, 1986, **24**, 49.
147. Yagci, Y., Onen, A., and Schnabel, W., *Macromolecules*, 1991, **24**, 4620.
148. Hazar, B., and Baysal, B.M., *Polymer*, 1986, **27**, 961.
149. O'Driscoll, K.F., and Bevington, J.C., *Eur. Polym. J.*, 1985, **21**, 1039.
150. Choi, K.Y., and Lee, G.D., *AIChE J.*, 1987, **33**, 2067.
151. Walz, R., Bomer, B., and Heitz, W., *Makromol. Chem.*, 1977, **178**, 2527.
152. Stewart, M.J., in 'New methods of Polymer Synthesis' (Ed. Ebdon, J.R.), p. 107 (Blackie: Glasgow 1991).
153. Schue, F., in 'Comprehensive Polymer Science' (Eds. Eastmond, G.C., Ledwith, A., Russo, S., and Sigwalt, P.), Vol. 4, p. 359 (Pergamon: London 1989).
154. Eastmond, G.C., *Pure & Appl. Chem.*, 1981, **53**, 657.
155. Abadie, M.J.M., and Ourahmoune, D., *Br. Polym. J.*, 1987, **19**, 247.
156. Riess, G., and Reeb, R., *ACS Symp. Ser.*, 1981, **166**, 477.
157. Abadie, M.J.M., Ourahmoune, D., and Mendjel, H., *Eur. Polym. J.*, 1990, **26**, 515.

158. Ren, Q., Zhang, H., Zhang, X., and Huang, B., *J. Polym. Sci., Part A: Polym. Chem.*, 1993, **31**, 847.
159. Tung, L.H., Lo, G.Y.S., and Griggs, J.A., *J. Polym. Sci., Polym. Chem. Ed.*, 1985, **23**, 1551.
160. Lindsell, W.E., Service, D.M., Soutar, I., and Richards, D.H., *Br. Polym. J.*, 1987, **19**, 255.
161. Cunliffe, A.V., Hayes, G.F., and Richards, D.H., *J. Polym. Sci., Polym. Lett. Ed.*, 1976, **14**, 483.
162. Abadie, M.J.M., Schue, F., Souel, T., and Richards, D.H., *Polymer*, 1981, **22**, 1076.
163. Abadie, M., Burgess, F.J., Cunliffe, A.V., and Richards, D.H., *J. Polym. Sci., Polym. Lett. Ed.*, 1976, **14**, 477.
164. Bamford, C.H., Eastmond, G.C., Woo, J., and Richards, D.H., *Polym. Commun.*, 1982, **23**, 643.
165. Bamford, C.H., Dyson, R.W., and Eastmond, G.C., *Polymer*, 1969, **10**, 885.
166. Eastmond, G.C., Parr, K.J., and Woo, J., *Polymer*, 1988, **29**, 950.
167. Eastmond, G.C., and Woo, J., *Polymer*, 1990, **31**, 358.
168. Eastmond, G.C., and Grigor, J., *Makromol. Chem., Rapid Commun.*, 1986, **7**, 375.
169. Alimoglu, A.K., Bamford, C.H., Ledwith, A., and Mullik, S.U., *Polym. Sci. USSR (Engl. Transl.)*, 1980, **21**, 2651.
170. Bamford, C.H., Middleton, I.P., Al-Lamee, K.G., and Paprotny, J., *Br. Polym. J.*, 1987, **19**, 269.
171. Meijs, G.F., and Rizzardo, E., *J. Macromol. Sci., Rev. Macromol. Chem. Phys.*, 1990, **C30**, 305.
172. Capek, I., and Akashi, M., *J. Macromol. Sci., Rev. Macromol. Chem. Phys.*, 1993, **C33**, 369.
173. Rempp, P.F., and Franta, E., *Adv. Polym. Sci.*, 1984, **58**, 1.
174. Cacioli, P., Hawthorne, D.G., Laslett, R.L., Rizzardo, E., and Solomon, D.H., *J. Macromol. Sci., Chem.*, 1986, **A23**, 839.

7
Controlling Polymerization

7.1 Introduction

Free radical polymerization is often the preferred mechanism for forming polymers and most commercial polymer materials involve free radical chemistry at some stage of their production cycle. From both economic and practical viewpoints, the advantages of free radical over other forms of polymerization are many (see Chapter 1). However, one of the often-cited "problems" with free radical polymerization is a perceived lack of control over the process and the range of undefined defect structures and other forms of "structure irregularity" that may be present in polymers prepared by this mechanism. Much research over the last decade has been directed at providing answers for problems of this nature.

Minor (by weight or mole percent) functionality is introduced into polymers as a consequence of the initiation or termination processes (see Chapters 3 and 5 respectively). These groups may either be at the chain ends (as a result of initiation, transfer, disproportionation) or they may be part of the backbone (as a consequence of termination by combination or the copolymerization of by-products or impurities). In section 7.2 we consider two polymers (PS and PMMA) and discuss the types of defect structure that may be present, their influence on polymer properties, and the prospects for controlling these properties.

Structural irregularities may also be introduced in the propagation step either through a lack of regio- or stereochemical specificity in free radical addition to monomer or by rearrangement of the propagating species (see Chapter 4). In section 7.3, the influence of the reaction media and added reagents on the rate and outcome of radical polymerization is explored. With this knowledge, prospects exist for controlling polymer structure and properties by appropriate choice of reaction conditions (solvent, temperature, pressure) or through the use of complexing agents and templates to direct the course of polymerization.[1]

All polymer chains in a given sample are not the same. This is true irrespective of the polymerization mechanism and is a consequence of the statistics of the polymerization process. In section 7.4 we describe how it is possible to predict the compositional heterogeneity of copolymers. With this ability, the degree of heterogeneity can be controlled or minimized by careful selection of the monomers, initiators, transfer agents and by taking into account the reactivities (see Chapters 3 and 5) and the functionality of each when designing the polymerization conditions.

Living polymerization mechanisms offer polymers of controlled architecture and molecular weight distribution. In particular, they provide a route to narrow polydispersity homopolymers, high purity block copolymers and end-functional polymers. Traditional methods of living polymerization are based on ionic, coordination or group transfer mechanisms. In section 7.5, mechanisms for living radical polymerization (also known as quasi- or pseudo-living radical polymerization) are described. Control over the termination process is achieved through the use of reagents that reversibly terminate polymer chains.

7.2 Controlling Structural Irregularities

The functional groups introduced into polymer chains as a consequence of the initiation or termination processes can be of vital importance in determining certain polymer properties. Some such functionality is unavoidable. However, the types of functionality can be controlled and should not be ignored.

Such functionality becomes of great practical importance since functional initiators, transfer agents, *etc.* are applied to prepare end-functional polymers (see 6.3.1) or block or graft copolymers (see 6.3.2, 7.5). In these cases maximizing the fraction of chains which contain the reactive or other desired functionality is of obvious importance. However, there are also well-documented cases where "weak links" formed in the initiation or termination processes impair the thermal or photochemical stability of polymers (see 7.2.1 & 7.2.2). It is often possible to control these "weak links" by selecting the initiator (chapter 3), transfer agents (section 5.3) and reaction conditions according to their chemistry.

Thus, it is important to know and understand the kinetics and mechanism of the entire polymerization process so that desirable aspects of the polymer structure can be maximized while those reactions that lead to an impairment of properties or a less than ideal functionality can be avoided or minimized. A corollary is that it is important to know how a particular polymer was prepared before using it in a critical application.

7.2.1 *"Defect Structures" in Polystyrene*

There is a substantial literature on the thermal and photochemical degradation of PS and it is well-established that polymer properties are sensitive to the manner in which a particular sample of PS is prepared. For example, it has been reported that PS prepared by anionic polymerization shows enhanced stability with respect to that prepared by a radical mechanism.[2-8] This has often been attributed to the presence of "weak links" in the latter polymers. However, the nature of the "weak links" remains the subject of some controversy. The situation is further confused by PS prepared by radical mechanisms often being considered as a class without reference to the particular polymerization conditions employed in their preparation. In many cases, the polymers are "commercial samples" with details of the method of preparation incomplete or unstated.

In some cases the "weak links" in radical PS may be peroxidic linkages. Such groups may become incorporated in polymers formed by radical polymerization through copolymerization of adventitious oxygen (see 5.4.2). However, there is also evidence that thermal behavior depends on the particular radical initiator or reaction conditions (solvent, temperature, conversion) employed in polymer preparation.

BPO is commonly used as an initiator for S polymerizations and copolymerizations and it has been reported that its use can lead to yellowing and impaired stability in PS.[9,10] The initiation and termination pathways observed for S polymerization when BPO is used as initiator are discussed in section 3.2. These give rise to end groups as follows (Scheme 1).

NMR studies[11,12] on polymers prepared with ^{13}C-labeled BPO have shown that the primary benzoyloxy and phenyl end groups formed by tail addition to monomer are thermally stable under conditions where the polymer degrades (they persist to > 50% weight loss at 300°C under nitrogen). These groups are unlikely to be directly responsible for the poor thermal stability of PS prepared with BPO as initiator. On the other hand, the secondary benzoate end groups formed by head addition or transfer to initiator appear extremely labile under these conditions (their half life at 300°C is <5 min).

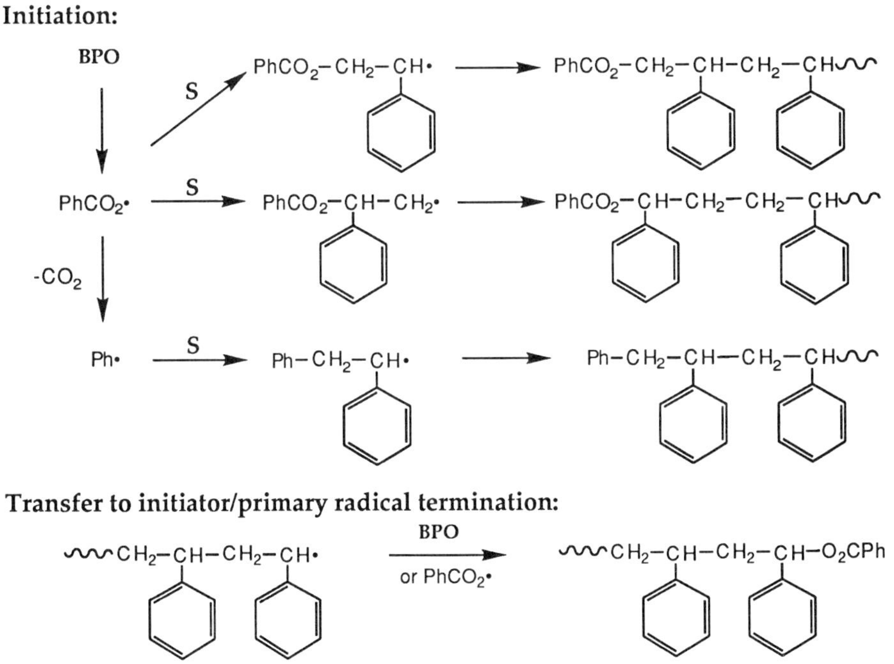

Initiation:

Transfer to initiator/primary radical termination:

Scheme 1

Studies with model compounds show that secondary benzoates eliminate benzoic acid to form unsaturated chain ends (see Scheme 2).[11] Unsaturation has long been thought to be a "weak link" in PS.[4,13] It is known that at high conversion most chain termination may be by way of transfer to initiator or primary radical termination.[14]

Thus, if these groups are responsible for initiating the chain degradation process, it provides a plausible explanation for high conversion PS formed with BPO initiator being less thermally stable than either a similar low conversion polymer or a polymer prepared with a different initiator.

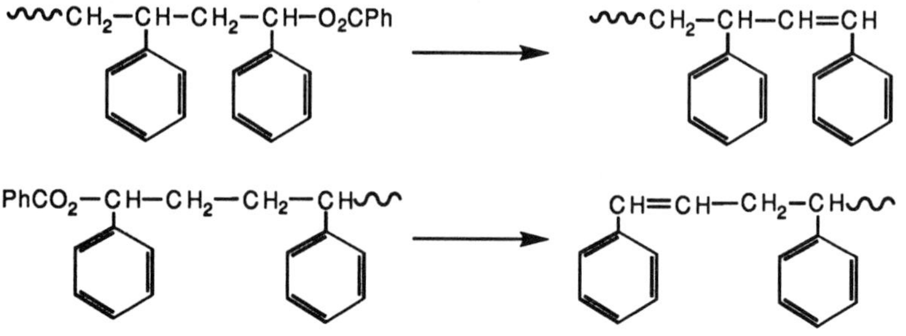

Scheme 2

This example shows how initiator selection can be critical in determining the properties of PS prepared by free radical polymerization. If thermal stability is of importance, then, since some initiator-derived ends cannot be avoided, a preferred initiator would be one which give rises to end groups which cannot readily eliminate. End groups formed with AIBN initiator appear stable with respect to the polymer backbone,[15] other systems remain to be studied.

7.2.2 "Defect Structures" in Poly(methyl methacrylate)

Unstable structures can also arise in the chain termination process. Mechanisms for radical-radical termination in MMA polymerization have been discussed in sections 5.2.1 and 5.2.1.1 and these are summarized in Scheme 3. Both disproportionation and combination occur to substantial extents.

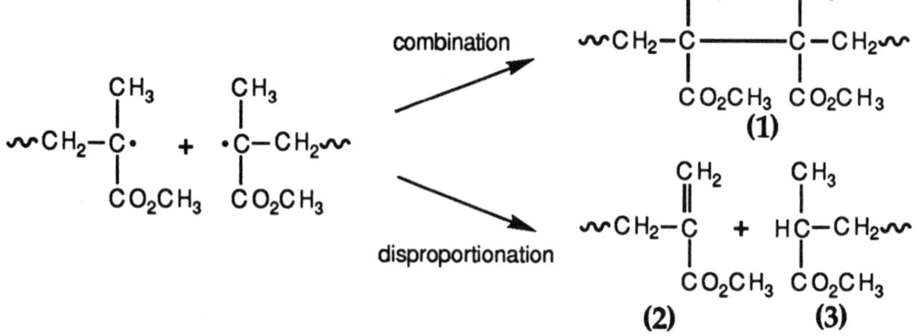

Scheme 3

It has been shown that the head-to-head linkages (1) and the unsaturated chain ends (2) constitute "weak links" in PMMA.[16-22] The presence of these groups is thought to account for PMMA formed by radical polymerization being significantly less stable than that formed by anionic polymerization. If terminated by a proton source, anionic PMMA will have saturated chain ends (3).

PMMA degrades by unzipping (*i.e.* the reverse of the polymerization process). Initiation of this process requires generation of a propagating radical. The head-to-head linkage (1) is thermally unstable at temperatures above 180°C and may undergo spontaneous scission to form two propagating radicals (Scheme 4).

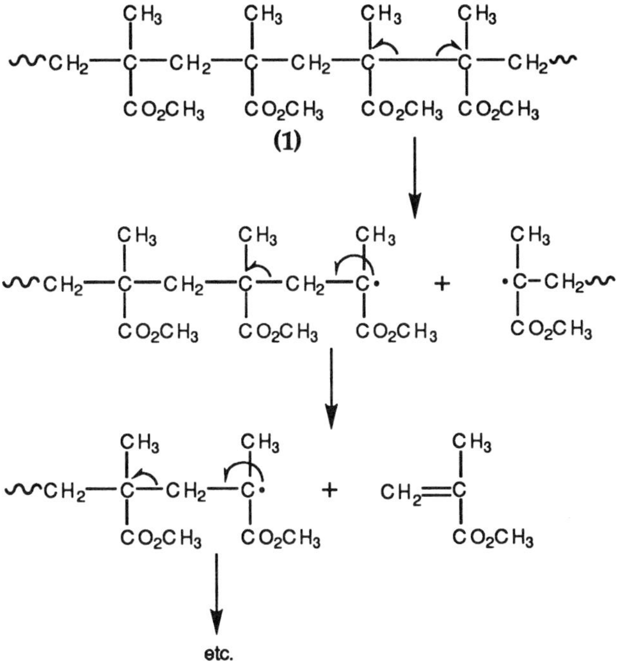

Scheme 4

The bond β- to the double bond of the unsaturated disproportionation product (2) may also be weaker than other backbone bonds. However it is believed that the instability of unsaturated linkages is due to a radical-induced decomposition mechanism (Scheme 5).[19] This mechanism for initiating degradation is analogous to the addition-elimination chain transfer observed in polymerizations carried out in the presence of (2) at lower temperatures (see 5.3.1.2 and 6.3.2.3).

To avoid these stability problems, it is necessary to minimize the proportion of chains that terminate by radical-radical reaction. One way of achieving this is by conducting the polymerization in the presence of an appropriate chain transfer agent. For example, if polymerization is performed with added thiol chain transfer agent, then conditions can be chosen such that most chains terminate by hydrogen-atom transfer with the thiol. These polymer chains will then possess the more thermally stable, saturated end groups (3, see Scheme 3).

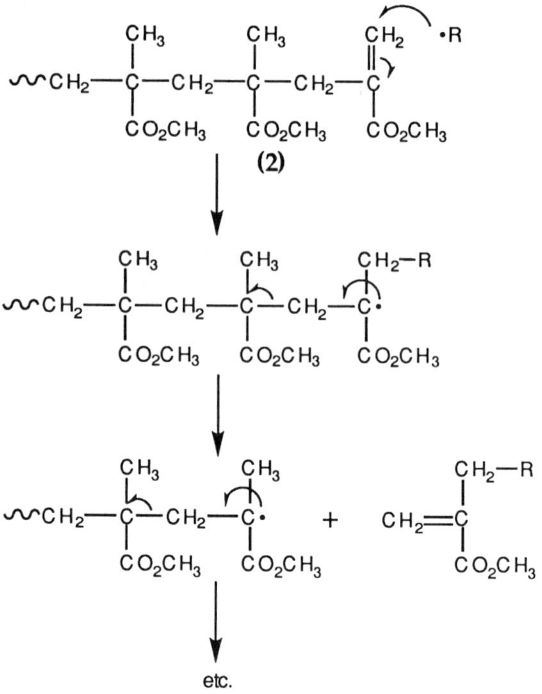

Scheme 5

There are other sources of unsaturated chain ends in PMMA formed by radical polymerization:

(a) End groups (2) are formed in chain transfer to certain addition-fragmentation transfer agents (*e.g.* allyl sulfides, see 5.3.2.6) or cobalt chain transfer agents (see 5.2.3.7).
(b) Unsaturated chain ends can arise during chain initiation if transfer to monomer occurs. The hydrogens of the α-methyl and, to a lesser extent, the ester methyl of MMA may be abstracted. If the monomer-derived radicals initiate polymerization, the polymer will contain end groups (4) and (5). The *t*-butoxy and other *t*-alkoxy radicals show a propensity for this chemistry (see 3.4.2.1).[23-25]

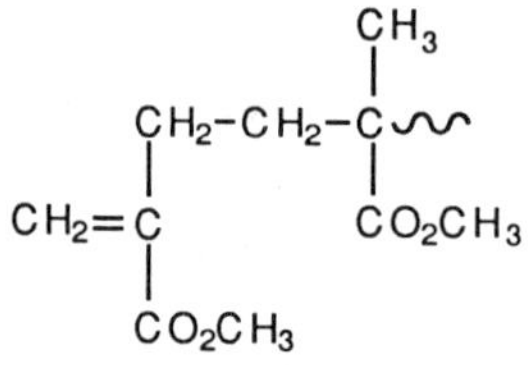

(4) 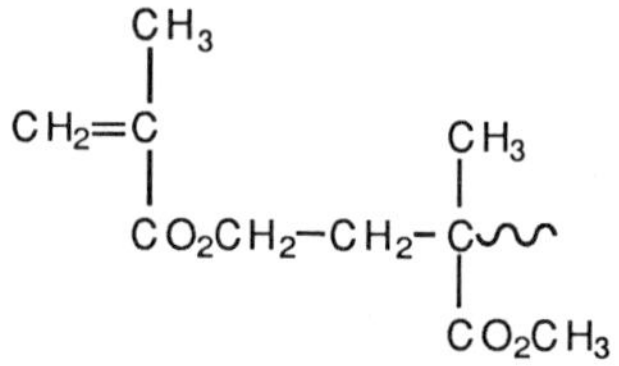   (5)

Note that chain ends (4, 5) may give different chemistry to those formed in termination by disproportionation (3, see Scheme 3). Chain scission β to the double bond will not lead to a MMA propagating species. It is not established whether the presence of these ends will give impaired thermal stability.

However, the presence of unsaturated chain ends can have other consequences for polymer properties:

(a) Those chains initiated by abstraction products will not contain an initiator residue.[25] Experiments which depend on determination of initiator-derived chain ends may be in error and some literature data may need to be reinterpreted in this light.[26] Syntheses of telechelic or end-functional polymers based on the use of functional initiators will also be detrimentally affected (see 6.3.1.1).
(b) The unsaturated end groups (2, 4, and 5) may be reactive under polymerization conditions (*i.e.* the polymer chains can be considered as macromonomers) and may copolymerize leading to graft or crosslink formation (see 6.3.2.3).[27] The end groups (2) may also give chain transfer by an addition-fragmentation mechanism (see 5.3.1.2 and 6.3.2.3).

7.3 Controlling Propagation

Given the important role that steric and polar factors play in determining the rate and regiospecificity of radical additions (see 2.2), it should be anticipated that reagents which coordinate with the propagating radical and/or the monomer and thereby modify the effective size, polarity, or inherent stability of that species, could alter the outcome of propagation.

These reagents used for controlling polymer structure may be low molecular weight (*e.g.* the solvent - see 7.3.1, Lewis acids - see 7.3.2) or polymeric (*e.g.* template polymers - see 7.3.3) in nature. For greatest effect propagation involving the complexed species should dominate over normal propagation. For this to occur one of the following should apply:

(a) The monomer or propagating species is completely complexed. This requires that concentration of the reagent whether mono- or polymeric to be at least stoichiometric with the species to be complexed throughout the polymerization.
(b) The reactivity of the complexed species is many-fold greater than that of any remaining uncomplexed species. It is also required that the equilibrium and rate constants associated with complex formation are high.

Bearing these requirements in mind, the more desirable way of controlling propagation would appear to be to complex the propagating radical (P•). Whereas the initial monomer concentrations are typically in the range 2-10 M, the typical "steady state" concentration of P• is usually very low (~10^{-7} M) (see Scheme 6). Therefore, only a small concentration of a catalytic reagent is required to complex all radicals. However, for this to occur, the reagent should interact specifically with P• and not associate strongly with either the monomer or the polymer. In any competitive equilibrium, the difference in concentrations (up to 10^8-fold) will clearly favor interaction with monomer or polymer over P•.

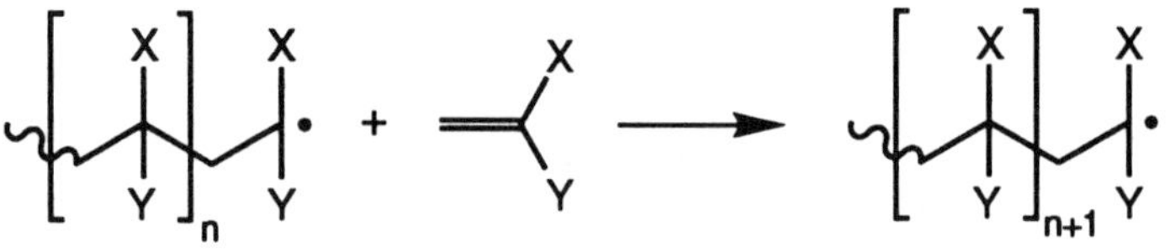

$$[P\bullet] \sim 10^{-7} M \qquad [M]_o\ 2\text{-}10\ M$$

Scheme 6

In seeking a suitable complexing agent for the propagating species, one approach is to consider the various species (X•) which are known to reversibly add carbon-centered radicals (*e.g.* Scheme 7). Many such reagents have been described in the organic literature. Notable examples are nitroxides[28,29] and various organometallic complexes.[30] These species react with carbon-centered radicals at near diffusion controlled rates[31,32] yet the X–R bonds in the adducts are relatively weak. The bond strength depends on the nature of R and the functionality on X. On heating, or on irradiation with UV light, the X-R bond may undergo reversible homolysis allowing monomer insertion by a radical mechanism. Could the proximity of X influence the course of propagation?

Scheme 7

Most of the studies on polymerization have been concerned with studying the utility of these reagents in living polymerization and controlling the rate of polymerization and polydispersity of polymers formed (see 7.5.3 and 7.5.4). Only a few have explicitly looked for effects on polymer structure and to date there is no definitive evidence that complexation of radicals by the reagents discussed above influences the regio- or stereospecificity of radical addition.

A report[33] that the stereoregularity of MMA propagation is influenced by Co-porphyrin has not been confirmed by subsequent studies. Giese *et al.*[34] reported that cyclohexyl radicals generated from alkylcobaloximes and cyclohexyl radicals generated from other sources show different specificity in atom transfer reactions. However, they[34] and Clarke and Jones[35] have also provided evidence that the radicals generated from square planar cobalt complexes behave as "normal" free radicals in simple radical additions. The utility of cobalt complexes as complexing agents in controlling propagation is limited by side reactions which give chain transfer (these may be used to advantage - see 5.3.2.7). There is some evidence that the importance of these reactions might be controlled by limiting the application to monosubstituted monomers (see 7.5.4) or by changing the ligands on cobalt.

7.3.1 Solvent

Solvent effects on radical reactions often appear insignificant when compared with the effects of several orders of magnitude often observed in reactions of ionic species.[36] Nonetheless, small, yet easily discernible, effects of solvent on radical reactivity have been reported even for reactions involving neutral radicals and monomers. An attractive feature of using the solvent as an agent to control propagation in solution polymerizations, is that the solvent is often present in very large excess in relation to the species to be complexed. Of course, economic, solubility, toxicity, waste disposal, and other considerations limit the range of solvents that can be employed in an industrial polymerization.

Most literature on solvent effects on the propagation step of radical polymerization deals with influences of the medium on reaction rate. Where monomers or radicals are charged, readily ionizable, or capable of forming hydrogen bonds, mechanisms whereby the solvent could affect radical reactivity seem obvious. For other systems mechanisms are still a matter of some controversy even in the case of small radicals (see 2.3.6).

There are at least three mechanisms whereby the solvent might modify the outcome of a radical process:

(a) Formation of a monomer or radical complex with different reactivity and/or specificity than the uncomplexed species.
(b) Solvation of the transition state or intermediate which may have polar character.
(c) Preferential solvation leading to local concentrations of the various reactants being different from those in the medium as a whole.

At least three forms of radical-solvent interaction should be considered:

(a) Reversible addition to the solvent molecule. For example, formation of a cyclohexadienyl radical in the case of aromatic solvents.
(b) Formation of a charge transfer complex.
(c) Orbital interaction with a C-H σ-bond or a π-system but without development of charge separation or bond formation.[37]

Solvent effects on radical reactions are discussed in general terms in Chapter 2 (see 2.2.5.2 & 2.3.6). In this section, examples of solvent effects which involve a reaction of the solvent as a transfer agent or a comonomer will not be considered. These properties of the solvent have been discussed in Chapters 3 and 5. In this section we concentrate on effects of the solvent on the rate and specificity of the propagation step of free radical polymerization.

7.3.1.1 Homopolymerization

The values of the rate parameters in certain homopolymerizations have been reported to be solvent dependent.[38-41] Large solvent effects have been reported for monomers which are ionizable (e.g. MAA, AA), give precipitation polymerization (AN), or contain hydroxy or amide groups (HEA, Am). For monomers not in these classes, the most extensively studied homopolymerizations are those of vinyl esters,[42,43] methacrylate derivatives,[38,44] and S.[45] These studies have focused wholly on the polymerization kinetics and only a few have examined the microstructures of the polymers formed. Some of the rate data in this area should be treated circumspectly because of the difficulties associated in separating effects of solvent on k_p, k_t and initiation rate and efficiency.

One of the most dramatic examples of a solvent effect on propagation is for vinyl acetate polymerization. Kamachi et al.[42] reported a ca. 80-fold reduction in k_p (30°C) on changing from ethyl acetate to benzonitrile solvent (see Table 7.2). Hatada et al.[46] conducted a ^{1}H NMR study on the structure of the PVAc formed in various solvents. They found that PVAc ($\bar{M}_n$~20000) produced in ethyl acetate solvent has ~0.7 branches/chain while that formed in aromatic solvents is essentially unbranched.

The solvent effects on k_p in S[45,47] and MMA[38,44,47] polymerization are small by comparison (<20%, see Table 7.1). Several groups have investigated the effect of aromatic solvents on k_p for MMA.[38,44] While the absolute numbers are not in close agreement the observed trends are similar. The values of k_p(30°C) for MMA reported by Kamachi et al.[44] are shown in Table 7.2. These values for aromatic solvents are all high in relation to the currently accepted bulk value suggesting that all aromatic solvents enhance k_p for this monomer.

Very large solvent effects are observed for systems where the monomers can aggregate either with themselves or another species. Micelle-forming surfactants,[49] for example, vinyl pyridinium salts and alkyl salts of dimethylaminoalkyl methacrylates in aqueous solution show this behavior. Rates of polymerization for these systems are dramatically higher in water than they are in non-aqueous media, though this does not necessarily mean a higher value for k_p. The heterogeneity of the medium needs to be considered. The effective concentration of double bonds in the vicinity of the propagating species can be up to 100-fold greater than the concentration of monomer in the solution considered as a whole. The number of surfactant molecules per micelle can determine the molecular weight. The microstructure (tacticity) of the polymer chains was claimed to be the same as that obtained in bulk polymerization.

The heterogeneity of the reaction medium is also important in determining the molecular weight and k_p in solution polymerization of macromonomers[50] and in template polymerizations (see 7.3.3). The magnitude of the effect varies according to the solvent quality. PS macromonomer chains in good solvents (e.g. toluene) have an extended conformation whereas in poor solvents (e.g. methylcyclohexane) chains are tightly coiled.[50] As a consequence, the radical center may see an environment which is medium dependent.

Table 7.1 Solvent Effect on Propagation Rate Constants at 30°C

Monomer	Solvent	$k_p \times 10^{-2}$ (M^{-1} s^{-1})
VAc (2.00 M)[42]	benzonitrile	8
	phenyl acetate	37
	anisole	48
	chlorobenzene	61
	ethyl benzoate	37
	fluorobenzene	97
	benzene-d_6	113
	benzene	117
	ethyl acetate	637
MMA (2.00 M)[44]	benzonitrile	6.1
	anisole	5.1
	chlorobenzene	5.0
	fluorobenzene	4.5
	benzene	4.5
	bulk[a]	3.7

[a] For bulk MMA.[48]

7.3.1.2 Copolymerization

The effects of solvent on radical copolymerization have been covered by a number of reviews.[40,41,51] For copolymerizations involving monomers which are ionizable or form hydrogen bonds (Am, MAm, HEA, HEMA, MAA, *etc*.) solvent effects on reactivity ratios can be dramatic. Some data for the MAA-MMA system is shown in Table 7.2.[52]

Table 7.2 Solvent Dependence of Reactivity Ratios for MMA-MAA Copolymerization at 70°C[52]

solvent	r_{MMA}	r_{MAA}
toluene	0.10	1.06
dioxane	0.12	1.33
acetonitrile	0.27	0.03
acetone	0.31	0.63
DMSO	0.78	0.23
isopropanol	0.78	0.33
ethanol	0.80	0.60
acetic acid	0.80	0.78
DMF	0.98	0.68
water	2.61	0.43

For MMA-MAA copolymerizations carried out in the more hydrophobic solvents (toluene, dioxane), MAA is the more reactive towards both propagating species while in water MMA is the more reactive. In solvents of intermediate polarity (alcohols, dipolar aprotic solvents), there is a tendency towards alternation. For these systems, choice of solvent could offer a means of controlling copolymer structure.

For non-protic monomers solvent effects are less marked. Indeed, early work suggested that the reactivity ratios in copolymerizations involving only non-protic monomers (*e.g.* S, MMA, AN, VAc, *etc.*) showed no substantial solvent dependence.[53,54] More recent studies on these and other systems (*e.g.* AN-S,[55-58] E-VAc,[59] MAN-S,[60] MMA-S,[61-63] MMA-VAc[64]) indicate small yet significant solvent effects (some recent data for AN-S copolymerization are shown in Table 7.3). However, the origin of the solvent effect is not clear. There have been various attempts to rationalize solvent effects on copolymerization by establishing correlations between radical reactivity and various solvent and monomer properties.[40,41,51,52] None has been entirely successful.

Table 7.3 Solvent Dependence of Penultimate Model Reactivity Ratios for S-AN Copolymerization at 60°C[56]

solvent	r_{SS}	r_{AS}	r_{SA}	r_{AA}
bulk	0.232	0.566	0.087	0.036
toluene	0.242	0.566	0.109	0.133
acetonitrile	0.322	0.621	0.105	0.052

Recently, Harwood proposed that the solvent need not directly affect monomer reactivity, rather it may influence the way the polymer chain is solvated.[56,65-69] Evidence for this proposal is the finding that, for certain copolymerizations, while the terminal model reactivity ratios appear solvent dependent, copolymers of the same overall composition have the same monomer sequence distribution. Harwood proposed that the choice of solvent may determine the monomer concentrations in the vicinity of the reactive chain end. He called this phenomenon "the bootstrap effect".[65,66] A partition coefficient K was defined as follows:

$$K = \frac{[M_A]/[M_B]}{[M_A^o]/[M_B^o]}$$

where $[M_A]/[M_B]$ is the ratio of monomer concentrations in the vicinity of the reactive chain end and $[M_A^o]/[M_B^o]$ is the global ratio. The conditional probabilities which determine the triad fractions are dependent on $[M_A]/[M_B]$ rather than $[M_A^o]/[M_B^o]$. The value of $[M_A]/[M_B]$ is determined by the polymer composition and the solvent. Recent work has shown that the effects of solvent on many copolymerizations can be interpreted in terms of this theory, including those of S with MAH,[67] S with MMA,[67,68] and S with AN.[56,69] It must be pointed out that, for these latter systems, while the experimental data are consistent with the bootstrap effect, it is not necessary to invoke the bootstrap effect for data interpretation.

Studies on the reactions of small model radicals with monomers provide indirect support for the bootstrap effect.[70] Krstina et al.[70] recently demonstrated that reactivity of MMA and MAN model radicals towards MMA, S and VAc were independent of solvent. However, significant solvent effects on reactivity ratios have been reported for MMA/VAc[64] and MMA/S[67,68] copolymerizations. For the model systems, since there is no polymer coil to solvate there should be no bootstrap effect. Reactivities are determined by the global monomer ratio $[M_A]/[M_B]$.[70]

Other phenomena which may be related to the bootstrap effect include the dependence of copolymer composition on molecular weight in certain copolymerizations[71] and the observation of significant solvent effects in macromonomer copolymerization. Tsukahara et al.[72] found that when copolymerizing macromonomers, the choice of solvent has a substantial influence on the reactivity ratios, the molecular weight of the polymer, and the particle size distribution of the final product. They interpreted their data in terms of the effects of solvent on the degree of interpenetration between unlike polymer chains.

7.3.2 Lewis Acids and Inorganics

Lewis acids are known to form complexes both with monomers and with propagating species. Their addition to a polymerization medium, even in catalytic amounts, can bring about dramatic changes in rate constants in homopolymerization (see 7.3.2.1) and reactivity ratios in copolymerization (see 7.3.2.2). This area has been reviewed most recently by Bamford[73] and Barton and Borsig.[40]

7.3.2.1 Homopolymerization

In 1957, Bamford et al.[74] reported that the addition of small amounts of lithium chloride brought about a significant (up to two-fold) enhancement in the rate of polymerization of AN in DMF and led to a higher molecular weight polymer. Subsequent studies have shown this to be a more general phenomenon for polymerizations involving, in particular, acrylic and vinylheteroaromatic monomers in the presence of a variety of Lewis acids.[40]

For the case of polymerization of AN in DMF, measurements of the absolute rate constants associated with the polymerizations indicated that the rate of initiation (by AIBN) was not significantly affected by added lithium salts. The enhancement in the rate of polymerization was largely attributed to an increase in k_p. The value of k_t remains essentially unchanged except when very high concentrations of the Lewis acid are employed.

Zubov et al.[75] suggested that during MMA polymerization in the presence of Lewis acids ($AlBr_3$) complexation occurs preferentially with the propagating radical rather than with monomer. They suggested a mechanism in which the metal ion is transferred to the incoming monomer in the transition state for addition so as to remain with the active chain end. It is known that Lewis acids can bring about significant changes in the appearance of the EPR spectra of MMA propagating radicals and related species.[76]

Although it is clear that added Lewis acids affect the rate of polymerization and the molecular weight of homopolymers formed in their presence,[40] it is not yet firmly established whether they modify the polymer structure. There are reports that Lewis acids affect the tacticity of PMMA.[40] Otsu and Yamada[77] found a slightly greater proportion of isotactic placements for the bulk polymerization of a 1:1 complex of MMA with zinc chloride than for a similar polymerization of MMA alone. However, for polymerizations carried out in solution or in the presence of lesser amounts of zinc chloride, no effect was observed.[77] It is also possible that complexation of monomer or propagating species could influence the regiospecificity of addition. However, since the effect is likely to be an enhancement of the usual tendency for head-to-tail addition it perhaps is not surprising that such effects have not been reported.

7.3.2.2 Copolymerization

The kinetics of copolymerization and the microstructure of copolymers can be markedly influenced by the addition of Lewis acids. In particular, Lewis acids are effective in enhancing the tendency towards alternation in copolymerization of donor-acceptor monomer pairs and give dramatic enhancement in the rate of copolymerization. Copolymerizations of acrylic monomers (*e.g.* MA, MMA) and S have been the most widely studied.[75,78-81] Strictly alternating copolymers of MMA and S can be prepared in the presence of, for example, diethylaluminum sesquichloride (reactivity ratios in absence of Lewis acid are ca. 0.51 and 0.49 - see 6.2.1.1).

Three basic mechanisms (not mutually exclusive) for the influence of Lewis acid on copolymerization have been proposed:

(a) The Lewis acid forms a binary complex with the acceptor monomer. The electron deficiency of the double bond is enhanced by complexation with the Lewis acid and thus its reactivity towards nucleophilic radicals is greater.
(b) A ternary complex is formed between acceptor, donor, and Lewis acid. An alternating polymer may be formed by homopolymerization of such a complex.
(c) Complexation of the propagating radical to create a species with selectivity different to that of the normal propagating species.

In an effort to distinguish these mechanisms studies on model propagating species have been carried out.[82,83] For S-MMA polymerization initiated by AIBMe-α-^{13}C (see Scheme 8) it has been established by end group analysis that extremely small amounts of ethyl aluminum sesqichloride (<10^{-3} M with 1.75 M monomers) are sufficient to cause a substantial enhancement in specificity for adding S in the initiation step. The result suggests that complexation of the propagating radical may be sufficient to induce alternating copolymerization but does not rule out other hypotheses.

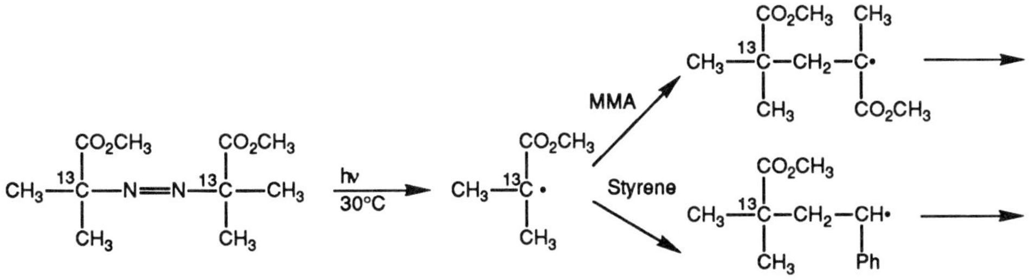

Scheme 8

7.3.3 Template Polymers

The possibility of using a template polymer to organize the monomer units prior to their being "zipped up" by the attack of a radical species has long attracted interest and has been the subject of a number of reviews.[84-86] Template polymerization, as its name suggests, involves the formation of a *daughter* polymer on a preformed *parent* polymer. The interest in this area may be seen to stem from the biological area where the phenomenon is well known and accounts for the regularity in the structure of natural proteins and polynucleotides.

The literature distinguishes two limiting forms of template polymerization:[84-86]

(a) Where the monomer is associated with the template and, ideally, initiation, propagation, and termination all occur on the template. The polymerization may be viewed as a ladder polymerization.
(b) Where only the propagating chain associates with the template. The rate of polymerization is limited by the rate at which monomer is attached from the bulk solution.

The interaction of the template with monomer and/or the propagating radical may involve solely Van der Waals forces or it may involve charge transfer complexation, hydrogen bonding, or ionic forces (see 7.3.3.1). Other systems involve formal covalent bonding (see 7.3.3.2).

7.3.3.1 Non-covalently bonded templates

In 1972 Buter et al.[87] reported that polymerization of MMA in the presence of isotactic PMMA leads to a greater than normal predominance of syndiotactic sequences during the early stages of polymerization. Other investigations of this system supporting[88] and disputing[89] this finding have appeared. The mechanism of the template polymerization is thought to involve initial stereocomplex formation between the oligomeric PMMA propagating radical (predominantly syndiotactic) and the isotactic template polymer with subsequent monomer additions being directed by the environment of the template. Isotactic and syndiotactic PMMA have been shown to form a 1:2 stereocomplex.[88]

The nature of the interaction between the monomer and the template is more obvious in cases where specific ionic or hydrogen bonding is possible. For example, N-vinylimidazole has been polymerized along a PMAA template[90,91] and acrylic acid has been polymerized on a N-vinylpyrrolidone template.[92] The daughter PAA had similar X_n to the template and had more isotactic triads than the PAA formed in the absence of the template.

It is well known that rates of polymerizations can increase markedly with the degree of conversion or with the polymer concentration. Some workers have attributed this solely or partly to a template effect. It has been proposed[93] that adventitious template polymerization occurs during polymerizations of AA, MAA and AN and that the gel or Trommsdorff effect observed during polymerizations of these monomers is linked to this phenomenon. However, it is difficult to separate possible template effects from the more generic effects of increasing solution viscosity and chain entanglement at high polymer concentrations on rates of termination and initiator efficiency (see 5.2.1).

There are also reports of template effects on reactivity ratios in copolymerization. For example, Polowinski[94] has reported that both kinetics and reactivity ratios in MMA-MAA copolymerization in benzene are affected by the presence of a PVA template.

$$CH_2=CH-\overset{O}{\underset{\|}{C}}-O-CH_2-CH_2-O-\overset{O}{\underset{\|}{C}}-\underset{NO_2}{\overset{NO_2}{\bigcirc}}$$

(6)

A template polymer may allow the use of monomers which do not otherwise undergo polymerization. An example is the dinitrobenzoate derivative (6). The dinitrobenzene derivatives are usually thought of as radical inhibitors (see 5.4.5) thus radical polymerization of such monomers is unlikely to be successful. Polymerization of (6) on a poly(N-vinylcarbazole) template succeeded in producing a high molecular weight polymer.[95] It can be envisaged that the monomer (6) forms a charge transfer complex with the electron donating carbazole group.

7.3.3.2 *Covalently bonded templates*

Template polymerizations where the monomer is covalently bound to the template clearly have limitations if polymers of high molecular weight or large quantities are required. However, their use offers much greater control over daughter polymer structure. The product in such cases is a ladder polymer and this may be viewed as a special case of cyclopolymerization (see 4.4.2).

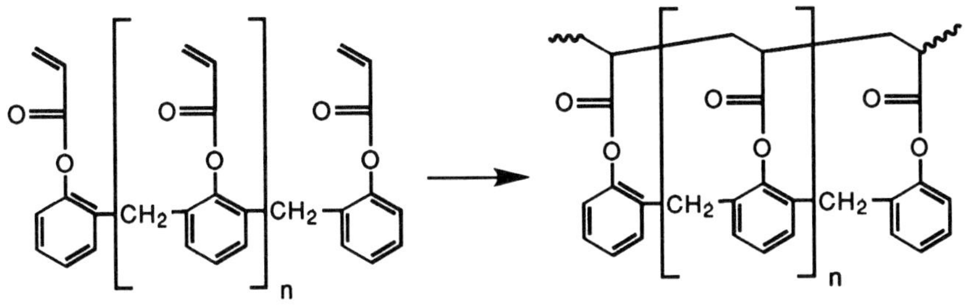

Scheme 9

Kämmerer et al.[96-98] have conducted extensive studies on the template polymerization of acrylate or methacrylate derivatives of polyphenolic oligomers (7) with $\bar{X}_n \leq 5$ (Scheme 9). Under conditions of low "monomer" and high initiator concentration they found that $\bar{X}_n$ for the daughter polymer was the same as $\bar{X}_n$ for the parent. The possibility of using such templates to control microstructure was considered but not reported.

Feldman et al.[99,100] and Wulff et al.[101] have examined other forms of template controlled oligomerization of acrylic monomers. The template (8) has initiator and transfer agent groups attached to a rigid template of precisely defined structure.[99,100]

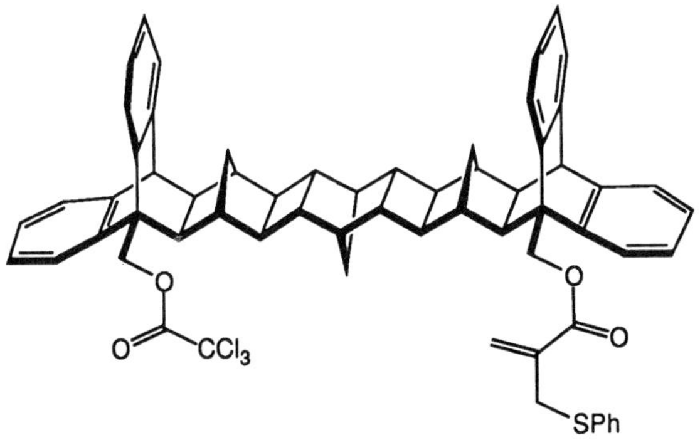

Polymerization of MMA in the presence of (8) gave a 3 unit oligo(MMA) as ca. 66% of the polymeric product. The stereochemistry of the oligomer was reported to be "different" from that of atactic PMMA.

Wulff et al.[101] attached vinyl groups to a large chiral sugar based template molecule and then copolymerized this substrate with various monomers. With MMA and MAN they achieved an unspecified degree of optical induction.

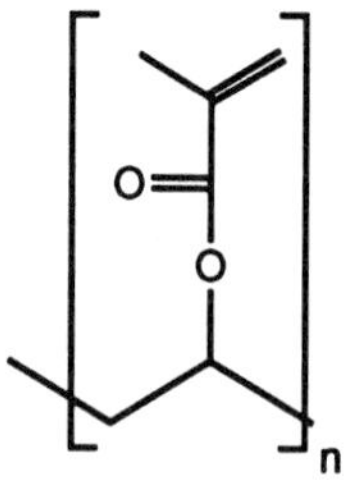

(9)

This approach has been extended in studies of higher molecular weight systems.[102-105] In this work, PVA was esterified with methacryloyl chloride to give a "multimethacrylate" (9) and polymerized to give a ladder polymer. "Multimethacrylates" based on PHEMA were also described. The daughter polymer was hydrolyzed to PMMA but only characterized in terms of molecular weight. The value of $\bar{X}_n$ for the daughter polymer was greater than $\bar{X}_n$ for the parent template indicating some inter-template reaction. These workers also examined the copolymerization of partially methacrylated PVA with MMA. It has not been established whether the tacticity of parent PVA or the presence of head-to-head and tail-to-tail linkages has an effect on the microstructure of the daughter polymer.[102]

A new form of template polymerization based on ring-opening polymerization of a 4-methylenedioxalane units has been reported by Endo and coworkers (Scheme 10).[106,107] For this system, the monomer is covalently bound and the daughter polymer is released from the template in the polymerization process.

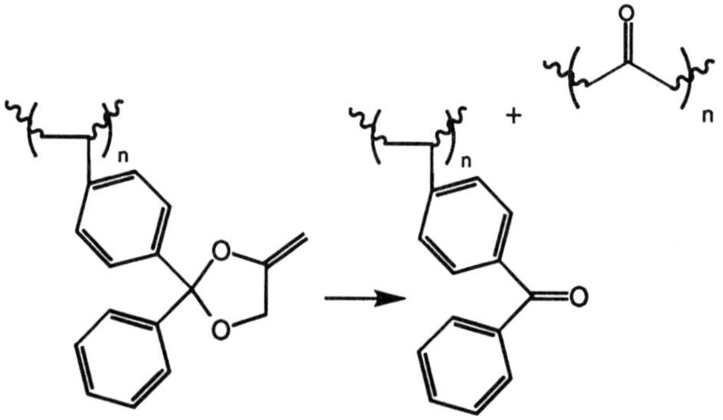

Scheme 10

7.4 Compositional Heterogeneity in Copolymers

Chain compositional heterogeneity is of particular relevance in the synthesis of functional copolymers which find widespread use in the coatings and adhesives industries.[108,109] Stockmayer[110] was one of the first to report on the problem of compositional heterogeneity and presented formulae for calculating the instantaneous copolymer composition as a function of chain length. Others[111-115] have examined the variation in copolymer composition with chain length by computer simulation.

Various factors are important in determining the composition and molecular weight distribution of multicomponent copolymers (*e.g.* monomer reactivity ratios, reaction conditions). However, the possible influence of selectivity in the initiation and termination steps on the distribution of monomer units within the copolymer chain is often ignored. Galbraith *et al.*[114] provided the first detailed analysis of these factors. They applied Monte Carlo simulation to examine the influence of the initiation and termination steps on the compositional heterogeneity and molecular weight distribution of binary and ternary copolymers. Spurling *et al.*[115] extended this treatment to consider the effects of conversion on compositional heterogeneity.

The ends of polymer chains are often not representative of the overall chain composition. This arises because the initiator and transfer agent derived radicals can show a high degree of selectivity for reaction with a particular monomer type (see 3.4). Similarly, there is specificity in chain termination. Transfer agents show a marked preference for particular propagating species (see 5.3.2). The kinetics of copolymerization are such that the probability for termination of a given chain by radical-radical reaction also has a marked dependence on the nature of the last added units (see 5.2.3.5).

The effect of the initiation and termination processes on compositional heterogeneity can be seen in data presented in Figures 1 and 2. The data come from a computer simulation of the synthesis of a hydroxy functional oligomer prepared from S, BA, and HEA with a thiol chain transfer agent. The recipe is similar to those used in some coatings applications.

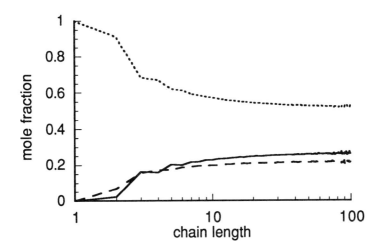

Fig. 1 Distribution of monomers [HEA(— — —), BA(———), S (-------)] within chains as a function of chain length for a HEA:BA:S copolymer prepared with butanethiol chain transfer agent.[115]

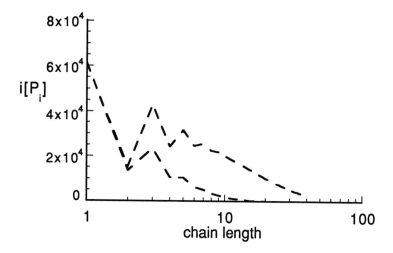

Fig. 2 Molecular weight distributions for HEA:BA:S copolymer prepared with butanethiol chain transfer agent: (a) all chains (———); (b) chains without HEA (— — —).[115]

In this copolymerization, most termination is by chain transfer and most chains are initiated by transfer agent-derived radicals. The thiyl radicals generated from the transfer agent react faster with S than they do with acrylate esters (Scheme 11).

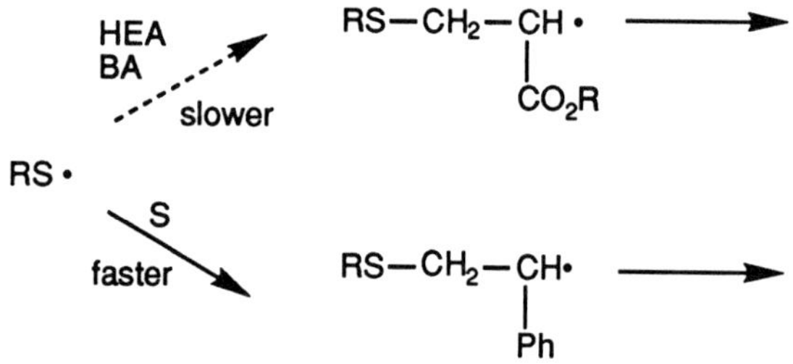

Scheme 11

The thiol shows a preference to react with propagating radicals with a terminal S unit (Scheme 12). This selectivity is due both to chemospecificity in the reaction with thiol and to the relative concentrations of the various propagating species (determined by the reactivity ratios).

A preponderance of chains that both begin and end in S results and this means that short chains are much richer in S than in the acrylic monomers (see Figure 1). This also has an influence on the fraction of chains which contain the functional monomer (see Figure 2). The fraction of HEA in very short chains is much less than that in the polymer as a whole and a significant fraction of these short chains contain no functional monomer.

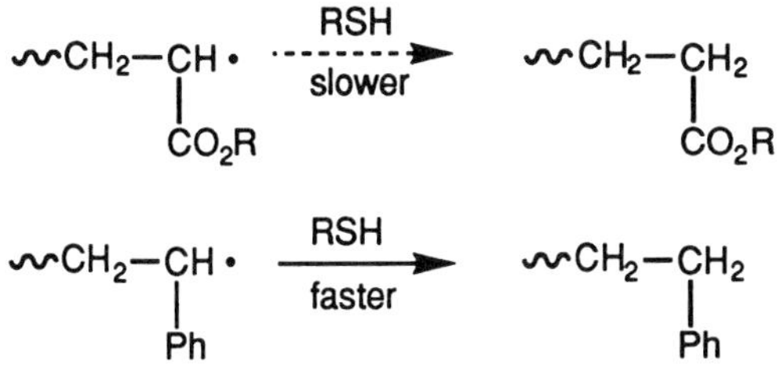

$$\text{\textasciitilde CH}_2\text{-CH}\cdot\text{(CO}_2\text{R)} \xrightarrow[\text{slower}]{\text{RSH}} \text{\textasciitilde CH}_2\text{-CH}_2\text{(CO}_2\text{R)}$$

$$\text{\textasciitilde CH}_2\text{-CH}\cdot\text{(Ph)} \xrightarrow[\text{faster}]{\text{RSH}} \text{\textasciitilde CH}_2\text{-CH}_2\text{(Ph)}$$

Scheme 12

In this copolymerization, the reactivity ratios are such that there is a tendency for styrene and the acrylic monomers to alternate in the chain. This, in combination with the above-mentioned specificity in the initiation and termination steps, causes chains with an odd number of units to dominate over those with an even number of units.

It is possible to exercise control over this form of compositional heterogeneity (*i.e.* the functionality distribution) by careful selection of the functional monomer and/or the transfer agent taking into account the reactivities of the radical species, monomers, and transfer agents, and the functionality of each.[114,115]

7.5 Agents For Controlling Termination

In conventional living polymerization (by anionic, group transfer or coordination mechanisms) all chains are initiated at the beginning of the reaction and grow until all monomer is consumed. Living radical polymerization (also known as quasi- or psuedo-living radical polymerization) becomes possible in the presence of species which reversibly terminate chains. These reagents serve to reduce the concentration of propagating species and the incidence of radical-radical termination. This chemistry has been reviewed by Moad *et al.*,[29] Kuchanov,[116] Greszta *et al.*,[117] and Georges *et al.*[118] A general mechanism is shown in Scheme 13.

The initiators used for this form of living radical polymerization have been called iniferters (*ini*tiator - trans*fer* agent - chain *ter*minator) or initers (*ini*tiator - chain *ter*minator). These terms were coined by Otsu and Yoshida[119] based on the similar terminology introduced by Kennedy[120] to cover analogous cationic systems. To be successful as initiators of living radical polymerization the iniferters or initers should possess the following attributes:

(a) One (or both) of the radicals formed on initiator decomposition is long lived and unable (or slow) to initiate polymerization.
(b) Primary radical termination and transfer to initiator should be the only significant mechanisms for the interruption of chain growth. Primary radical termination should occur exclusively by combination. Transfer to initiator should occur exclusively by group transfer.

(c) The bond to the end group (X) formed by these mechanisms must be thermally or photochemically labile. Reversible homolysis under the reaction conditions regenerates the propagating radical.

Propagation:

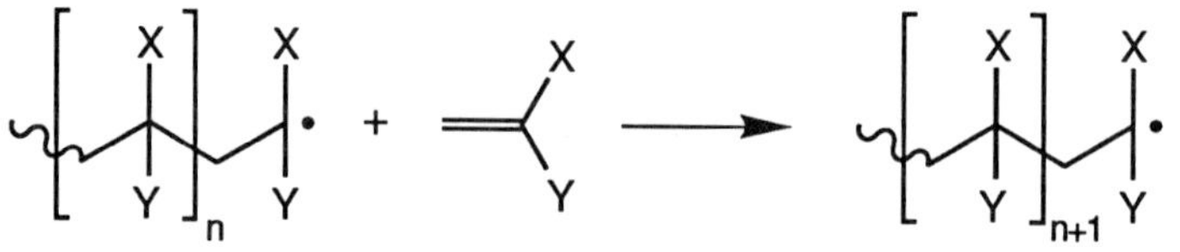

Primary radical termination/ Initiation:

Transfer to initiator:

Scheme 13

Since the concentration of propagating species is lower than in conventional polymerization, rates of polymerization are often relatively slow. The applications of the living radical polymerization are in the synthesis of:

(a) End-functional homopolymers.
(b) Block and graft copolymers.
(c) Polymers of narrow polydispersity[121-123] and chemical homogeneity.[124]
(d) Polymers of controlled molecular weight.

In the following discussion, iniferters or initers have been classed according to the type of stable radical produced. Those described include sulfur-centered (see 7.5.1), carbon-centered (see 7.5.2), oxygen-centered (7.5.3), and organocobalt species (7.5.4).

7.5.1 Organosulfur iniferters

The first detailed study of dialkyldithiuram disulfides as initiators in polymerizations of MMA and S was reported by Werrington and Tobolsky in 1955.[125] They observed that the transfer constant to the disulfide was high and also found significant retardation. The potential of this and other disulfides as initiators of living radical polymerization was recognized by Otsu and Yoshida in 1982.[119] A range of disulfides has now been investigated in this context with varying degrees of success. These include diaryl disulfides (10),[126,127] dibenzoyl disulfide (11),[119] tetraethyldithiuram disulfide (12),[119,128,129] and bis(isopropylxanthogen) disulfide (13).[130] To date, the dithiuram disulfides (12) are the most studied and the most successful.

The mechanism of polymerization with dithiuram disulfide initiators (12) is shown in Scheme 14. The initiator may be decomposed thermally or photochemically to give dithiocarbamyl radicals (14). These radicals add monomer only slowly and relatively high reaction temperatures (typically >80°C) appear necessary even if the initiator is decomposed photochemically. Transfer to initiator (by transfer of a dithiocarbamyl group) is probably the most important mechanism for the termination of polymer chains. Transfer constants of tetraethyldithiuram disulfide (12) are reported to be ca. 0.5 in S and MMA polymerizations.[131,132] The end groups formed by transfer to initiator are indistinguishable from those formed by primary radical termination. The dithiocarbamyl end groups are thermally stable but photochemically labile. Thus, only photo-initiated polymerizations should be described as living polymerizations.

The chemistry has been shown to be compatible with a range of monomers (S, MMA, MA, VAc) though degree of livingness depends on the particular monomer. Studies on model compounds have shown that the primary dithiocarbamate end groups formed by addition to monomer are much less susceptible to photodissociation than benzyl or tertiary derivatives. Therefore, depending on the reaction conditions, the only "living" ends are those formed by primary radical termination or transfer to initiator and polymerization of MA and VAc are less successful than those of MMA and S.

The direct use of the disulfide iniferters [e.g. (12)] suffers from the disadvantage that both of the radicals produced on decomposition of the macroinitiator may add monomer. This is unavoidable since the experiment relies on the one radical species to both initiate polymerization and terminate chains. Various side reactions likely to lead to a slow loss of "living" ends have also been described.[129,133]

Initiation:

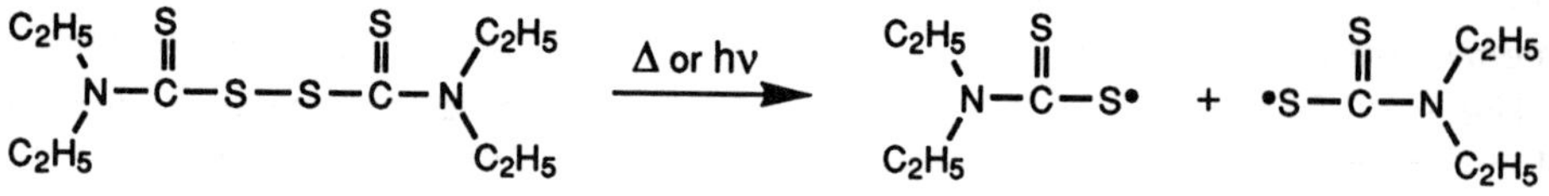

(12) → (14) + (14) [Δ or hν]

Chain transfer:

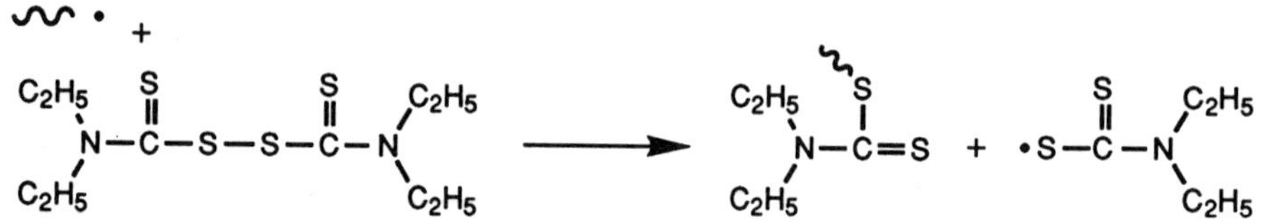

~• + (12) → + (14)

Primary radical termination/reinitiation:

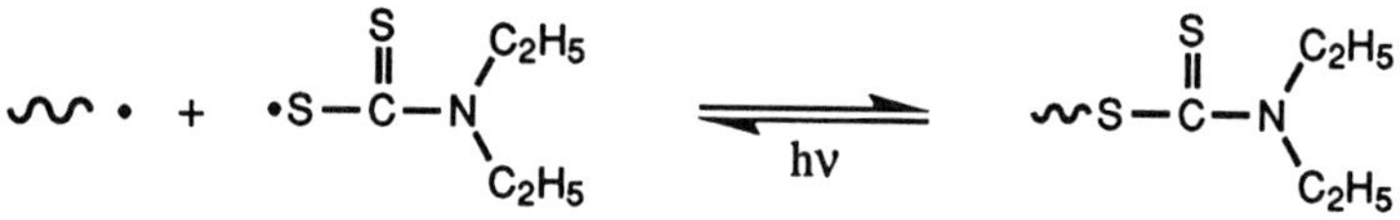

~• + (14) ⇌ [hν]

Scheme 14

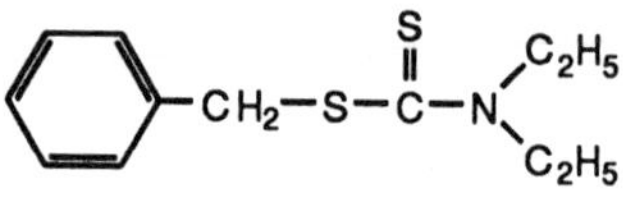

(15) 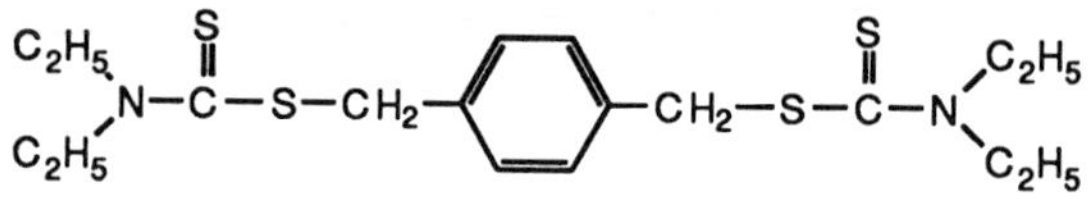   (16)

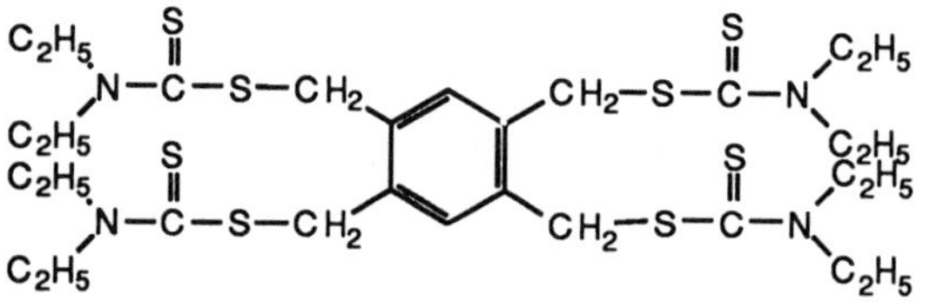

(17) 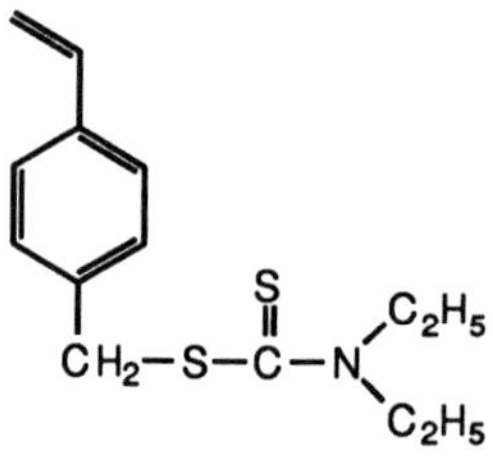   (18)

Benzylic dithiocarbamates (15-18) may also be used as photoiniferters. Photodissociation of the C–S bond affords a reactive alkyl radical (to initiate polymerization) and a less reactive sulfur-centered radical (to undergo primary radical termination).[126,134,135] Since the experiment is no longer reliant on the dithiocarbamyl radical to initiate chains, lower reaction temperatures may be used (where the dithiocarbamyl radical is slower to add monomer) and better control over the polymerization process is obtained. The transfer constants for the benzyl dithiocarbamates are low, thus primary radical termination becomes the main chain termination mechanism. Doi et al.[136] have recently described a two part initiator system comprising a mixture of a benzyl dithiocarbamate and a dithiuram disulfide.

The use of benzyl dithiocarbamate derivatives offers greater scope for designing polymer architectures. The use of the mono-, di- and multifunctional initiators, (15), (16) and (17), in the production of block or star polymers has been demonstrated. Homopolymers of (18) or copolymers of (18) with S or MMA have also been successfully used in photoinitiated graft polymerization of S or MMA.[137-139] The monomer (18) has reactivity ratios similar to those of other *para*-substituted styrenes.

Since the dithiocarbamyl end groups are thermally stable, only photo-initiated polymerizations should be described as living polymerizations. The dithiocarbamates are used to prepare end-functional polymers, for example, (19) and (20), have been used to form telechelics[140] and blocks[141] respectively by non-living thermally-initiated polymerization.

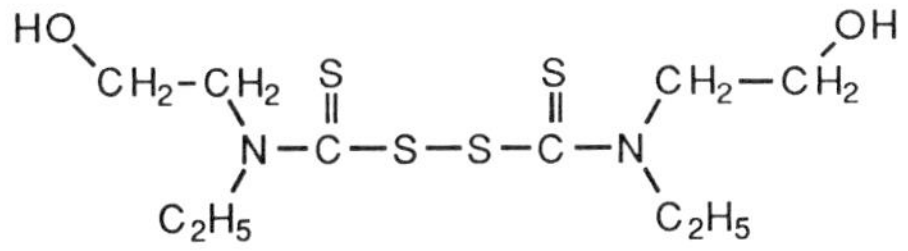

(19)

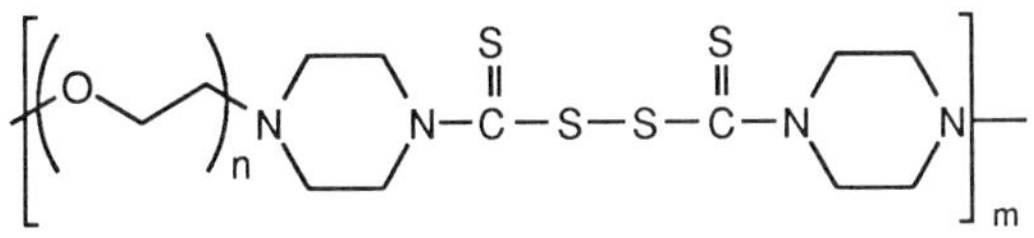

(20)

7.5.2 Hexasubstituted ethanes and azo compounds

Stable carbon-centered radicals, for example, diphenylalkyl and triarylmethyl radicals, are also effective in reversibly coupling with propagating radicals. The first use of 1,2-disubstituted-1,1,2,2-tetraphenylethanes [*e.g.* (21)] as initiators of living polymerization was reported by Bledzki et al.[142,143] Related initers based on silylated pinacols [*e.g.* (21)] have been investigated by Crivello et al.[144-147]

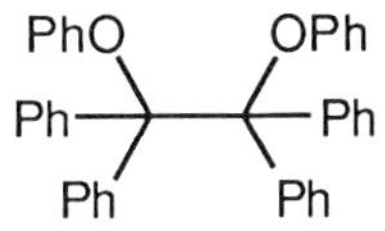

(21)

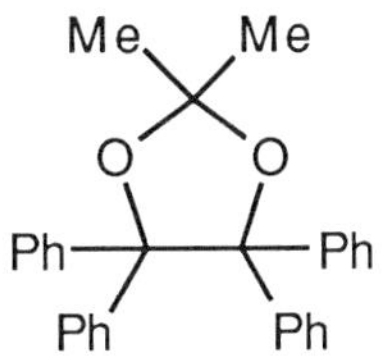

(22)

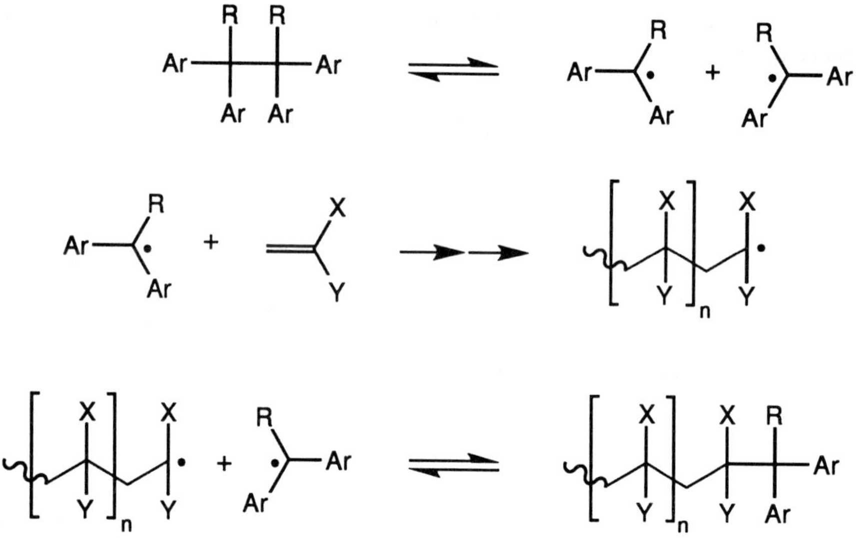

Scheme 15

The proposed mechanism is shown in Scheme 15. Thermal decomposition of the tetraarylethane derivative affords diphenylalkyl radicals which can initiate polymerization or combine with propagating species (primary radical termination) to form an oligomeric macroinitiator. Otsu and Tazaki[148] have reported on the use of triphenylmethylazobenzene (23) as an initiator of living radical polymerization. In this case phenyl radical initiates polymerization and the triphenylmethyl radical reacts mainly by primary radical termination. The chemistry of the macroinitiators formed is then analogous to those formed from the hexasubstituted ethanes (*i.e.* Scheme 15).

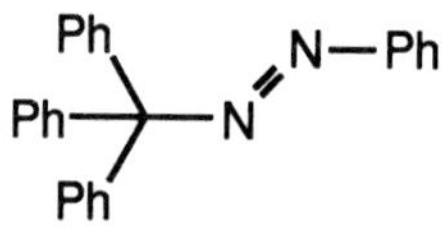

(23)

There is a gradual loss of "living" ends with these systems as the di- or triarylmethyl radicals produced from the macroinitiator can add monomer, *albeit* slowly.[149,150] A further problem with these iniferters is loss of "living" ends through primary radical termination by disproportionation. The ratio of k_{td}/k_{tc} reported for the cross reaction between radicals (24) and (25) at 110°C is 0.61.[151]

It is of interest to speculate on the structure of the macroiniter species in these polymerizations. The recent work of Engel *et al.*[151] suggests the likelihood of the quinonoid intermediate (26, see Scheme 16) at least for the polymerizations involving triphenylmethyl radical (25). The early report that (25) does not initiate MMA polymerization may be rationalized in terms of the low rate of polymerization in the presence of the initer.[152]

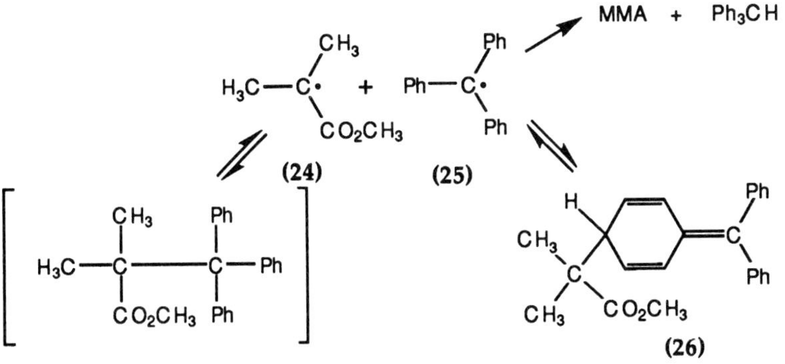

Scheme 16

The rates of decomposition of hexasubstituted ethanes and the derived macroinitiators vary according to the degree of steric crowding about the C–C bond undergoing homolysis.[153] These initers are most suited for the polymerization of 1,1-disubstituted monomers (*e.g.* MMA,[143,154] other methacrylates and MAA,[155] and AMS[156]). Polymerizations of monosubstituted monomers are not living. Dead end polymerization is observed with S at usual polymerization temperatures (<100°C).[157] Monosubstituted monomers may be used in the second stage AB block copolymer synthesis (formation of the B block).[154] However the non-living nature of the polymerization limits the length of the B block that can be formed.

7.5.3 Alkoxyamines and related species

Rizzardo *et al.*[28,158] pioneered the use of alkoxyamines as initiators of living radical polymerization. These compounds are formed from the reaction between carbon-centered radicals and appropriate nitroxides. Examples include (27), from decomposition of DBPOX in S in the presence of di-*t*-butyl nitroxide, and (28) formed by decomposition of AIBN in the presence of TEMPO.[28,159] The alkoxyamine initiators may either be produced *in-situ* in the polymerization medium[122,123,158,160] or, where they are stable at room temperature, they can be isolated and characterized.[158,159]

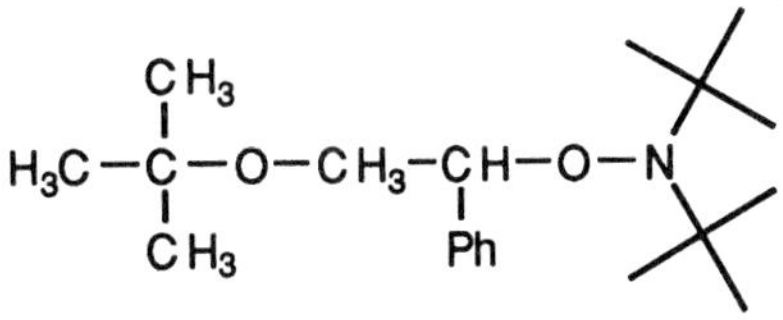

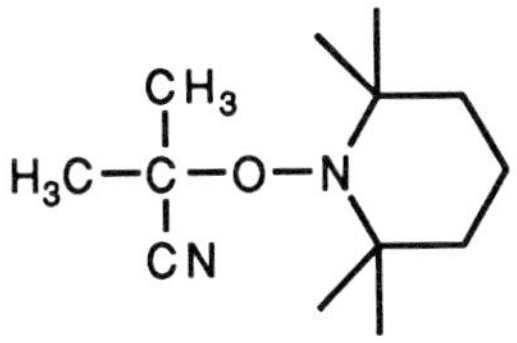

The C–O bond of alkoxyamines and similar species is relatively weak and undergoes reversible homolysis on heating to afford an alkyl radical and a stable nitroxide (see Scheme 17).

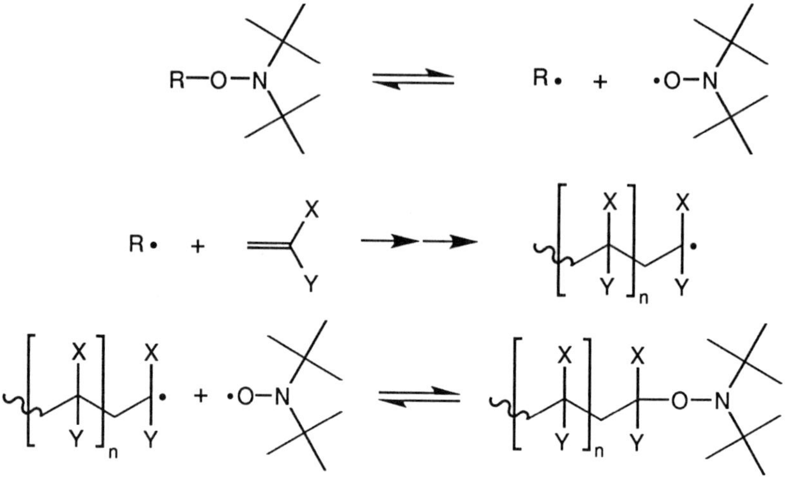

Scheme 17

The reactive carbon-centered radical initiates polymerization while the nitroxide reacts with the propagating radical by primary radical termination to form a new oligo- or polymeric alkoxyamine initiator. The method has been applied to make a wide range of block and graft copolymers[118,158,159] and functional and narrow polydispersity homopolymers.[121-123]

Various polymers, including PMMA[158] and PS,[122] with $\bar{M}_w/\bar{M}_n \leq 1.5$ have been prepared using alkoxyamine-initiated living radical polymerization (conventional radical polymerization with termination by chain transfer or disproportionation will give $\bar{M}_w/\bar{M}_n \sim 2.0$). Computer modeling studies show that the polymerization rates and molecular weight distributions obtained in these polymerizations depend strongly on the rate of homolysis of alkoxyamine initiator and derived macro-initiators. If these are sufficiently fast, polydispersities can approach those obtained by anionic living polymerization.[121]

The effectiveness of alkoxyamine initiators in this context depends on the structure of the alkoxyamine. Experimental work[158,159] and theoretical studies[161] have been carried out to establish a structure-property correlation and give further understanding of the mechanism. The rate of homolysis of the alkoxyamines (R-ON<) is determined by a combination of the degree of steric compression around the C-O bond, the stabilities of the radicals formed, and polar factors.[118,158,159,161,162] The rates of homolysis vary with the structure of both the reactive radical and the nitroxide fragment. For example, rates of C-O bond homolysis have been shown to increase in the series where the nitroxide is:[162]

and where the radical is:

$$\text{\textasciitilde}CH_2-\overset{H}{\underset{CO_2CH_3}{C}}\cdot \quad < \quad \text{\textasciitilde}CH_2-\overset{H}{\underset{Ph}{C}}\cdot \quad < \quad \text{\textasciitilde}CH_2-\overset{CH_3}{\underset{CO_2CH_3}{C}}\cdot \quad < \quad \text{\textasciitilde}CH_2-\overset{CH_3}{\underset{Ph}{C}}\cdot$$

It has been shown that trends in reactivity may be qualitatively predicted through molecular orbital calculations.[161]

A major advantage of the alkoxyamines over the other initiators described in this section is that the nitroxides used appear completely inert towards most monomers under normal polymerization conditions. Nitroxides do not directly initiate radical polymerization. Alkoxyamines are also not susceptible to induced decomposition under typical reaction conditions. Disproportionation between propagating species and the nitroxide, with consequent loss of "living" ends, is not observed except with tertiary radicals (*e.g.* that from MMA) but then is only a minor pathway.

The chemistry can also be used for graft copolymer synthesis.[158] Two routes have been reported for synthesis of the required macroinitiator. The first involves copolymerization with a monomer containing pendant alkoxyamine functionality [*e.g.* (29)].[158]

(29)

In the second route alkoxyamine functionality is first grafted onto a pre-existing polymer (see Scheme 18). This can be achieved by heating the polymer with a nitroxide and a t-butoxy radical source.[158] The polymeric alkoxyamine is then used to initiate living polymerization. This procedure has been applied to polybutadiene and used to prepare poly(B-graft-MA).[158]

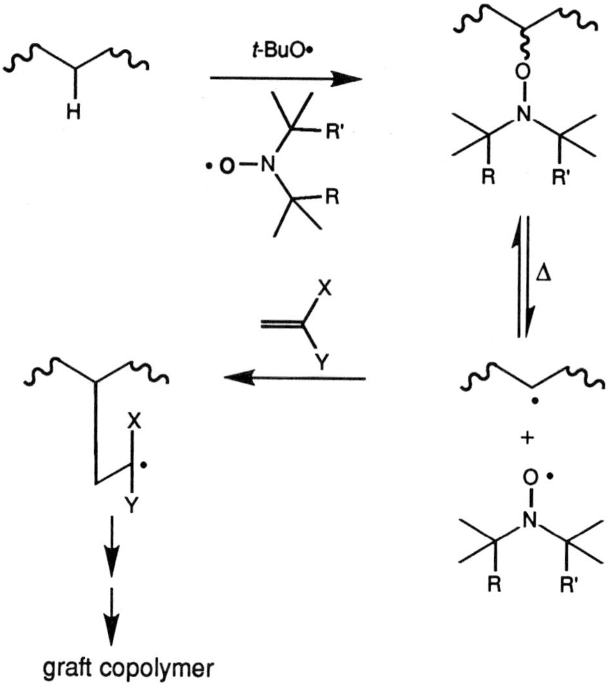

Scheme 18

graft copolymer

A number of related chemistries have been reported.[160,163] Druliner[160] reported on living polymerization systems where inorganic species, ⁻O-N=NO• and NCO•, or (**30**) are the stable radical. Homolysis rates for these systems were suitable for polymerizations at ambient temperature. Yamada et al.[163] examined block polymerization of S and MMA initiated by derivatives of the triphenylverdazyl radical (**31**). However, yields were relatively low.

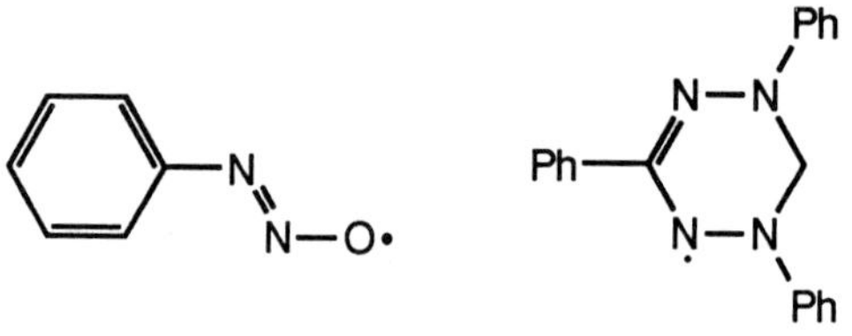

(**30**)   (**31**)

A living radical polymerization mechanism has also been proposed for the polymerization initiated by certain aluminum complexes in the presence of nitroxides.[164,165] The system has been used for VAc polymerization, however, the mechanism has not been completely defined. It was proposed that a carbon-aluminum bond is formed in the reversible termination step.[164,165]

7.5.4 Organocobalt complexes

Oganova et al. observed that certain cobalt (II) porphyrin complexes reversibly inhibit BA polymerization[166-169] presumably with formation of a cobalt (III) intermediate (33). Thus, these species may function as initers in living radical polymerization (Scheme 19).[170,171] Wayland et al.[171] have used the complex (34) to synthesize high molecular weight PMA with very narrow polydispersities (1.1-1.3).

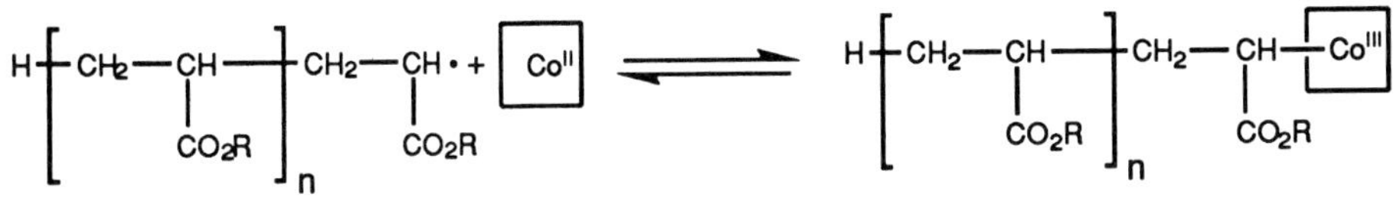

Scheme 19

(32) ⇌ (33)

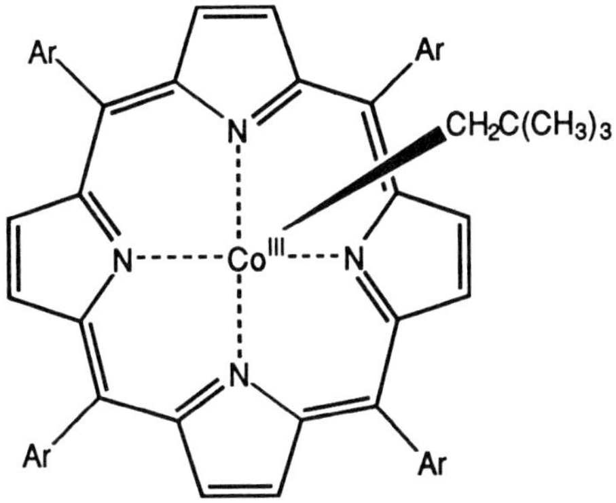

(34)

The most important side reactions are disproportionation between the cobalt(II) complex (32) and the propagating species and/or β-elimination of an alkene from the cobalt(III) intermediate (35) (see Scheme 20).

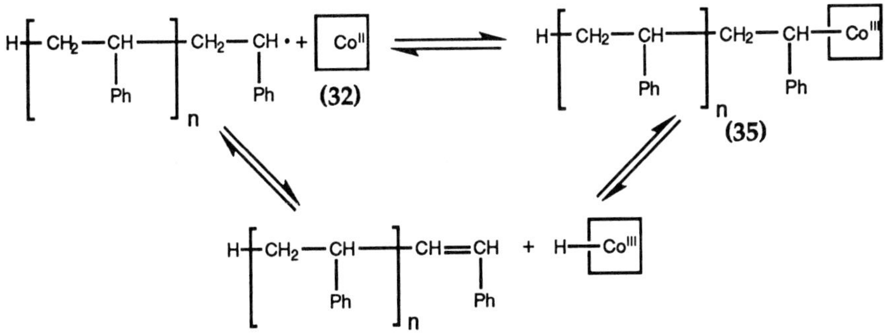

Scheme 20

(32) ⇌ (35)

Both pathways appear unimportant in the case of acrylate ester polymerizations but are of major importance with methacrylate esters and S; at least with the cobalt complexes explored thus far. This chemistry (Scheme 20), while precluding living polymerization, has led to the development of cobalt complexes as catalytic chain transfer agents (see 5.3.2.7).

It has also been suggested that alkyl cobalt intermediate may be subject to radical-induced decomposition.[168]

References

1. Moad, G., and Solomon, D.H., *Aust. J. Chem.*, 1990, **43**, 215.
2. Lehrle, R.S., Peakman, R.E., and Robb, J.C., *Eur. Polym. J.*, 1982, **18**, 517.
3. Rudin, A., Samanta, M.D., and Reilly, P.M., *J. Appl. Polym. Sci.*, 1979, **24**, 171.
4. Cameron, G.G., Meyer, J.M., and McWalter, I.T., *Macromolecules*, 1978, **11**, 696.
5. Wall, L.A., Straus, S., Florin, R.E., and Fetters, L.J., *J. Res. Nat. Bur. Stand.*, 1973, **77**, 157.
6. Cascaval, C.N., Straus, S., Brown, D.W., and Florin, R.E., *J. Polym. Sci., Polym. Symp.*, 1976, **57**, 81.
7. Singh, M., and Nandi, U.S., *J. Polym. Sci., Polym. Lett. Ed.*, 1979, **17**, 121.
8. Grassie, N., and Kerr, W.W., *Trans. Faraday Soc.*, 1959, 1050.
9. Schildknecht, C.E., in 'Polymerization Processes' (Eds. Schildknecht, C.E., and Skeist, I.), p. 88 (Wiley: New York 1977).
10. Boundy, R.H., and Boyer, R.F., Styrene (Reinhold: New York 1952).
11. Krstina, J., Moad, G., and Solomon, D.H., *Eur. Polym. J.*, 1989, **25**, 767.
12. Moad, G., Solomon, D.H., and Willing, R.I., *Macromolecules*, 1988, **21**, 855.
13. Cameron, G.G., and MacCallum, J.R., *J. Macromol. Sci., Rev. Macromol. Chem.*, 1967, **C1**, 327.
14. Moad, G., Solomon, D.H., Johns, S.R., and Willing, R.I., *Macromolecules*, 1984, **17**, 1094.
15. Krstina, J., Moad, G., Willing, R.I., Danek, S.K., Kelly, D.P., Jones, S.L., and Solomon, D.H., *Eur. Polym. J.*, 1993, **29**, 379.
16. Hodder, A.N., Holland, K.A., and Rae, I.D., *J. Polym. Sci., Polym. Lett. Ed.*, 1983, **21**, 403.
17. Manring, L.E., *Macromolecules*, 1988, **21**, 528.
18. Manring, L.E., Sogah, D.Y., and Cohen, G.M., *Macromolecules*, 1989, **22**, 4652.
19. Manring, L.E., *Macromolecules*, 1989, **22**, 2673.
20. Cacioli, P., Moad, G., Rizzardo, E., Serelis, A.K., and Solomon, D.H., *Polym. Bull. (Berlin)*, 1984, **11**, 325.
21. Kashiwagi, T., Kirata, T., and Brown, J.E., *Macromolecules*, 1985, **18**, 131.
22. Meisters, A., Moad, G., Rizzardo, E., and Solomon, D.H., *Polym. Bull. (Berlin)*, 1988, **20**, 499.
23. Rizzardo, E., and Solomon, D.H., *J. Macromol. Sci., Chem.*, 1979, **A13**, 1005.
24. Rizzardo, E., and Solomon, D.H., *Polym. Bull. (Berlin)*, 1979, **1**, 529.

25. Bednarek, D., Moad, G., Rizzardo, E., and Solomon, D.H., *Macromolecules*, 1988, **21**, 1522.
26. Aliwi, S.M., and Bamford, C.H., *J. Chem. Soc., Faraday Trans. 1*, 1977, **73**, 776.
27. Bamford, C.H., and White, E.F.T., *Trans. Faraday Soc.*, 1958, **54**, 268.
28. Rizzardo, E., *Chem. Aust.*, 1987, **54**, 32.
29. Moad, G., Rizzardo, E., and Solomon, D.H., in 'Comprehensive Polymer Science' (Eds. Eastmond, G.C., Ledwith, A., Russo, S., and Sigwalt, P.), Vol. 3, p. 141 (Pergamon: London 1989).
30. Ng, F.T.T., Rempel, G.L., and Halpern, J., *J. Am. Chem. Soc.*, 1982, **104**, 621.
31. Beckwith, A.L.J., Bowry, V.W., and Moad, G., *J. Org. Chem.*, 1988, **53**, 1632.
32. Sanayei, R.A., and O'Driscoll, K.F., *J. Macromol. Sci. Chem.*, 1989, **A26**, 1137.
33. Ozerhwskii, B.V., and Reshchupkin, V.P., *Dokl. Phys. Chem. (Engl. Transl.)*, 1981, **254**, 731.
34. Giese, B., Ghosez, A., Gobel, T., Hartung, J., Huter, O., Koch, A., Kroder, K., and Springer, R., in 'Free Radicals in Chemistry and Biology' (Ed. Minisci, F.), p. 97 (Kluwer: Dordrecht 1989).
35. Clark, A.J., and Jones, K., *Tetrahedron Lett.*, 1989, **30**, 5485.
36. Reichardt, C., Solvent Effects in Organic Chemistry, p. 110 (Verlag Chemie: Weinheim 1978).
37. Brumby, S., *Polym. Commun.*, 1989, **30**, 13.
38. Bamford, C.H., and Brumby, S., *Makromol. Chem.*, 1967, **105**, 122.
39. Kamachi, M., *Adv. Polym. Sci.*, 1981, **38**, 55.
40. Barton, J., and Borsig, E., Complexes in Free Radical Polymerization (Elsevier: Amsterdam 1988).
41. Gromov, V.F., and Khomiskovskii, P.M., *Russ. Chem. Rev. (Engl. Transl.)*, 1979, **48**, 1040.
42. Kamachi, M., Liaw, D.J., and Nozakura, S.-I., *Polym. J. (Tokyo)*, 1979, **11**, 921.
43. Kamachi, M., Satoh, J., and Nozakura, S., *J. Polym. Sci., Polym. Chem. Ed.*, 1978, **16**, 1789.
44. Kamachi, M., Liaw, D.J., and Nozakura, S., *Polym. J. (Tokyo)*, 1981, **13**, 41.
45. Burnett, G.M., Cameron, G.G., and Joiner, S.N., *Trans. Faraday Soc.*, 1973, **69**, 322.
46. Hatada, K., Terawaki, Y., Kitayama, T., Kamachi, M., and Tamaki, M., *Polym. Bull. (Berlin)*, 1981, **4**, 451.
47. Morrison, B.R., Piton, M.C., Winnik, M.A., Gilbert, R.G., and Napper, D.H., *Macromolecules*, 1993, **26**, 4368.
48. Mahabadi, H.K., and O'Driscoll, K.F., *J. Macromol. Sci., Chem.*, 1977, **A11**, 967.
49. Paleos, C.M., and Malliaris, A., *J. Macromol. Sci., Rev. Macromol. Chem.*, 1988, **C28**, 403.
50. Tsutsumi, K., Tsukahara, Y., and Okamota, Y., *Polym. J. (Tokyo)*, 1994, **26**, 13.
51. Plochocka, K., *J. Macromol. Sci., Rev Macromol. Chem.*, 1981, **C20**, 67.
52. Georgiev, G.S., and Dakova, I.G., *Macromol. Chem. Phys.*, 1994, **195**, 1695.

53. Lewis, F.M., Walling, C., Cummings, W., Briggs, E.R., and Mayo, F.R., *J. Am. Chem. Soc.*, 1948, 1519.
54. Price, C.C., and Walsh, J.G., *J. Polym. Sci.*, 1951, **6**, 239.
55. Pichot, C., Zaganiaris, E., and Guyot, A., *J. Polym. Sci., Polym. Symp.*, 1975, **52**, 55.
56. Hill, D.J.T., Lang, A.P., Munro, P.D., and O'Donnell, J.H., *Eur. Polym. J.*, 1992, **28**, 391.
57. Hill, D.J.T., Lang, A.P., Munro, P.D., and O'Donnell, J.H., *Eur. Polym. J.*, 1989, **28**, 391.
58. Asakura, J., Yoshihara, M., Matsubara, Y., and Maeshima, T., *J. Macromol. Sci., Chem.*, 1981, **A15**, 1473.
59. Van Der Meer, R., Aarts, M.W.A.M., and German, A.L., *J. Polym. Sci., Polym. Chem. Ed.*, 1980, **18**, 1347.
60. Cameron, G.G., and Esslemont, G.F., *Polymer*, 1972, **13**, 435.
61. Ito, T., and Otsu, T., *J. Macromol. Sci. Chem*, 1969, **A3**, 197.
62. San Roman, J., and Madruga, E.L., *Angew. Makromol. Chem.*, 1980, **86**, 1.
63. Bonta, G., Gallo, B.M., and Russo, S., *Polymer*, 1975, **16**, 429.
64. Busfield, W.K., and Low, R.B., *Eur. Polym. J.*, 1975, **11**, 309.
65. Park, K.Y., Santee, E.R., and Harwood, H.J., *Eur. Polym. J.*, 1989, **25**, 651.
66. Harwood, H.J., *Makromol. Chem., Macromol. Symp.*, 1987, **10/11**, 331.
67. Klumperman, B., and O'Driscoll, K.F., *Polymer*, 1993, **34**, 1032.
68. Davis, T.P., *Polym. Commun.*, 1990, **31**, 442.
69. Klumperman, B., and Kraeger, I.R., *Macromolecules*, 1994, **27**, 1529.
70. Krstina, J., Moad, G., and Solomon, D.H., *Eur. Polym. J.*, 1992, **28**, 275.
71. Semchikov, Y.D., *Polym. Sci. USSR (Engl. Transl.)*, 1990, **32**, 177.
72. Tsukahara, Y., Hayashi, N., Jiang, X.L., and Yamashita, Y., *Polym. J. (Tokyo)*, 1989, **21**, 377.
73. Bamford, C.H., in 'Alternating Copolymers' (Ed. Cowie, J.M.G.), p. 75 (Plenum: New York 1985).
74. Bamford, C.H., Jenkins, A.D., and Johnston, R., *Proc. R. Soc., London, Ser. A,*, 1957, **241**, 364.
75. Zubov, V.P., Valuev, L.I., Kabanov, V.A., and Kargin, V.A., *J. Polym. Sci., Part A-1*, 1971, **9**, 833.
76. Tanaka, T., Kato, H., Sakai, I., Sato, T., and Ota, T., *Makromol. Chem., Rapid Commun.*, 1987, **8**, 223.
77. Otsu, T., and Yamada, B., *J. Macromol. Sci. Chem*, 1966, **A1**, 61.
78. Momtaz-Afchar, J., Polton, A., Tardi, M., and Sigwalt, P., *Eur. Polym. J.*, 1985, **21**, 1067.
79. Rogueda, C., Polton, A., Tardi, M., and Sigwalt, P., *Eur. Polym. J.*, 1989, **25**, 1259.
80. Rogueda, C., Polton, A., Tardi, M., and Sigwalt, P., *Eur. Polym. J.*, 1989, **25**, 1251.
81. Rogueda, C., Tardi, M., Polton, A., and Sigwalt, P., *Eur. Polym. J.*, 1989, **25**, 885.

82. Lyons, R.A., Moad, G., and Senogles, E., *Eur. Polym. J.*, 1993, **29**, 389.
83. Krstina, J., Moad, G., and Solomon, D.H., *Polym. Bull. (Berlin)*, 1992, **27**, 425.
84. Bamford, C.H., *Chem. Aust.*, 1982, **49**, 341.
85. Tan, Y.Y., in 'Recent Advances in Mechanistic and Synthetic Aspects of Polymerization', p. 281 (Dordrecht: Reidel 1987).
86. Tan, Y.Y., in 'Comprehensive Polymer Science' (Eds. Eastmond, G.C., Ledwith, A., Russo, S., and Sigwalt, P.), Vol. 3, p. 245 (Pergamon: London 1989).
87. Buter, R., Tan, Y.Y., and Challa, G., *J. Polym. Sci., Part A-1*, 1972, **10**, 1031.
88. Schomaker, E., and Challa, G., *Macromolecules*, 1988, **21**, 3506.
89. Matsuzaki, K., Kanai, T., Ichijo, C., and Yuzawa, M., *Makromol. Chem.*, 1984, **185**, 2291.
90. van de Grampel, H.T., Tan, Y.Y., and Challa, G., *Macromolecules*, 1991, **24**, 3767.
91. van de Grampel, H.T., Tan, Y.Y., and Challa, G., *Macromolecules*, 1991, **24**, 3773.
92. Ferguson, J., Al-Alawi, S., and Grannmayeth, R., *Eur. Polym. J.*, 1985, **19**, 475.
93. Chapiro, A., *Pure Appl. Chem.*, 1981, **53**, 643.
94. Polowinski, S., *Eur. Polym. J.*, 1983, **19**, 679.
95. Natansohn, A., *Polym. Prepr. (Am. Chem. Soc., Div. Polym. Chem)*, 1984, **25(2)**, 65.
96. Kämmerer, H., *Angew. Chem., Int. Ed. Engl.*, 1965, **4**, 952.
97. Kern, W., and Kämmerer, H., *Pure Appl. Chem.*, 1967, **15**, 421.
98. Kämmerer, H., and Onder, N., *Makromol. Chem.*, 1968, **111**, 67.
99. Feldman, K.S., Bobo, J.S., Ensel, S.M., Lee, Y.B., and Weinreb, P.H., *J. Org. Chem.*, 1990, **55**, 474.
100. Feldman, K.S., and Lee, Y.B., *J. Am. Chem. Soc.*, 1987, **109**, 5850.
101. Wulff, G., Kemmerer, R., and Vogt, B., *J. Am. Chem. Soc.*, 1987, **109**, 7449.
102. Jantas, R., *J. Polym. Sci., Part A: Polym. Chem.*, 1990, **28**, 1973.
103. Jantas, R., and Polowinski, S., *J. Polym. Sci., Polym. Chem. Ed.*, 1986, **24**, 1819.
104. Jantas, R., *J. Polym. Sci., Part A: Polym. Chem.*, 1994, **32**, 295.
105. Bamford, C.H., in 'Developments in Polymerization 2' (Ed. Howard, R.N.), Vol. 49, p. 215 (Applied Science: London 1979).
106. Sugiyama, J.-I., Yokozawa, T., and Endo, T., *Macromolecules*, 1994, **27**, 1987.
107. Sugiyama, J.-I., Yokozawa, T., and Endo, T., *J. Am. Chem. Soc.*, 1993, **115**, 2041.
108. Hill, L.W., and Wicks, Z.W., *Prog. Org. Coat.*, 1982, **10**, 55.
109. Spinelli, H.J., *Am. Chem. Soc., Org. Coat. Plas. Chem., Reprints.*, 1982, 529.
110. Stockmayer, W.H., *J. Chem. Phys.*, 1945, **13**, 199.
111. Fueno, T., and Fukawa, J., *J. Polym. Sci., Part A*, 1964, **2**, 3681.
112. Mirabella, F.M., Jr., *Polymer*, 1977, **18**, 705.
113. O'Driscoll, K.F., *J. Coat. Technol.*, 1983, **55**, 57.
114. Galbraith, M.N., Moad, G., Solomon, D.H., and Spurling, T.H., *Macromolecules*, 1987, **20**, 675.
115. Spurling, T.H., Deady, M., Krstina, J., and Moad, G., *Makromol. Chem., Macromol. Symp.*, 1991, **51**, 127.
116. Kuchanov, S.I., in 'Comprehensive Polymer Science' (Eds. Agarwal, S.L., and Russo, S.), Vol. Suppl. 1, p. 23 (Pergamon: London 1992).

117. Greszta, D., Mardare, D., and Matyjaszewski, K., *Macromolecules*, 1994, **27**, 638.
118. Georges, M.K., Veregin, R.P.N., Kazmaier, P.M., and Hamer, G.K., *Trends Polym. Sci.*, 1993, **2**, 66.
119. Otsu, T., and Yoshida, M., *Makromol. Chem., Rapid Cummun.*, 1982, **3**, 127.
120. Kennedy, J.P., *J. Macromol. Sci. Chem.*, 1979, **A13**, 695.
121. Johnson, C.H.J., Moad, G., Solomon, D.H., Spurling, T.H., and Vearing, D.J., *Aust. J. Chem.*, 1990, **43**, 1215.
122. Georges, M.K., Veregin, R.P.N., Kazmaier, P.M., and Hamer, G.K., *Macromolecules*, 1993, **26**, 2987.
123. Veregin, R.P.N., Georges, M.K., Kazmaier, P.M., and Hamer, G.K., *Macromolecules*, 1993, **26**, 5316.
124. Kuchanov, S.I., and Olenin, A.V., *Polym. Bull. (Berlin)*, 1992, **28**, 449.
125. Ferington, T.E., and Tobolsky, A.V., *J. Am. Chem. Soc.*, 1955, **77**, 4510.
126. Otsu, T., and Kuriyama, A., *J. Macromol. Sci., Chem.*, 1984, **A21**, 961.
127. Shefer, A., Grodzinsky, A.J., Prime, K.L., and Busnel, J.P., *Macromolecules*, 1993, **26**, 2240.
128. Otsu, T., and Yoshida, M., *Polym. Bull. (Berlin)*, 1982, **7**, 197.
129. Turner, S.R., and Blevina, R.W., *Macromolecules*, 1990, **23**, 1856.
130. Niwa, M., Matsumoto, T., and Izumi, H., *J. Macromol. Sci. Chem.*, 1987, **A24**, 567.
131. Nair, C.P.R., Clouet, G., and Chaumont, P., *J. Polym. Sci.*, 1982, **27**, 1795.
132. Staudner, E., Kysela, G., Beniska, J., and Mikolaj, D., *Eur. Polym. J.*, 1978, **14**, 1067.
133. Lambrinos, P., Tardi, M., Polton, A., and Sigwalt, P., *Eur. Polym. J.*, 1990, **26**, 1125.
134. Otsu, T., and Kuriyama, A., *Polym. J. (Tokyo)*, 1985, **17**, 97.
135. Otsu, T., Matsunaga, T., Kuriyama, A., and Yoshioka, M., *Eur. Polym. J.*, 1989, **25**, 643.
136. Doi, T., Matsumoto, A., and Otsu, T., *J. Polym. Sci., Part A: Polym. Chem.*, 1994, **32**, 2911.
137. Otsu, T., Yamashita, K., and Tsuda, K., *Macromolecules*, 1986, **19**, 287.
138. Yamashita, K., Ito, K., Tsuboi, H., Takahama, S., Tsuda, K., and Otsu, T., *J. Appl. Polym. Sci.*, 1990, **40**, 1445.
139. Yamashita, K., Kanamori, T., and Tsuda, K., *J. Macromol. Sci., Chem.*, 1990, **A27**, 897.
140. Nair, C.P.R., and Clouet, G., *J. Macromol. Sci., Rev. Macromol. Chem. Phys.*, 1991, **C31**, 311.
141. Nair, C.P.R., Chaumont, P., and Clouet, G., *J. Macromol. Sci., Chem*, 1990, **A27**, 791.
142. Bledzki, A., Braun, D., and Titzschkau, K., *Makromol. Chem.*, 1983, **184**, 745.
143. Bledzki, A., and Braun, D., *Makromol. Chem.*, 1981, **182**, 1047.
144. Crivello, J.V., Lee, J.L., and Conlon, D.A., in 'Advances in Elastomers and Rubber Elasticity' (Ed. Lal, J.), p. 157 (Plenum: New York 1986).

145. Crivello, J.V., Lee, J.L., and Conlon, D.A., *J. Polym. Sci., Polym. Chem. Ed.*, 1986, **24**, 1251.
146. Crivello, J.V., Lee, J.L., and Conlon, D.A., *Polym. Bull. (Berlin)*, 1986, **16**, 95.
147. Crivello, J.V., Lee, J.L., and Conlon, D.A., *J. Polym. Sci., Polym. Chem. Ed.*, 1986, **24**, 1197.
148. Otsu, T., and Tazaki, T., *Polym. Bull. (Berlin)*, 1986, **16**, 277.
149. Moad, G., Rizzardo, E., and Solomon, D.H., *Macromolecules*, 1982, **15**, 909.
150. Marvel, C.S., Dec, J., and Corner, J.O., *J. Am. Chem. Soc.*, 1945, **67**, 1855.
151. Engel, P.S., Chen, Y., and Wang, C., *J. Org. Chem.*, 1991, **56**, 3073.
152. McElvain, S.M., and Aldridge, C.L., *J. Am. Chem. Soc.*, 1953, **75**, 3987.
153. Rüchardt, C., *Top. Curr. Chem.*, 1980, **88**, 1.
154. Otsu, T., Matsumoto, A., and Tazaki, T., *Polym. Bull. (Berlin)*, 1987, **17**, 323.
155. Bledzki, A., Braun, D., Menzel, W., and Titzschkau, K., *Makromol. Chem.*, 1983, **184**, 287.
156. Bledzki, A., and Braun, D., *Polym. Bull. (Berlin)*, 1986, **16**, 19.
157. Bledzki, A., and Braun, D., *Makromol. Chem.*, 1986, **187**, 2599.
158. Solomon, D.H., Rizzardo, E., and Cacioli, P., Eur. Pat. Appl. EP135280 (Chem. Abstr. (1985) 102: 221335q).
159. Rizzardo, E., and Chong, Y.K., in '2nd Pacific Polymer Conference, Preprints', p. 26 (Pacific Polymer Federation: Tokyo 1991).
160. Druliner, J.D., *Macromolecules*, 1991, **24**, 6079.
161. Moad, G., and Rizzardo, E., in 'Pacific Polymer Conference Preprints', Vol. 3, p. 651 (Polymer Division, Royal Australian Chemical Institute: Brisbane 1993).
162. Moad, G., and Rizzardo, E., 1995, manuscript in preparation.
163. Yamada, B., Tanaka, H., Konishi, K., and Otsu, T., *J. Macromol. Sci., Chem.*, 1994, **A31**, 351.
164. Mardare, D., Matyjaszewski, K., and Coca, S., *Macromol. Rapid Commun.*, 1994, **15**, 37.
165. Mardare, D., and Matyjaszewski, K., *Polym. Prepr. (Am. Chem. Soc., Div. Polym. Chem)*, 1993, **34(2)**, 566.
166. Oganova, A.G., Smirnov, B.R., Ioffe, N.T., and Enikolopyan, N.S., *Bull. Acad. Sci. USSR*, 1983, 1837.
167. Oganova, A.G., Smirnov, B.R., Ioffe, N.T., and Kim, I.P., *Bull. Acad. Sci. USSR*, 1984, 1154.
168. Oganova, A.G., Smirnov, B.R., Ioffe, N.T., and Enikopyan, N.S., *Doklady Akad. Nauk SSR (Engl. Transl.)*, 1983, **268**, 66.
169. Morozova, I.S., Oganova, A.G., Nosova, V.S., Novikov, D.D., and Smirnov, B.R., *Bull. Acad. Sci. USSR*, 1987, 2628.
170. Smirnov, V.R., *Polym. Sci. USSR (Engl. Transl.)*, 1990, **32**, 583.
171. Wayland, B.B., Poszmik, G., Mukerjee, S.L., and Fryd, M., *J. Am. Chem. Soc.*, 1994, **116**, 7943.

Abbreviations

AA	acrylic acid	I•	initiator-derived radical
AAm	acrylamide	K	degrees Kelvin
AIBN	azobisisobutyronitrile	k_d	rate constant for initiator decomposition
AIBMe	azobis(methyl isobutyrate)		
AN	acrylonitrile	k_H	rate constant for head addition to monomer
B	butadiene		
BA	n-butyl acrylate	k_i	rate constant for initiator-derived radical adding to monomer
BMA	n-butyl methacrylate		
BPB	t-butyl perbenzoate	k_p	rate constant for propagation
BPO	benzoyl peroxide	k_{prt}	rate constant for primary radical termination
C_I	transfer constant to initiator		
C_M	transfer constant to monomer	k_T	rate constant for tail addition to monomer
C_P	transfer constant to polymer	k_t	rate constant for radical-radical termination
C_S	transfer constant to solvent or added transfer agent		
		k_{tc}	rate constant for radical-radical termination by combination
EA	ethyl acrylate		
EMA	ethyl methacrylate	k_{td}	rate constant for radical-radical termination by disproportionation
EPR	electron paramagnetic resonance (spectroscopy)		
DBPOX	di-t-butyl peroxalate	k_{tr}	rate constant for reaction with chain transfer agent
DPPH	diphenylpicrylhydrazyl		
DMF	dimethylformamide	k_z	rate constant for reaction with inhibitor
DMSO	dimethylsulphoxide		
E	ethylene	L	liter(s)
f	initiator efficiency	LPO	lauroyl (dodecanoyl) peroxide
f_X	instantaneous mole fraction of monomer X in monomer feed during copolymerization	M	mole(s)
		m-	meta-
		MA	methyl acrylate
F_X	instantaneous mole fractions of monomer X in a copolymer	MAA	methacrylic acid
		MAAm	methacrylamide
GPC	gel permeation chromatography	MAH	maleic anhydride
		MAN	methacrylonitrile
HEA	hydroxyethyl acrylate	MMA	methyl methacrylate
HEMA	hydroxyethyl methacrylate		

Author Index

Aarts, M.W.A.M., **59**, 326
Abadie, M., **163**, 305
Abadie, M.J.M., **155**, 304; **157**, 304; **162**, 305
Abbas, K.B., **51**, 156; **226**, 253
Abe, K., **294**, 95, 96
Abe, M., **165**, 72; **295**, 95, 96; **302**, 95, 96; **408**, 114, 117
Abuin, E., **123**, 27
Abul'khanov, A.G., **333**, 98, 101
Acar, M.H., **194**, 175, 176, 177, 178
Achilias, D.S., **29**, 51, 62, 63, 71
Adelman, R.L., **44**, 155; **253**, 258, 259
Adler, M.G., **58**, 59
Aerdts, A.M., **37**, 287, 295; **38**, 287
Agarwal, S.H., **255**, 258
Ahn, K.-D., **160**, 172; **166**, 172
Akashi, M., **172**, 306
Al-Alawi, S., **92**, 330
Al-Lamee, K.G., **272**, 90, 91; **170**, 305
Alberti, J.M., **32**, 286
Aldridge, C.L., **152**, 341
Alexander, I.J., **263**, 88
Alfassi, Z.B., **126**, 28, 33, 34
Alfrey, T., **33**, 13; **26**, 285; **100**, 295
Alhadeff, E.S., **158**, 34
Alimoglu, A.K., **169**, 305
Aliwi, S.M., **26**, 321
Allen, J.L., **404**, 114
Allen, N.S., **186**, 76
Allen, P.E.M., **254**, 187
Allen, P.W., **115**, 230
Altschul, R., **217**, 251, 254
Ameduri, B., **180**, 244
Amiya, S., **46**, 155, 185
Amone, M.J., **191**, 175
Amrich, M.J., Jr., **131**, 28, 35; **67**, 221, 222

Anderson, D.R., **444**, 122
Anderson, F.E., III, **109**, 26
Ando, S., **113**, 166
Ando, W., **113**, 66
Angelo, R.J., **91**, 163
Anh, N.T., **72**, 18
Anisimov, Y.N., **151**, 71
Anselme, J.-P., **42**, 57, 60
Aoki, S., **186**, 246; **191**, 246
Arfaei, P.Y., **210**, 177
Arias, C., **71**, 292
Arnaud, R., **48**, 15; **71**, 18; **74**, 18
Arnett, L.M., **109**, 229
Arnoldi, A., **17**, 48, 98, 118; **267**, 260, 264
Arnoldi, C., **393**, 112; **394**, 112
Aronson, M.T., **306**, 191; **309**, 191
Asahara, T., **150**, 236
Asakura, J., **58**, 326
Asmus, K.-D., **311**, 97
Aso, C., **113**, 166
Atherton, J.N., **45**, 214, 217
Athey, R.D., **100**, 65
Atkinson, W.H., **132**, 232
Avci, D., **268**, 190
Avila, D.V., **118**, 26; **371**, 107, 108
Awaya, H., **58**, 157
Axelson, D.E., **238**, 184; **237**, 256
Ayrey, G., **77**, 61, 125; **99**, 65; **319**, 98; **99**, 229, 230; **112**, 229, 230; **115**, 230; **121**, 230, 231; **123**, 231
Ayscough, P.B., **46**, 57
Azuma, K., **205**, 176; **208**, 177
Bacon, R.G.R., **211**, 79, 89
Baer, B., **142**, 31
Baignee, A., **375**, 108
Bailey, B.E., **124**, 231

The number in bold type is the reference number and the following numbers are the page(s) on which the reference appears. References appear at the end of each Chapter.

Bailey, W.J., **146**, 171; **149**, 172, 175, 177, 178, 180, 181, 183; **190**, 175; **191**, 175; **195**, 175, 177; **198**, 175; **199**, 175; **200**, 175; **203**, 176; **204**, 176; **207**, 177; **208**, 177; **209**, 177; **210**, 177; **211**, 177, 181; **214**, 178; **219**, 179, 180; **222**, 181; **224**, 183; **226**, 183; **227**, 183; **230**, 183; **140**, 301
Baitis, F., **445**, 122
Bajpai, U.D.N., **265**, 89
Baker, C.A., **95**, 228
Balke, S.T., **33**, 213
Ballard, M.J., **154**, 238
Bamford, C.H., **1**, 8, 21; **266**, 89, 90; **267**, 90; **269**, 90; **272**, 90, 91; **428**, 118; **8**, 209; **16**, 211, 213; **22**, 213; **23**, 213; **24**, 213; **25**, 213; **26**, 213; **54**, 217; **89**, 228, 231, 232; **92**, 228, 231; **102**, 229, 231; **118**, 230; **132**, 232; **153**, 236, 244; **157**, 239; **158**, 239; **159**, 239, 240; **167**, 241; **260**, 260; **285**, 266; **108**, 297; **109**, 297; **110**, 297; **121**, 299; **164**, 305; **165**, 305; **169**, 305; **170**, 305; **26**, 321; **27**, 321; **38**, 324; **73**, 327; **74**, 327; **84**, 329; **105**, 332
Bandermann, F., **308**, 191
Banthia, A.K., **400**, 112, 124
Barantsevich, E.N., **128**, 300
Barbe, W., **134**, 29; **49**, 59, 63; **77**, 225
Barclay, L.R.C., **383**, 109; **212**, 178
Barone, V., **48**, 15; **71**, 18
Barreles-Rienda, J.M., **71**, 292
Barson, C.A., **321**, 98; **136**, 234; **148**, 236
Bartlett, P.D., **130**, 28, 34; **137**, 30; **54**, 59; **128**, 68, 75; **135**, 70, 72; **66**, 221, 222; **217**, 251, 254; **231**, 254
Barton, J., **38**, 54, 60, 61, 66, 70, 71, 72, 79; **80**, 225, 227, 231; **40**, 324, 325, 326, 327, 328
Barton, J.M., **145**, 170
Basahel, S.N., **158**, 239; **159**, 239, 240; **167**, 241
Bassi, G.L., **243**, 84
Bastianelli, U., **126**, 231
Bauer, D.R., **452**, 123
Bausch, M.J., **103**, 25
Bawn, C.E.H., **125**, 68
Baxter, H.N.III., **110**, 26; **111**, 26
Baxter, J.E., **258**, 87, 115; **419**, 115, 120
Baysal, B., **95**, 64, 71; **211**, 251

Baysal, B.M., **148**, 302
Beckhaus, H.D., **145**, 32
Beckwith, A.L.J., **21**, 5; **12**, 10, 16, 24, 30, 33; **21**, 12, 23; **53**, 16, 20; **54**, 16; **55**, 16, 56, 16, 23, 24; **76**, 18; **96**, 24; **98**, 24; **307**, 97, 120; **384**, 109; **438**, 121; **441**, 121; **442**, 121; **94**, 163; **97**, 164, 167, 168, 176; **123**, 167; **134**, 168; **141**, 169, 170; **153**, 172, 173, 174; **157**, 172, 173; **184**, 174; **213**, 178; **246**, 185; **317**, 193; **264**, 260, 261; **265**, 260; **31**, 322
Bednarek, D., **66**, 18; **21**, 48, 102, 107, 120, 122; **25**, 320, 321
Behari, K., **316**, 98
Behrman, E.J., **214**, 80
Bel'govskii, I.M., **203**, 249
Belgovskii, I.M., **200**, 249, 250
Bell, B., **184**, 246
Benedikt, G.M., **55**, 156, 186; **221**, 252, 253
Beniska, J., **132**, 337
Benson, S.W., **119**, 26; **141**, 31; **203**, 79; **204**, 79; **18**, 211
Benyon, K.I., **121**, 166
Benzing, E.P., **128**, 68, 75
Beranek, I., **14**, 10, 15, 19; **314**, 98
Berenbaum, M.B., **56**, 59; **63**, 60
Bergamini, F., **313**, 98
Berger, K.C., **150**, 71, 110; **114**, 229; **164**, 240, 242, 243, 245, 251, 255
Berger, P.A., **16**, 150; **17**, 150
Berlin, K.D., **120**, 166
Bernardi, R., **334**, 98, 101
Berndt, A., **149**, 32
Berner, G., **249**, 84
Bertrand, M.P., **121**, 26; **376**, 108; **377**, 109
Beshah, K., **14**, 150
Bessiere, J.-M., **418**, 114; **110**, 229
Bethea, T.W., **79**, 160
Bevington, J.C., **2**, 1; **26**, 50, 99; **96**, 65, 125; **102**, 65, 99; **163**, 72; **164**, 72; **316**, 98; **317**, 98, 102, 126; **318**, 98; **319**, 98; **320**, 98, 99; **321**, 98; **322**, 98; **324**, 98; **328**, 98, 123; **329**, 98; **337**, 99; **342**, 101; **347**, 102, 109, 110, 126; **348**, 102; **379**, 109; **423**, 117; **424**, 117; **455**, 123, 126; **456**, 123; **461**,

124; **475**, 125; **476**, 125; **477**, 125; **478**, 126; **98**, 229, 230, 232, 255; **101**, 229; **107**, 229, 230; **125**, 231; **278**, 264; **149**, 303
Bhanu, V.A., **271**, 262
Bhardwaj, I.S., **147**, 71
Biçak, N., **146**, 302
Bickel, A.F., **76**, 225
Bickley, H.T., **134**, 69, 70, 75
Bielecki, A., **112**, 166
Bieringer, H., **113**, 229
Billingham, N.C., **115**, 166
Binegar, G.A., **45**, 57
Bischoff, C., **193**, 77
Bitai, I., **291**, 191; **292**, 191
Bizilj, S., **138**, 30, 35; **86**, 63; **55**, 217, 219, 222, 223
Bledzki, A., **142**, 339; **143**, 339, 341; **155**, 341; **156**, 341; **157**, 341
Blevina, R.W., **129**, 337
Blitz, J.P., **233**, 184
Blomquist, A.T., **127**, 68
Bobo, J.S., **99**, 331
Bömer, B., **79**, 225, 227; **125**, 299; **151**, 303
Bonacic-Koutecky, V., **30**, 13
Bond, W.C., Jr., **119**, 166
Bonifacic, M., **311**, 97
Bonta, G., **63**, 326
Borchardt, J.K., **105**, 296, 297
Borsig, E., **38**, 54, 60, 61, 66, 70, 71, 72, 79; **40**, 324, 325, 326, 327, 328
Bothe, H., **152**, 172
Bottle, S., **166**, 72, 120, 121
Bottle, S.E., **412**, 114, 115, 120; **57**, 290
Boudevska, H., **116**, 230
Bouillion, G., **122**, 67, 74
Boukhors, A., **229**, 82
Boundy, R.H., **16**, 47; **10**, 317
Boutevin, B., **418**, 114; **110**, 229; **140**, 234; **168**, 241; **169**, 241; **170**, 241; **180**, 244; **114**, 298, 300; **132**, 300; **135**, 300; **136**, 300;
Bovey, F.A., **1**, 146, 149, 150, 152, 160, 161; **3**, 146, 149; **7**, 150; **12**, 150; **18**, 150; **22**, 150, 151; **24**, 150; **51**, 156; **57**, 157, 185; **239**, 184; **216**, 251, 259; **226**, 253; **238**, 256; **262**, 260; **69**, 292; **94**, 295

Bower, D.I., **9**, 150, 151, 152
Bowmer, T.N., **235**, 184
Bowry, V.W., **438**, 121; **440**, 121; **441**, 121; **157**, 172, 173; **178**, 173; **263**, 260; **264**, 260, 261; **265**, 260; **31**, 322
Boyer, R.F., **16**, 47; **10**, 317
Bradley, J.N., **139**, 31
Brady, R.F., Jr., **147**, 171
Braks, J.G., **101**, 65; **105**, 229; **120**, 230
Brame, E.G., **83**, 161; **84**, 161
Brame, E.G., Jr., **66**, 158
Brandup, G., **164**, 240, 242, 243, 245, 251, 255
Branston, R.E. **83**, 294
Braun, D., **80**, 62; **4**, 278; **142**, 339; **143**, 339, 341; **155**, 341; **156**, 341; **157**, 341
Bray, N.F., **136**, 169
Bredeweg, C.J., **198**, 78; **199**, 78
Breitenbach, J.W., **129**, 28; **286**, 94; **64**, 219, 220, 221; **93**, 228, 229; **113**, 229
Bresler, L.S., **128**, 300
Bresler, S.E., **279**, 190; **280**, 190
Breuer, S.W., **102**, 65, 99; **322**, 98; **348**, 102
Briggs, E.R., **288**, 94; **53**, 326
Britten, C., **176**, 172, 173
Brooks, B.R., **46**, 57
Brosse, J.-C., **227**, 81
Brosse, J.C., **117**, 298
Brown, A.S., **34**, 286
Brown, C.E., **118**, 26; **371**, 107, 108
Brown, D.W., **94**, 64; **6**, 316
Brown, J.E., **459**, 124; **51**, 217, 230; **21**, 318
Brown, L.R., **21**, 150
Bruch, M.D., **18**, 150; **24**, 150
Brudzynski, R.J., **391**, 111
Brumby, S., **37**, 323; **38**, 324
Brutchkov, C., **116**, 230
Buback, M., **83**, 62; **123**, **450**, 123; **451**, 123; **272**, 190, 191; **273**, 190, 191; **312**, 192; **3**, 209; **4**, 209; **40**, 213
Bugada, D.C., **41**, 152; **240**, 184; **239**, 256; **243**, 256; **249**, 257, 258; **250**, 257, 258
Buhle, E.L., **137**, 169
Bullock, R.M., **155**, 172, 173

Bunce, N.J., **107**, 26, 29
Buncel, E., **79**, 62
Burchard, W., **223**, 253
Burczyk, A.F., **201**, 249, 250
Burgess, F.J., **163**, 305
Burke, A.L., **91**, 294
Burnett, G.M., **91**, 228, 229; **45**, 324
Buruiana, E.C., **221**, 179
Busfield, W.K., **94**, 24; **95**, 24; **112**, 66, 103, 108, 120, 121; **166**, 72, 120, 121; **350**, 103, 108, 120; **360**, 105, 120; **361**, 105, 120; **368**, 106; **369**, 106; **437**, 120; **256**, 187; **258**, 187, 188, 189; **64**, 326, 327
Busnel, J.P., **127**, 337
Buter, R., **87**, 329
Butler, G.B., **16**, 4; **18**, 4; **19**, 4; **20**, 4; **79**, 20; **86**, 162, 164, 166, 167; **87**, 162, 166; **88**, 162; **89**, 162; **90**, 162; **91**, 163; **92**, 163; **109**, 166; **116**, 166; **117**, 166; **119**, 166; **120**, 166; **131**, 167; **145**, 170; **230**, 254; **232**, 254
Butler, P.E., **170**, 172
Buzanowski, W.C., **285**, 94, 95
Byrikhina, N.A., **336**, 99
Bywater, S., **51**, 59, 63; **89**, 63; **260**, 188; **263**, 188; **264**, 188; **73**, 223
Cacioli, P., **7**, 3; **2**, 44; **50**, 217; **56**, 217, 247, 249, 256; **174**, 306; **20**, 318; **158**, 341, 342, 343
Cais, R.E., **13**, 4; **11**, 150; **24**, 150; **51**, 156; **57**, 157, 185; **60**, 157; **63**, 157; **64**, 157, 158; **65**, 157; **216**, 251, 259; **226**, 253; **48**, 288, 289
Cameron, G.G., **4**, 316, 317; **13**, 317; **45**, 324; **60**, 326
Canadell, E., **27**, 13, 15; **49**, 15, 18
Capek, I., **80**, 225, 227, 231; **172**, 306
Caraculacu, A.A., **49**, 156
Cardenas, J.N., **27**, 213; **161**, 239
Carduner, K.R., **452**, 123
Carlblom, L.H., **253**, 86
Caronna, T., **427**, 118
Carrat, G.M., **41**, 213
Carswell, T.G., **281**, 190, 191; **286**, 191
Carter, R.O., **452**, 123
Cary, L.W., **111**, 66

Cascaval, C.N., **94**, 64; **6**, 316
Casinos, I., **276**, 91
Castaldi, G., **393**, 112
Castleman, J.K., **129**, 68, 79; **206**, 79
Catalina, F., **186**, 76
Chalfont, G.R., **159**, 72; **160**, 72; **421**, 117; **283**, 266; **284**, 266
Challa, G., **87**, 329; **88**, 329, 330; **90**, 330; **91**, 330
Chang, K.H.S., **222**, 253
Chapel, H.L., **444**, 122
Chapin, E.C., **145**, 170
Chapiro, A., **93**, 330
Chateauneuf, J., **346**, 102, 109, 110; **439**, 121; **269**, 261
Chatterjee, S.R., **139**, 234; **213**, 251
Chattopadhyay, A.K., **253**, 86
Chaudhuri, A.K., **119**, 230, 231
Chaumont, P., **287**, 94; **163**, 240; **131**, 337; **141**, 339
Chawla, O.P., **224**, 81, 112
Chen, C.Y., **133**, 233
Chen, P.Y., **198**, 175; **210**, 177
Chen, S., **282**, 264
Chen, S.-C., **198**, 175; **210**, 177
Chen, Y., **128**, 28; **151**, 340
Cheng, J., **269**, 190
Chenier, J.H.B., **43**, 14
Chiang, Y.S., **230**, 82
Chiantore, O., **135**, 169
Chiao, W.-B., **198**, 175
Chiriac, A.P., **32**, 52
Chiriac, M.V., **32**, 52
Chiu, W.Y., **41**, 213
Cho, I., **160**, 172; **166**, 172; **167**, 172; **168**, 172; **171**, 172; **180**, 174; **186**, 175; **187**, 175; **188**, 175; **196**, 175, 177; **197**, 175, 178; **218**, 179, 180
Choe, S., **169**, 72, 73
Choi, K.Y., **217**, 80; **150**, 303
Choi, S.Y., **186**, 175
Chong, Y.K., **284**, 93, 118; **159**, 341, 342
Chou, J.L., **190**, 175; **191**, 175; **200**, 175
Christie, D.I., **77**, 18; **25**, 285
Chujo, R., **25**, 150; **26**, 151; **27**, 151
Chung, D.C., **299**, 95

Author Index

Chung, H., **221**, 252, 253
Chung, R.P.-T., **92**, 63
Church, D.F., **39**, 14, 22, 23, 27; **413**, 114
Citterio, A., **19**, 11; **17**, 48, 98, 118; **313**, 98; **334**, 98, 101; **339**, 101, 118; **393**, 112; **394**, 112; **427**, 118; **267**, 260, 264
Clark, A.J., **35**, 322
Clarke, J.T., **165**, 240, 243, 244
Clay, P.A., **77**, 18; **155**, 238; **20**, 285
Clouet, G., **264**, 89; **287**, 94; **144**, 236; **163**, 240; **177**, 242; **131**, 337; **140**, 339; **141**, 339
Coca, S., **164**, 344
Coco, J.H., **97**, 65, 100; **218**, 252; **133**, 300
Cohen, F., **59**, 59
Cohen, G.M., **18**, 318
Cohen, S.G., **59**, 59; **60**, 59; **175**, 73
Coleman, M.L., **81**, 160
Coleman, M.M., **83**, 161; **84**, 161
Coleman, R.J., **172**, 241
Colletti, R.F., **112**, 166
Collot, J., **262**, 188, 190
Comanita, E., **239**, 83; **141**, 302
Conlon, D.A., **144**, 339; **145**, 339; **146**, 339; **147**, 339
Cook, R.E., **261**, 188
Cook, W.D., **261**, 88
Cooke, H.G., Jr., **73**, 158
Corfield, G.C., **122**, 166
Corley, R.C., **124**, 28, 31, 32, 34; **62**, 219, 221; **84**, 226
Corner, J.O., **150**, 340
Corner, T., **141**, 234, 235, 241; **130**, 300
Corpart, P., **287**, 94
Costa, L., **135**, 169
Costanza, A.J., **172**, 241
Cowan, J.C., **72**, 158; **73**, 158
Cowie, J.M.G., **5**, 278, 288; **40**, 288; **41**, 288
Cox, R.A., **79**, 62
Coxon, J.M., **447**, 122
Craddock, J., **230**, 82
Crano, J., **180**, 74
Craven, I., **34**, 286
Crawford, T.H., **136**, 169
Crawshaw, A., **20**, 4; **92**, 163

Crivello, J.V., **144**, 339; **145**, 339; **146**, 339; **147**, 339
Crosato-Arnaldi, A., **224**, 253
Crowther, M.W., **23**, 150, 151
Cudby, M.E.A., **236**, 184
Cummings, W., **53**, 326
Cunliffe, A.V., **161**, 305; **163**, 305
Curci, R., **148**, 71; **223**, 81
Curtis, H.C., **137**, 70, 109
Cuthbertson, M.C., **351**, 103, 111
Cuthbertson, M.J., **15**, 10, 15; **177**, 74, 103, 110, 117, 120; **358**, 105, 106, 120; **359**, 105, 106, 120
Cutler, D.J., **236**, 184
Cywar, D.A., **328**, 98, 123; **329**, 98; **456**, 123; **20**, 284, 285
Czerwinski, W.K., **80**, 62; **4**, 278
Dais, V.A., **184**, 246
Dakova, I.G., **52**, 325, 326
Dambatta, B.B., **167**, 72
Danek, S.K., **50**, 59, 63, 98, 100, 127; **92**, 63; **316**, 193; **83**, 225; **15**, 318
Dannenberg, J.J., **90**, 23; **142**, 31; **244**, 184, 185
Danusso, F., **128**, 232
Darricades-Llauro, M.F., **52**, 156
Das, S.K., **213**, 251
Davidson, R.S., **258**, 87, 115; **419**, 115, 120
Davies, A.G., **119**, 66
Davis, T.P., **293**, 191; **294**, 191; **295**, 191; **296**, 191; **305**, 191; **204**, 249; **106**, 296, 297; **107**, 296; **68**, 326, 327
Davis, W.H., Jr., **86**, 22
Dawkins, J.V., **94**, 228, 229
de Gennes, P.G., **39**, 213
de Haan, J.W., **37**, 287, 295
de Hargrave, C.V., **463**, 124
de Pooter, M., **297**, 95
Deady, M., **70**, 60, 63, 127; **304**, 191, 192, 193; **21**, 211, 214; **44**, 214; **10**, 278, 300; **115**, 332, 333, 334, 335
Deb, P.C., **150**, 71, 110
Dec, J., **73**, 158; **150**, 340
Deibert, S., **308**, 191
Delano, G., **223**, 81

Delbecq, F., **72**, 18
Derouet, D., **227**, 81; **117**, 298
Dewar, M.J.S., **73**, 18
Di Silvestro, G., **405**, 114, 124
Dickerman, S.C., **341**, 101
Dickson, J.K., Jr., **223**, 181
Dietrich, H.J., **114**, 166
DiFuria, F., **223**, 81
Dinan, F.J., **39**, 287, 295; **70**, 292, 295
Dinerstein, R.J., **445**, 122
Dixon, D.W., **235**, 83
Dixon, W.T., **231**, 82
Dodgson, K., **46**, 288
Doi, T., **136**, 339
Dong, L., **19**, 150, 152
Donners, W.A.B., **45**, 155; **256**, 259
Dorigo, A.E., **91**, 23
Douady, J., **74**, 18
Douzou, P., **229**, 82
Drewer, R.J., **76**, 61
Druliner, J.D., **152**, 33; **109**, 66; **58**, 290; **160**, 341, 344
Drumright, R.E., **191**, 77, 78
Duever, T.A., **91**, 294
Duisman, W., **57**, 59; **64**, 60
Dulog, L., **105**, 65
Dumitriu, S., **239**, 83; **141**, 302
Dumont, E., **145**, 71
Dunn, A.S., **248**, 257, 258
Dünnebacke, D., **135**, 29
Dusek, K., **227**, 81; **117**, 298
Duynstee, E.F.J., **179**, 74
Dyball, C.J., **163**, 72; **164**, 72
Dyson, R.W., **89**, 228, 231, 232; **165**, 305
Dzhabiyeva, Z.M., **336**, 99
Eastmond, G.C., **268**, 90; **269**, 90; **7**, 209; **89**, 228, 231, 232; **118**, 230; **132**, 232; **138**, 234, 260; **154**, 304; **164**, 305; **165**, 305; **166**, 305; **167**, 305; **168**, 305
Easton, C.J., **53**, 16, 20; **96**, 24; **98**, 24
Eaton, D.R., **285**, 191
Eaves, D.E., **125**, 231

Ebdon, J.R., **167**, 72; **226**, 81; **455**, 123, 126; **461**, 124; **477**, 125; **46**, 288; **113**, 298
Edelmann, D., **129**, 300
Edge, D.J., **380**, 109, 110
Edwards, J.O., **148**, 71; **214**, 80; **223**, 81
Eigenmann, H.K., **82**, 21
Eisenstein, O., **27**, 13, 15
El Idrissi, A., **168**, 241; **135**, 300
Elghoul, A.M.R., **146**, 236
Elias, H.G., **34**, 152
Elliasyan, M.A., **132**, 168
Elliot, J.J., **170**, 172
Elson, I.H., **6**, 9; **23**, 49, 105, 107
Encina, M.V., **365**, 106, 107
Endo, T., **150**, 172; **162**, 172; **163**, 172; **164**, 172; **165**, 172; **172**, 172, 174, 183; **173**, 172, 183; **174**, 172; **175**, 172; **177**, 172, 183; **179**, 173, 174; **181**, 174; **182**, 174; **185**, 174; **193**, 175, 179, 180; **194**, 175, 176, 177, 178; **198**, 175; **205**, 176; **206**, 176, 177; **208**, 177; **210**, 177; **215**, 178, 180; **216**, 179, 180; **217**, 179, 180; **220**, 179; **224**, 183; **225**, 183; **227**, 183; **228**, 183; **140**, 301; **106**, 332; **107**, 332
Engel, P.S., **128**, 28; **40**, 57, 60, 61; **151**, 340
Engelbrecht, R., **17**, 284
Englin, B.A., **151**, 236; **152**, 236
Enikolopyan, N.S., **200**, 249, 250; **207**, 250; **166**, 345; **168**, 345, 346
Ensel, S.M. **99**, 331
Epaillard, F., **227**, 81; **117**, 298
Errede, L.A., **189**, 175
Esser, M.L., **179**, 74
Esslemont, G.F., **60**, 326
Evans, C.A., **124**, 68, 72
Evans, H.E., **46**, 57
Evans, M.G., **83**, 21
Faldi, A., **85**, 62; **29**, 213; **42**, 214
Farina, M., **405**, 114, 124; **467**, 124; **6**, 148, 150; **137**, 234
Farmer, R.G., **48**, 288, 289
Fedorova, E.F., **161**, 172
Feix, C., **57**, 17
Feldman, K.S., **99**, 331; **100**, 331
Feng, P.-Z., **200**, 175; **222**, 181

Author Index

Feng, R.H.C., **197**, 78
Ferguson, J.,**92**, 330
Ferguson, R.C., **32**, 151; **44**, 155; **66**, 158; **253**, 258, 259
Ferington, T.E., **176**, 242; **125**, 336
Fernandez-Garcia, M., **71**, 292
Ferris, A., **127**, 68
Fessenden, R.W., **224**, 81, 112
Fetters, L.J., **5**, 316
Field, N.D., **101**, 166
Fineman, M., **76**, 293
Finestone, A.B.,
Finestone, A.B., **63**, 60; **61**, 219
Fink, J.K., **78**, 61
Firestone, R.A., **66**, 60
Fischer, H., **14**, 10, 15, 19; **133**, 28; **160**, 34, 35; **314**, 98; **323**, 98, 99, 100, 124; **327**, 98; **331**, 98, 124; **386**, 111, 124; **387**, 111; **464**, 124; **465**, 124; **69**, 221, 222
Fischer, J.P., **63**, 291
Fiske, T.R., **93**, 64, 65, 100; **340**, 101
Flemming, I., **78**, 18
Florin, R.E., **94**, 64; **5**, 316; **6**, 316
Flory, P.J., **3**, 1; **282**, 93; **43**, 155; **134**, 234; **135**, 234
Fomichev, V.N., **279**, 190
Fong, C.W., **68**, 18
Foote, R.S., **87**, 63; **71**, 222, 223
Fossey, J., **54**, 156
Foster, F.J., **145**, 236
Fox, T.G., **10**, 150; **31**, 151; **246**, 257
Fraenkel, G., **146**, 32; **68**, 221
Francis, A.P., **82**, 294
Franta, E., **173**, 306
Fraser-Reid, B., **223**, 181
Freed, D.J., **57**, 157, 185; **216**, 251, 259
Freidlina, R.K., **152**, 236
Fridd, P.F., **423**, 117; **424**, 117
Friedlander, H.N., **127**, 300
Fristad, W.E., **395**, 112
Fryd, M., **171**, 345
Fueno, T., **111**, 332
Fuhrman, J., **89**, 294
Fujimora, K., **34**, 286

Fujimori, K., **121**, 67
Fujimoto, H., **25**, 13; **144**, 31, 33; **82**, 225
Fukawa, J., **111**, 332
Fukuda, T., **81**, 62, 120; **275**, 190; **47**, 215; **48**, 215; **24**, 285; **28**, 286, 287
Fukuda, W., **124**, 167, 169
Fukui, K., **25**, 13; **144**, 31, 33; **82**, 225
Fukuoka, K., **69**, 60
Fuller, D.L., **41**, 14, 22
Funt, B.L., **127**, 231
Furakawa, J., **43**, 288
Galbraith, M.N., **7**, 278; **114**, 332, 333, 335
Gall, E.J., **102**, 166
Gallo, B.M., **63**, 326
Gallopo, A.R., **223**, 81
Gapud, B., **198**, 175; **140**, 301
Garcia-Rubio, L.H., **142**, 71, 123; **143**, 71, 109; **449**, 123; **272**, 190, 191; **3**, 209; **84**, 294
Gasparro, F.P., **60**, 291
Gaylord, N.G., **303**, 96
Geckle, M.J., **146**, 32; **68**, 221
Geers, B.N., **413**, 114
Georges, M.K., **118**, 335, 342; **122**, 336, 341, 342; **123**, 336, 341, 342
Georgiev, G.S., **52**, 325, 326
Gerlock, J.L., **452**, 123
German, A.L., **32**, 286; **37**, 287, 295; **38**, 287; **86**, 294; **87**, 294; **88**, 294; **59**, 326
Ghanem, N.A., **278**, 264
Ghatge, N.D., **122**, 299; **123**, 299
Ghirardini, M., **427**, 118
Ghosez, A., **34**, 322
Ghosh, N.N., **397**, 112; **472**, 124
Giacometti, G., **387**, 111
Gibian, M.J., **124**, 28, 31, 32, 34; **62**, 219, 221; **84**, 226
Giese, B., **9**, 10, 13; **10**, 10, 13, 19, 20, 21; **13**, 10; **17**, 10, 11; **29**, 13; **36**, 14, 17; **37**, 14; **57**, 17; **58**, 17, 25; **65**, 18; **305**, 97, 98; **312**, 97; **429**, 119; **430**, 119; **435**, 120; **17**, 284; **34**, 322
Gilbert, B.C., **104**, 25
Gilbert, R.D., **261**, 260, 264

Gilbert, R.G., **77**, 18; **35**, 53; **82**, 62; **272**, 190, 191; **273**, 190, 191; **274**, 190, 191; **287**, 191; **298**, 191; **302**, 191; **303**, 191; **315**, 193; **3**, 209; **4**, 209; **5**, 209; **154**, 238; **155**, 238; **25**, 285; **47**, 324
Gilliom, R.D., **85**, 22, 23
Gilman, L.B., **463**, 124
Giordano, C., **393**, 112
Gippert, G.P., **21**, 150
Glaser, D.M., **186**, 76
Gleicher, G.J., **413**, 114
Gleixner, G., **129**, 28; **17**, 211, 212; **64**, 219, 220, 221
Gobel, T., **34**, 322
Gol'dfein, M.D., **268**, 260
Golan, D.R., **155**, 33
Goldfinger, G., **261**, 260, 264; **26**, 285
Golubev, V.A., **443**, 121
Golubev, V.B., **281**, 264
Gong, M.-S., **218**, 179, 180
Gonzalez, C., **28**, 13, 18
Gonzalez-Gomez, J.A., **435**, 120
Goodrich, S.D., **97**, 229
Gopalan, A., **138**, 169, 170; **139**, 169, 170
Görlitz, V.M., **62**, 157
Gostowski, R., **103**, 25
Gould, C.W., **207**, 79
Graham, J.D., **285**, 94, 95
Graham, P.J., **137**, 169
Grannmayeth, R., **92**, 330
Grant, R.D., **116**, 26; **117**, 26; **20**, 48, 49, **104**, 106, 107, 120; **236**, 83, 120; **344**, 102, 103, 104, 106, 107, 120; **345**, 102, 103, 111, 120
Grassie, N., **8**, 316
Gratch, S., **246**, 257
Greber, G., **260**, 87
Greenley, R.Z., **12**, 281, 282, 293; **13**, 281, 282; **101**, 296; **103**, 296; **104**, 296
Gregg, R.A., **149**, 71; **103**, 229, 251
Greszta, D., **117**, 335
Grice, D.I., **94**, 24; **95**, 24; **368**, 106; **369**, 106
Gridnev, A.A., **202**, 249, 251
Griffiths, P.G., **116**, 26; **11**, 46, 102, 103, 106, 120; **20**, 48, 49, 104, 106, 107, 120; **352**, 104, 105, 120
Griggs, J.A., **159**, 304, 305

Grigor, J., **268**, 90; **168**, 305
Griller, D., **100**, 24; **104**, 25; **125**, 28; **147**, 32, 34; **148**, 32; **150**, 32; **18**, 48; **262**, 88; **370**, 107; **383**, 109; **212**, 178; **20**, 211
Grinshpun, V., **240**, 256; **241**, 256
Grodzinsky, A.J., **127**, 337
Gromov, V.F., **61**, 17; **41**, 324, 325, 326
Gronski, W., **72**, 292
Grossi, L., **139**, 70
Grow, R.H., **45**, 57
Gruber, H., **260**, 87
Gu, J.M., **195**, 175, 177
Guadalupe-Fasano, C., **103**, 25
Guaita, M., **126**, 167; **135**, 169
Guillot, J., **272**, 190, 191; **3**, 209; **35**, 286
Gunderson, H.J., **45**, 57
Gupta, B.S., **274**, 91
Gupta, I., **185**, 76
Gupta, S.N., **185**, 76
Guth, W., **60**, 219, 225, 227; **124**, 299, 300
Guthrie, J.T., **273**, 91
Guyot, A., **425**, 117; **52**, 156; **35**, 286; **55**, 326
Haber, F., **228**, 82
Haddleton, D.M., **204**, 249
Hageman, H.J., **247**, 84; **254**, 86; **256**, 86; **258**, 87, 115; **419**, 115, 120; **275**, 263
Hajeck, J., **112**, 26
Hakozaki, J., **100**, 229
Hale, W.F., **63**, 60
Halford, R.G., **125**, 68
Hall, H.K., Jr., **280**, 92, 95, 96; **293**, 95, 96; **300**, 95; **151**, 172
Halou, O.-M., **377**, 109
Halpern, J., **30**, 322
Ham, G.E., **16**, 283; **27**, 285
Hamashima, M., **178**, 243
Hamer, G.K., **118**, 335, 342; **122**, 336, 341, 342; **123**, 336, 341, 342
Hamielec, A.E., **296**, 95; **272**, 190, 191; **285**, 191; **3**, 209; **12**, 211; **15**, 211, 213; **33**, 213
Hamilton, C.J., **210**, 79
Hammond, G.S., **24**, 13; **91**, 63
Hanna, G.M., **87**, 22

Author Index

Hansen, J.L., **108**, 229
Hara, H., **154**, 72
Hardy, S.J., **186**, 76
Hargis, J.H., **133**, 69, 70
Harris, D.O., **347**, 102, 109, 110, 126
Harrison, D., **196**, 247
Hartless, R.L., **57**, 157, 185; **216**, 251, 259
Hartung, J., **34**, 322
Hartzler, H.D., **25**, 50, 99
Harwood, H.J., **37**, 152; **97**, 229; **162**, 239; **62**, 291; **74**, 292; **75**, 292; **65**, 326; **66**, 326
Hasha, D.L., **297**, 95
Hatada, K., **458**, 123; **459**, 124; **460**, 124; **8**, 150; **25**, 150; **26**, 151; **27**, 151; **38**, 152; **51**, 217, 230; **122**, 231; **257**, 259; **46**, 324
Hautus, F.L.M., **87**, 294; **88**, 294
Haward, R.N., **143**, 169, 170
Hawkins, E.G.R., **120**, 66
Hawthorne, D.G., **17**, 4; **22**, 5; **5**, 9; **94**, 163; **96**, 163, 166; **133**, 168; **56**, 217, 247, 249, 256; **210**, 251; **174**, 306
Hayama, S., **219**, 80
Hayashi, N., **72**, 327
Hayashi, R., **206**, 176, 177
Hayashi, S., **38**, 14
Hayes, G.F., **161**, 305
Haynes, A.C., **99**, 65; **121**, 230, 231
Hazar, B., **148**, 302
He, J., **13**, 10; **29**, 13
Heatley, F., **247**, 257
Hebeish, A., **273**, 91
Heberger, K., **323**, 98, 99, 100, 124; **327**, 98
Heicklein, J.P., **354**, 105
Heilmann, S.M., **220**, 80; **221**, 80
Heine, H.-G., **252**, 84
Heitz, W., **60**, 219, 225, 227; **79**, 225, 227; **85**, 226, 227, 234; **115**, 298, 300; **119**, 298, 300; **124**, 299, 300; **125**, 299; **145**, 302; **151**, 303
Heldoorn, G.M., **155**, 33
Henderson, J.N., **78**, 160
Henderson, R.W., **40**, 14, 22; **46**, 14, 23
Hendra, P.J., **236**, 184
Hendrickson, W.H., Jr., **170**, 72

Hendry, D.G., **82**, 21
Henrici, G., **245**, 256
Henrici-Olivé, G., **68**, 60; **106**, 229; **117**, 230
Hensley, D.R., **97**, 229
Herk, L., **326**, 98
Hershberger, J., **434**, 120
Hershberger, S., **434**, 120
Heseltine, E.N.J., **102**, 65, 99
Heuts, J.P.A., **77**, 18; **25**, 285
Hey, D.H., **159**, 72; **160**, 72; **283**, 266; **284**, 266
Hiatt, R., **123**, 67, 73; **129**, 68, 79; **131**, 68; **184**, 76; **189**, 77; **201**, 78; **202**, 78; **205**, 79; **206**, 79; **207**, 79
Hikichi, K., **20**, 150, 151, 152; **35**, 152
Hilborn, J.W., **378**, 109
Hill, C.L., **431**, 119
Hill, D., **272**, 190, 191; **3**, 209
Hill, D.J.T., **103**, 65, 122; **275**, 91; **19**, 150, 152; **273**, 190, 191; **281**, 190, 191; **286**, 191; **4**, 209; **22**, 284, 285, 286, 295; **30**, 286, 295; **33**, 286, 288, 295; **42**, 288; **48**, 288, 289; **51**, 289; **93**, 295; **56**, 326; **57**, 326
Hill, L.W., **9**, 278; **108**, 332
Hill, R.J., **62**, 291
Hinkelmann, F., **291**, 191
Hiraguri, Y., **185**, 174; **193**, 175, 179, 180; **215**, 179, 180; **216**, 179, 180; **217**, 179, 180
Hirota, M., **196**, 78
Hirota, N., **138**, 70
Hjertberg, T., **50**, 156; **251**, 186
Ho, T.-L., **270**, 90
Hochster, H.S., **132**, 28; **70**, 222
Hodder, A.N., **16**, 318
Holland, K.A., **16**, 318
Holt, T., **142**, 169, 170
Horii, F., **27**, 151
Horsfield, A., **421**, 117
Hosomi, A., **38**, 14; **42**, 14, 22; **363**, 106
Hoss, W.P., **44**, 14, 22
Houk, K.N., **29**, 13; **69**, 18; **75**, 18; **91**, 23; **122**, 27
House, D.A., **213**, 80, 81
Howard, J.A., **23**, 13; **43**, 14; **82**, 21; **100**, 24; **262**, 88; **343**, 101; **375**, 108; **401**, 113; **403**, 113
Howard, R.O., **165**, 240, 243, 244

Howell, J.A., **65**, 291
Hu, S.S., **330**, 98
Huang, B., **158**, 304
Huang, J., **72**, 60, 63
Huang, R.Y.M., **101**, 65; **105**, 229; **120**, 230
Huang, X.L., **90**, 23; **244**, 184, 185
Huckerby, T.N., **102**, 65, 99; **226**, 81; **316**, 98; **317**, 98, 102, 126; **318**, 98; **319**, 98; **320**, 98, 99; **321**, 98; **322**, 98; **328**, 98, 123; **329**, 98; **337**, 99; **348**, 102; **455**, 123, 126; **456**, 123, **461**, 124
Huckestein, B., **83**, 62, 123; **450**, 123; **451**, 123; **312**, 192; **40**, 213
Huie, R.E., **402**, 113; **272**, 262
Humphrey, M.J., **123**, 231
Hunter, D.S., **286**, 191
Hunter, T.C., **226**, 81
Hutchinson, R.A., **306**, 191; **309**, 191
Huter, O., **34**, 322
Hutovic, J., **77**, 18; **25**, 285
Hutton, N.W.E., **317**, 98, 102, 126; **318**, 98; **337**, 99
Huyser, E.S., **45**, 14; **62**, 17, 25; **130**, 68, 78; **197**, 78; **198**, 78; **199**, 78; **373**, 107
Hwang, E.F.J., **8**, 3; **3**, 44
Ichihashi, T., **125**, 167
Ichijo, C., **89**, 329
Icli, S., **150**, 32
Idage, B.B., **122**, 299
Igeta, S.-I., **153**, 72
Iizawa, T., **140**, 169, 170
Ilavsky, D., **72**, 18
Imoto, M., **143**, 31; **169**, 72, 73; **173**, 241, 242
Inaba, A., **459**, 124; **51**, 217, 230
Inagaki, H., **81**, 62, 120; **275**, 190; **47**, 215; **48**, 215; **24**, 285; **28**, 286, 287
Ingold, K.U., **21**, 5; **21**, 12, 23; **99**, 24; **107**, 26, 29; **113**, 26; **118**, 26; **147**, 32, 34; **156**, 34; **18**, 48; **24**, 49, 99; **139**, 70; **346**, 102, 109, 110; **357**, 105; **371**, 107, 108; **383**, 109; **439**, 121; **440**, 121; **441**, 121; **156**, 172, 173; **178**, 173; **212**, 178; **246**, 185; **263**, 260; **265**, 260; **266**, 260, 262; **269**, 261
Ingram, J.E., **111**, 166
Inoue, T., **31**, 52, 64
Ioffe, N.T., **166**, 345; **167**, 345; **168**, 345, 346

Irwin, K.C., **129**, 68, 79; **202**, 78; **206**, 79; **207**, 79
Isa, K., **127**, 167
Ishigaki, H., **195**, 77, 113
Ishiguro, S., **468**, 124
Ishiwata, H., **31**, 52, 64
Issari, B., **200**, 175; **230**, 183
Ito, D., **253**, 186
Ito, K., **88**, 63; **38**, 213; **59**, 219, 226; **72**, 222; **87**, 227; **138**, 339
Ito, O., **67**, 18; **406**, 114; **407**, 114; **409**, 114, 115; **414**, 114; **415**, 114; **416**, 114; **417**, 114
Ito, T., **342**, 101; **61**, 326
Ittel, S.D., **205**, 249
Ivanchev, S.S., **151**, 71
Ivanov, B.E., **333**, 98, 101
Ivin, K.J., **255**, 187; **256**, 187; **260**, 188; **261**, 188; **265**, 188
Iwanami, K., **67**, 158; **68**, 158; **71**, 158; **144**, 170
Iwatsuki, S., **301**, 95
Iyoda, S., **88**, 63; **72**, 222; **87**, 227
Izu, M., **62**, 291; **65**, 291
Izumi, H., **130**, 337
Jacobine, A., **186**, 76
Jaffe, A.B., **90**, 63; **81**, 225
Jankauskas, K.J., **373**, 107
Janousek, Z., **273**, 262
Jansen, L.G.J., **256**, 86
Jantas, R., **102**, 332; **103**, 332; **104**, 332
Janzen, E.G., **124**, 68, 72
Jenkins, A.D., **34**, 13; **35**, 13; **428**, 118; **115**, 166; **92**, 228, 231; **102**, 229, 231; **124**, 231; **285**, 266; **98**, 295, 297; **108**, 297; **109**, 297; **110**, 297; **111**, 297; **121**, 299; **74**, 327
Jenkins, I.D., **94**, 24; **95**, 24; **112**, 66, 103, 108, 120; **121**; **166**, 72, 120, 121; **350**, 103, 108, 120; **360**, 105, 120; **361**, 105, 120; **368**, 106; **369**, 106; **437**, 120
Jenkins, R.F., **255**, 258
Jiang, W., **330**, 98
Jiang, X.L., **72**, 327
Johns, S.R., **15**, 4; **22**, 5; **5**, 9; **6**, 45, 52, 64, 65, 71, 99, 100, 123; **8**, 46, 47, 71, 109; **28**, 51, 59, 61, 62,

Author Index

63, 64, 127; **457,** 123; **15,** 150; **30,** 151, 152; **88,** 227, 229; **111,** 229; **126,** 299; **14,** 317
Johnsen, A., **72,** 292; **73,** 292
Johnson, C.H.J., **43,** 214; **121,** 336, 342
Johnson, D.H., **104,** 229
Johnson, K.A., **153,** 33; **155,** 33
Johnson, M., **347,** 102, 109, 110, 126
Johnston, L.J., **446,** 122
Johnston, R., **428,** 118; **92,** 228, 231; **285,** 266; **108,** 297; **121,** 299; **74,** 327
Joiner, S.N., **45,** 324
Jones, K., **35,** 322
Jones, M.J., **18,** 10, 14, 15, 23; **362,** 106, 120
Jones, R.G., **162, 72**
Jones, S.A., **432,** 119; **19,** 284, 285; **54,** 290; **56,** 290
Jones, S.L., **50,** 59, 63, 98, 100, 127; **316,** 193; **83,** 225; **15,** 318
Jovanovich, S., **242,** 184; **243,** 184
Jug, K., **304,** 96
Julia, M., **51,** 16; **52,** 16
Jumangat, K., **319,** 98
Jumonville, S., **41,** 14, 22
Juranicova, V., **80,** 225, 227, 231
Kabanov, V.A., **229,** 254; **75,** 327, 328
Kable, S.H., **77,** 18; **25,** 285
Kadokawa, J., **201,** 175; **202,** 175
Kageoka, M., **288,** 191, 192; **289,** 191
Kaizerman, S., **271,** 90
Kajiwara, A., **411,** 114, 115; **77,** 159, 161
Kakiuchi, H., **124,** 167, 169
Kale, L.T. **70,** 292, 295; **84,** 294
Kamachi, M., **60,** 17; **335,** 99; **411,** 114, 115; **77,** 159, 161; **283,** 191; **290,** 191; **257,** 259; **258,** 259; **39,** 324; **42,** 324, 325; **43,** 324; **44,** 324, 325; **46,** 324
Kamath, V.R., **36,** 54; **374,** 107, 108
Kameshima, T., **276,** 263
Kamide, K., **35,** 152
Kamiya, Y., **27,** 50, 57, 60, 62; **183,** 76
Kamlet, M.J., **68,** 18
Kammerer, H., **96,** 331; **97,** 331; **98,** 331
Kanabus-Kaminske, J.M., **104,** 25
Kanai, H., **29,** 151

Kanai, T., **36,** 152; **89,** 329
Kanamori, T., **139,** 339
Kanda, N., **181,** 174
Kanoya, A., **153,** 72; **154, 72**
Karad, P., **53,** 289
Kargin, V.A., **75,** 327, 328
Karmilova, L.V., **203,** 249
Kashiwagi, T., **459,** 124; **51,** 217, 230; **21,** 318
Kastl, P.E., **191,** 77, 78
Kato, E., **190,** 246
Kato, H., **76,** 328
Kaufmann, H.F., **286,** 94; **113,** 229
Kawa, H., **197,** 247, 256
Kawaguchi, N., **68,** 158
Kawahara, F., **277,** 263, 264
Kawai, W., **125,** 167
Kawamura, T., **36,** 152; **39,** 152, **40,** 152
Kaye, H., **130,** 167
Kazbekov, E.N., **279,** 190; **280,** 190
Kazmaier, P.M., **118,** 335, 342; **122,** 336, 341, 342; **123,** 336, 341, 342
Keana, J.F.W., **445,** 122
Kelen, T., **79,** 293; **80,** 293; **92,** 294
Kelly, D.P., **138,** 30, 35; **50,** 59, 63, 98, 100, 127; **86,** 63; **316,** 193; **55,** 217, 219, 222, 223; **58,** 219, 222, 223; **83,** 225; **15,** 318
Kemmerer, R., **101,** 331
Kennedy, J.P., **170,** 172; **120,** 335
Kennedy, V.H., **155,** 33
Kenyon, W.O., **251,** 258
Kern, W., **474,** 125; **97,** 331
Kerr, J.A., **140,** 31
Kerr, W.W., **8,** 316
Keute, J., **444,** 122
Keys, R.T., **91,** 63
Khachaturov, A.S., **161,** 172
Khanarian, G., **11,** 150
Khanna, R.K., **330,** 98
Kharasch, M.S., **277,** 263, 264
Kharitonova, O.A., **144,** 71
Khomiskovskii, P.M., **61,** 17; **41,** 324, 325, 326
Kida, S., **118,** 166
Kiefer, H., **61,** 59, 65; **110,** 66, 76

Kikuta, M., **69**, 158
Kim, B.-G., **218**, 179, 180
Kim, C.-B., **218**, 179, 180
Kim, I.P., **167**, 345
Kim, J.-B., **180**, 174
Kim, S.-K., **187**, 175; **197**, 175, 178
Kim, Y.H., **222**, 80
Kimura, H., **192**, 77
Kimura, S., **131**, 167
Kimura, T., **178**, 243
King, C., **172**, 241
King, J., **9**, 150, 151, 152
Kinoshita, Y., **173**, 241, 242
Kiparissides, C., **29**, 51, 62, 63, 71
Kirata, T., **21**, 318
Kirchmayr, R., **249**, 84
Kirchner, K., **281**, 92, 94; **298**, 95
Kirtman, B., **154**, 33
Kishore, K., **271**, 262
Kita, S., **171**, 72, 117
Kitagawa, K., **126**, 68, 73
Kitamaru, R., **26**, 151; **27**, 151
Kitamura, T., **109**, 166
Kitayama, T., **458**, 123; **459**, 124; **460**, 124; **8**, 150; **25**, 150; **26**, 151; **27**, 151; **51**, 217, 230; **122**, 231; **257**, 259; **46**, 324
Klein, P., **105**, 65
Klemm, E., **192**, 175; **229**, 183; **137**, 300
Klesper, E., **72**, 292; **73**, 292
Klos, R., **260**, 87
Klumperman, B., **67**, 326, 327; **69**, 326
Knipper, M., **144**, 236
Kobatake, S., **186**, 246; **190**, 246; **191**, 246
Kobayashi, S., **201**, 175; **202**, 175
Koch, A., **34**, 322
Koch, T.H., **444**, 122
Kochi, J.K., **6**, 9; **120**, 26, 30; **23**, 49, 105, 107; **136**, 70; **157**, 72, 109, 118; **355**, 105, 107, 118; **380**, 109, 110; **381**, 109, 118
Kodaira, K., **88**, 63; **72**, 222
Kodaira, T., **127**, 167; **128**, 167; **87**, 227; **178**, 243
Koenig, J.L., **4**, 146, 152; **68**, 291
Koenig, T., **43**, 57, 66

Koga, G., **42**, 57, 60
Koga, N., **42**, 57, 60
Köhler, K.H., **79**, 225, 227; **125**, 299
Kohno, M., **283**, 191; **290**, 191
Koizumi, T., **182**, 174
Kokubo, T., **301**, 95
Kolthoff, I.M., **132**, 68; **108**, 229; **262**, 260
Komai, T., **126**, 68, 73
Kometani, J.M., **13**, 4; **11**, 150; **60**, 157; **63**, 157; **65**, 157
Komori, T., **153**, 72; **154**, 72
Konishi, K., **163**, 344
Konishi, Y., **411**, 114, 115
Konter, W., **79**, 225, 227; **125**, 299
Kopecky, K.R., **159**, 34
Korth, H.-G., **367**, 106
Kotnour, T.A., **221**, 80
Kotyk, J.J., **16**, 150; **17**, 150
Koutecky, J., **30**, 13
Kovacic, P., **158**, 72, 109, 111
Kozlov, Y.N., **443**, 121
Kozlowski, S.A., **18**, 150
Kraeger, I.R., **69**, 326
Kraeutler, B., **33**, 52
Krasavina, N.B., **144**, 71
Kremminger, P., **276**, 190; **277**, 190
Krepski, L.R., **220**, 80
Kretzschmar, G., **65**, 18
Kroder, K., **34**, 322
Kronfli, E.B., **115**, 166
Krstina, J., **13**, 47, 64, 72, 127; **50**, 59, 63, 98, 100, 127; **53**, 59, 60, 63; **70**, 60, 63, 127; **75**, 61; **315**, 98, 127; **316**, 193; **44**, 214; **83**, 225; **11**, 317; **15**, 318; **70**, 327; **83**, 328; **115**, 332, 333, 334, 335
Krusic, P.D., **152**, 33; **109**, 66; **58**, 290
Kubo, K., **48**, 215; **24**, 285
Kuchanov, S.I., **215**, 251; **1**, 278; **143**, 302; **116**, 335; **124**, 336
Kuchta, F.-D., **83**, 62, 123; **312**, 192
Kuindersma, M.E., **310**, 191
Kuki, M., **253**, 186
Kumar, M.V., **229**, 254
Kunitake, T., **366**, 106; **113**, 166

Author Index

Kuo, J.F., **133**, 233
Kurbatov, V.A., **279**, 92
Kuriyama, A., **126**, 337, 338; **134**, 338; **135**, 338
Kuruganti, V., **200**, 175
Kuruganti, Y., **198**, 175
Kurz, M.E., **158**, 72, 109, 111
Kusefoglu, S.H., **111**, 166; **268**, 190
Kuwae, Y., **335**, 99; **283**, 191; **290**, 191
Kuwata, K., **411**, 114, 115
Kuzayev, A.I., **207**, 250
Kysela, G., **132**, 337
Lachhein, S., **17**, 10, 11; **312**, 97
Lai, Y., **79**, 20
Laidler, K.J., **74**, 158
Lambrinos, P., **133**, 337
Landers, J.P., **107**, 26, 29
Lane, J., **382**, 109, 117
Lang, A.P., **22**, 284, 285, 286, 295; **93**, 295; **56**, 326; **57**, 326
Langhals, H., **133**, 28; **69**, 221, 222
Laslett, R.L., **56**, 217, 247, 249, 256; **196**, 247; **174**, 306
Lasswell, L.D., **278**, 92, 93, 94
Laurie, D., **183**, 174
Laurier, G.C., **14**, 282, 295, 296
Le, T.P.T., **198**, 248
Ledwith, A., **250**, 84, 85; **392**, 112; **169**, 305
Lee, C.Y., **217**, 80
Lee, G.D., **150**, 303
Lee, J.-Y., **167**, 172
Lee, J.L., **144**, 339; **145**, 339; **146**, 339; **147**, 339
Lee, M.-H., **187**, 175; **188**, 175
Lee, Y.B., **99**, 331; **100**, 331
Leech, J., **163**, 72; **164**, 72
Leffler, J.E., **58**, 59
Lefort, D., **140**, 70
LeFort, D., **245**, 184
Lefour, J.M., **72**, 18
Legeay, G., **227**, 81; **117**, 298
Lehr, G.F., **152**, 33; **109**, 66; **58**, 290
Lehrle, R.S., **2**, 316
Leicht, R., **89**, 294
Leinhos, U., **451**, 123

Lelj, F., **48**, 15
Leonard, J., **265**, 188
Leutner, F.S., **43**, 155
Levin, Y.A., **333**, 98, 101
Levitt, F.G., **112**, 229, 230
Levy, G.C., **23**, 150, 151; **238**, 184; **237**, 256
Lewis, F.M., **11**, 280, 293; **53**, 326
Lewis, R.N., **178**, 74
Lewis, T.D., **96**, 65, 125
Liang, C.K., **63**, 219
Liang, K.S.Y., **159**, 72; **160**, 72; **283**, 266; **284**, 266
Liaw, D.J., **258**, 259; **42**, 324, 325; **44**, 324, 325
Lick, C., **122**, 67, 74
Lightsey, A.K., **193**, 246
Lim, D., **235**, 255
Lin, J., **31**, 286
Lin, T.H., **46**, 14, 23
Lin, Y.-N., **198**, 175; **140**, 301
Lindsell, W.E., **160**, 305
Lingnau, J., **289**, 94; **290**, 94; **291**, 94
Linssen, H.N., **32**, 286; **86**, 294; **87**, 294; **88**, 294
Lipscomb, N.T., **255**, 86
Lishanskii, I.S., **161**, 172
Lissi, E., **123**, 27
Lissi, E.A., **365**, 106, 107
Litt, M., **49**, 289
Litt, M.H., **222**, 253
Lo, G.Y.S., **159**, 304, 305
Lodge, T.P., **85**, 62; **29**, 213; **42**, 214
Londero, D.I., **281**, 190, 191
Lopez-Gonzalez, M.M.C., **71**, 292
Lorand, J.P., **22**, 13; **310**, 97; **313**, 192
Loubet, O., **110**, 229
Lovell, P.A., **247**, 257
Low, R.B., **64**, 326, 327
Lowry, G.G., **64**, 291; **81**, 294
Lu, J., **311**, 191
Ludwig, B., **450**, 123
Lusinchi, J.-M., **169**, 241; **132**, 300
Lusztyk, J., **107**, 26, 29; **113**, 26; **118**, 26; **139**, 70; **346**, 102, 109, 110; **371**, 107, 108; **439**, 121; **178**, 173; **269**, 261
Lutz, P.J., **112**, 298

Lyons, R.A., **73**, 61; **74**, 61; **324**, 98; **82**, 328
Ma, Y.-D., **81**, 62, 120; **275**, 190; **47**, 215; **48**, 215; **24**, 285; **28**, 286, 287
MacCallum, J.R., **13**, 317
Mackie, J.S., **73**, 223
Mackrodt, W.C., **107**, 296
Maddams, W.F., **9**, 150, 151, 152
Madruga, E.L., **62**, 326
Madruga, E.M., **267**, 190; **71**, 292
Mae, Y., **128**, 167
Maeshima, T., **58**, 326
Mahabadi, H.K., **338**, 100; **278**, 190, 192; **307**, 191; **10**, 211, 213; **48**, 325
Mahadevan, V., **212**, 251
Maillard, B., **24**, 49, 99; **156**, 172, 173; **266**, 260, 262
Majima, T., **259**, 87, 115
Makimoto, T., **29**, 151
Makishima, T., **150**, 236
Malatesta, V., **97**, 24; **99**, 24
Malek, J., **26**, 13
Malliaris, A., **49**, 324
Mandal, B.M., **397**, 112; **398**, 112; **399**, 112, 124; **400**, 112, 124
Mandelkern, L., **238**, 184; **237**, 256
Manka, M.J., **157**, 34
Manring, L.E., **53**, 217; **17**, 318; **18**, 318; **19**, 318, 319
Mao, S.W., **6**, 9; **23**, 49, 105, 107
Marchenko, A.P., **207**, 250
Marcus, N.L., **93**, 23
Mardare, D., **117**, 335; **164**, 344; **165**, 344
Marechal, E., **116**, 298
Margaritova, M.F., **161**, 72
Marriott, P.R., **100**, 24; **148**, 32; **262**, 88
Marten, F.L., **15**, 211, 213
Martin, J., **244**, 256
Martin, J.C., **63**, 17, 25; **55**, 59; **133**, 69, 70
Maruthamuthu, P., **389**, 111, 113
Marvel, C.S., **72**, 158; **73**, 158; **102**, 166; **103**, 166; **172**, 241; **150**, 340
Masnovi, J., **155**, 172, 173
Masson, J.C., **39**, 54

Masterova, M.N., **229**, 254
Masuda, E., **458**, 123; **459**, 124; **460**, 124; **51**, 217, 230; **122**, 231
Mather, R.R., **148**, 236
Matheson, M.S., **149**, 71; **103**, 229, 251
Mathias, L.J., **105**, 166, 167; **111**, 166; **112**, 166; **268**, 190; **193**, 246; **233**, 254
Matkovskii, P.Y., **336**, 99
Matsoyan, S.G., **132**, 168
Matsubara, Y., **58**, 326
Matsuda, M., **67**, 18; **294**, 95, 96; **406**, 114; **407**, 114; **414**, 114; **415**, 114; **416**, 114; **417**, 114
Matsugo, S., **187**, 77
Matsumoto, A., **67**, 158; **68**, 158; **69**, 158; **70**, 158; **166**; **71**, 158; **109**, 166; **144**, 170; **136**, 339; **154**, 341
Matsumoto, S., **36**, 152
Matsumoto, T., **130**, 337
Matsumura, Y., **202**, 175
Matsunaga, T., **135**, 338
Matsuyama, K., **192**, 77
Matsuzaki, K., **36**, 152; **39**, 152; **40**, 152; **89**, 329
Mattice, W.L., **231**, 184; **250**, 186
Matyjaszewski, K., **117**, 335; **164**, 344; **165**, 344
Mau, A.W.H., **304**, 191, 192, 193; **21**, 211, 214; **10**, 278, 300
Mauldin, R.L., **225**, 81
Maxwell, I.A., **35**, 53; **315**, 193; **38**, 287
May, J.A., Jr., **98**, 65
Mayer, G., **120**, 230
Mayer, Z., **6**, 3
Mayo, F.R., **149**, 71; **205**, 79; **283**, 93; **49**, 215, 216; **103**, 229, 251; **143**, 235, 238, 245; **149**, 236; **181**, 245; **11**, 280, 293; **53**, 326
Mazza, R.J., **112**, 229, 230
McAskill, N.A., **390**, 111; **396**, 112
McBay, H.C., **173**, 73
McBride, J.M., **132**, 28; **137**, 30; **90**, 63; **70**, 222; **81**, 225
McCallion, D., **238**, 83
McCrackin, F.L., **239**, 184; **238**, 256
McCurdy, K.G., **74**, 158
McDowell, D.J., **274**, 91

McDowell, W.H., **251**, 258
McElvain, S.M., **152**, 341
McFaddin, D.C., **233**, 184
McFarlane, R.C., **90**, 294
McGinniss, V.D., **248**, 84, 85
McGuinness, J.A., **89**, 22; **364**, 106
McMillan, A.M., **275**, 91
McNeill, I.C., **5**, 3
McWalter, I.T., **4**, 316, 317
Medsker, R.E., **162**, 239
Megna, I.S., **341**, 101
Mehl, W., **13**, 10
Mehta, J., **142**, 71, 123
Meijs, G.F., **453**, 123; **454**, 123; **182**, 246, 248; **183**, 246, 248; **185**, 246; **192**, 246, 248; **194**, 246, 248; **195**, 246; **196**, 247; **198**, 248; **199**, 248; **134**, 300; **171**, 306
Meister, J., **36**, 14, 17
Meisters, A., **466**, 124; **52**, 217; **22**, 318
Meixner, J., **37**, 14
Melville, H.W., **249**, 185; **98**, 229, 230, 232, 255; **107**, 229, 230; **220**, 252; **278**, 264
Mendenhall, G.D., **115**, 26; **104**, 65; **106**, 65; **107**, 65; **108**, 66; **111**, 66
Mendjel, H., **157**, 304
Menzel, W., **155**, 341
Merenyi, R., **273**, 262
Merrett, F.M., **115**, 230
Merz, E., **26**, 285
Meyer, J.M., **4**, 316, 317
Meyer, V.E., **81**, 294
Meyerhoff, G., **84**, 62; **150**, 71, 110; **289**, 94; **290**, 94; **291**, 94; **292**, 94; **114**, 229
Michejda, C.J., **44**, 14, 22
Michel, A., **52**, 156
Middleton, I.P., **272**, 90, 91; **170**, 305
Mignani, S., **273**, 262
Mikawa, H., **45**, 288
Mikolaj, D., **132**, 337
Milford, G.N., **104**, 166
Mill, T., **82**, 21; **129**, 68, 79; **205**, 79; **206**, 79
Miller, I.K., **132**, 68
Miller, W.L., **20**, 4; **92**, 163

Milovanovic, J., **146**, 71
Minagawa, M., **76**, 159
Minato, T., **25**, 13; **144**, 31, 33; **82**, 225
Minisci, F., **19**, 11; **17**, 48, 98, 118; **339**, 101, 118; **394**, 112; **426**, 118; **427**, 118; **267**, 260, 264
Minke, R., **62**, 157
Mino, G., **271**, 90
Mirabella, F.M., Jr., **112**, 332
Mirau, P.A., **12**, 150; **22**, 150, 151
Mishra, M.K., **244**, 84
Misra, N., **265**, 89; **398**, 112; **399**, 112, 124
Mita, I., **9**, 3; **4**, 44
Mitani, K., **58**, 157
Miyakawa, T., **158**, 172
Miyake, T., **107**, 166
Moad, G., **7**, 3; **11**, 3; **14**, 4; **15**, 416, 10, 11; **18**, 10, 14, 15, 23; **55**, 16; **66**, 18; **116**, 26; **127**, 28, 34, 35; **1**, 43, 64; **2**, 44; **6**, 45, 52, 64, 65, 71, 99, 100, 123; **7**, 45, 46, 47, 101, 102, 103, 107, 109, 120, 122; **8**, 46, 47, 71, 109; **9**, 46, 101, 102, 103, 109, 110, 120; **10**, 46, 102; **13**, 47, 64, 72, 127; **20**, 48, 49, 104, 106, 107, 120; **21**, 48, 102, 107, 120, 122; **28**, 51, 59, 61, 62, 63, 64, 127; **34**, 52, 109, 120; **37**, 54, 57, 66, 67, 127; **50**, 59, 63, 98, 100, 127; **53**, 59, 60, 63; **70**, 60, 63, 127; **73**, 61; **74**, 61; **75**, 61; **152**, 72; **155**, 72, 120, 121; **177**, 74, 103, 110, 117, 120; **277**, 92, 122; **315**, 98, 127; **349**, 102, 104, 109, 110, 120; **362**, 106, 120; **438**, 121; **442**, 121; **448**, 122; **466**, 124; **479**, 126; **480**, 126; **13**, 150; **15**, 150; **30**, 151, 152; **106**, 166; **134**, 168; **141**, 169, 170; **153**, 172, 173, 174; **176**, 172, 173; **184**, 174, 272, 190, 191; **273**, 190, 191; **304**, 191, 192, 193; **316**, 193; **317**, 193; **3**, 209; **4**, 209; **21**, 211, 214; **43**, 214; **44**, 214; **50**, 217; **52**, 217; **83**, 225; **86**, 226, 232; **88**, 227, 229; **111**, 229; **156**, 238, 239; **264**, 260, 261; **7**, 278; **10**, 278, 300; **15**, 283; **23**, 284, 295; **29**, 286; **36**, 287, 294, 295; **96**, 295; **97**, 295; **126**, 299; **1**, 315; **11**, 317; **12**, 317; **14**, 317; **15**, 318; **20**, 318; **22**, 318; **25**, 320, 321; **29**, 322, 335; **31**, 322; **70**, 327; **82**, 328; **83**, 328; **114**, 332, 333, 335; **115**, 332, 333, 334, 335; **121**, 336, 342; **149**, 340; **161**, 342, 343; **162**, 342
Momtaz-Afchar, J.P., **78**, 328

Monks, H.H., **122,** 166
Monteiro, M.J., **95,** 24; **369,** 106
Moore, C.G., **99,** 229, 230; **115,** 230
Morariu, S., **221,** 179
Morawetz, H., **75,** 159; **57,** 218
Mori, M., **31,** 13; **28,** 151
Morishima, Y., **411,** 114, 115; **77,** 159, 161; **248,** 185; **219,** 252, 258; **252,** 258
Morkved, E.H., **134,** 69, 70, 75
Morozova, I.S., **169,** 345
Morrison, B.R., **35,** 53; **303,** 191; **315,** 193; **47,** 324
Mortimer, G.A., **77,** 293; **78,** 293, 294
Morton, T.C., **453,** 123; **192,** 246, 248; **196,** 247; **198,** 248; **134,** 300
Mossoba, M.M., **141,** 70
Mujica, C., **123,** 27
Mukerjee, S.L., **171,** 345
Mukherjee, A.R., **139,** 234
Mulcahy, M.F.R., **209,** 79
Mullik, S.U., **169,** 305
Mun, G.A., **281,** 264
Münger, K., **331,** 98, 124
Munro, P.D., **93,** 295; **56,** 326; **57,** 326
Murahashi, S., **118,** 166; **219,** 252, 258; **252,** 258
Murakami, K., **13,** 211; **14,** 211
Murakami, S., **366,** 106
Myshkin, V.E., **64,** 17
Nagase, S., **69,** 18; **84,** 22
Nair, C.P.R., **264,** 89; **163,** 240; **131,** 337; **140,** 339; **141,** 339
Nakahama, T., **154,** 72
Nakamura, T., **190,** 77
Nakano, T., **31,** 13; **28,** 151
Nakao, M., **124,** 167, 169
Nakashio, Y., **196,** 78
Nakata, T., **385,** 111
Nambu, Y., **194,** 175, 176, 177, 178
Nandi, U.S., **14,** 47; **7,** 316
Nangia, P.S., **203,** 79
Napper, D.H., **35,** 53; **82,** 62; **272,** 190, 191; **298,** 191; **302,** 191; **303,** 191; **315,** 193; **3,** 209; **154,** 238; **47,** 324
Naravane, S.R., **248,** 257, 258

Narita, N., **237,** 83
Natansohn, A., **95,** 330
Nate, K., **205,** 176; **208,** 177
Navaratnam, S., **186,** 76
Navolokina, R.A., **144,** 71
Nayatani, K., **282,** 190, 191; **175,** 241
Nazran, A.S., **234,** 83; **238,** 83
Neale, R.S., **93,** 23
Neckers, D.C., **185,** 76
Nedelec, J.Y., **140,** 70; **54,** 156; **245,** 184
Neel, J., **31,** 286
Nefedov, A.G., **333,** 98, 101
Neff, D.L., **91,** 63
Nelsen, S.F., **130,** 28, 34; **54,** 59; **66,** 221, 222
Neta, P., **402,** 113; **272,** 262
Neuman, R.C., Jr., **131,** 28, 35; **158,** 34; **45,** 57; **67,** 221, 222
Neumann, W.P., **135,** 29; **136,** 29; **74,** 223
Newcomb, M., **19,** 48, 120, 121; **154,** 172, 173
Ng, F.T.T., **30,** 322
Nguyen, H.A., **116,** 298
Ni, Z., **198,** 175; **203,** 176; **204,** 176; **207,** 177; **209,** 177; **210,** 177; **140,** 301
Nicolini, M., **313,** 98
Niki, E., **27,** 50, 57, 60, 62; **183,** 76
Nilsson, W.B., **153,** 33; **154,** 33
Nishi, Y., **124,** 68, 72
Nishikubo, T., **140,** 169, 170
Nitta, M., **140,** 169, 170
Niwa, M., **130,** 337
Nojima, Y., **182,** 174
Nonhebel, D.C., **183,** 174
Norman, R.O.C., **231,** 82; **233,** 82, 111, 112
North, A.M., **1,** 209, 212, 213, 217; **18,** 211; **45,** 214, 217; **46,** 214, 217; **91,** 228, 229
Nosova, V.S., **169,** 345
Novikov, D.D., **169,** 345
Nozaki, K., **135,** 70, 72
Nozakura, S., **118,** 166; **248,** 185; **283,** 191; **290,** 191; **219,** 252, 258; **252,** 258; **258,** 259; **42,** 324, 325; **43,** 324; **44,** 324, 325
Nudenberg, W., **277,** 263, 264
Nuyken, O., **138,** 301; **139,** 301; **142,** 302; **144,** 302

O'Connor, P.R., **108**, 229
O'Donnell, J.H., **103**, 65, 122; **275**, 91; **19**, 150, 152; **235**, 184; **281**, 190, 191; **286**, 191; **22**, 284, 285, 286, 295; **30**, 286, 295; **33**, 286, 288, 295; **42**, 288; **48**, 288, 289; **51**, 289; **93**, 295; **56**, 326; **57**, 326
O'Driscoll, K., **240**, 256
O'Driscoll, K.F., **72**, 60, 63; **338**, 100; **272**, 190, 191; **273**, 190, 191; **278**, 190, 192; **293**, 191; **294**, 191; **295**, 191; **296**, 191; **307**, 191; **310**, 191; **2**, 209, 210, 211, 214; **3**, 209; **4**, 209; **10**, 211, 213; **27**, 213; **101**, 229; **161**, 239; **201**, 249, 250; **206**, 250; **208**, 250; **209**, 251; **8**, 278; **14**, 282, 295, 296; **60**, 291; **62**, 291; **65**, 291; **70**, 292, 295; **84**, 294; **85**, 294; **90**, 294; **95**, 295; **149**, 303; **32**, 322; **48**, 325; **67**, 326, 327; **113**, 332
O'Shaughnessy, B., **30**, 213; **31**, 213
O'Sullivan, P.W., **103**, 65, 122; **22**, 284, 285, 286, 295; **30**, 286, 295; **33**, 286, 288, 295; **42**, 288; **51**, 289
Odian, G., **6**, 209
Oganova, A.G., **166**, 345; **167**, 345; **168**, 345, 346; **169**, 345
Ogata, T., **58**, 157
Ohanessian, G., **27**, 13, 15
Ohkawa, H., **219**, 80
Ohta, N., **27**, 50, 57, 60, 62
Ohtani, H., **468**, 124; **234**, 184
Ohya, T., **99**, 165, 166, 170; **108**, 166, 170
Oiwa, M., **67**, 158; **68**, 158; **69**, 158; **70**, 158, 166; **71**, 158; **109**, 166; **144**, 170
Okada, H., **195**, 77, 113
Okajima, K., **35**, 152
Okamota, Y., **50**, 324
Okamoto, Y., **31**, 13; **28**, 151
Okawara, M., **208**, 177
Okumura, K., **124**, 167, 169
Okumura, M., **127**, 167
Olaj, O.F., **129**, 28; **286**, 94; **272**, 190, 191; **276**, 190, 277, 190; **291**, 191; **292**, 191; **297**, 191; **299**, 191; **300**, 191; **301**, 191; **3**, 209; **11**, 211; **17**, 211, 212; **19**, 211; **64**, 219, 220, 221; **93**, 228, 229; **113**, 229
Olenin, A.V., **215**, 251; **124**, 336
Olivé, S., **68**, 60; **106**, 229; **117**, 230; **245**, 256

Olivella, S., **48**, 15; **71**, 18; **73**, 18
Onder, N., **98**, 331
Onen, A., **147**, 302
Onishchenko, T.A., **151**, 236; **152**, 236
Onishi, **196**, 78
Ono, A., **85**, 161
Oosterhoff, P., **275**, 263
Osei-Twum, E.Y., **238**, 83
Oster, G., **245**, 84
Ota, T., **69**, 60; **252**, 186; **253**, 186; **197**, 247, 256; **274**, 263; **276**, 263; **18**, 284; **21**, 284; **76**, 328
Otsu, T., **12**, 46, 106, 117; **52**, 59, 60; **165**, 72; **171**, 72, 117; **295**, 95; **302**, 95, 96; **408**, 114, 117; **99**, 165, 166, 170; **108**, 166, 170; **266**, 190; **269**, 190; **282**, 190, 191; **288**, 191, 192; **289**, 191; **173**, 241, 242; **174**, 241, 242; **175**, 241; **187**, 246; **188**, 246; **189**, 246; **190**, 246; **61**, 326; **77**, 328; **119**, 335, 336, 337; **126**, 337, 338; **128**, 337; **134**, 338; **135**, 338; **136**, 339; **137**, 339; **138**, 339; **148**, 340; **154**, 341; **163**, 344
Ouchi, T., **143**, 31
Ourahmoune, D., **155**, 304; **157**, 304
Ovenall, D.W., **32**, 151; **33**, 151, 155; **42**, 152; **61**, 157; **254**, 258, 259; **259**, 259
Overberger, C.G., **56**, 59; **63**, 60; **61**, 219
Overeem, T., **254**, 86; **258**, 87, 115; **419**, 115, 120; **275**, 263
Ozerhwskii, B.V., **33**, 322
Pace, R.J., **287**, 191
Padden-Row, M.N., **69**, 18
Padias, A.B., **293**, 95, 96; **300**, 95
Padwa, A., **92**, 23, 24; **372**, 107, 108
Pajaro, G., **128**, 232
Pak, H., **223**, 181
Paleos, C.M., **49**, 324
Palit, S.R., **400**, 112, 124; **469**, 124; **470**, 124; **119**, 230, 231; **139**, 234; **166**, 240; **213**, 251
Palma, G., **53**, 156; **225**, 253
Pan, C.-Y., **198**, 175; **210**, 177; **219**, 179, 180; **140**, 301
Pandya, A., **300**, 95
Pang, S., **242**, 256
Pannicucci, R., **238**, 83

Pappas, N., **137**, 169
Pappas, S.P., **240**, 84; **241**, 84; **242**, 84; **251**, 84; **253**, 86
Pappiaonnou, C.G., **146**, 71
Paprotny, J., **170**, 305
Parisi, J.P., **168**, 241; **135**, 300
Park, G.S., **30**, 52, 64, 123; **56**, 156; **130**, 232; **131**, 232; **227**, 253
Park, K.Y., **74**, 292; **65**, 326
Park, Y.-C., **218**, 179, 180
Parr, K.J., **166**, 305
Parshall, G.W., **205**, 249
Parsons, B.J., **186**, 76
Pascal, P., **298**, 191; **302**, 191
Paskia, W., **127**, 231
Pastorino, R.L., **178**, 74
Pastravanu, M., **239**, 83; **141**, 302
Patai, S., **115**, 66
Patel, R.D., **143**, 71, 109
Patino-Leal, H., **85**, 294
Patrick, C.R., **254**, 187
Patron, L., **126**, 231
Pattsalides, E., **447**, 122
Paul, H., **465**, 124
Paulrajan, S., **138**, 169, 170; **139**, 169, 170
Peakman, R.E., **2**, 316
Pearce, E.M., **8**, 3; **3**, 44
Pearson, J.M., **32**, 13
Pechatnikov, Y.L., **336**, 99
Penelle, J., **262**, 188, 190
Penenory, A., **135**, 29
Penlidis, A., **91**, 294
Perkins, M.J., **159**, 72; **160**, 72; **420**, 116, 117; **421**, 117; **283**, 266; **284**, 266
Peterson, J.H., **109**, 229
Peterson, J.R., **395**, 112
Petiaud, R., **52**, 156
Petit, A., **31**, 286
Petrov, A.N., **443**, 121
Petukhow, G.G., **174**, 73
Pham, Q.T., **52**, 156
Piasecki, M.L., **155**, 33
Pichot, C., **425**, 117; **55**, 326

Pickering, T.L., **171**, 241, 242, 243
Pierson, R.M., **172**, 241
Pietrasanta, Y., **169**, 241; **170**, 241; **132**, 300; **136**, 300
Pike, P., **434**, 120
Pincock, J.A., **378**, 109
Pincock, R.E., **128**, 68, 75
Pinto, J.A., **88**, 22, 23
Piszkiewicz, L., **82**, 21
Piton, M.C., **77**, 18; **293**, 191; **294**, 191; **295**, 191; **296**, 191; **298**, 191; **303**, 191; **25**, 285; **47**, 324
Pittman, C.U., **50**, 289
Platchkova, S., **116**, 230
Platz, K.-H., **193**, 77
Plaumann, H.P., **83**, 294
Plitz, I.M., **30**, 52, 64, 123; **57**, 157, 185; **131**, 232; **216**, 251, 259
Plochocka, K., **51**, 325, 326
Plotnikov, V.D., **207**, 250
Poblet, J.M., **27**, 13, 15
Pocius, A.V., **221**, 80
Pogosyan, G.M., **132**, 168
Polanyi, M., **83**, 21
Poller, R.C., **123**, 231
Polman, R.J., **275**, 263
Polowinski, S., **94**, 330; **103**, 332
Polton, A., **78**, 328; **79**, 328; **80**, 328; **81**, 328; **133**, 337
Polyansky, V.I., **128**, 300
Pomery, P.J., **275**, 91; **281**, 190, 191; **286**, 191
Ponec, R., **26**, 13; **112**, 26
Ponomarev, G.V., **200**, 249, 250; **203**, 249
Popielarz, R., **177**, 242
Porter, N.A., **200**, 78
Porter, R.S., **255**, 258
Postlethwaite, D., **46**, 214, 217
Poszmik, G., **171**, 345
Poutsma, M.L., **81**, 21
Pramanick, D., **166**, 240
Pramanik, A., **472**, 124
Prementine, G.S., **432**, 119; **19**, 284, 285; **54**, 290
Price, C.C., **33**, 13; **279**, 264; **100**, 295; **54**, 326

Priddy, D.B., **191**, 77, 78; **285**, 94, 95; **297**, 95; **184**, 246
Prime, K.L., **127**, 337
Priola, A., **126**, 167
Pritchard, G.O., **153**, 33; **154**, 33; **155**, 33
Pross, A., **84**, 22
Protasiewicz, J., **106**, 65
Pryor, W.A., **39**, 14, 22, 23, 26, 27; **41**, 14, 22; **46**, 14, 23; **86**, 22; **93**, 64, 65, 100; **97**, 65, 100; **134**, 69, 70, 75; **170**, 72; **278**, 92, 93, 94; **340**, 101; **171**, 241, 242, 243; **218**, 252; **133**, 300
Purmal, A.P., **443**, 121
Pyszora, H., **9**, 150, 151, 152
Qiu, X.-Y., **145**, 302
Quach, C., **92**, 63
Quinga, E.M.Y., **104**, 65; **108**, 66
Quinn, M.J., **62**, 291
Rabinovitch, B.S., **139**, 31
Rae, I.D., **16**, 318
Ranby, B., **422**, 117; **462**, 124
Randall, J.C., **2**, 146, 150; **66**, 291, 295
Raner, K.D., **113**, 26
Rao, K.V., **139**, 169, 170
Rapo, A., **162**, 239
Rasmussen, E., **271**, 90
Rasmussen, J.K., **216**, 80; **218**, 80; **220**, 80; **221**, 80
Rätzsch, M., **47**, 288
Rawlinson, D.J., **156**, 72; **208**, 79; **212**, 79
Raymond, M.A., **114**, 166
Razuvaev, G.A., **174**, 73
Read, D.H., **279**, 264
Reeb, R., **156**, 304
Reed, S.F., **120**, 299
Reichardt, C., **105**, 25, 29; **36**, 323
Reilly, P.M., **14**, 282, 295, 296; **84**, 294; **85**, 294; **90**, 294; **3**, 316
Reinmöller, M., **31**, 151
Remmp, P.F., **112**, 298; **173**, 306
Rempel, G.L., **201**, 249, 250; **30**, 322
Remsen, E.E., **16**, 150; **17**, 150
Ren, Q., **158**, 304
Reshchupkin, V.P., **33**, 322
Reynolds, J.L., **404**, 114

Richards, D.H., **269**, 90; **160**, 305; **161**, 305; **162**, 305; **163**, 305; **164**, 305
Richards, J.R., **306**, 191; **309**, 191
Richards, S.N., **204**, 249
Richou, M.C., **163**, 240
Riedel, S., **80**, 225, 227, 231
Riederle, K., **281**, 92, 94
Riess, G., **156**, 304
Riesz, P., **141**, 70
Riga, J., **273**, 262
Rigo, A., **53**, 156; **225**, 253
Rinaldi, P.L., **97**, 229
Risbood, P.A., **238**, 83
Rist, G., **249**, 84
Ritter, H., **129**, 300
Rivera, M., **365**, 106, 107
Rivera, W.H., **136**, 169
Rizk, N.A., **280**, 264
Rizzardo, E., **14**, 4; **15**, 4; **15**, 10, 15; **16**, 10, 11; **18**, 10, 14, 15, 23; **66**, 18; **116**, 26; **117**, 26; **7**, 45, 46, 47, 101, 102, 103, 107, 109, 120, 122; **9**, 46, 101, 102, 103, 109, 110, 120; **10**, 46, 102; **11**, 46, 102, 103, 106, 120; **20**, 48, 49, 104, 106, 107, 120; **21**, 48, 102, 107, 120, 122; **28**, 51, 59, 61, 62, 63, 64, 127; **34**, 52, 109, 120; **62**, 59, 65, 66, 102, 103, 107, 108, 120; **112**, 66, 103, 108, 120, 121; **155**, 72, 120, 121; **166**, 72, 120, 121; **177**, 74, 103, 110, 117, 120; **236**, 83, 120; **277**, 92, 122; **284**, 93, 118; **344**, 102, 103, 104, 106, 107, 120; **345**, 102, 103, 111, 120; **349**, 102, 104, 109, 110, 120; **350**, 103, 108, 120; **351**, 103, 111, 120; **352**, 104, 105, 120; **353**, 104; **358**, 105, 106, 120; **359**, 105, 106, 120; **360**, 105, 120; **361**, 105, 120; **362**, 106, 120; **436**, 120; **437**, 120; **442**, 121; **448**, 122; **453**, 123; **454**, 123; **457**, 123; **466**, 124; **471**, 124; **473**, 124, 125; **15**, 150; **106**, 166; **176**, 172, 173; **317**, 193; **50**, 217; **52**, 217; **56**, 217, 247, 249, 256; **182**, 246, 248; **183**, 246, 248; **185**, 246; **192**, 246, 248; **194**, 246, 248; **195**, 246; **196**, 247; **199**, 248; **270**, 261; **134**, 300; **171**, 306; **174**, 306; **20**, 318; **22**, 318; **23**, 320; **24**, 320; **25**, 320, 321; **28**, 322; **341**; **29**, 322, 335; **149**, 340; **158**, 341, 342, 343; **159**, 341, 342; **161**, 342, 343; **162**, 342

Ro, N., **143**, 71, 109
Robb, J.C., **148**, 236; **2**, 316
Roberts, C., **101**, 25; **102**, 25
Robin, J.-J., **169**, 241; **132**, 300
Robison, J., **100**, 165, 166
Roedel, M.J., **241**, 184
Rogers, C.L., **119**, 166
Rogers, S.C., **106**, 296, 297; **107**, 296
Rogueda, C., **79**, 328; **80**, 328; **81**, 328
Rondan, N.G., **69**, 18
Rosen, S.L., **32**, 213
Rosenkranz, H.-J., **252**, 84
Rosenthal, I., **141**, 70
Ross, A.B., **402**, 113; **272**, 262
Ross, S.D., **76**, 293
Roth, H.K., **388**, 111, 124
Rounsefell, T.D., **50**, 289
Roy, K.K., **166**, 240
Rozantsev, E.G., **268**, 260
Rüchardt, C., **8**, 9, 12; **20**, 12, 21, 28, 31, 32; **134**, 29; **145**, 32; **49**, 59, 63; **57**, 59; **64**, 60; **65**, 60, 97; **77**, 225; **153**, 341
Rudin, A., **215**, 80; **41**, 152; **240**, 184; **96**, 228; **209**, 251; **239**, 256; **240**, 256; **241**, 256; **242**, 256; **243**, 256; **249**, 257, 258; **250**, 257, 258; **95**, 295; **3**, 316
Rudolf, H., **252**, 84
Rudolf, P.R., **297**, 95
Ruffland, G., **262**, 188, 190
Ruland, W., **145**, 302
Rumack, M.S., **95**, 295
Rusakova, A., **161**, 72
Russell, G.A., **80**, 21, 22, 23; **106**, 25, 29; **330**, 98; **433**, 120
Russell, G.T., **82**, 62; **83**, 62, 123; **273**, 190, 191; **312**, 192; **4**, 209; **28**, 213; **40**, 213
Russell, K.E., **241**, 256
Russell, P.J., **392**, 112
Russo, N., **48**, 15; **71**, 18
Russo, S., **63**, 326
Sack, R., **84**, 62
Saegusa, K., **77**, 159, 161
Saito, I., **187**, 77
Saito, J., **55**, 290

Sakai, I., **21**, 284; **76**, 328
Sakai, S., **143**, 31
Sakuragi, H., **138**, 70
Sakurai, H., **38**, 14; **42**, 14, 22; **363**, 106
Saleem, M., **56**, 156; **227**, 253
Salem, L., **30**, 13
Saltiel, J., **137**, 70, 109
Samanta, M.C., **215**, 80
Samanta, M.D., **3**, 316
Samsel, E.G., **155**, 172, 173
San Roman, J., **62**, 326
Sanayei, R.A., **206**, 250; **208**, 250; **32**, 322
Sanda, F., **172**, 172, 174, 183; **173**, 172, 183; **174**, 172; **175**, 172; **177**, 172, 183; **179**, 173, 174
Sangster, D.F., **390**, 111; **396**, 112
Santee, E.R. **74**, 292; **65**, 326
Santee, E.R., Jr., **37**, 152; **59**, 157
Santhappa, M., **212**, 251
Santi, R., **313**, 98
Saremi, A.H., **30**, 52, 64, 123; **131**, 232
Sargent, J.D., Jr., **374**, 107, 108
Sarnecki, J., **308**, 191
Sarraf, L., **418**, 114
Sasai, K., **276**, 263; **18**, 284; **21**, 284
Sataka, Y., **272**, 90, 91
Satake, M., **267**, 190; **189**, 246
Sato, H., **26**, 151; **27**, 151; **38**, 152; **82**, 161; **85**, 161
Sato, T., **12**, 46, 106, 117; **165**, 72; **171**, 72, 117; **295**, 95; **302**, 95, 96; **408**, 114, 117; **252**, 186; **253**, 186; **197**, 247, 256; **274**, 263; **276**, 263; **18**, 284; **21**, 284; **76**, 328
Satoh, J., **43**, 324
Savedoff, L.G., **422**, 117
Sawada, H., **257**, 187; **59**, 291
Sawaki, Y., **181**, 74; **194**, 77, 107
Sawant, S., **75**, 159
Scaiano, J.C., **23**, 13; **97**, 24; **100**, 24; **107**, 26, 29; **115**, 26; **24**, 49, 99; **107**, 65; **262**, 88; **332**, 98, 101; **343**, 101; **370**, 107; **375**, 108; **446**, 122; **266**, 260, 262
Scammell, M.V., **241**, 256
Schank, K., **122**, 67, 74
Schellekens, R., **179**, 74

Author Index

Schiesser, C.H., **76**, 18
Schildknecht, C.E., **15**, 47; **9**, 317
Schilling, F.C., **30**, 52, 64, 123; **18**, 150; **51**, 156; **57**, 157, 185; **239**, 184; **131**, 232; **216**, 251, 259; **226**, 253; **238**, 256
Schlapkohl, H., **298**, 95
Schlegel, H.B., **28**, 13, 18; **70**, 18
Schluter, A.-D., **152**, 172
Schlüter, K., **149**, 32
Schmid, E., **83**, 62, 123; **312**, 192
Schnabel, W., **257**, 87; **259**, 87, 115; **410**, 114, 115; **411**, 114, 115; **147**, 302
Schnecko, H.W., **10**, 150
Schneider, C., **53**, 289
Schnöll-Bitai, I., **276**, 190; **277**, 190; **297**, 191; **299**, 191; **300**, 191; **301**, 191
Schomaker, E., **88**, 329, 330
Schreck, V.A., **65**, 219, 220, 221, 222
Schue, F., **153**, 304; **162**, 305
Schuh, H., **160**, 34, 35
Schulz, G.V., **84**, 62; **117**, 230; **234**, 255; **245**, 256
Schulze, T., **192**, 175; **229**, 183
Schweer, J., **308**, 191
Scott, G.P., **404**, 114; **145**, 236; **146**, 236; **147**, 236
Scott, R.J., **263**, 88
Sebastiano, R., **313**, 98
Seiner, J.A., **49**, 289
Selmarten, D., **103**, 25
Semchikov, Y.D., **99**, 295; **71**, 327
Sengputa, P.K., **472**, 124
Senogles, E., **73**, 61; **74**, 61; **324**, 98; **328**, 98, 123; **329**, 98; **456**, 123; **82**, 328
Sensfuss, S., **137**, 300
Senyek, M.L., **80**, 160
Serelis, A.K., **53**, 16, 20; **138**, 30, 35; **62**, 59, 65, 66, 102, 103, 107, 108, 120; **86**, 63; **50**, 217; **55**, 217, 219, 222, 223; **58**, 219, 222, 223; **65**, 219, 220, 221, 222; **75**, 224, 225; **78**, 225, 226, 227; **86**, 226, 232; **20**, 318
Service, D.M., **160**, 305
Sewell, P.R., **249**, 185; **220**, 252
Seyferth, D., **100**, 165, 166
Shadrin, V.N., **279**, 190; **280**, 190

Shaffer, S.E., **198**, 175; **210**, 177; **211**, 177, 181; **140**, 301
Shah, T.H., **247**, 257
Shaik, S.S., **49**, 15, 18
Shefer, A., **127**, 337
Sheldon, R.A., **120**, 26, 30; **136**, 70; **188**, 77, 78
Shelton, J.R., **63**, 219
Shen, J., **272**, 190, 191; **273**, 190, 191; **284**, 191; **3**, 209; **4**, 209
Sheppard, C.S., **36**, 54; **41**, 57; **114**, 66
Shero, E., **285**, 94, 95
Shetty, S., **449**, 123
Shiga, T., **229**, 82
Shiraishi, H., **462**, 124
Shirota, Y., **44**, 288; **45**, 288
Shoda, S., **201**, 175
Shostenko, A.G., **64**, 17
Sianesi, D., **128**, 232
Sidney, L., **198**, 175; **210**, 177
Sidney, L.N., **211**, 177, 181
Sigwalt, P., **78**, 328; **79**, 328; **80**, 328; **81**, 328; **133**, 337
Sikkema, K.D., **184**, 246
Simamura, O., **385**, 111
Simionescu, B.C., **221**, 179
Simionescu, C., **141**, 302
Simionescu, C.I., **32**, 52; **239**, 83
Simpson, W., **142**, 169, 170
Singer, L.A., **182**, 74
Singh, M., **14**, 47; **7**, 316
Singhae, R.K., **147**, 71
Sivaram, S., **147**, 71
Skeist, I., **6**, 278, 294
Skell, P.S., **110**, 26; **111**, 26
Skinner, D.L., **119**, 166
Skinner, K.J., **132**, 28; **90**, 63; **70**, 222; **81**, 225
Skorobogatova, M.S., **333**, 98, 101
Skoultchi, M.M., **341**, 101
Sloane, N.J.A., **64**, 157, 158
Sloane, T.M., **391**, 111
Smart, B.E., **47**, 15
Smirnov, B.R., **200**, 249, 250; **203**, 249; **207**, 250; **166**, 345; **167**, 345; **168**, 345, 346; **169**, 345

Smirnov, V.R., **170**, 345
Smith, D.G., **130**, 232
Smith, H.K., **216**, 80; **218**, 80; **220**, 80
Smith, P., **463**, 124; **184**, 246
Smith, P.A.S., **44**, 57
Smith, W.B., **98**, 65
Sogah, D.Y., **18**, 318
Soh, S.K., **9**, 210, 213; **34**, 213; **35**, 213, **36**, 213
Solé, A., **48**, 15; **71**, 18
Solomon, D.H., **7**, 3; **10**, 3; **11**, 3; **14**, 4; **15**, 4; **17**, 4; **22**, 5; **5**, 9; **15**, 10, 15; **16**, 10, 11; **18**, 10, 14, 15, 23; **66**, 18; **116**, 26; **117**, 26; **127**, 28, 34, 35; **138**, 30, 35;**1**, 43, 64; **2**, 44; **5**, 44; **6**, 45, 52, 64, 65, 71, 99, 100, 123; **7**, 45, 46, 47, 101, 102, 103, 107, 109, 120, 122; **8**, 46, 47, 71, 109; **9**, 46, 101, 102, 103, 109, 110, 120; **10**, 46, 102; **11**, 46, 102, 103, 106, 120; **13**, 47, 64, 72, 127; **20**, 48, 49, 104, 106, 107, 120; **21**, 48, 102, 107, 120, 122; **28**, 51, 59, 61, 62, 63, 64, 127; **34**, 52, 109, 120; **37**, 54, 57, 66, 67, 127; **50**, 59, 63, 98, 100, 127; **53**, 59, 60, 63; **62**, 59, 65, 66, 102, 103, 107, 108, 120; **75**, 61; **86**, 63; **92**, 63; **112**, 66, 103, 108, 120, 121; **152**, 72; **155**, 72, 120, 121; **166**, 72, 120, 121; **177**, 74, 103, 110, 117, 120; **236**, 83, 120; **277**, 92, 122; **284**, 93, 118; **315**, 98, 127; **344**, 102, 103, 104, 106, 107, 120; **345**, 102, 103, 111, 120; **349**, 102, 104, 109, 110, 120; **350**, 103, 108, 120; **352**, 104, 105, 120; **358**, 105, 106, 120; **359**, 105, 106, 120; **360**, 105, 120; **361**, 105, 120; **362**, 106, 120; **436**, 120; **437**, 120; **442**, 121; **448**, 122; **457**, 123; **466**, 124; **471**, 124; **473**, 124, 125; **15**, 150; **30**, 151, 152; **93**, 163, 166; **94**, 163; **95**, 163, 164; **96**, 163, 166; **133**, 168; **272**, 190, 191; **316**, 193; **317**, 193; **3**, 209; **43**, 214; **50**, 217; **52**, 217; **55**, 217, 219, 222, 223; **56**, 217, 247, 249, 256; **58**, 219, 222, 223; **65**, 219, 220, 221, 222; **75**, 224, 225; **83**, 225; **86**, 226, 232; **88**, 227, 229; **111**, 229; **270**, 261; **7**, 278; **15**, 283; **29**, 286; **36**, 287, 294, 295; **82**, 294; **126**, 299; **174**, 306; **1**, 315; **11**, 317; **12**, 317; **14**, 317; **15**, 318; **20**, 318; **22**, 318; **23**, 320; **24**, 320; **25**, 320, 321; **29**, 322, 335; **70**, 327; **83**, 328; **114**, 332, 333, 335; **121**, 336, 342; **149**, 340; **158**, 341, 342, 343

Solomon, S., **60**, 59
Song, K.Y., **196**, 175, 177
Song, S.S., **168**, 172; **171**, 172
Soong, C.C., **404**, 114
Soong, D.S., **41**, 213
Sorvik, E., **50**, 156; **251**, 186
Sosa, C., **28**, 13, 18; **70**, 18
Sosnovsky, G., **156**, 72; **208**, 79; **212**, 79
Souel, T., **162**, 305
Soutar, I., **160**, 305
Soutif, J.-C., **227**, 81; **117**, 298
Sozzani, P., **405**, 114, 124
Sparrow, D.B., **175**, 73
Spellmeyer, D.C., **69**, 18; **75**, 18
Spinelli, H.J., **109**, 332
Spirin, Y., L., **59**, 17
Spitz, R., **425**, 117
Springer, R., **34**, 322
Spurling, T.H., **70**, 60, 63, 127; **30**, 151, 152; **304**, 191, 192, 193; **21**, 211, 214; **43**, 214; **44**, 214; **86**, 226, 232; **7**, 278; **10**, 278, 300; **15**, 283; **29**, 286; **36**, 287, 294, 295; **82**, 294; **114**, 332, 333; **115**, 332, 333, 334, 335; **121**, 336, 342
Spychaj, T., **296**, 95
Stackman, R.W., **116**, 166; **117**, 166
Stanley, J.P., **46**, 14, 23
Stannett, V.T., **274**, 91
Stansbury, J.W., **110**, 166; **148**, 171, 172, 182, 183
Stapel, R., **135**, 29; **136**, 29; **74**, 223
Stark, E.J., **297**, 95
Starks, C.M., **314**, 193; **142**, 234, 236, 243, 244; **131**, 300
Starnes, W.H., Jr., **30**, 52, 64, 123; **47**, 156, 186; **48**, 156; **51**, 156; **55**, 156, 186; **57**, 157, 185; **131**, 232; **216**, 251, 259; **221**, 252, 253; **226**, 253; **228**, 253
Staudinger, H., **1**, 1, 2
Staudner, E., **132**, 337
Stefani, A., **326**, 98
Stefani, A.P., **151**, 33
Stein, D.J., **214**, 251, 253; **234**, 255
Stein, S.E., **157**, 34
Steven, J.R., **209**, 79
Stevens, R.D., **463**, 124

Stewart, C.A., **81,** 160
Stewart, L.C., **115,** 26; **107,** 65; **332,** 98, 101; **375,** 108
Stewart, M.J., **152,** 304
Stewen, U., **135,** 29
Stickler, M., **71,** 60, 63, 64; **145,** 71; **291,** 94; **292,** 94; **270,** 190; **271,** 190; **272,** 190, 191; **3,** 209; **90,** 228, 230, 239
Stille, J.K., **299,** 95
Stockmayer, W.H., **165,** 240, 243, 244; **110,** 332
Stoiljkovich, D., **242,** 184; **243,** 184
Stone, R.A., **36,** 287, 294, 295
Strachan, W.M.J., **131,** 68
Straus, S., **94,** 64; **5,** 316; **6,** 316
Strong, W.A., **172,** 73
Subbaratnam, N.R., **138,** 169, 170; **139,** 169, 170
Subra, R., **48,** 15; **74,** 18
Suckling, C.J., **183,** 174
Suddaby, K.G., **209,** 251
Suehiro, T., **153,** 72; **154,** 72
Suga, K., **162,** 172; **163,** 172; **164,** 172; **165,** 172
Sugihara, Y., **190,** 77
Sugiyama, J.-I., **220,** 179; **225,** 183; **106,** 332; **107,** 332
Sumiyoshi, T., **257,** 87; **410,** 114, 115
Sun, R.L., **146,** 171
Sundberg, D.C., **9,** 210, 213; **34,** 213; **35,** 213; **36,** 213
Surzur, J.-M., **121,** 26; **376,** 108; **377,** 109
Sustmann, R., **367,** 106
Suyama, S., **190,** 77; **194,** 77, 107; **195,** 77, 113
Suzuki, T., **37,** 152
Swern, D., **116,** 66; **117,** 66; **118,** 66
Szeverenyi, N.M., **23,** 150, 151
Szwarc, M., **32,** 13; **325,** 98; **326,** 98
Tabb, D.L., **83,** 161
Tabner, B.J., **382,** 109, 117; **423,** 117; **424,** 117
Taft, R.W., **68,** 18
Tagoshi, H., **228,** 183
Takada, Y., **129,** 167
Takahama, S., **138,** 339
Takahara, S., **138,** 70
Takahashi, A., **303,** 96

Takahashi, H., **252,** 186
Takahashi, T., **158,** 172; **159,** 172; **169,** 172
Takata, T., **172,** 172, 174, 183; **173,** 172, 183; **174,** 172; **175,** 172; **177,** 172, 183; **179,** 173, 174
Takayama, S., **232,** 184; **237,** 184; **236,** 256
Takebayashi, K., **82,** 161
Takeishi, M., **219,** 80
Takeshita, T., **81,** 160
Talamini, G., **53,** 156; **129,** 232; **224,** 253; **225,** 253
Talat-Erben, M., **51,** 59, 63; **89,** 63
Tamaki, M., **257,** 259; **46,** 324
Tan, Y.Y., **85,** 329; **86,** 329; **87,** 329; **90,** 330; **91,** 330
Tanaka, H., **69,** 60; **300,** 95; **252,** 186; **253,** 186; 197, 247, 256; **274,** 263; **276,** 263; **18,** 284; **21,** 284; **163,** 344
Tanaka, M., **468,** 124
Tanaka, T., **37,** 152; **76,** 328
Tanaka, Y., **26,** 151; **27,** 151; **38,** 152; **82,** 161; **85,** 161
Tang, F.Y., **39,** 14, 22, 23, 27
Tang, R.H., **39,** 14, 22, 23, 27
Tang, Y., **225,** 81
Tanko, J.M., **109,** 26; **111,** 26
Tardi, M., **78,** 328; **79,** 328; **80,** 328; **81,** 328; **133,** 337
Tarshiani, Y., **255,** 86
Tate, D.P., **79,** 160
Tate, F.A., **231,** 254
Taylor, C.K., **110,** 26
Taylor, R.P., **98,** 229, 230, 232, 255; **107,** 229, 230
Tazaki, T., **148,** 340; **154,** 341
Tedder, J.M., **12,** 4; **2,** 8, 10, 12, 13, 15, 20, 21, 27; **3,** 8, 21; **7,** 9, 10, 12, 15; **11,** 10, 20; **306,** 97; **308,** 97; **309,** 97; **179,** 243
Tencer, M., **446,** 122
Terada, T., **70,** 158, 166
Teraoka, Y., **274,** 263
Terawaki, Y., **25,** 150; **27,** 151; **257,** 259; **46,** 324
Terman, L.M., **174,** 73
Tezuka, T., **237,** 83
Thaler, W., **22,** 49, 107
Thang, S., **166,** 72, 120, 121

Thang, S.H., **112**, 66, 103, 108, 120, 121; **350**, 103, 108, 120; **360**, 105, 120; **361**, 105, 120; **437**, 120; **453**, 123; **454**, 123; **106**, 166; **176**, 172, 173; **185**, 246; **192**, 246, 248; **194**, 246, 248; **195**, 246; **196**, 247; **134**, 300
Thankachan, C., **150**, 32
Thomas, C.B., **384**, 109; **213**, 178
Thompson, P.E., **58**, 219, 222, 223
Thompson, R.D., **268**, 190; **193**, 246
Thorn, R.P., **225**, 81
Tian, Y., **284**, 191; **285**, 191
Tidwell, P.W., **77**, 293; **78**, 293, 294
Tidwell, T., **150**, 32
Tiers, G.V.D., **7**, 150
Tighe, B.J., **210**, 79
Timberlake, J.W., **47**, 60; **55**, 59
Tirrell, D.A., **328**, 98, 123; **329**, 98; **432**, 119; **456**, 123; **2**, 278, 291; **3**, 278; **19**, 284, 285; **20**, 284, 285; **54**, 290; **55**, 290; **56**, 290
Tirrell, M., **85**, 62; **272**, 190, 191; **3**, 209; **29**, 213; **37**, 213; **42**, 214
Titzschkau, K., **142**, 339; **155**, 341
Tobolsky, A.V., **95**, 64, 71; **104**, 229; **176**, 242; **211**, 251; **118**, 298; **125**, 336
Tobolsky, S., **48**, 59
Tokumaru, K., **138**, 70; **385**, 111
Tolman, C.A., **152**, 33; **109**, 66; **58**, 290
Tomari, Y., **58**, 157
Tomono, T., **52**, 289
Tonellato, U., **41**, 14, 22
Tonelli, A.E., **5**, 146, 149; **11**, 150; **67**, 291, 292, 295
Tonge, M.P., **287**, 191
Toren, P.E., **221**, 80
Toshima, N., **40**, 152
Towns, C.R., **46**, 288
Trapp, O.D., **91**, 63
Trautvetter, W., **62**, 157
Traylor, T.G., **61**, 59, 65; **110**, 66, 76; **184**, 76
Trecker, D.J., **87**, 63; **71**, 222, 223
Trossarelli, L., **126**, 167
Troth, H.G., **26**, 50, 99
Trotman-Dickenson, A.F., **140**, 31
Trubnikov, A.V., **268**, 260

Tsai, L., **282**, 264
Tsang, R., **223**, 181
Tsuboi, H., **138**, 339
Tsuchida, E., **52**, 289
Tsuda, K., **174**, 241, 242; **137**, 339; **138**, 339; **139**, 339
Tsuda, T., **105**, 166, 167
Tsuge, S., **468**, 124; **234**, 184
Tsukahara, Y., **50**, 324; **72**, 327
Tsuruta, T., **29**, 151
Tsutsumi, K., **50**, 324
Tucker, J.A., **91**, 23
Tucker, O., **173**, 73
Tüdos, F., **79**, 293; **80**, 293; **92**, 294
Tulig, T.J., **37**, 213
Tunca, U., **146**, 302
Tung, L.H., **159**, 304, 305
Turcsányi, B., **79**, 293
Turkevich, J., **230**, 82
Turner, S.R., **129**, 337
Turro, N.J., **33**, 52
Uebel, J.J., **39**, 287, 295; **70**, 292, 295
Uetsuki, M., **46**, 155, 185
Urushisaki, M., **127**, 167
Uryu, T., **36**, 152; **39**, 152
Usami, T., **232**, 184; **234**, 184; **237**, 184; **236**, 256
Uschold, R.E., **61**, 157; **259**, 259
Ute, K., **8**, 150
Uyama, H., **201**, 175; **202**, 175
Vaidya, R.A., **233**, 254
Valuev, L.I., **75**, 327, 328
Van Damme, F., **297**, 95
van de Grampel, H.T., **90**, 330; **91**, 330
Van Der Hoff, B.M.E., **215**, 80
Van Der Meer, R., **32**, 286; **86**, 294; **59**, 326
van der Werf, S., **275**, 263
Van Sickle, D.E., **176**, 73
Van-Hook, J.P., **48**, 59
VanScoy, R., **130**, 68, 78
Vanscoy, R.M., **199**, 78
Varma, S.C., **102**, 65, 99; **320**, 98, 99
Vaughn, A., **103**, 25

Author Index

Vearing, D.J., **43**, 214; **15**, 283; **29**, 286; **121**, 336, 342
Velazquez, A., **55**, 156, 186
Venkatarao, K., **138**, 169, 170
Venkatasuryanarayana, C., **111**, 26
Verbist, J., **273**, 262
Vercauteren, F.F., **45**, 155; **256**, 259
Veregin, R.P.N., **118**, 335, 342; **122**, 336, 341, 342; **123**, 336, 341, 342
Vernekar, S.P., **122**, 299; **123**, 299
Verravalli, M.S., **32**, 213
Vest, R.D., **103**, 166
Vialle, J., **35**, 286
Vidotto, G., **224**, 253
Viehe, H.G., **273**, 262
Villacorta, G.M., **30**, 52, 64, 123; **131**, 232
Vismara, E., **19**, 11; **334**, 98, 101; **339**, 101, 118
Viswanadhan, V.N., **231**, 184; **250**, 186
Vogl, O., **37**, 152; **98**, 164, 165, 166, 170; **47**, 288
Vogt, B., **101**, 331
Völkel, T., **138**, 301; **139**, 301
von Meewwall, E., **85**, 62; **42**, 214
Wadgaonkar, P.P., **123**, 299
Wagner, H.L., **239**, 184; **238**, 256
Wagner, P.J., **114**, 26; **246**, 84
Waits, H.P., **146**, 71
Walbiner, M., **327**, 98
Wall, L.A., **5**, 316
Walling, C., **4**, 1; **4**, 9; **89**, 22; **92**, 23, 24; **108**, 26, 34; **114**, 26; **22**, 49, 107; **146**, 71; **168**, 72, 73; **232**, 82; **288**, 94; **356**, 105, 107; **364**, 106; **372**, 107, 108; **49**, 215, 216; **160**, 239, 240; **53**, 326
Walsh, J.G., **54**, 326
Walsh, M.R., **93**, 23
Walton, D.R.M., **115**, 166
Walton, J.C., **7**, 9, 10, 12, 15; **11**, 10, 20; **101**, 25; **102**, 25; **308**, 97; **309**, 97; **156**, 172, 173; **183**, 174
Walton, R., **223**, 181
Walz, R., **151**, 303
Wanatabe, Y., **194**, 77, 107; **195**, 77, 113
Wang, C., **128**, 28; **151**, 340
Wang, C.H., **59**, 59; **60**, 59
Wang, G., **284**, 191
Wang, J.C., **147**, 236
Wang, L.-H., **103**, 25
Ward, J.C., **209**, 79
Ward, R.D., Jr., **40**, 14, 22
Warkentin, J., **234**, 83; **238**, 83
Watanabe, M., **162**, 172; **164**, 172; **165**, 172
Watanabe, Y., **190**, 77
Waters, W.A., **76**, 225
Waton, H., **52**, 156
Wayland, B.B., **171**, 345
Wehrli, F.W., **72**, 292
Weidner, R., **142**, 302; **144**, 302
Weinreb, P.H., **99**, 331
Weisgerber, G., **62**, 157
Weiss, J.J., **228**, 82
Whang, B.Y.C., **154**, 238
White, E.F.T., **54**, 217; **27**, 321
White, K.E., **138**, 30, 35; **86**, 63; **55**, 217, 219, 222, 223
Whitesides, G.M., **431**, 119
Whittaker, A.K., **19**, 150, 152
Whittle, D., **118**, 230
Wichterle, O., **235**, 255
Wicks, Z.W. **9**, 278; **108**, 332
Wilbourn, A.H., **247**, 185
Wiley, R.H., **136**, 169
Williams, R.J.P., **95**, 228
Willing, R.I., **15**, 4; **22**, 5; **5**, 9; **6**, 45, 52, 64, 65, 71, 99, 100, 123; **8**, 46, 47, 71, 109; **28**, 51, 59, 61, 62, 63, 64, 127; **50**, 59, 63, 98, 100, 127; **152**, 72; **457**, 123; **15**, 150; **30**, 151, 152; **316**, 193; **83**, 225; **88**, 227, 229; **111**, 229; **96**, 295; **126**, 299; **12**, 317; **14**, 317; **15**, 318
Willis, H.A., **236**, 184
Wilson, C.W., III, **59**, 157
Wilt, J.W., **50**, 16
Wine, P.H., **225**, 81
Winnik, M.A., **272**, 190, 191; **273**, 190, 191; **293**, 191; **294**, 191; **295**, 191; **296**, 191; **298**, 191; **302**, 191; **303**, 191; **3**, 209; **4**, 209; **47**, 324
Winzor, C.L., **281**, 190, 191; **286**, 191; **156**, 238, 239
Wirthlin, T., **73**, 292
Wittmer, P., **61**, 291

Witzel, T., **435**, 120
Wojciechowski, B.J., **47**, 156, 186; **55**, 156, 186; **228**, 253
Wolf, B., **93**, 228, 229
Wolf, C., **223**, 253
Wolf, R.A., **67**, 60
Wong, P.C., **370**, 107
Woo, J., **269**, 90; **164**, 305; **166**, 305; **167**, 305
Worsfield, D.J., **264**, 188
Wu, S.-R., **198**, 175; **203**, 176; **204**, 176; **207**, 177; **209**, 177; **210**, 177; **140**, 301
Wu, T.K., **42**, 152
Wu, Z., **219**, 179, 180
Wu, Z.Z., **133**, 233
Wulff, G., **101**, 331
Wunsche, P., **388**, 111, 124
Xi, F., **98**, 164, 165, 166, 170
Yagci, Y., **146**, 302; **147**, 302
Yako, N., **205**, 176
Yamabe, S., **25**, 13; **144**, 31, 33; **82**, 225
Yamada, B., **52**, 59, 60; **266**, 190; **269**, 190; **288**, 191, 192; **289**, 191; **186**, 246; **187**, 246; **188**, 246; **189**, 246; **190**, 246; **191**, 246; **77**, 328; **163**, 344
Yamada, M., **126**, 68, 73
Yamada, N., **100**, 229
Yamamoto, K., **194**, 175, 176, 177, 178
Yamamoto, N., **198**, 175
Yamamoto, T., **196**, 78
Yamashita, I., **158**, 172; **159**, 172
Yamashita, K., **137**, 339; **138**, 339; **139**, 339
Yamashita, Y., **301**, 95; **72**, 327
Yamataka, H., **84**, 22
Yamauchi, S., **138**, 70
Yamauchi, T., **153**, 72
Yamazaki, H., **35**, 152
Yamazaki, N., **198**, 175; **210**, 177; **140**, 301
Yang, M., **284**, 191
Yang, N., **245**, 84
Yang, Y., **311**, 191
Yassin, A.A., **280**, 264
Yasuda, M., **20**, 150, 151, 152
Yasukawa, T., **13**, 211; **14**, 211
Yeadon, G., **94**, 228, 229

Yee, W., **261**, 260, 264
Yen, I.F., **202**, 175
Yeung, M.-Y., **159**, 34
Ykman, P.J., **151**, 172
Yokata, K., **129**, 167
Yokona, H., **208**, 177
Yokozawa, T., **150**, 172; **162**, 172; **164**, 172; **165**, 172; **206**, 176, 177; **220**, 179; **225**, 183; **106**, 332; **107**, 332
Yonezawa, K., **198**, 175; **140**, 301
Yoshida, M., **119**, 335, 336, 337; **128**, 337
Yoshihara, M., **58**, 326
Yoshihira, K., **31**, 52, 64
Yoshinaga, A., **140**, 169, 170
Yoshioka, M., **135**, 338
Young, L.J., **102**, 296
Yu, J., **30**, 213; **31**, 213
Yurzhenko, A.I., **151**, 71
Yuzawa, M., **89**, 329
Zaganiaris, E., **55**, 326
Zak, A.G., **161**, 172
Zarkadis, A.K., **135**, 29
Zavitsas, A.A., **87**, 22; **88**, 22, 23
Zetie, R.J., **142**, 169, 170
Zhang, H., **311**, 191; **158**, 304
Zhang, X., **158**, 304
Zheng, Z.-F., **226**, 183
Zhou, L.-L., **195**, 175, 177; **198**, 175; **200**, 175; **214**, 178
Zhu, S., **285**, 191; **12**, 211
Zifferer, G., **11**, 211; **17**, 211, 212; **19**, 211
Zilberman, E.N., **144**, 71
Zinbo, M., **452**, 123
Zipse, H., **29**, 13
Zubov, V.P., **229**, 254; **281**, 264; **75**, 327, 328

Subject Index

Abstraction (see abstraction vs. addition, backbiting, hydrogen atom transfer, chain transfer)
Abstraction vs. addition
  by alkoxy radicals 26, 27, 101-108
  by alkoxycarbonyloxy radicals 110
  by alkyl radicals 26, 27, 99, 107
  by arenethiyl radicals 114
  by aryl radicals 27, 102
  by benzoyloxy radicals 27, 46, 103, 109
    with MMA 46, 103
  by t-butoxy radicals 27, 46-49, 101-107
    solvent effects 48, 49, 107
    with alkenes 106
    with allyl acrylates 106
    with AMS 103, 107
    with BMA 47, 106
    with MA 103
    with MAN 104
    with MMA 46-49, 101, 103, 320
    with VAc 104
    with vinyl ethers 106
  by carbon-centered radicals 26, 27
  by cumyloxy radicals 103, 107
  by α-cyanoalkyl radicals 99
  by heteroatom-centered radicals 26, 114
  by hydroxy radicals 27, 103, 111, 112
    with AMS 103, 111
    with MA 103
    with MAA 112
    with MMA 103, 111
  by isopropoxycarbonyloxy radicals 101, 103, 110
    with MMA 101, 103
  by oxygen-centered radicals 26, 27
  by sulfate radical anion 112
  prediction
    from bond dissociation energies 26
    from frontier molecular orbital theory 27
    from radical polarity 27
  solvent effects 48, 49, 107
Acceptor monomers
  alternating copolymerization 288
  interaction with Lewis acids 328
  list 288
  spontaneous initiation 95
Acrylamide
  polymerization
    head vs. tail addition 159

k_p solvent effects 324
Acrylate esters (see also n-butyl acrylate, methyl acrylate)
  polymerization
    chain transfer to polymer 257
    combination vs. disproportionation 231
    head vs. tail addition 158
    spontaneous initiation 94
    template 331
  reaction with radicals 10, 11, 14, 103, 106
Acrylic acid
  copolymerization with S
    spontaneous initiation 95
  polymerization
    k_p solvent effects 324
    tacticity 150, 152
    template 330
    thermodynamics 188
  reaction with carbon-centered radicals
    rate constants 98
Acrylic polymers (see also poly(acrylic acid), polyacrylonitrile, poly(methacrylic acid), polymethacrylonitrile, poly(methyl acrylate), poly(methyl methacrylate))
  head-to-head linkages 158
  long chain branching 256
Acrylonitrile
  copolymerization
    penultimate unit effects 284-286
    Q-e values 296
    reactivity ratios 282
  copolymerization with S
    bootstrap effect 326
    mechanism 284, 285, 290
    solvent effects 326
    spontaneous initiation 95
  polymerization
    chain transfer to halocarbons 243
    chain transfer to monomer 251
    chain transfer to solvent 245
    chain transfer to thiols 240
    combination vs. disproportionation 231
    head vs. tail addition 159
    k_p - effect of Lewis acid 327
    tacticity 152
    thermodynamics 188
  reaction with alkyl radicals
    penultimate unit effects 284
    rate constants 98

381

reaction with aryl radicals
  rate constants 98
reaction with heteroatom-centered radicals
  rate constants 114
reaction with oxygen-centered radicals
  rate constants 102
  specificity 104
Activation energy
  for hydrogen atom transfer 21
  for initiator decomposition
    azo-compounds 59
    peroxides 68
  for propagation 191
  for propagation in copolymerization 285
  for radical addition 13
    calculation 18
  for radical-radical termination 208, 221
  small radicals 28
Activation entropy (see Arrhenius A factor)
Acyl peroxides (see diacyl peroxides, dibenzoyl peroxide, dilauroyl peroxide)
Acyloxy radicals (see also benzoyloxy radicals)
  from α-acylperoxydiazenes 83
  aliphatic
    fragmentation to alkyl radicals 67, 69, 97, 109
    from diacyl peroxides 67-71
    from diacyl hyponitrites 65
    from diacyl peroxides 67-73, 108
    from peresters 74-76, 108
Addition (see radical addition)
Addition vs. abstraction (see abstraction vs. addition)
Addition-abstraction polymerization 186
Addition-fragmentation chain transfer
  mechanism 238, 246, 247
  template polymerization 331
  to allyl halides 246
  to allyl peroxides 246
  to allyl silanes 246
  to allyl sulfides 246, 248, 300, 331
  to allyl sulfones 246
  to benzyl vinyl ethers 248, 300
  to PMMA macromonomer 247, 256, 257, 306, 320
  to thiohydroxamic esters 248
  to thionoesters 248
  to VC 156, 247
  transfer constants 247, 248
Alkanethiyl radicals
  from allyl sulfides 246
  from disulfides 241, 242
  from thiols 240, 241
  polarity 241
  reaction with monomers 114
ω-Alkenyl radicals (see also hex-5-enyl radicals)
  cyclization

but-3-enyl radicals 16, 174
hept-6-enyl radicals 5, 16, 170
hex-5-enyl radicals 5, 16, 48, 164, 168
pent-4-enyl radicals 16
Alkoxy radicals (see also t-butoxy, cumyloxy radicals)
  abstraction vs. addition 26, 27, 101-108
  combination vs. disproportionation 33
  fragmentation
    substituent effects 107, 108
    from alkyl hydroperoxides 78
    from dialkyl hyponitrites 65
    from dialkyl peroxides 77
    from dialkyl peroxydicarbonates 73
    from peresters 74-76
  polarity 22, 27, 106
  primary and secondary 27, 33, 108
  tertiary 27, 101-108
Alkoxyamines
  formation 120, 121
  in living polymerization 341-344
  stability 122, 342
Alkoxycarbonyloxy radicals (see also isopropoxycarbonyloxy radicals)
  abstraction vs. addition 110
  fragmentation to alkoxy radicals 110
  from dialkyl peroxydicarbonates 73, 108
Alkyl hydroperoxides (see also t-butyl hydroperoxide, α-hydroperoxydiazenes)
  as initiators 56, 68, 78, 79
  as transfer agents 79
  induced decomposition 79
  kinetic data for decomposition 68
  non-radical decomposition 79
Alkyl radicals (see also benzyl, n-butyl, t-butyl, cyclohexyl, n-hexyl, hex-5-enyl, methyl, 1-phenylalkyl, undecyl radicals)
  abstraction vs. addition 26, 27
  combination vs. disproportionation 32, 34
  disproportionation 30
  from diacyl peroxides 69
  from dialkyldiazenes 57
  from fragmentation of acyloxy radicals 67, 69, 97, 109
    rate constant 109
  head vs. tail addition 97
  hydrogen abstraction by 99
  polarity 13, 22, 27, 97
  reaction with inhibitors 116-122, 260-266
    nitroxides 120, 260
    oxygen 49, 50, 113, 260
    rate constants 260
  reaction with monomers 97-100
    effect of temperature 17
    penultimate unit effects 284
    rate constants 98
  reaction with oxygen 49, 50, 113
  reviews 97

Subject Index

Alkylperoxy radicals
  epoxidation by 113
  from alkyl hydroperoxides 79, 113
  from alkyl radicals and oxygen 50, 99, 113
  in autoxidation processes 113
  reaction with monomers 113
Allyl acrylate
  cyclopolymerization 167
  reaction with t-butoxy radicals 105
Allyl amines
  polymerization
    chain transfer to monomer 167, 254
Allyl esters
  polymerization
    chain transfer to monomer 251, 254
    head *vs.* tail addition 158
Allyl halides
  chain transfer to 246, 251
Allyl methacrylate
  cyclopolymerization 167
  reaction with t-butoxy radicals 105
Allyl monomers
  polymerization
    chain transfer to monomer 167, 254
    head *vs.* tail addition 158, 165
Allyl peroxides
  chain transfer to 246
Allyl polymers
  head-to-head linkages 158, 165
Allyl silanes
  chain transfer to 246
Allyl sulfides
  chain transfer constants 248
  chain transfer to 243, 246, 248
  synthesis of end-functional polymers 300
Allyl sulfones
  chain transfer to 246
Alternating copolymerization
  Lewis acid induced 328
  mechanisms for 288-290
  monomers for 288
  of divinyl ether with MAH 170
  of MAH with S 288
  of MMA with S 328
  spontaneous initiation 95, 289
t-Amyl peroxides
  as initiators 108
Anionic polymerization
  transformation to radical polymerization 304
Antepenultimate unit effect (see penpenultimate unit effect)
Arenethiyl radicals
  from disulfides 237, 241, 242
  from thiols 240
  polarity 114
  reaction with monomers
    AMS 18
    MMA 114

S 114
Aromatic substitution
  by benzoyloxy radicals
    of benzene 109
    of PS 109
    of S 46, 109
    reversibility 109
  by chlorine atoms
    of benzene 26
  by hydroxy radicals
    of S 111
  by isopropoxycarbonyloxy radicals
    of S 111
  by phenyl radicals
    of S 46, 101
Arrhenius A factor (see also activation entropy)
  for hydrogen atom transfer 23
  for initiator decomposition
    azo-compounds 59
    peroxides 68
  for propagation 191
  for propagation in copolymerization 285
  for radical addition 13
    calculation 18
Aryl radicals (see also phenyl radicals)
  abstraction *vs.* addition 27
  from fragmentation of aroyloxy radicals 100
  head *vs.* tail addition 101
  polarity 27, 101
  reaction with monomers
    rate constants 98, 101
    specificity 101
Azeotropic copolymer composition
  multicomponent 283
  penultimate model 286
  terminal model 283
Azo-compounds (see also dialkyldiazenes, dialkyl hyponitrites, azobisisobutyronitrile, azobis(methyl isobutyrate), azonitriles)
  as initiators 53-66
  classes 54
  kinetic data for decomposition 59
Azo-peroxides
  decomposition mechanism 83
  synthesis of block & graft copolymers 302
Azobis(methyl isobutyrate)
  ^{13}C-labeled 127, 328
  advantages 64
  as initiator 57, 59, 60, 64, 126, 127, 328
  chain transfer to 64
  effect of solvent on k_d 60
  kinetic data for decomposition 59
  synthesis of end-functional polymers 299
Azobisisobutyronitrile
  ^{13}C-labeled 127, 229
  ^{14}C-labeled 229
  cage reaction 52, 62-64
  cage return 57, 60

chain transfer to 65
decomposition mechanism 51, 52, 57, 63, 226
diazenyl radicals from 57
effect of viscosity on k_d 60, 62
initiator efficiency 62-65
initiator of MMA polymerization 44
initiator of S polymerization 44, 64, 100, 318
ketenimine from 51, 63, 99, 100, 224-226
kinetic data for decomposition 59
MAN from 51, 52, 64
solvent effect on decomposition 57, 60
synthesis of end-functional polymers 299
toxicity 64
Azonitriles (see also azobisisobutyronitrile)
as initiators 57-59
water soluble 57
Backbiting
addition-abstraction polymerization 186
chain transfer to polymer 184-186, 254, 255
effect of monomer concentration 186
effect of temperature 184, 186
in E propagation 184
in VAc propagation 185
in VC propagation 185
transition state 23, 24, 184
Benzene
chain transfer to 245
solvent effect on VAc polymerization 259
Benzoin derivatives
acyl phosphine oxides 87, 115
acyl phosphinates 115
acyl phosphonates 87
as photoinitiators 84-87
benzil monooxime 84
benzoin esters 86
benzoin ethers
α-alkyl 85
α-ether 86
photodecomposition mechanism 85
shelf life 86
α-sulfonyl 86
polymerizable 87
Benzoyl peroxide (see dibenzoyl peroxide)
Benzoyloxy radicals
abstraction vs. addition 27
aromatic substitution
of benzene 109
of PS 109
of S 46, 103, 109
fragmentation to phenyl radicals 46, 48
rate constant 109
from BPO 44, 69-73
head vs. tail addition 103,104,109
polarity 27, 109
reaction with monomers
AN 102, 104, 109
MA 10, 102, 103
MMA 10, 46, 103, 109

rate constants 102, 110
S 4, 45, 46, 102, 103, 109, 121, 317
specificity 4, 10, 45, 46, 103, 104, 109, 121
VAc 102, 104, 109
Benzyl radicals
from toluene 48
pathways for combination 29
radical-radical reaction 221
reaction with monomers 98
MMA 48
reaction with nitroxides 121
Benzyl vinyl ethers
chain transfer constants 248
chain transfer to 246, 248, 300
synthesis of end-functional polymers 300
Bicyclobutanes
ring-opening polymerization 172
Bimolecular termination (see combination,
    disproportionation, primary radical
    termination, radical-radical reaction,
    radical-radical termination)
Block copolymers
definition 277
synthesis 297, 302-307, 336, 339, 341, 342
    by living radical polymerization 241, 336,
        339, 341, 342
    with end-functional polymers 297
    with macromonomers 306
    with multifunctional initiators 83, 218,
        302, 303
    with transformation reactions 303
Bond dissociation energies
addition vs. abstraction 26, 27
alkoxyamines 342
and fragmentation of t-alkoxy radicals 107
and hydrogen atom transfer 21, 26, 27
and living radical polymerization 322, 336,
    342
and radical addition 9, 15
and radical reactions 8
effect of fluorine substituents 15
substituent effects
    C-C bond 15, 26
    C-H bond 21, 26
    C-O bond 15, 26, 342
    O-H bond 26
Bond strengths (see bond dissociation energies)
Bootstrap effect
in copolymerization 292, 326
Branched polymers (see graft copolymers, long
    chain branches, short chain branches)
Butadiene
copolymerization
    penultimate unit effects 286
polymerization
    1,2- vs. 1,4-addition 159-161
    k_p 191
t-Butoxy radicals

Subject Index

abstraction vs. addition 27
fragmentation 47, 48
from DBPOX 44, 75
Hammett parameters
  reaction with substituted styrenes 14
  reaction with substituted toluenes 14
polarity 14, 15, 27, 106
reaction with ethers 24
reaction with monomers
  alkenes 106
  allyl acrylate 105
  allyl methacrylate 105
  AMS 103, 107
  AN 104, 290
  BMA 106
  fluoro-olefins 10, 15
  MA 10, 103
  MAN 104
  MMA 10, 46, 101, 103
  rate constants 102
  S 45, 103, 117
  specificity 10, 103, 104
  VAc 104, 290
  VF2 9, 10
  vinyl ethers 106
reaction with toluene 48, 102
specificity of hydrogen transfer 25, 103-106
n-Butyl acrylate
  copolymerization
    Q-e values 296
n-Butyl methacrylate
  polymerization
    combination vs. disproportionation 223, 231
    k_p 191
    thermodynamics 188
  reaction with t-butoxy radicals 106
n-Butyl radicals
  reaction with monomers 290
t-Butyl hydroperoxide
  as initiator 68, 78, 79, 113
  kinetic data for decomposition 68
t-Butyl Methacrylate
  polymerization
    k_p 191
t-Butyl perbenzoate
  as initiator 68, 74
  kinetic data for decomposition 68
t-Butyl radicals
  abstraction vs. addition 27
  combination vs. disproportionation 33, 34
    effect of reaction conditions 35
  Hammett parameters
    reaction with substituted styrenes 14
    reaction with substituted toluenes 14
  polarity 27
  reaction with monomers
    effect of temperature 17

    rate constants 98
  reaction with oxygen 113
t-Butylperoxy radicals
  from t-butyl hydroperoxide 79, 113
  Hammett parameters
    reaction with substituted styrenes 14
    reaction with substituted toluenes 14
  reaction with monomers 113
Cage reaction
  by-products from 51
    AIBN 52, 63
    BPO 69
    dialkyl hyponitrites 66
  cage return
    AIBN 57, 60
    BPO 52, 70
    diacyl peroxides 71
    dialkyldiazenes 57
  definition 51
  effect of initiator structure 55
  effect of magnetic field 52
  effect of solvent 76
  effect of viscosity 52, 61, 62, 66, 71, 76
    AIBN 60, 62
    DBPOX 76
    diacyl peroxides 70, 71
    dialkyl hyponitrites 66
    dialkyldiazenes 60-62
  effect on initiator efficiency 51, 52, 71
  effect on k_d 52, 71
  effect on product distribution 35
  initiator decomposition
    AIBMe 63
    AIBN 51, 52, 62, 63
    BPO 52, 69-71
    DBPOX 75, 76
    diacyl peroxides 71
    dialkyl hyponitrites 66
    dialkyl peroxide 78
    dialkyl peroxydicarbonates 74
    dialkyldiazenes 61, 62
  photodecomposition
    BPO 70
Captodative olefins
  as inhibitors 262
  copolymerization 262
2-Carboalkoxyprop-2-yl radicals
  combination vs. disproportionation 222
2-Carbomethoxyprop-2-yl radicals
  combination vs. disproportionation 222
  from AIBMe 60, 63, 64, 126, 328
  reaction with monomers 98, 126, 328
  reaction with PS• 226
Carbon monoxide
  copolymerization with depropagation 291
Carbon-centered radicals (see also alkyl radicals, cyanoalkyl radicals, aryl radicals)
Catalytic chain transfer

limitations 251
mechanism 249
to cobalt complexes 249-251, 346
to PMMA macromonomer 256
Cationic polymerization
transformation to radical polymerization 305
Ceiling temperature
in copolymerization 290
in homopolymerization 188-190
in ring-opening polymerization 173
table 188
Ceric ions
reaction with 1,2-Diols 91
redox initiation 90, 91
Chain length dependence
combination vs. disproportionation
for PE• 226
for PMAN• 224
for PS• 220
of chain transfer constant 236, 237, 244
of chain transfer to polymer 255-257
PE 256
PMMA 257
of copolymer composition 327
of k_p 187, 192
of k_t 208-214
Chain transfer (see also addition-fragmentation chain transfer, backbiting, hydrogen atom transfer)
by atom or group transfer 237
control of
compositional heterogeneity 333
defect groups 319
molecular weight distribution 334
definition 207
in VAc polymerization 155
mechanisms 237
penultimate unit effects 236, 237, 241, 244
reviews 234
structural irregularities from 3
synthesis of end-functional polymers 234, 241, 246, 248-250, 300
termination by 207, 234
to disulfides 241
polar effects 242
substituent effects 242
to functional transfer agents 300
to halocarbons 243
catalysis 244
penultimate unit effects 244
to initiator 52
AIBN 64, 65
aliphatic diacyl peroxides 69, 109
alkyl hydroperoxides 50, 79
BPO
in S polymerization 52, 71, 110, 317
in VAc polymerization 72
in VC polymerization 72

mechanism 53
diacyl peroxides 52, 71
dialkyl peroxides 78
dialkyl peroxydicarbonates 74
dialkyldiazenes 62, 64
disulfides 89, 241, 242
dithiocarbamates 337
dithiuram disulfides 89, 242, 337
effect of conversion 71
effect on initiator efficiency 53
effect on k_d 53, 70
α–hydroperoxydiazenes 83
ketenimine from AIBN 63, 65
peresters 76
persulfate 80
to monomer 251-254
allyl amines 167, 254
allyl esters 254
allyl monomers 167, 251, 254
MA 251
mechanisms 251
MMA 251
S 251, 252
transfer constants 251
VAc 251-253
VC 156, 247, 251, 253
to polymer 254-259
mechanisms 254
molecular weight dependence 255
PE 256
PMMA 256, 257
PMMA macromonomer 217, 247, 256, 257
poly(alkyl acrylates) 257
poly(alkyl methacrylates) 256
PVAc 257-259
PVC 259
PVF 259
transfer constants 255
to solvent 244
to sulfides 243
to thiols
penultimate unit effects 241
polar effects 241
substituent effects 241
transfer constants 240
uses 234
Chain transfer constants
chain length dependence 236, 244, 250
definition 235
effect of conversion 239
for allyl sulfides 248
for benzyl vinyl ethers 248
for disulfides 242
for monomers 251
for polymers 255
for thiols 240
ideal 235
Mayo equation 235

Subject Index

measurement 238
prediction 237
Chlorine atoms
  1,2-shift 156
  aromatic substitution 26
  hydrogen atom transfer to
    solvent effects 25
    specificity 23
  in VC polymerization 157
  polarity 22
Chloroprene
  polymerization
    1,2- vs. 1,4-addition 160, 161
Cobalt complexes
  catalytic chain transfer 249-251, 346
  control of propagation 322
  in living polymerization 345
Comb polymers (see graft copolymers)
Combination (see also combination vs.
  disproportionation)
  pathways for 28
    benzyl radicals 29, 221, 222
    α–cyanoalkyl radicals 29, 99, 100, 224-226, 231
    α–ketoalkyl radicals 29
    PAN• 231
    PMAN• models 224-226
    PMMA• models 223
    PS• models 222
    triphenylmethyl radicals 29
  termination by 207
Combination vs. disproportionation
  for 1-phenylalkyl radicals 34, 220
  for 2-carboalkoxy-2-propyl radicals 222
  for alkoxy radicals 33, 66
  for alkyl radicals 32, 34
  for t-butyl radicals 33, 34
  for cumyl radicals 32, 220, 221
  for cyanoisopropyl radicals 100, 224
  for ethyl radicals 33
  for ethyl with fluoromethyl radicals 33
  for nitroxides with carbon-centered radicals 121
  for PAMS•
    model studies 220, 221
  for PAN• 231
  for PBMA• 231
    model studies 222
  for PBMA• with PMAN• 225, 227
  for PBMA• with PMMA• 222, 223
  for PEMA• 231
    model studies 222
  for PE•
    chain length dependence 226
    model studies 226
  for PE• with PMAN• 225, 227
  for PMAN• 231
    chain length dependence 224

    model studies 224-226
  for PMAN• with PS• 225, 227
  for PMA• 231
  for PMMA• 217, 230, 231
    effect of temperature 222, 223, 230
    model studies 222-224
  for PMMA• with PS• 226, 232
  for PS• 229
    effect of temperature 220, 221, 229
    model studies 219-222
  for PVAc• 231
  for PVC• 232
  guidelines 35, 233
  in copolymerization 217-219, 222, 223, 225-227, 232, 233
  in radical-radical reactions 31-35, 217-233
  influence of reaction conditions 34
  measurement
    kinetics 228
    model studies 218, 219
    polymer analysis 228
  statistical factors 31
  stereoelectronic factors 33
  synthesis of end-functional polymers 299, 300
  temperature dependence 31, 34, 35, 208, 220-223, 229, 230
Computer modeling
  compositional heterogeneity 333
  of chain end functionality 300
  of molecular weight distribution 191, 214
  of radical-radical termination 232
Concentration
  initiator
    effect on k_d 69, 74, 80
  monomer
    effect on backbiting 184, 186
    effect on ceiling temperature 189
    effect on ring-opening polymerization 172
Conversion
  effect on branching in PVAc 258
  effect on chain transfer constant 239
  effect on copolymer composition 278
  effect on initiation 48
  effect on initiator efficiency 214
    AIBN in S polymerization 62, 64
    BPO in MMA polymerization 71
  effect on propagation rate constant 214
  effect on radical-radical termination 210, 213, 214
Copolymerization 277-307
  complex dissociation model
    mechanism 289
    model studies 290
  complex participation model
    mechanism 289
    model studies 290
  composition drift with conversion 278

compositional heterogeneity 278, 315, 332-334
  control 315, 332
  effect of initiation 333
  effect of termination 333
  Monte Carlo simulation 333
depropagation model 290, 291
  AMS 291
  carbon monoxide 291
  ceiling temperature 291
  sulfur dioxide 291
effect of conversion 294
effect of Lewis acids 328
effect of molecular weight 327
effect of solvent 279, 292, 294, 325-327
estimation of reactivity ratios 293-295
mechanisms 279-291
model discrimination 287, 294, 295
  by composition data 294
  by monomer sequence distribution 295
monomer complex models 288-290
monomer sequence distribution 278, 291-293, 295, 326
  effect of solvent 292, 326
  effect of tacticity 292
  penultimate model 292
of AN with S 284, 326
of macromonomers 215, 306
of MAH with S 288
of MMA with S
  penultimate unit effects 287
  propagation rate constant 216
  radical-radical termination 216, 226, 232
of quinones 264
penultimate model 284-287
  AN 286, 326
  B 286
  bootstrap effect 326
  instantaneous composition equation 286
  MAH 286
  mechanism 286
  model discrimination 287
  reactivity ratios 285
  VC 286
prediction of reactivity ratios 295-297
  Patterns of reactivity scheme 297
  Q-e scheme 13, 295-297
radical-radical termination 214-217, 222, 223, 225, 226, 227, 232
  chemical control model 216
  combination vs. disproportionation 217, 222, 223, 225, 226, 227, 232
  diffusion control model 214, 217
  model studies 219, 222, 223, 225, 226, 227
spontaneous initiation 289
template effects 330
terminal model 280-283
  assumptions 280
  azeotropic composition 283
  bootstrap effect 292, 326
  implications 282
  limitations 216
  Mayo-Lewis equation 280, 281
  reactivity ratios for common monomers 282
  termination kinetics 215
Copolymers (see block, graft copolymers)
  compositional heterogeneity 278, 315, 332-334
  monomer sequence distribution 278, 291-293
  terminology 277, 278
Critical point (see azeotropic composition)
Cross termination
  ethyl and fluoromethyl radicals 33
  in copolymerization
    model studies 219
    of PBMA• with PMAN• 225, 227
    of PBMA• with PMMA• 222, 223
    of PE• with PMAN• 225, 227
    of PMAN• with PS• 225, 227
    of PMMA• with PS• 216, 226, 232
  specificity 216
Cumyl radicals
  combination vs. disproportionation 32, 220, 221
  from azocumene 96
  pathways for combination 222
Cumyloxy radicals
  from cumyl hydroperoxide 78
  from dicumyl hyponitrite 65
  from dicumyl peroxide 77
  reaction with monomers
    rate constants 102
    specificity 103, 108
  solvent effects on fragmentation 107, 108
α–Cyanoalkyl radicals (see also cyanoisopropyl radicals)
  from dialkyldiazenes 57-59, 99
  pathways for combination 29, 99, 100, 224-226
  polarity 99
  reaction with monomers 99
Cyanoisopropyl radicals
  abstraction vs. addition 99
  combination vs. disproportionation 100, 224
  from AIBN 44
  ketenimine formation from 99, 225
  mechanism of formation from AIBN 51
  radical-radical reaction 224
  reaction with monomers 99, 100
    MAN 100
    MMA 46
    rate constants 98
    S 45
    specificity 98
    VAc 99
  reaction with oxygen 50, 99
Cyclization (see also cyclopolymerization)

of but-3-enyl radicals 16, 174
of hept-6-enyl radicals 5, 16, 170
of hex-5-enyl radicals 5, 16, 48, 164, 168
    radical clock 48
    substituent effects 16, 164, 168
of pent-4-enyl radicals 16
stereoelectronic effects 16
Cyclobutylmethyl radicals
    ring-opening of 174
Cyclo-copolymerization
    of divinyl ether with maleic anhydride 170
Cyclohexyl radicals
    disproportionation 30
    Hammett parameters
        reaction with substituted styrenes 14
        reaction with substituted toluenes 14
    from cyclohexylmercury salts 119
    reaction with monomers
        effect of temperature 17
        MA 10, 98
        MMA 10, 98
        rate constants 98
Cyclopentadiene
    reaction with t-butoxy radicals 107
Cyclopolymerization 162-170
    1,4-dienes 169
        divinyl ether 169
    1,5-dienes 169
        o-divinylbenzene 169
        vinyl acrylate 169
        vinyl methacrylate 169
    1,6-dienes 4, 5, 9, 163-169
        1,7-cyclization 165
        allyl acrylamide 167
        allyl acrylate 167
        allyl methacrylamide 167
        allyl methacrylate 167
        diacrylic anhydride 164
        diallyl monomers 4, 163-166
        diallylammonium salts 163
        dimethacrylic anhydride 164
        dimethacrylic imides 165
        dimethallyl monomers 164
        kinetic $vs.$ thermodynamic control 5, 9, 163
        o-isopropenylstyrene 167
        propagation kinetics 167
        ring size 4, 163-166
        stereochemistry of ring closure 166
        stereoelectronic effects 162
        substituent effects 163-166
        symmetrical 166
        table 166
        unsymmetrical 167
    1,7-dienes 169
        bisacryloylhydrazine 170
        ethylene glycol divinylether 170
    1,8-dienes - methylene-bis-acrylamide 170
    1,9-dienes - o-dimethacryloylbenzene 170
    1,11-dienes - diallyl o-phthalate 170
    triene monomers 168
        double ring closure 168
        triallylamine 168
Cyclopropylmethyl radicals
    ring-opening 171-173
    rate constant for 172
    reversibility 173
Dead-end polymerization
    definition 298
    synthesis of end-functional polymers 298
Decomposition (see photochemical, thermal decomposition)
Defect groups (see also structural irregularities)
    control 315
    effect on polymer properties 3, 44, 316
    from radical-radical termination 217
    in PMMA 47
        anionic $vs.$ radical initiation 3, 319
        from addition-fragmentation chain transfer 320
        from t-butoxy radical initiation 47, 320
        from radical-radical termination 318
        head-to-head linkages 318
        unsaturated chain ends 47, 318, 320
    in PS 47
        anionic $vs.$ radical initiation 316
        benzoate end groups 317
        peroxide linkages 317
        prepared with AIBN initiator 318
        prepared with BPO initiator 317
        unsaturated end groups 317
    in PVC
        evidence for 3
        formation 156
    in PVA/PVAc 257
Degradative chain transfer (see also retardation)
    definition 208
    to allyl monomers 167, 254
Depropagation
    effect of temperature 188-190
    in copolymerization 290
Di-t-butyl hyponitrite
    kinetic data for decomposition 59
Di-t-butyl peroxalate
    decomposition 75
    initiator of MMA polymerization 44
    initiator of S polymerization 44
    kinetic data for decomposition 68
Di-t-butyl peroxide
    from DBPOX decomposition 76
    induced decomposition 78
    initiator efficiency 78
    kinetic data for decomposition 68
Di-t-butylmethyl radical
    persistent radical 32

Diacyl peroxides (see also dibenzoyl peroxide, dilauroyl peroxide)
  alkyl radicals from 69, 108
  as initiators 55, 68-73, 108
  cage decomposition products 69
  cage return 71
  chain transfer to 52, 53, 71
  concerted decomposition 69
  decomposition rate
    k_d values 68
    effect of solvent & viscosity 70, 71
    substituent effects 69
  explosive decomposition 69
  induced decomposition 52, 71, 72
  kinetic data for decomposition 68
  non-radical decomposition 71
  reaction with nitroxides 121
  redox decomposition 72, 73
  photodecomposition 70
  synthesis of end-functional polymers 300
  thermal decomposition 67
Dialkyl hyponitrites (see also di-t-butyl hyponitrite, dicumyl hyponitrite)
  as initiators 55, 65, 66
  induced decomposition 66
  kinetic data for decomposition 59
Dialkyl peroxalates (see also di-t-butyl peroxalate)
  as initiators 55
  decomposition mechanism 82
Dialkyl peroxides
  alkoxy radicals from 77
  as initiators 68, 77, 78
  decomposition mechanisms 77
  induced decomposition 78
  kinetic data for decomposition 68
Dialkyl peroxydicarbonates (see dialkyl peroxydicarbonates, di-i-propyl peroxydicarbonate)
  alkoxy radicals from 73
  as initiators 55, 68, 73, 74
  effect of concentration k_d 74
  effect of solvent on k_d 74
  induced decomposition 74
  kinetic data for decomposition 68
  synthesis of end-functional polymers 300
Dialkyl peroxyketals
  decomposition mechanism 78, 83
  initiator efficiency 78
Dialkylamino radicals
  polarity 22
Dialkyldiazenes (see also azobisisobutyronitrile, azobis(methyl isobutyrate), azonitriles)
  as initiators 55, 57-65, 299
  cage return 57
  cis-trans isomerization 61
  decomposition rate

k_d values 59
  effect of Lewis acids 60
  effect of transition metals salts 60
  effect of solvent 57, 60
  substituent effects 60
decomposition mechanisms 57
initiator efficiency 61, 62
kinetic data for decomposition 59
photochemical decomposition 61
selection 57
solubility 58
synthesis of end-functional polymers 299
transfer to initiator 64
unsymmetrical 57
water soluble 57
Diallyl monomers (see cyclopolymerization)
Diazenes (see dialkyldiazenes, azobisisobutyronitrile)
Dibenzoyl disulfide
  in living radical polymerization 337
  radicals from 114
Dibenzoyl peroxide
  ^{13}C-labeled 127, 229, 317
  ^{14}C-labeled 125
  cage return 52, 70
  chain transfer to 53, 110
  decomposition mechanism
    photochemical 70
    redox 72, 73
    thermal 69
  induced decomposition 72
    by nitroxides 72, 122
  initiator efficiency 69-71
  initiator of MMA polymerization 44, 71
  initiator of S polymerization 44, 52, 71, 72, 317
  initiator of VAc polymerization 72
  initiator of VC polymerization 72
  kinetic data for decomposition 68
  photodecomposition 70
  redox systems
    dimethylaniline 72, 73
    transition metals 72
Dicumyl hyponitrite
  kinetic data for decomposition 59
Diene monomers (see butadiene, chloroprene, cyclopolymerization, isoprene)
Diffusion
  control of radical-radical termination
    copolymerization 217
    homopolymerization 208
    small radicals 28
  mechanisms
    reaction diffusion 210, 213
    reptation 210
    segmental motion 210
Diisopropyl peroxydicarbonate
  t-amine redox couple 74

Subject Index

kinetic data for decomposition 68
radicals from 73
Dilauroyl peroxide
  cage return 70
  decomposition mechanism 69
  kinetic data for decomposition 68
Dimethylamino radicals
  Hammett parameters
    reaction with substituted styrenes 14
    reaction with substituted toluenes 14
Disproportionation (see also combination vs. disproportionation)
  end groups from
    PMAN 224
    PMMA 222, 223, 231, 318
  of alkoxy radicals 66
  of alkyl radicals 30
  pathways for 28, 30
    statistical factors 30
    stereoelectronic control 31
  specificity of hydrogen atom transfer 30, 222, 223, 224, 231
  termination by 207
  transition state 33
Disulfides (see also dibenzoyl disulfide, dithiuram disulfides)
  as initiators 56
  chain transfer constants 242
  chain transfer to 241
  in living radical polymerization 89, 241, 336-339
  in synthesis of telechelics 89
  radicals from 114
Dithiocarbamates
  as photoinitiators 89, 337-339
  chain transfer to 337, 338
  in living radical polymerization 337-339
Dithiocarbamyl radicals
  from dithiocarbamates 337, 338
  primary radical termination of PMMA• 337
  reaction with S 337
Dithiuram disulfides
  as initiators 336-339
  as photoinitiators 89, 337, 338
  block and end-functional polymers 338, 339
  chain transfer to 89, 242, 337
  in living radical polymerization 336-339
Donor monomers
  alternating copolymerization 288
  interaction with Lewis acids 328
  list 288
  spontaneous initiation 95
Double ring-opening polymerization
  see ring-opening polymerization 182
DPPH
  inhibition by 64, 261
Electron transfer
  pathway for radical-radical reaction 28

Electrophilicity (see polarity)
Emulsion polymerization
  initiation 53
End group determination (see techniques for end group determination)
End groups
  ester from aliphatic diacyl peroxides 109
  in persulfate initiated PS 112
  in polymers initiated with dialkyl peroxides 77
  in PVAc 155, 259
  initiator-derived 47
  peroxy groups 77
  solvent-derived 48
End-functional polymers 298-302
  methods of synthesis 298
  synthesis
    by living radical polymerization 336
    with functional inhibitors 302
    with functional initiators 298
    with functional monomers 301
    with functional transfer agents 241, 248, 300
    with halocarbon transfer agents 243
Energy of activation (see activation energy)
Entropy of activation (see activation entropy, Arrhenius A factor)
Entropy of polymerization 187-190
  steric effects 189
  table of 188
EPR spectroscopy
  initiation mechanisms 116-118, 124
  measurement of k_p 190
  spin trapping 116-118
ESR spectroscopy (see EPR spectroscopy)
Ethyl methacrylate
  polymerization
    combination vs. disproportionation 223, 231
    k_p 191
    thermodynamics 188
Ethylene
  copolymerization with VAc
    solvent effects 326
  polymerization
    backbiting 184
    chain transfer to polymer 256
    combination vs. disproportionation 226
    k_p solvent effects 17
    reaction with t-butoxy radicals 10
    reaction with methyl radicals 10
    reaction with trichloromethyl radicals 10
    reaction with trifluoromethyl radicals 10
Evans-Polanyi equation
  applied to hydrogen atom transfer 21
Fenton's reagent
  reaction of organic substrates with 82

Fluoro-olefins (see also vinyl fluoride, vinylidene fluoride, trifluoroethylene)
  reaction with radicals 9-11, 14, 15, 19, 105
Fragmentation
  of acyloxy to alkyl radicals 67, 69, 97
    rate constant 109
  of alkoxy to alkyl radicals 77
    substituent effects 107, 108
  of alkoxycarbonyloxy to alkoxy radicals 110
  of benzoyloxy to phenyl radicals 46, 48
    as radical clock 110, 126
    photogenerated 70
    rate constant 109
  of t-butoxy to methyl radicals 47, 108
    as radical clock 48
    solvent effects 107
  of cumyloxy to methyl radicals
    rate constant 108
  of hept-6-enoyloxy to hex-5-enyl radicals 48
  of initiator-derived radicals 47
  of isopropoxycarbonyloxy to isopropoxy radicals 74
  ring-opening polymerization 172
Free radical addition (see radical addition)
Free radical polymerization (see radical polymerization)
Free radical reactions (see radical reactions)
Frontier molecular orbital theory
  abstraction vs. addition 27
  radical addition 18, 19
Fumarodinitrile
  reaction with alkyl radicals
    effect of temperature 17
Galvinoxyl
  inhibition by 64, 261
Gel effect
  and k_t 210
  and template polymerization 330
  control of with chain transfer agents 234
  in acrylate polymerization 257
Graft copolymers
  definition 277
  nomenclature 297
  synthesis 84, 91, 297, 298, 302-307, 336, 339, 342, 343
    by living radical polymerization 336, 339, 342, 343
    terminology 298
    with ceric ion redox initiation 91
    with end-functional polymers 297
    with macromonomers 306
    with multifunctional initiators 83, 302
    with photoinitiation 84
    with transformation reactions 303-306
Group transfer polymerization
  transformation to radical polymerization 305
Guidelines
  for initiator selection 55

for predicting
  combination vs. disproportionation 35, 233
  outcome of hydrogen atom transfer 27
  outcome of radical addition 20
Half lives
  for initiator decomposition
    azo-compounds 59
    peroxides 68
    typical values 54
Halo-olefins (see also vinyl chloride, vinyl fluoride, vinylidene fluoride, trifluoroethylene)
  reaction with t-butoxy radicals 105
Haloalkyl radicals (see also trichloromethyl, trifluoromethyl radicals)
  disproportionation with ethyl radicals 33
  from halocarbons 243, 244
Halocarbons
  chain transfer to 243
  effect of chain length on C_s 244
  in preparation of telomers 243
  polarity 244
Hammett parameters
  for radical reactions 13, 14, 22
    hydrogen atom transfer 22
    substituted styrenes 14
    substituted toluenes 14
  Patterns of reactivity scheme 297
Hammond postulate
  radical addition 13
Head addition (see head-to-head linkages, head vs. tail addition)
Head vs. tail addition
  definition 4, 8, 152, 160
  effect of temperature 99, 158
  initiation
    by alkyl radicals 97
    by aryl radicals 101
    by benzoyloxy radicals 4, 46, 103, 104, 109, 317
    by t-butoxy radicals 15, 103-106
    by cyanoisopropyl radicals 99
    by hydroxy radicals 103, 111
    with acrylate esters 10, 11, 14, 103
    with fluoro-olefins 9-11, 14, 15, 19, 105
  polar effects 15
  propagation 4-6, 145, 152-161
    acrylic monomers 158, 159
    allyl monomers 158
    diene monomers 159-161
    fluoro-olefins 4, 154, 157
    in cyclopolymerization 5, 162
    VAc 154, 155
    VC 156
  steric effects 14
Head-to-head linkages
  from captodative olefins 262
  from head addition 152-161

Subject Index

from radical-radical termination 28, 153, 217, 318
   in acrylic polymers 158
   in cyclopolymers 5
   in fluoro-olefin polymers 157
   in PMMA 158, 217, 318
     effect on thermal stability 318
   in PVAc 154, 259
   in PVF 4
Heat of polymerization 187-190
   steric effects 189
   table of 188
Hept-6-enoyloxy radicals
   fragmentation to hex-5-enyl radicals 48
n-Hexyl radicals
   Hammett parameters
     reaction with substituted styrenes 14
     reaction with substituted toluenes 14
   reaction with monomers
     effect of temperature 17
Heterogeneous polymerization (see emulsion, suspension polymerization)
Hex-5-enyl radicals
   cyclization 5, 16, 48, 164, 168
     as radical clock 48, 98
     substituent effects 16, 164, 168
   from hept-6-enoyl peroxide 48
   reaction with monomers 98
Hexasubstituted ethanes
   in living polymerization 339-341
Hydrogen abstraction (see addition *vs.* abstraction, hydrogen atom transfer)
Hydrogen atom transfer (see also abstraction *vs.* addition, backbiting, chain transfer)
   from cobalt hydrides 250, 251
   from initiator
     alkyl hydroperoxides 79, 113
     dialkyl hyponitrites 66
     dialkyl peroxydicarbonates 74
     dialkyldiazenes 62
     ketenimines 65
   from metal hydrides 119
   from polymer 254, 255
     PE 256
     poly(alkyl acrylates) 257
     PVAc 257-259
     PVC 259
     PVF 259
   from solvent 244-246
   from toluene 14, 22, 98, 245, 297
   from transfer agent
     halohydrocarbons 244
     thiols 236, 237, 240, 241
   Hammett parameters 14, 22
   in disproportionation 30, 31, 233
     from PMAN• 224
     from PMMA• 223, 224, 231, 251
   intramolecular 24, 184-186

rate and specificity 21-27
   bond dissociation energies 21, 26
   Evans-Polanyi equation 21
   guidelines 27
   polar effects 22, 23, 27
   reaction conditions 25
   stereoelectronic effects 23-25
   steric effects 21, 22, 27
   reactivity-selectivity principle 21
   to α–cyanoalkyl radicals 99
   to alkoxy radicals 24-27, 101-108
     solvent effects 26
   to alkyl radicals 27, 99
   to alkylperoxy radicals 50
   to aryl radicals 27, 101
   to benzoyloxy radicals 46, 103, 109
   to t-butoxy radicals 49
     from alkenes 106
     from allyl acrylates 106
     from amines 24
     from AMS 103, 107
     from BMA 47, 106
     from conformationally constrained compounds 25
     from cyclohexanes 25
     from ethers 24
     from MA 103
     from MAN 104
     from MMA 46-49, 103
     from toluene 14, 48, 102
     from VAc 104
     from vinyl ethers 106
   to chlorine atoms 23, 25, 26
     solvent effects 25, 26
     specificity 23
   to cumyloxy radicals 103, 107
   to cyanoisopropyl radicals 99
   to hydroxy radicals 23, 27, 103, 111, 112
     specificity 23
   to methyl radicals 23, 27, 97, 98
     specificity 23
   to nitroxides 122
   to phenyl radicals 23, 27, 98, 101
     specificity 23
   to sulfate radical anion 112
   transition state 21
Hydrogen peroxide
   as initiator 81, 82
   photodecomposition 81
   redox systems 82
   synthesis of end-functional polymers 300
   thermal decomposition 81
Hydroperoxides (see alkyl hydroperoxides, t-butyl hydroperoxide, α–hydroperoxydiazenes)
Hydroperoxy ketals
   decomposition mechanism 83
α–Hydroperoxydiazenes

decomposition mechanism 82
induced decomposition 83
Hydroxy radicals
   abstraction vs. addition 27
   from α–hydroperoxydiazenes 83
   from alkyl hydroperoxides 78
   from hydrogen peroxide 82
   from sulfate radical anion 112
   polarity 27, 111
   reaction with aliphatic esters 23
   reaction with monomers
      rate constants 102
      specificity 103, 111
α–Hydroxyalkyl radicals
   from ceric ion initiation 91
Hyponitrites (see dialkyl hyponitrites, di-t-butyl hyponitrite, dicumyl hyponitrite)
Induced decomposition of initiator (see also chain transfer to initiator)
   α–hydroperoxydiazenes 83
   alkyl hydroperoxides 79, 113
   diacyl peroxides 52, 71, 72, 122
   dialkyl hyponitrites 66
   dialkyl peroxides 78
   dialkyl peroxydicarbonates 74
   dialkyldiazenes 62
   ketenimines 65
   peresters 76
   persulfate 80, 81
Inhibition 208, 260-266
   definition 208
   mechanisms 260
   reversible 335-346
Inhibitors
   aromatic nitro-compounds 265
   captodative olefins 262
   DPPH 64, 261
   functional 302
   galvinoxyl 64, 261
   nitro-compounds 265
   nitrones and nitroso-compounds 116-118, 265
   nitroxides 64, 120-122, 260, 261, 341-344
   oxygen 260, 262
   phenols 263
   quinones 260, 264
   stable radicals 260-262
   styrene 260
   transition metal salts 118, 260, 266
   triphenylverdazyl radical 64, 261, 344
   use of to measure initiator efficiency 63
Iniferter (see living radical polymerization)
Initers (see living radical polymerization)
Initiation (see also initiator efficiency, initiators, radicals)
   by-products 51
   definition 43
   historical view 43
   in heterogeneous polymerization 53

initiator efficiency 50
initiator-derived radicals
   formation 43, 44
   fragmentation 47
   reaction with monomer 45, 47
   reaction with oxygen 49
   solvent effects 48
   primary radical termination 52
   structural irregularities from 2-4, 43-53
   transfer to initiator 52
Initiator efficiency
   definition 50
   effect of cage reaction 51, 63, 71
   effect of cage return 52
   effect of chain transfer to initiator 52
   effect of conversion 71, 213
   effect of initiator structure 55
   effect of primary radical termination 52
   effect of radicals formed 51
   effect of temperature 63
   effect of viscosity 51
   measurement 63, 126
   of AIBMe 63
   of AIBN 62, 63
   of BPO 69, 71
      photochemical decomposition 70
   of BPO/dimethylaniline redox couple 73
   of DBPOX 75
   of di-t-butyl peroxide 78
   of diacyl peroxides 70, 71
   of dialkyldiazenes 61, 62
   of LPO 69
   of persulfate redox systems 81
Initiators (see also under individual initiators, cage reaction, induced decomposition of initiator, photoinitiators, redox initiators, spontaneous initiation), 53-96
   ^{13}C-labeled 126, 127
   ^{14}C-labeled 125
   alkoxyamines 322, 341-344
   azo-compounds 53, 54-66
      dialkyl hyponitrites 65, 66
      dialkyldiazenes 57-65, 299
   disulfides 89, 241, 336-339
   functional 298-300
      alkoxyamines 343
      benzoin derivatives 87
      dialkyldiazenes 57, 299
      disulfides 89, 339
      peroxides 300
   half lives
      azo-compounds 59
      peroxides 68
      typical values 54
   hexasubstituted ethanes 339-341
   multifunctional initiators 82-84
   organometallics 90, 322, 345
   peroxides 53-56, 66-82

alkyl hydroperoxides 78, 79, 113
diacyl peroxides 67-73, 108, 121
dialkyl peroxides 77-78
dialkyl peroxydicarbonates 73, 74, 108, 300
hydrogen peroxide 81
inorganic peroxides 79-82
peresters 74-77
persulfate 80, 111
photoinitiators 61, 65, 70, 76, 84-89
rate constants for decomposition
azo-compounds 59
peroxides 68
typical values 54
redox initiators 89-91
reviews 54
selection 46, 54, 55
spontaneous initiation 92-96
Intramolecular rearrangement (see backbiting, cyclization, ring-opening)
Isoprene
polymerization
1,2- vs. 1,4-addition 160, 161
Isopropenyl acetate
reaction with carbon-centered radicals
rate constants 98
Isopropoxycarbonyloxy radicals
aromatic substitution by 111
fragmentation 74
from diisopropyl peroxydicarbonate 73, 108
reaction with monomers
MMA 101
specificity 103
reaction with S 111
Ketene acetals (see also 2-methylene-1,3-dioxolanes)
ring-opening copolymerization 171, 301
ring-opening polymerization 175-179
ring size effects 175
substituent effects 175
synthesis of end-functional polymers 171, 301
Ketenimine
from α–cyanoalkyl radicals 29, 99
from cyanoisopropyl radicals 63, 99, 225
from dimeric PMAN• 225
thermal stability 225
α–Ketoalkyl radicals
pathways for combination 29
Kinetic vs. thermodynamic control
in 1,6-diene cyclopolymerization 163
of cyclization of hex-5-enyl radicals 5
of cyclopolymerization 5
of radical addition 4
of radical reactions 8, 44
of radical ring-opening 178
of ring-opening polymerization 173
Ladder polymers (see also template polymerization)

from methacryloyl derivatives of poly(hydroxy compounds) 170
Lauroyl peroxide (see dilauroyl peroxide)
Lewis acids
control of propagation 321, 327, 328
copolymerization 328
head vs. tail addition 328
k_p 327
tacticity of PMMA 328
effect on decomposition of azo-compounds 60
Living polymerization (see living radical polymerization)
Living radical polymerization
definition 335
initiators for
alkoxyamines 341-344
alkyl dithiocarbamates 337
cobalt complexes 345
disulfides 89, 241, 336-339
dithiuram disulfides 337
hexasubstituted ethanes 339-341
organosulfur iniferters 336
triphenylmethylazobenzene 340
mechanisms 335
of MA 337-345
of MMA 337, 339, 341-343
of S 337, 339, 342
of VAc 337, 344
stable radicals
diarylmethyl 340
nitroxides 341
tripehenylverdazyl 344
triphenylmethyl 340
synthesis
of block copolymers 336, 342
of end functional polymers 336
of graft copolymers 336, 342, 343
of narrow polydispersity polymers 336, 342
Long chain branches
by transfer to polymer 254
in PE 256
in PVA 257
in PVAc 257
in PVC 259
in PVF 259
Macromonomers
acrylate 306
definition 306
PMMA
addition-fragmentation chain transfer 217, 247, 256, 306
copolymerization 306
polymerization 306, 307, 324
solvent effects 324
S 306
synthesis
by catalytic chain transfer 249

with functional transfer agents 300
synthesis of block & graft copolymer 306
Magnetic field
  effect on cage reaction 52
Maleic anhydride
  copolymerization
    penultimate unit effects 286
    reactivity ratios 282
  copolymerization with S 288
    bootstrap effect 326
    spontaneous initiation 95
  cyclo-copolymerization with divinyl ether 170
Mass spectrometry
  for end group determination 124
Mayo-Lewis equation
  in copolymerization 280, 281
Mercaptoethanol
  synthesis of end-functional polymers 300
Methacrylamide
  polymerization
    catalytic inhibition 250
Methacrylate esters (see also n-butyl methacrylate, methyl methacrylate)
  polymerization
    catalytic chain transfer 249
    chain transfer to polymer 256
    combination vs. disproportionation 222-224, 230
    head vs. tail addition 158
    spontaneous initiation 94
    template 331
Methacrylic acid
  copolymerization with MMA
    solvent effects 325
  polymerization
    k_p solvent effects 324
    thermodynamics 188
  reaction with carbon-centered radicals
    rate constants 98
Methacrylonitrile
  by-product from AIBN 52
  copolymerization
    initiator by-product 52
    Q-e values 296
  copolymerization with S
    solvent effects 326
  polymerization
    combination vs. disproportionation 224-226, 231
    tacticity 150, 152
    thermodynamics 188
  reaction with carbon-centered radicals
    rate constants 98
  reaction with cyanoisopropyl radicals
    rate constant 100
  reaction with heteroatom-centered radicals
    rate constants 114

reaction with oxygen-centered radicals
  rate constants 102
  specificity 104
Methyl α–chloroacrylate
  reaction with alkyl radicals
    effect of temperature 17
Methyl acrylate
  copolymerization
    Q-e values 296
    reactivity ratios 282
  polymerization
    addition-fragmentation chain transfer 248
    chain transfer to halocarbons 243
    chain transfer to monomer 251
    chain transfer to solvent 245
    chain transfer to thiols 236, 240
    combination vs. disproportionation 231
    head vs. tail addition 158
    living radical 337
    tacticity 152
    thermodynamics 188
  reaction with benzoyloxy radicals 10, 102, 103
  reaction with t-butoxy radicals 10, 102, 103, 105
  reaction with carbon-centered radicals
    rate constants 98
  reaction with cyclohexyl radicals 10, 98
  reaction with heteroatom-centered radicals
    rate constants 114
  reaction with hex-5-enyl radicals 18, 98
    solvent effects 18
  reaction with oxygen-centered radicals
    rate constants 102
    specificity 103
  reaction with phenyl radicals 10
Methyl crotonate
  reaction with radicals 10
Methyl ethacrylate
  polymerization
    thermodynamics 188
Methyl methacrylate
  copolymerization
    Q-e values 296
    reactivity ratios 282
  copolymerization with MAA
    solvent effects 325
  copolymerization with S
    bootstrap effect 326
    effect of Lewis acid 328
    penultimate unit effects 287
    propagation rate constant 216
  copolymerization with VAc
    solvent effects 326
  polymerization
    addition-fragmentation chain transfer 248

Subject Index

catalytic chain transfer 249, 250-251
catalytic inhibition 250
chain transfer to disulfides 242
chain transfer to halocarbons 243
chain transfer to monomer 251
chain transfer to polymer 256
chain transfer to solvent 244, 245
chain transfer to thiols 240
combination vs. disproportionation 222-224, 230
head vs. tail addition 158
k_p 191
k_p - effect of Lewis acid 327
k_p solvent effects 324, 325
living radical 337
spontaneous initiation 94
tacticity 150-152
template 329
thermodynamics 188
with AIBN initiator 44, 62
with BPO initiator 44, 71
with DBPOX initiator 44, 48
reaction with arenethiyl radicals 114
reaction with benzoyloxy radicals 10, 46, 102, 103
reaction with t-butoxy radicals 10, 46, 48, 101-103
reaction with carbon-centered radicals
 rate constants 98
reaction with cumyloxy radicals 102, 103, 108
reaction with cyanoisopropyl radicals 46
reaction with cyclohexyl radicals 10
reaction with heteroatom-centered radicals
 rate constants 114
reaction with hydroxy radicals 102, 103
reaction with isopropoxycarbonyloxy radicals 101, 103
reaction with methyl radicals 46
reaction with oxygen-centered radicals
 rate constants 102
 specificity 103
reaction with phenyl radicals 10, 46
 solvent effects 18
α–Methylstyrene
 copolymerization with depropagation 291
 polymerization
  combination vs. disproportionation 220
  depropagation 188, 190
  thermodynamics 188, 190
 reaction with arenethiyl radicals
  rate constants 114
  solvent effects 18
 reaction with t-butoxy radicals
  rate constants 102
  solvent effects 107
  specificity 103
 reaction with carbon-centered radicals
  rate constants 98

reaction with heteroatom-centered radicals
 rate constants 114
reaction with hydroxy radicals
 rate constants 102
 specificity 103, 111
α–Methylvinyl monomers (see also methacrylonitrile, methyl methacrylate, α–methylstyrene)
 abstraction vs. addition
  solvent effects 49
 ene reaction with nitroso-compounds 117
 polymerization
  combination vs. disproportionation 233
  thermodynamics 190
Methyl radicals
 abstraction vs. addition 27
 from fragmentation of t-butoxy radicals 47
 Hammett parameters
  reaction with substituted styrenes 14
  reaction with substituted toluenes 14
 polarity 15, 22, 27, 97
 reaction with aliphatic esters 23
 reaction with monomers 98
  fluoro-olefins 10, 15
  MMA 46, 98
  rate constants 98
  S 45, 98
  specificity 10
 reaction with propionic acid 23
2-Methylene-1,3-dioxolanes (see also ketene acetals)
 ring-opening polymerization 175-179, 181
2-Methylene-1,4-dioxanes
 ring-opening polymerization 180
2-Methylenetetrahydrofurans
 ring-opening polymerization 181, 182
2-Methylenetetrahydropyran
 ring-opening polymerization 181
4-Methylene-1,3-dioxolanes
 ring-opening polymerization 179-181
Molecular weight (see also chain length dependence)
 control 234-237, 316, 330-332, 336
  by living polymerization 316, 336
  by template polymerization 330-332
  with transfer agents 234-237
Molecular weight distribution (see polydispersity)
Multifunctional initiators
 applications 83, 302
 azo-peroxides 83, 302
 decomposition mechanisms 82
 dialkyl peroxalates 75, 82
 α–hydroperoxydiazenes 82
 peroxyketals 78, 83
 polymeric initiators 302
 synthesis of block & graft copolymers 218, 302, 304

Neoprene (see chloroprene)
Nitro-compounds
  inhibition by 265
Nitrones
  as spin traps 117
  inhibition by 265
Nitroso-compounds
  as spin traps 117
  inhibition by 265
Nitroxides
  control of propagation 322
  in living polymerization 122, 341-344
  inhibition by 64, 208, 261
  radical trapping 120-122
  reaction with diacyl peroxides 72, 121
  reaction with radicals 120-122
    combination vs. disproportionation 121
    rate constants 121, 260
NMR spectroscopy
  measurement of
    end groups 123, 126, 127
    monomer sequence distribution 295
    tacticity 150
  monomer reactivity correlation 296
Nucleophilicity (see polarity)
Oxygen
  as chain transfer facilitator 99, 262
  as comonomer 262
  effect on initiation 49
  inhibition by 208, 260, 262
  reaction with alkyl radicals 49, 50, 99, 113
  reaction with cyanoisopropyl radicals 50, 99
  reaction with polymeric anions 304
Oxygen radical anion
  abstraction vs. addition 27
  polarity 27
Patterns of reactivity scheme
  for prediction of reactivity ratios 13, 297
  for prediction of transfer constants 23, 237
Penpenultimate unit effects
  on tacticity 149, 285
Penultimate unit effects
  in copolymerizations 284
  in radical addition 284
  on chain transfer 236, 237
    to halocarbons 244
    to thiols 241
  on tacticity 149, 285
Peresters (see also dialkyl peroxalates)
  as initiators 55, 68, 74-77
  kinetic data for decomposition 68
  non-radical decomposition 76
  photochemical decomposition 76
  thermal decomposition 75
  transfer to initiator 76
Peroxalates (see dialkyl peroxalates, di-t-butyl peroxalate)
Peroxide linkages
  formation 49, 50, 262, 304
  in PS 262, 317
Peroxides (see also alkyl hydroperoxides, diacyl peroxides, dialkyl peroxides, dialkyl peroxydicarbonates, hydrogen peroxide, peresters, persulfate)
  and chain transfer 50
  as initiators 53-56, 66-82
  rate constants for decomposition 68
Peroxydicarbonates (see dialkyl peroxydicarbonates, di-i-propyl peroxydicarbonate)
Peroxyoxalates (see dialkyl peroxalates, di-t-butyl peroxalate)
Persistent radicals
  di-t-butylmethyl radical 32
  triisopropylmethyl radical 32
Persulfate
  as initiator 56, 68, 80, 81, 111, 112
  crown ether complexes 80, 111
  effect of reaction conditions on k_d 80
  kinetic data for decomposition 68
  non-radical decomposition 80
  photodecomposition 81
  redox systems 81
Phenols
  inhibition by 208, 263
Phenyl radicals
  abstraction vs. addition 27
  aromatic substitution of S 46
  effect on solvent on relative reactivity towards S and MMA 18
  from BPO 69, 70
  from fragmentation of benzoyloxy radicals 46, 48, 109
  polarity 27, 101
  reaction with MMA 340
  reaction with monomers 98
    MA 10
    MMA 10, 46
    S 45
  reaction with tolyl chloride 23
1-Phenylalkyl radicals (see also benzyl, 1-phenylethyl radicals)
  reaction with monomers 284
  self reaction 220
1-Phenylethyl radicals
  reaction with monomers 98
  reaction with PMMA• 226
  self reaction 220
Phosphinyl radicals
  from acyl phosphine oxides 87, 115
  from transfer to phosphines 115
  polarity 115
  reaction with monomers 114, 115
Photochemical decomposition
  of peresters 76

Subject Index

Photoinitiators (see also benzoin derivatives, photoredox initiators), 84-89
  aromatic carbonyl compounds 84
  BPO 70
  diacyl peroxides 70
  dialkyl hyponitrites 65
  dialkyldiazenes 61
  dithiuram disulfides 89
  hydrogen peroxide 81
  peresters 76
  persulfate 81
  polymerizable 87
  reviews 84
  visible light 88
Photoredox initiators
  carbonyl compound/tertiary amine 88
  metal complex/organic halide 90
Polar effects
  on chain transfer
    to disulfides 242
    to halocarbons 244
    to thiols 241
  on fragmentation of t-alkoxy radicals 107
  on hydrogen atom transfer 22, 23, 241
    abstraction vs. addition 27
  on radical addition 4, 13-15, 19, 20
    copolymerization 13, 295
    Hammett parameters 14
    head vs. tail addition 14, 106, 109, 157
    rate 13-15, 19, 20, 97, 99, 101, 106, 109, 114, 115, 264
    to acrylate esters 14
    to fluoro-olefins 14, 15, 106, 157
    to p-benzoquinone 264
  on radical reactions 8
  on radical-radical reactions
    combination vs. disproportionation 33, 222
    combination 225
Polarity
  Hammett parameters 13
  of α–cyanoalkyl radicals 99
  of alkanethiyl radicals 241
  of alkoxy radicals 22, 27
  of alkyl radicals 22, 27
  of aryl radicals 27, 101
  of benzoyloxy radicals 27, 109
  of t-butoxy radicals 15, 27, 106
  of t-butyl radicals 27
  of carbon-centered radicals 97
  of chlorine atoms 22
  of dialkylamino radicals 22
  of hydroxy radicals 27
  of methyl radicals 15, 22, 27
  of phenyl radicals 27, 101
  of phosphinyl radicals 115
  of selenium-centered radicals 114
  of sulfur-centered radicals 114
  of thiols 241
  of trichloromethyl radicals 15
  of trifluoromethyl radicals 15
Polyacrylamide
  head-to-head linkages 159
Poly(acrylic acid)
  tacticity 150
Polyacrylonitrile
  head-to-head linkages 159
  tacticity 152
Poly(allyl esters)
  head-to-head linkages 158
Polyamides
  from ring-opening polymerization 175
Polybutadiene
  microstructure 159-161
Polychloroprene
  microstructure 160, 161
Polyesters
  from ring-opening polymerization 171, 175
Polyethers
  from ring-opening polymerization 174
Polyethylene
  short chain branches 184
Polydispersity
  control
    by living radical polymerization 316, 336, 342
    by transfer constant 235
  effect on radical-radical termination 213, 228
    combination vs. disproportionation 228
Polyisoprene
  microstructure 160, 161
Polyketones
  from ring-opening polymerization 171
Poly(methacrylic acid)
  template polymer 330
Poly(methacrylonitrile)
  tacticity 150, 152
Poly(methyl acrylate)
  head-to-head linkages 158
  tacticity 152
Poly(methyl methacrylate)
  defect groups
    control 319
    head-to-head linkages 158, 217, 318
    unsaturated chain ends 47, 217, 318, 320
  tacticity 150-152, 329, 331
  thermal stability 3, 318, 319
Polystyrene
  aromatic substitution
    by benzoyloxy radicals 109
  defect groups 316
    benzoate end groups 317
    peroxide linkages 317
    unsaturated end groups 317
  tacticity 152
  thermal stability
    anionic vs. free-radical 316

prepared with AIBN 318
prepared with BPO 47, 72, 317
weathering 47
Poly(trifluoroethylene)
head-to-head linkages 157
Poly(vinyl acetate) (see also poly(vinyl alcohol)
end groups 155
head-to-head linkages 154, 155
temperature dependence 154
long chain branches 257
effect of reaction conditions 258
measurement 258
short chain branches 186
Poly(vinyl alcohol)
ceric ion initiated graft copolymerization 91
long chain branches 257
tacticity 150-152
Poly(vinyl chloride)
long chain branches 259
short chain branches 186
structural irregularities 156
tacticity 150-152
temperature dependence 152
thermal stability 3, 156
model compounds 3
Poly(vinyl fluoride)
head-to-head linkages 4, 157
long chain branches 259
tacticity 150
Poly(vinylidene fluoride)
head-to-head linkages 157
Polymer degradation (see defect groups, thermal stability)
Polymer properties (see also thermal stability, photochemical stability)
control 315
prediction 43
Polymer structure (see also structural irregularities, defect groups)
control 315-346
copolymers 277
compositional heterogeneity 332-334
monomer sequence distribution 291-293, 295, 326, 327, 328
types 278
end-groups
from chain transfer 234, 237, 238, 320
from radical-radical termination 2, 207, 217-233, 318-320
initiator-derived 2, 43-53, 316-318, 320
general formulae 1-4, 43
in chain groups
head-to-head linkages 4, 145, 152-162, 322, 328
rings 163-170
short chain branches 145, 184-186
tacticity 328-330

tail-to-tail linkages 145, 153-155
unsaturation 159-161, 171, 184
Staudinger concept 1
tacticity 145-152
Pressure effect
backbiting in E polymerization 184
on chain transfer to PVF 259
on radical-radical reactions 35
Primary radical termination
by sulfur-centered radicals 89
definition 52, 207
effect on initiator efficiency 52
in photoinitiation 86
model studies of radical-radical termination 219
of PBMA• with cyanoisopropyl radicals 227
of PE• with cyanoisopropyl radicals 227
of PMMA•
with 1-phenylethyl radicals 226
with dithiocarbamyl radicals 337
with nitroxides 341
with triphenylmethyl radicals 340
of PS•
with 2-carbomethoxyprop-2-yl radicals 226
with benzoyloxy radical 317
with cyanoisopropyl radicals 100, 227, 299
Primary radicals
definition 43
Propagating radical
complexation 321
with nitroxide 322
with organometallic reagent 322
Propagation 145-193
Arrhenius parameters 191
effect of solvent
mechanisms 323
head vs. tail addition 4, 145, 152-162
kinetics 190-193
k_p 190-193
chain length dependence 192
conversion dependence 192
measurement 190
values 191
polymerization thermodynamics 187-190
regiosequence isomerism 145, 152-161
acrylic polymers 158
allyl polymers 158
diene polymers 159-161
fluoro-olefin polymers 4, 157
PAN 159
PVAc 154
PVC 156
terminology 152
stereosequence isomerism/tacticity 145-152
structural irregularities 4, 145
structural isomerism 145, 162-186

Subject Index

addition-abstraction polymerization 186
backbiting 184-186
cyclopolymerization 162-170
ring-opening polymerization 171-184
Psuedo-living radical polymerization (see
living radical polymerization)
Q-e scheme
for prediction of reactivity ratios 13, 23, 295, 296
Q-e values 296
Quasi-living radical polymerization (see living radical polymerization)
Quinones
copolymerization 264
inhibition by 264
Radical addition (see also abstraction vs. addition, head vs. tail addition)
Hammett parameters 14
Hammond postulate 13
intramolecular 4, 5, 16
kinetic vs. thermodynamic control 4, 8
rate and specificity 8-20
bond strengths 15
guidelines 20
polar effects 4, 13, 15
product radical stability 4, 8-12
reaction conditions 16-18
stereoelectronic effects 15
steric effects 4, 12, 13
rate constants 9
reactivity-selectivity principle 17
theoretical studies 13, 18-20
transition state 9, 12, 16
Radical clock
cyclization of hex-5-enyl radicals 48, 98
fragmentation of benzoyloxy radicals 110, 126
fragmentation of t-butoxy radicals 48
reaction of carbon-centered radicals with metal hydrides 120
reaction of carbon-centered radicals with nitroxides 121
Radical cyclization (see cyclization)
Radical fragmentation (see Fragmentation)
Radical polarity (see Polarity)
Radical polymerization
benefits 1
component reactions 7
control 315
general mechanism 2, 234
historical review 1
Radical-radical reactions (see also combination, disproportionation, primary radical termination, radical-radical termination)
cage vs. encounter reaction 35, 218, 219
combination vs. disproportionation 28, 31-35, 218-227

cross termination 33, 219, 222, 223, 225, 226, 227
electron transfer 28
rate and specificity 28-35
diffusion control 28
guidelines 35
p-substituents 28, 29, 34, 35
polar effects 33, 223
pressure effect 31, 34
radical stability 34
rate constant 28
reaction conditions 34, 35
statistical factors 31, 32
stereoelectronic effects 30, 31, 33, 34
steric effects 32, 33, 35, 220, 223
temperature effect 31, 34, 35, 220-223
techniques 218, 219
transition states 31
Radical-radical termination (see also combination, disproportionation, primary radical termination, radical-radical reactions)
activation energy 208
chain length dependence
of combination vs. disproportionation 220, 224, 226
of rate constant 208, 213, 214
combination vs. disproportionation 217-233
in copolymerization 214
chemical control model 214-216, 232
combination vs. disproportionation 222, 223, 225, 226, 232, 233
diffusion control model 214, 217, 232
effect on compositional heterogeneity 333
model studies 218-227
pathways for 207, 208
rate constant 208-217
diffusion mechanisms 210
effect of conversion 210, 213, 214
effect of polydispersity 213
effect of solvent 208
effect of viscosity 213
gel or Trommsdorff effect 210
prediction 211
terminology 208
structural irregularities from 2, 153, 217
Radicals 96-115
carbon-centered radicals 96-101
α–aminoalkyl radicals 72, 88
α–cyanoalkyl radicals 99
alkyl radicals 27, 97
aryl radicals 27, 100
t-butyl radicals 27
methyl radicals 27
phenyl radicals 27
rate constants 98
heteroatom-centered radicals 101-115
initiator derived 55

classification 46
oxygen-centered radicals 101-113
  acyloxy radicals 108
  alkoxy radicals 27, 105-110
  alkoxycarbonyloxy radicals 73, 108, 110
  alkylperoxy radicals 113
  t-amyloxy radicals 107
  benzoyloxy radicals 27, 109
  t-butoxy radicals 27, 105-107
  t-butylperoxy radicals 113
  cumyloxy radicals 107
  ethoxy radicals 108
  formation 101
  hydroxy radicals 27, 111
  isopropoxy radicals 108, 110
  isopropoxycarbonyloxy radicals 73, 101, 110
  methoxy radicals 27, 108
  rate constants 102
  reviews 101
  specificity 102, 103
  sulfate radical anion 111-113
phosphorous-centered radicals 115
  phosphinyl radicals 115
  rate constants 114
polarity 13
selenium-centered radicals 114
sulfur-centered radicals 114
  alkanethiyl radicals 114
  arenethiyl radicals 114
  rate constants 114
Random copolymers (see statistical copolymers)
Rate constant
  for abstraction
    effect of bond dissociation energies 26
    from toluene 98, 102
  for fragmentation
    of benzoyloxy radicals 109
    of cumyloxy radicals 108
  for inhibitor-radical reaction (k_z)
    inhibitors and carbon-centered radicals 260
    nitroxides and carbon-centered radicals 121, 260
  for initiator decomposition (k_d)
    azo-compounds 59
    effect of cage return 52
    effect of chain transfer to initiator 53
    effect of conversion 214
    peroxides 68
    typical values 54
  for propagation (k_p) 190-193
    chain length dependence 192
    conversion dependence 192
    effect of conversion 214
    effect of Lewis acid 327
    effect of solvent 17
    measurement 190
    MMA/S copolymerization 216
    solvent effects 323-325
    table 191
  for radical addition (k_i)
    aryl radicals 101
    carbon-centered radicals 98
    cyanoisopropyl radicals 100
    effect of bond dissociation energies 26
    effect of solvent 17
    effect of temperature 17
    heteroatom-centered radicals 114
    oxygen-centered radicals 102
    to acrylate esters 10
    to halo-olefins 10
  for radical-radical termination (k_t)
    chain length dependence 208, 211
    definition 209
    prediction 211, 213
    small radicals 28
    typical values 209
  for ring-opening 172
    cyclobutylmethyl radicals 174
    cyclopropylmethyl radicals 172
    dioxalan-2-yl radicals 178
    in ring-opening polymerization 172, 174, 178
  use of radical clocks in calibration 48
Reaction conditions (see concentration, pressure, solvent, temperature, viscosity)
Reactivity ratios
  definition
    penultimate model 285, 286
    terminal model 281
  estimation
    from composition data 293-295
    from monomer sequence distribution 295
  for common monomers 282
  prediction
    NMR chemical shifts 296
    Patterns of reactivity scheme 297
    Q-e scheme 295-297
  solvent effects 17, 295, 296, 325-327
  substituent effects 284
  template effects 330
Reactivity-selectivity principle
  and hydrogen atom transfer 21
  and radical addition 17
Rearrangement (see fragmentation, cyclization)
  of radicals during polymerization 162
Redox initiators (see also photoredox initiators), 89-91
  BPO/N,N-dimethylaniline 72
  metal complex/organic halide 90
  with alkyl hydroperoxides 79
  with ceric ions 90
  with diacyl peroxides 72
  with hydrogen peroxide 82
  with persulfate 81

Subject Index

Redox systems
  with inorganic peroxides 79
Regiosequence isomerism (see head vs. tail addition)
Retardation (see also degradative chain transfer)
  definition 208
  mechanisms 260
  with transfer agents 234
    addition-fragmentation chain transfer 247
    solvents 244
    thiols 241
    VAc polymerization 244
Ring-opening
  of cyclobutylmethyl radicals 174
  of cyclopropylmethyl radicals 171-173
  of dioxolan-2-yl radicals
    rate constant 178
    reversibility 178
Ring-opening copolymerization
  of ketene acetals 171, 301
  of methylenecyclohexadiene spiro compounds 175
  of spiroorthocarbonates 182
  of spiroorthoesters 182
  synthesis of end-functional polymers 171
Ring-opening polymerization 171-184
  double ring-opening 182, 183
  effect of concentration 172
  effect of temperature 172, 176, 180
  of 2-methylene-1,3-dioxolanes 175, 176-179, 181
  of 2-methylene-1,3-dithiolanes 175
  of 2-methylene-1,4-dioxanes 180
  of 2-methylenetetrahydrofurans 181, 182
  of 2-methylenetetrahydropyrans 181
  of 4-methylene-1,3-dioxolanes 179-181
    effect of temperature 180
    polyketone synthesis 179
    substituent effects 179
  of α–cyclopropylstyrene 173
  of bicyclobutanes 172
  of ketene acetals 175-179
    effect of temperature 176
    polyester synthesis 175
    reversibility 178
    substituent effects 176
  of spiroorthocarbonates 182
  of spiroorthoesters 182
  of vinyl cyclobutanes 174
  of vinyl cyclopropanes 171-174
    rate constant for ring-opening 172
    reversibility 173
    substituent effects 173
  of vinyl oxiranes 174
  of vinyl sulfones 175
  of vinylcycloalkanes 172-175
  ring size effects 174, 175
  rate constant for ring-opening 172, 174, 178
  synthesis of polyesters 171, 175
  synthesis of polyketones 171, 179
  template polymerization 332
  volume expansion 171, 182
Scission (see fragmentation)
Secondary radicals
  definition 48
Segmented copolymers (see block copolymers)
Short chain branches
  by backbiting 254
Solvent
  chain transfer to 244, 245
Solvent effects (see also cage reaction)
  on chain transfer
    to PVAc 258, 259
  on copolymerization 281, 325-327
    bootstrap effect 292, 326
    model studies 327
    monomer sequence distribution 292
    of macromonomers 307
    reactivity ratios 17, 294, 325-327
  on fragmentation
    benzoyloxy radicals 110
    t-butoxy radicals 48, 49, 107
    cumyloxy radicals 107, 108
  on hydrogen atom transfer 25
    to alkoxy radicals 26
    to t-butoxy radicals 48, 107
    to chlorine atoms 25
  on initiation 48
  on initiator decomposition
    alkoxyamines 122
    DBPOX 76
    diacyl peroxides 70
    di-t-alkyl peroxides 78
    dialkyl peroxydicarbonates 74
    dialkyldiazenes 60
  on polymerization 323-325
    k_p 324, 325
    mechanisms 321, 323
    MMA 324, 325
    molecular weight 324
    propagation rate constant 17
    S 324
    tacticity 324
    VAc 17, 258, 259, 324, 325
  on radical addition 17, 323-327
  on radical reactivity
    t-butoxy radicals 49, 107
    mechanisms 323
    phosphinyl radicals 115
  on radical-radical termination 34, 35, 208, 210
    t-butyl radicals 35
    PMMA• 230
Spin traps

for initiation mechanism 116-118
Spiroorthocarbonates/esters
  ring-opening polymerization 182
Spontaneous initiation 92-96
  of acrylate esters 94
  of copolymerization 95
    of donor and acceptor monomers 289
  of MMA 94
  of perfluorostyrene 94
  of S 92-94
  unified mechanism 96
Stability (see photochemical stability, radical stability, thermal stability)
Stable radicals
  as radical traps
    for initiation mechanism 120-122
    for initiator efficiency 63
  in living radical polymerization 335, 339, 344
    diarylmethyl radicals 339-340
    nitroxides 340-344
    triphenylmethyl radicals 339-340
    triphenylverdazyl radical 344
  inhibition by 208, 260, 261
Star copolymer
  synthesis
    by living radical polymerization 339
Statistics of propagation
  1st order Markov 149, 151
  Bernoullian 149, 151
  Coleman-Fox 150
  random 149
Stereoelectronic effects
  on cyclopolymerization 162
  on hydrogen atom transfer 23-25
    backbiting 23, 24, 184, 186
    disproportionation 31, 33, 34
  on radical addition 15
  on radical cyclization 16
  on radical reactions 8
  on radical-radical reactions 31, 33, 34
  on ring-opening polymerization 173, 174, 176
Stereosequence isomerism (see tacticity)
Steric effects
  on chain transfer 236, 237, 247
    to halocarbons 244
  on fragmentation of t-alkoxy radicals 107
  on hydrogen atom transfer 21, 22
    abstraction vs. addition 27
    disproportionation 223
  on initiator decomposition
    alkoxyamines 342
    dialkyldiazenes 60
  on radical addition
    head vs. tail addition 4, 8, 12, 13, 14, 20, 101, 153, 154
    intramolecular 16, 164, 168
    polymerization thermodynamics 189, 190
    rate 12, 13
  on radical reactions 8
  on radical-radical reactions
    combination 28, 29, 225
    combination vs. disproportionation 32, 33, 35, 220, 222, 233
    disproportionation 30, 223
Structural irregularities (see also end groups, defect groups, head-to-head linkages, polymer structure)
  control 234, 315, 316, 321, 322
  definition 2 1-6
  from anomalous propagation 4, 5, 145, 146
  from chain transfer 3, 237, 238
  from initiation 2-4, 43, 53
  from radical-radical termination 2, 3, 217
Structural isomerism (see backbiting, cyclopolymerization, ring-opening polymerization)
Styrene
  aromatic substitution
    by benzoyloxy radicals 46, 109
    by hydroxy radicals 111
    by isopropoxycarbonyloxy radicals 111
    by phenyl radicals 46, 101
  copolymerization
    Q-e values 296
    reactivity ratios 282
    spontaneous initiation 95
  copolymerization with AN
    solvent effects 326
  copolymerization with MAH
    bootstrap effect 326
  copolymerization with MAN
    solvent effects 326
  copolymerization with MMA
    bootstrap effect 326
    effect of Lewis acid 328
  inhibition of VAc polymerization 260
  polymerization
    addition-fragmentation chain transfer 248
    catalytic chain transfer 250
    chain transfer to disulfides 242
    chain transfer to halocarbons 236, 243
    chain transfer to monomer 251, 252
    chain transfer to solvent 245
    chain transfer to thiols 236, 240
    combination vs. disproportionation 219, 220-222, 229
    effect of oxygen 260, 262
    k_p 191
    k_p solvent effects 324
    living radical 337
    spontaneous initiation 92-94
    thermodynamics 188
    with AIBN initiator 44, 62, 100, 318
      chain transfer to initiator 64
    with BPO initiator 44, 71, 317

with DBPOX initiator 44
reaction with alkyl radicals
  penultimate unit effects 284
  rate constants 98
reaction with alkylperoxy radicals 113
reaction with arenethiyl radicals 114
reaction with benzoyloxy radicals 4, 45, 46, 102, 103, 109, 121
reaction with t-butoxy radicals 45, 102, 103, 117
reaction with cumyloxy radicals 103, 108
reaction with cyanoisopropyl radicals 45, 98
reaction with heteroatom-centered radicals
  rate constants 114
reaction with hydroxy radicals 103, 111
reaction with isopropoxycarbonyloxy radicals 111
reaction with methyl radicals 45, 98
reaction with oxygen-centered radicals 103
reaction with phenyl radicals 45
  rate constants 98
  solvent effects 18
reaction with sulfate radical anion 112
ring-opening copolymerization 171, 175, 301
Substituent effects
  on alkoxyamine decomposition 342
  on bond dissociation energies
    C-C 15, 26
    C-H 26
    C-O 15, 26, 342
    O-H 26
  on ceiling temperature 188
  on chain transfer
    to disulfides 242
    to thiols 241
  on combination $vs.$ disproportionation
    for 2-carboalkoxy-2-propyl radicals 222
    for phenylethyl radicals 220
  on cyclopolymerization of 1,6-dienes 163
  on fragmentation of alkoxy radicals 107, 108
  on hydrogen atom transfer from toluene 14
  on k_d
    diacyl peroxides 69
    dialkyldiazenes 60
  on monomer reactivity ratios 284
  on radical addition
    to acrylate esters 10
    to alkenes 13
    to halo-olefins 10
    to styrene 14
  on radical cyclization 16
  on ring-opening polymerization
    of 2-methylene-1,3-dioxolanes 176
    of 4-methylene-1,3-dioxolanes 179
    of vinyl cyclobutanes 174
    of vinyl cyclopropanes 173
    of vinyl oxiranes 174
Sulfate radical anion
  effect of pH 112
  hydrolysis to hydroxy radicals 112
  pathways for reaction with monomers 111
Sulfur compounds
  as photoinitiators 89
Sulfur dioxide
  copolymerization with depropagation 291
Suspension polymerization
  initiation 53
Tacticity 145-152
  definition 146
  dyad composition 148
  effect of Lewis acids 152, 328
  effect of solvent 152, 324
  effect of temperature 151, 152
  effect of template polymer 329, 331
  effect on backbiting 186
  effect on monomer sequence distribution 292
  measurement 150
  of diene polymers 160
  of PAA 150
  of PAN 152
  of PMA 152
  of PMAN 150, 152
  of PMMA 150-152
  of PS 152
  of PVA 150-152
  of PVC 150-152
  of PVF 150
  penpenultimate unit effects 151
  penultimate unit effects 149, 284
  reviews 146
  statistics 149, 151
  terminology 146
Tail-to-tail linkages
  in PVAc 259
  occurrence with head-to-head linkages 153
Techniques
  NMR 126
Techniques for measurement
  of branching 184, 255
  of chain transfer constants 239, 240, 255
  of combination $vs.$ disproportionation 218, 219, 228, 229
    model studies 218, 219
  of end groups 123, 127
  of initiation mechanism 115-127
    chemical methods 124
    EPR 116
    radical trapping 118-122
    radiolabeling 125, 126
    spectroscopic methods 123, 124, 126, 127
    spin trapping 116-118
  of initiator efficiency 63, 126
  of k_p 190
  of k_t 209
  of monomer reactivities 116-122, 124-127
  of monomer sequence distribution 291, 295

of reactivity ratios 293-295
of tacticity 150
Telechelic polymers
  methods of synthesis 298
  synthesis
    by living radical polymerization 89
    with functional transfer agents 241, 248
Temperature effect
  backbiting in E polymerization 184
  on chain transfer
    backbiting 184, 186
    C_s 236
    to PVAc 258, 259
    to PVF 259
    VAc polymerization 258
  on combination vs. disproportionation
    for t-butyl radicals 34, 35
    for PMMA• 223, 230
    for PS• 221, 229
  on copolymerization
    ceiling temperature 290, 291
  on hydrogen abstraction 25
  on inhibition
    by nitroxides 261
  on initiation 16, 17, 49, 107
  on propagation
    allyl monomers 158
    backbiting 184, 186
    ceiling temperature 188
    diene monomers 160, 161
    k_p 236
    ring-opening polymerization 172, 176, 180
    tacticity 151
    VAc 155
    VF 157
Template polymerization
  and gel or Trommsdorff effect 330
  covalently bound templates 330
  ladder polymerization 170, 331
  mechanisms 329
  non-covalently bound templates 329
  of MMA 329
  of multimethacylate 331
  optical induction 331
  ring-opening polymerization 332
Terminal model
  for copolymerization 280
Termination (see chain transfer, combination, disproportionation, inhibition, primary radical termination, radical-radical reactions, radical-radical termination)
  control 335
  pathways for 207, 208
Theoretical studies
  of radical addition 13, 18, 19
  on 1,2-chlorine shifts 156
  on reactivity ratios in copolymerization 296
Thermal initiation (see spontaneous initiation)

Thermal stability
  of PMMA 3, 318-320
  of PS 47, 72, 316-318
    prepared with AIBN initiator 318
    prepared with BPO 47, 72, 317
  of PVC 3, 156
Thermodynamic control (see kinetic vs. thermodynamic control)
Thiols
  chain transfer constants 240
  chain transfer to 240, 241
  polarity 241
  polymeric 304
  synthesis of end-functional polymers 300
Thiophenol
  radicals from 114
Toluene
  chain transfer to 245
  reaction with radicals
    t-butoxy radicals 48, 102
    rate constants 98, 102
    substituent effects 14
Transfer (see chain transfer, hydrogen atom transfer)
Transformation reactions
  anion-radical 304
  cationic-radical 305
  group transfer-radical 305
  synthesis of block & graft copolymers 303-306
Transition metal salts/complexes
  decomposition of azo-compounds 60
  inhibition by 266
  reaction with carbon-centered radicals 118
  reaction with dialkyldiazenes 60
  reaction with hydrogen peroxide 82
  reaction with persulfate 81
  redox initiation 72, 79, 81, 82, 90
Transition state
  for combination 31
  for disproportionation 31, 33
  for hydrogen atom transfer 21
  for radical addition 9, 12
  for radical cyclization 16
  for radical reactions 8
  for radical-radical reactions 31
Trichloromethyl radicals
  Hammett parameters
    reaction with substituted styrenes 14
    reaction with substituted toluenes 14
  polarity 15
  reaction with monomers
    fluoro-olefins 10, 15
    specificity 10
Trifluoroethylene
  polymerization
    head vs. tail addition 157, 158
    reaction with t-butoxy radicals 10
    reaction with methyl radicals 10

reaction with trichloromethyl radicals 10
reaction with trifluoromethyl radicals 10
Trifluoromethyl radicals
  polarity 15
  reaction with monomers
    fluoro-olefins 10, 15
    specificity 10
Triisopropylmethyl radical
  persistent radical 32
Triphenylmethyl radicals
  abstraction vs. addition 27
  pathways for combination 29
  polarity 27
  reaction with MMA 340
  reaction with PMMA• 340
Triphenylmethylazobenzene
  as initiator 58
  in living polymerization 340
  kinetic data for decomposition 59
Trommsdorff effect (see gel effect)
Undecyl radicals
  from LPO 69
  Hammett parameters
    reaction with substituted styrenes 14
    reaction with substituted toluenes 14
Unsaturated chain ends
  from addition-fragmentation chain transfer 246-248
  from catalytic chain transfer 249-251
  from radical-radical termination 217
  from transfer to monomer 251-254
  in PMAN
    from disproportionation 224
  in PMMA
    effect on thermal stability 318-320
    effect on C_p 257
    chain transfer to 247, 256, 257, 306, 320
    from catalytic chain transfer 249-251
    from disproportionation 222, 223, 231, 257, 318
    from initiation 47
  in PS
    effect on thermal stability 317, 318
    from benzoate chain ends 317
    from disproportionation 219-221, 229, 230
Verdazyl radicals
  in living radical polymerization 344
  inhibition by 64, 261
Vinyl acetate
  copolymerization
    Q-e values 296
    reactivity ratios 282
  copolymerization with E
    solvent effects 326
  copolymerization with MMA
    solvent effects 326
  polymerization

addition-fragmentation chain transfer 248
backbiting 185
chain transfer to disulfides 242
chain transfer to halocarbons 236, 243
chain transfer to monomer 251-253
chain transfer to polymer 257-259
chain transfer to solvent 244, 245
chain transfer to thiols 240
combination vs. disproportionation 231
effect of oxygen 262
head vs. tail addition 154, 155
inhibition by styrene 260
k_p 191
k_p solvent effects 17, 324, 325
living radical 337
retardation 244
solvent effects 259, 324
thermodynamics 188
with AIBN initiator 65
reaction with benzoyloxy radicals 102, 104
reaction with t-butoxy radicals 102, 104, 105
reaction with carbon-centered radicals
  rate constants 98
reaction with heteroatom-centered radicals
  rate constants 114
reaction with oxygen-centered radicals
  rate constants 102
  specificity 104
Vinyl chloride
  copolymerization
    penultimate unit effects 286
    Q-e values 296
    reactivity ratios 282
  polymerization
    1,2-chlorine shifts 156
    backbiting 185
    chain transfer to halocarbons 236
    chain transfer to monomer 251, 253
    chain transfer to polymer 259
    combination vs. disproportionation 232
    head vs. tail addition 154, 156, 157
    tacticity 150-152
    thermodynamics 188
    transfer to monomer 157
  reaction with carbon-centered radicals
    rate constants 98
Vinyl ethers
  chain transfer to 246
  reaction with t-butoxy radicals 106
Vinyl fluoride
  polymerization
    chain transfer to polymer 259
    head vs. tail addition 157, 158
    tacticity 150, 152
  reaction with t-butoxy radicals 10
  reaction with methyl radicals 10
  reaction with trichloromethyl radicals 10

reaction with trifluoromethyl radicals 10
Vinyl monomers
　polymerization
　　combination vs. disproportionation 233
Vinyl sulfones
　ring-opening polymerization 175
Vinylcycloalkanes
　ring-opening polymerization 172-175
　　ring size effects 174
　　vinylcyclobutanes 174
　　vinylcyclopropanes 172-174
Vinylidene fluoride
　polymerization
　　head vs. tail addition 157, 158
　radical addition to 15
　reaction with t-butoxy radicals 9, 10, 105
　reaction with methyl radicals 10
　reaction with trichloromethyl radicals 10
　reaction with trifluoromethyl radicals 10
N-vinylimidazole
　induced decomposition of BPO 72
　template polymerization 330
Vinyloxiranes
　ring-opening polymerization 174
N-vinylpyrrolidone
　induced decomposition of BPO 72
　polymerization
　　with AIBN initiator 62
Viscosity effect
　on cage reaction 52
　　AIBN 60, 62
　　DBPOX 76
　　diacyl peroxides 70, 71
　　dialkyl hyponitrites 66
　　dialkyldiazenes 60-62
　on initiator efficiency 51, 213
　on k_d 70
　on macromonomer polymerization 307
　on radical-radical termination 35, 213
Volume expansion
　in double ring-opening polymerization 182
　in ring-opening polymerization 171
Weak links (see defect groups, structural irregularities)